Principles of Terahertz Time-Domain Spectroscopy

Principles of Terahertz Time-Domain Spectroscopy

Jean-Louis Coutaz
Frédéric Garet
Vincent Wallace

Pan Stanford Publishing

Published by

Pan Stanford Publishing Pte. Ltd.
Penthouse Level, Suntec Tower 3
8 Temasek Boulevard
Singapore 038988

Email: editorial@panstanford.com
Web: www.panstanford.com

British Library Cataloguing-in-Publication Data
A catalogue record for this book is available from the British Library.

Principles of Terahertz Time-Domain Spectroscopy

ISBN 978-981-4774-56-7 (Hardcover)
ISBN 978-0-203-70969-6 (eBook)

Contents

Preface

The purpose of this book is to describe the basic principles of terahertz time-domain spectroscopy (THz-TDS) and to explain how to extract the electromagnetic response in the far infrared of materials and devices from THz-TDS data.

Since its introduction, THz-TDS has become a rich scientific research topic, which allows one to obtain spectral parameters from the measurements of pulses, i.e., traces of the electric field recorded in the time domain. This technique is unique and very useful, and it is a perfect pedagogical example of the application of the Fourier transform to physics. It addresses very modern research studies and a myriad of applications, in a spectral range which is still difficult to investigate because of the lack of efficient, compact, and low-cost sources and receivers, which is mainly due to the technique's reliance on mode-locked femtosecond pulsed lasers.

In addition, the THz range of the electromagnetic spectrum, lying between infrared and microwave regions, is itself remarkable. Indeed, it corresponds to frequencies for which the dielectric response (i.e., matter seen from the point of view of optics) and electrical conductivity (which commonly appears in metallic conductors as employed in electrical engineering and electronics) contribute almost equally to the electromagnetic properties of matter. Simultaneously, the period of the THz pulses, which is of the order of the picosecond or shorter, corresponds to the time constant of the fastest classical[a] phenomena occurring in matter, like thermalization, carrier recombination, carrier

[a]By "classical," we mean that we exclude phenomena related to nuclear physics or particles physics.

scattering, molecular oscillations, fastest chemical reactions, and so on.

Both aspects of the THz response of matter makes this scientific domain fascinating, and opens the route towards many fruitful applications that need a precise characterization of the THz response of materials and devices.

The origins of this technique lie in the seminal works of David H. Auston and Kin P. Cheung, and prior equivalent works in the microwave region by Alexander Murray Nicholson and Gerald F. Ross. In these early papers most of the sample characterization was performed in transmission. Nowadays, almost all different kinds of possible experimental schemes have been used, including reflection, attenuated total reflection, scattering, polarimetry, optical pump and THz probe, etc. This has not only been done with home-made THz-TDS systems which are widely spread in laboratories worldwide, but several companies sell high-performing, reliable instruments, which can easily be used by scientists, and not only physicists. Therefore, we thought a book dedicated to THz-TDS would be useful for many students and researchers working in this field, to go beside the other excellent textbooks and manuals that describe THz science and technology more generally.

The content of this book is not only based on our own experience of having worked in the field for about 25 years, but also draws on the extensive literature which now exists and much of which is by leaders in the field, with whom we have interacted with over many years, and who we can call friends that have provided the inspiration for this book.

This book includes a description of the far-infrared spectrum, of the related response of matter and of the classical techniques that are employed to explore it. Then the principle of THz-TDS is explained, followed by a presentation of the common THz-TDS instrumentation. The major part of the book is dedicated to the extraction issue, i.e., how to derive the parameters of matter and devices (mostly the index of refraction and the coefficient of absorption) from the THz-TDS data. Most of the experimental schemes are described as well as the relevant extraction techniques. The treatment of noise and uncertainty is largely addressed in this book. Finally, a comparison with competing and complementary techniques, like Fourier transform (FTIR) and Raman spectroscopies, is given.

Jean-Louis Coutaz
Frédéric Garet
Vincent Wallace

Chapter 1

Introduction

Terahertz time-domain spectroscopy (THz-TDS) is an extremely useful technique for characterizing the far-infrared properties of matter, namely its complex permittivity (or complex refractive index) and permeability. The advantages of THz-TDS over traditional characterization methods [1], mainly Fourier transform infrared spectroscopy (FTIR), i.e., fast recordings at room temperature, high dynamic range, magnitude and phase recording, and broad bandwidth contributed to the revival of far-infrared studies at the end of the 1980s [2–4]. Today, still based on its excellent performance and advantages, THz-TDS is being used for the characterization of many different materials and devices, like anisotropic crystals, scattering heterogeneous materials, magneto-optical materials, metamaterials, and so on. The development and interest in the technique is so significant that several companies now sell advanced THz-TDS systems. This is also due to the numerous fundamental and applied research that can be performed in the THz spectral domain, as described in several excellent books [5–9]. On the fundamental side, weak resonances in matter can be excited, like vibration of molecules, optical phonons in crystals, etc., and any phenomenon with a picosecond/sub-picosecond characteristic response time. On the applied side, promising applications are foreseen in various

Principles of Terahertz Time-Domain Spectroscopy
Jean-Louis Coutaz, Frédéric Garet, and Vincent Wallace

ISBN 978-981-4774-56-7 (Hardcover), 978-0-203-70969-6 (eBook)
www.panstanford.com

domains like security and defense, telecoms, imaging for the industry and medicine, sensors, and the environment.

The principle of THz-TDS was derived from pioneering original works in the microwave domain by Nicolson and coworkers [10, 11]. The basic idea is to measure the temporal shape of a short electromagnetic pulse, which in turn can be converted to the frequency domain using a Fourier transform. This is a perfect pedagogical example of the application of the Fourier transform in physics. As soon as the pulse duration is of the order of picoseconds, the corresponding spectrum enters the THz range. Thus, the method combines the advanced technology that is required to deliver and measure the ultrashort pulses, with 19th-century mathematics. The necessary picosecond or, even better, femtosecond pulses are easily produced using mode-locked ultrafast lasers. It is the availability of highly reliable and stable trains of repetitive femtosecond laser pulses, typically from titane–sapphire or fiber lasers, that makes THz-TDS so successful. In addition, getting the temporal shape of the electromagnetic pulses through time-equivalent methods requires a delay line where the optical components need only to move over a few centimeters, which is technically easy, as opposed to the microwaves or infrared domains. Indeed, producing infrared pulses (greater than a few terahertz) requires electromagnetic pulses of a few tens of femtoseconds, which has been demonstrated but is rather inefficient and is still a challenge. In the microwave domain, the size of the quasi-optical setup is rather big (due to the longer wavelengths), and the delay lines need to be moved over tens of centimeters, which can be tricky. Moreover, pulses are preferably produced and time-delayed by electronic means, for which there are jitter issues, which are more troublesome than with mode-locked lasers.

Extracting the complex refractive index materials, without any need of a Kramers–Kronig transform, is also a great advantage of THz-TDS. Since the reports by D. Auston [2] in the late 1980s, this subject has been addressed by many research teams worldwide. The extraction procedure is not straightforward: one has to find the zero of a complex function, which is, most of the time, oscillating in the complex plane. This complexity may lead to some erroneous analysis of the experimental results, or if simplified extraction procedures are used, to an imprecise determination of the material parameters.

The goal of this book is to present and explain all the main THz-TDS extraction methods that have been developed to characterize any type of material and device. We focus on the different experimental cases to be treated, on the uncertainty and precision of the techniques, and on materials that require special attention, like anisotropic crystal, metamaterials, and so on. In most of the topics treated in this book, we derive the main useful equations, give experimental examples extracted from our work or from the literature, and advise the reader about the best procedure to calculate the desired parameters. The book is dedicated to master's and PhD students, as well as to researchers who are just entering the field of THz spectroscopy. Therefore, in all the chapters, we have made an effort to ensure that all basic principles and equations are given.

The book is organized along 13 chapters. Chapter 2 presents the THz range of the electromagnetic spectrum and its peculiarities. Chapter 3 is devoted to other techniques aimed at investigating the far-infrared properties of matter and devices. Chapter 4 describes the most common THz-TDS instrumentation. Chapters 5 and 6 are dedicated to the extraction of the complex refractive index of THz-TDS data from samples in transmission and reflection, respectively. Chapter 7 addresses pump-and-probe TDS techniques, in view of obtaining dynamical THz responses. Chapter 8 presents THz-TDS with waveguides, metamaterials, and plasmonic devices. Chapter 9 is devoted to THz-TDS polarimetry for the study of anisotropic samples. Chapter 10 treats scattering in heterogeneous materials. Noise in THz-TDS is explained in Chapter 11, and precision and uncertainties are described in Chapter 12. Finally, Chapter 13 presents some advanced topics, including comparison of Raman spectroscopy and THz-TDS, comparison of CW spectroscopy and THz-TDS, and simultaneous extraction of permittivity and permeability. A large annex introduces the fundamentals of electromagnetism and optics, and the basics of Fourier transform mathematics.

References

1. Houghton, J. T., and Smith, S. D. (1966). *Infra-red Physics*, Clarendon Press, Oxford.
2. Cheung, K. P., and Auston, D. H. (1986). A novel technique for measuring far-infrared absorption and dispersion, *Infrared Phys.*, **26**, pp. 23–27.
3. Pastol, Y., Arjavalingam, G., Halbout, J.-M., and Kopcsay, G. V. (1989). Dielectric properties of uniaxial crystals measured with optoelectronically generated microwave transient radiation, *Appl. Phys. Lett.*, **55**, pp. 2277–2279.
4. van Exter, M., Fattinger, Ch., and Grischkowsky, D. (1989). Terahertz time-domain spectroscopy of water vapour, *Opt. Lett.*, **14**, pp. 1128–1130.
5. Mittleman, D. (ed.) (2003). *Sensing with Terahertz Radiation*, Springer-Verlag Berlin Heidelberg.
6. Sakai, K. (ed.) (2005). *Terahertz Optoelectronics*, Springer-Verlag Berlin Heidelberg.
7. Dexheimer, S. L. (ed.) (2007). *Terahertz Spectroscopy: Principles and Applications*, Optical Science and Engineering Series, CRC Press, Boca Raton.
8. Zhang, X.-C., and Xu, J. (2009). *Introduction to THz Wave Photonics*, Springer-Verlag US.
9. Lee, Y.-S. (2009). *Principles of Terahertz Science and Technology*, Springer-Verlag US.
10. Nicolson, A. M. (1968). Broad-band microwave transmission characteristics from a single measurement of the transient response, *IEEE Trans. Instrum. Meas.*, **IM-17**, pp. 395–402.
11. Nicolson, A. M., and Ross, G. F. (1970). Measurement of the intrinsic properties of materials by time-domain techniques, *IEEE Trans. Instrum. Meas.*, **19**, pp. 377–382.

Chapter 2

The THz Frequency Range

2.1 Definitions

The terahertz (THz) frequency range is the part of the electromagnetic spectrum located between the infrared and the microwave regions, and until the 1980s, it was called the far infrared [1]. The borders of this spectral range are not precisely defined, but it is generally agreed within the community that the wavelength limits are 30 μm and 3 mm. The wavelength λ of an electromagnetic wave is related to the frequency f:

$$\lambda = \frac{c}{f}, \tag{2.1}$$

where $c \approx 3 \times 10^8$ m/s (see annex) is the velocity of light in vacuum. Therefore, the THz range spreads from 100 GHz to 10 THz. Let us also define the angular frequency[a] ω:

$$\omega = 2\pi f = \frac{2\pi c}{\lambda}. \tag{2.2}$$

The energy of a THz photon is given by

$$E = hf = \hbar\omega = \frac{hc}{\lambda} \simeq \frac{1.98 \times 10^{-19}}{\lambda_{\mu m}} \simeq 6.62 \times 10^{-22} f_{\mathrm{THz}}. \tag{2.3}$$

[a] Also named pulsation. In many publications, for the sake of simplicity, ω is called frequency, which can lead to some confusion.

Principles of Terahertz Time-Domain Spectroscopy
Jean-Louis Coutaz, Frédéric Garet, and Vincent Wallace

ISBN 978-981-4774-56-7 (Hardcover), 978-0-203-70969-6 (eBook)
www.panstanford.com

Here the energy is given in joule (h is Planck's constant: $h \approx 6.62 \times 10^{-34}$ J·s, $\hbar = h/2\pi$). This energy, for submillimeter waves, is extremely small. So it is preferable to use the electron-volt (eV) as unity (1 eV $= 1.6 \times 10^{-19}$ J):

$$E \simeq 4.125 \times 10^{-3} \, f_{\mathrm{THz}} \simeq \frac{1.237}{\lambda_{\mu\mathrm{m}}} \text{ eV}. \tag{2.4}$$

Thus, the energy range of THz waves spreads from 0.41 to 41 meV. The wavenumber ν, with the units cm^{-1}, is commonly employed in spectroscopy because the first spectrometers deviated the light beam with an angle proportional to ν:

$$\nu = \frac{1}{\lambda} = \frac{f}{c}. \tag{2.5}$$

Hence the THz wavenumbers range between 3 cm^{-1} and 300 cm^{-1}. Another interesting feature is that the wavenumber is directly proportional to the energy of the photons ($E = hc\nu$). Finally, the THz range can also be defined in terms of equivalent temperature T, with regard to the photon energy:

$$E = hf = k_{\mathrm{B}} T \quad \Rightarrow \quad T = \frac{h}{k_{\mathrm{B}}} f = \frac{hc}{k_{\mathrm{B}}} \frac{1}{\lambda} = \frac{hc}{k_{\mathrm{B}}} \nu. \tag{2.6}$$

Here, $k_{\mathrm{B}} \approx 1.38 \times 10^{-23}$ J/K is the Boltzmann constant. The equivalent temperatures of the THz waves lie between 4.8 and 480 K. All these different ways of defining the THz regions are scaled in Fig. 2.1, which shows the correspondence between the different units.

The THz domain separates the optical infrared region from the microwaves one not only in terms of frequencies but also in terms of the electromagnetic response of the matter. In fact, the electrical response of matter to any electromagnetic excitation is clearly depicted by the Maxwell–Faraday equation (relation A.2 in the annex)

$$\vec{\nabla} \times \vec{H} = \frac{\partial \vec{D}}{\partial t} + \vec{J}, \tag{2.7}$$

in which $\vec{D}$ is the displacement field (induced by the movement of the charges bound to the atoms or to the molecules), $\vec{H}$ the magnetic field strength, and $\vec{J}$ the current density (movement of the free

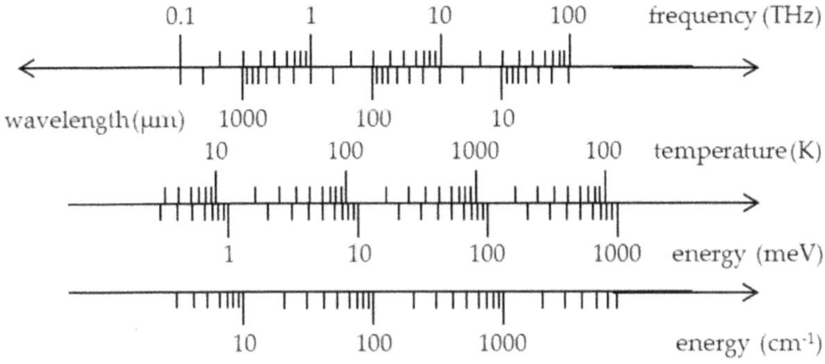

Figure 2.1 The THz range of the electromagnetic spectrum.

charge carriers). As long as the electric field $\vec{E}$ is weak compared to atomic fields, both $\vec{D}$ and $\vec{J}$ vary linearly with $\vec{E}$:

$$\vec{D} = \varepsilon_0 \varepsilon \vec{E}, \quad \vec{J} = \sigma \vec{E}. \tag{2.8}$$

σ is the material conductivity, and ε_0 and ε are respectively the permittivity of vacuum and the relative permittivity of the material (see relation A.1 in the annex). We substitute these expressions in Eq. 2.7:

$$\vec{\nabla} \times \vec{H} = j\omega\varepsilon_0\varepsilon\vec{E} + \sigma\vec{E} = j\omega\varepsilon_0\left(\varepsilon - j\frac{\sigma}{\omega\varepsilon_0}\right)\vec{E} = j\omega\varepsilon_0\varepsilon_{\text{eff}}\vec{E}. \tag{2.9}$$

In a material where both bound outer shell electrons and free charges are excited by an electromagnetic wave, both displacement current $\partial\vec{D}/\partial t$ and conduction current $\vec{J}$ contribute to an overall response described by the effective permittivity ε_{eff}:

$$\varepsilon_{\text{eff}} = \varepsilon - j\frac{\sigma}{\omega\varepsilon_0}. \tag{2.10}$$

ε gives the contribution of bound electrons, i.e., the dielectric response of the matter, while $\sigma/\omega\varepsilon_0$ is related to the free carriers, i.e., to the conduction of electricity. Usually, in the visible and infrared regions, $\varepsilon \gg \sigma/\omega\varepsilon_0$, and thus the response, is purely dielectric. On the other hand, in the microwave and radio frequency domains, only the conduction effect is relevant. Let us consider, as an example, silicon doped with arsenic (n-doping), at a concentration of 10^{16} atoms per cm^3. We will determine the frequency of the E field for which $\varepsilon = \sigma/\omega\varepsilon_0$, i.e., at the point both bound and free electrons contribute equally. If we neglect the losses, i.e., the regarded values are real, for Si, $\varepsilon \approx 11.6$ whatever the THz frequency, and for the considered doping concentration, $\sigma \approx 188\,\mathrm{m}^{-1}\,\Omega^{-1}$, leading to $f = 0.29$ THz. This means that, for this material, the THz range is at the transition between a pure dielectric behavior and a pure conductive one. At these THz frequencies, both bound and free charges contribute almost equally to the electric response of the matter. Let us notice that some ambiguities related to expression 2.10. If the dielectric term ε exhibits loss, i.e., it is a complex value, its imaginary part may be interpreted as a conduction current. Similarly, the imaginary part of $\vec{J}$ may be taken as dielectric losses.

From now on, we will write any complex value with a tilde. Because $\tilde{\varepsilon}_{\text{eff}}$ or $\tilde{\varepsilon}$ are complex values, the refractive index $\tilde{n}$ of the material (see expression A.6 in the Annex) is also complex. We note it (see A.9):

$$\tilde{n} = \sqrt{\tilde{\varepsilon}_{\text{eff}}\tilde{\mu}} = n - j\,\kappa, \tag{2.11}$$

$\tilde{\mu}$ is the permeability of the material and $j^2 = -1$.

The corresponding coefficient of absorption of energy α is

$$\alpha = 2\frac{\omega}{c}\,\kappa, \tag{2.12}$$

which has units per unit length, e.g., cm^{-1}.

2.2 Distinguishing Features of THz Wave-Matter Interactions

When considering energy, the THz photon is very weak, since its energy is of the order of the meV. Below 6.27 THz, this energy is even weaker than the thermal energy $E_{\text{th}} = k_{\text{B}}T$ at room temperature ($T = 300$ K, $E_{\text{th}} \simeq 26$ meV). This has two consequences: (i) in THz experiments, temperature and thermal radiation are sources of noise and can completely hide the THz signal, and (ii) the interaction of THz waves with matter will only involve weak resonances, like phonons or rotations and vibrations of molecules. Therefore, some materials could exhibit absorption spectra with fingerprints that could permit their unambiguous identification. This could be complementary to other spectroscopic methods, for instance, performed in the visible or in the infrared (e.g., Raman Spectroscopy). In several cases, these THz spectral signatures are unique, and THz spectroscopy gives information that cannot be obtained from other spectral domains. This is the case of transparent polymers made of low-weight molecules. X-rays are unable to "see" or interact with the lightest atoms, typically lighter than oxygen. If the polymers are transparent in the visible part of the spectrum, imaging is unable to distinguish them. Moreover, some polymeric molecules are IR-inactive due to their symmetry, and thus IR spectroscopy cannot identify them.

In gases, molecules are almost free to rotate and vibrate. The rotational energy levels of a rigid molecule, when rotation occurs around a principal axis x, are given by

$$E(J) = E_B J (J + 1). \tag{2.13}$$

J is a quantum integer number ($J \geq 0$) and $E_B = \hbar/4\pi I_x$ is the rotational constant. For diatomic molecules, the moment of inertia perpendicular to the molecule axis x is $I_x = \mu_{mass} R_{at}^2$, where R_{at} is the distance between the two atoms, and μ_{mass} is the reduced mass. Quantum mechanics selection rules impose $\Delta J = J_a - J_b = \pm 1$ when the molecule energy evolves from the b state to the a state. Hence, the frequency of the photon emitted or absorbed during this energetic transition is

$$f_{ab} = \left| \frac{E(J_a) - E(J_b)}{h} \right| = \frac{2E_B}{h}(1 + \max(J_a, J_b)). \tag{2.14}$$

Thus, we expect to observe lines periodically arranged in the rotational spectrum of such diatomic molecules. As an example, let us consider the rotational spectrum of a hydrogen chloride (HCl) molecule, for which $\mu_{mass} = 1.62 \times 10^{-27}$ kg. The frequencies f_{ab} are listed in the table below.

As seen in Table 2.1, the calculated frequencies are within the THz range. The frequency interval between 2 lines is about

Table 2.1 First rotation frequencies of HCl [2]

$J_a = J_b+1$	1	2	3	4	5	6	7	8	9	10
J_b	0	1	2	3	4	5	6	7	8	9
f(THz)	0.626	1.251	1.876	2.500	3.122	3.742	4.360	4.976	5.588	6.196

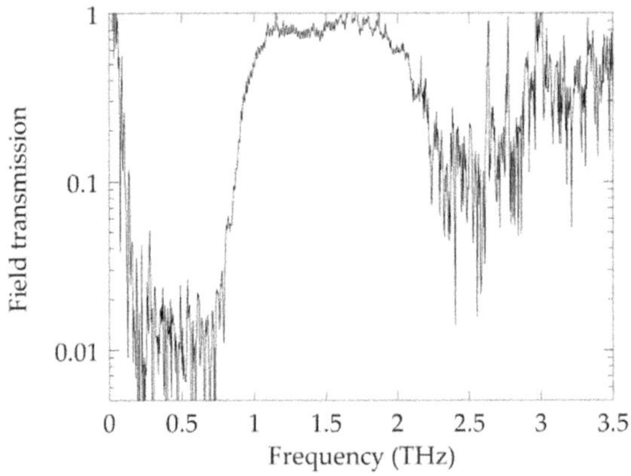

Figure 2.2 THz absorption spectrum of CF_4 measured by THz-TDS.

0.62 THz. This value leads to $R_{at} \approx 0.14$ nm, while the actual value is $R_{at} \approx 0.1275$ nm for the lowest energy levels [3]. In fact, relation 2.13 originates from a model that is too simplistic. Additional effects, like centrifugal distortion, must be incorporated in the model, resulting in the expression

$$E(J) = E_B J (J+1) - E_D J^2 (J+1)^2, \qquad (2.15)$$

where E_D is the centrifugal distortion constant, typically three orders of magnitude smaller than $\hbar/4\pi I_x$. For polyatomic molecules, the spectra are much more complicated. First, they originate from rotations in three dimensions, so they are described through three rotational constants, E_B, and five distortion constants, E_D. Second, they are coupled with vibrations of the whole molecule structure. Thus they exhibit an absorption spectrum in the THz range that includes many lines. Figure 2.2 shows the transmission spectrum of carbon tetrafluoride (CF_4) gas. This spectrum has been obtained by THz-TDS with a 70 cm long cell in which the gas was at atmospheric pressure. The dense number of lines, for example between 0.1 and 1 THz, are not resolved by classical THz-TDS, and thus this window appears incorrectly as an opaque region.

In the liquid or solid states, the vibration of each molecule is coupled to the neighboring ones. In pure polyatomic crystals, this vibration propagates from one atom to the other as an acoustic wave. This wave must be studied within the formalism of quantum

Table 2.2 Frequency of optical phonons in several crystals at room temperature, mostly from the Ioffe Institute database (http://www.ioffe.ru/SVA/NSM/). Some data have been determined through THz-TDS [6–8]

Crystal	GaSe	$SrTiO_3$	AgCl	CdTe	NaCl	InSb	ZnTe
f_{phonon} (THz)	0.59	2.7	3.09	4.2	4.92	5.21	5.32
crystal	$BaTiO_3$	GaAs	GaP	ZnO	Si	DAST	
f_{phonon} (THz)	5.35	8.05	10.96	12.4	13.7–14.9	1.1//a; 0.8,1.15,1.45//b	

mechanics. It follows that the energy of a given vibration mode is quantified. The quantum of energy is called a phonon. Only transverse phonons, for which the atoms oscillate perpendicularly to the wavevector, can be excited by an electromagnetic wave. Moreover, for such an excitation, two neighboring atoms must oscillate in opposition, which is known as an optical phonon. The phonon energy in most common crystals is listed in Table 2.2.

In polycrystalline materials, the phonon lines broaden to form absorption bumps due to microscopic disorder. The coherent propagation of the mechanical waves is no longer possible. The vibration is limited to some neighboring atoms or molecules. The Fourier transform of such spatially limited waves explains the broadening of the spectral features. Sometimes, especially in amorphous materials like glasses, this localized vibration leads to an absorption peak, with regard to the classical Debye model, whose temperature dependence resembles the one of a harmonic oscillator, which obeys Bose–Einstein statistics. Therefore this peak is called the Boson peak [4, 5].

Conductive materials are usually fairly well described by the Drude model. In this simple model, the free charges form a gas that does not interact with the crystal lattice. Also, one ignores the interactions between free charges. The finite conductivity of the material is depicted by a damping force, resulting from the averaged effect of collisions against structural defects, impurities, or phonons (especially at room temperature). The Drude permittivity is given by

$$\tilde{\varepsilon} = \varepsilon_{\infty} - \frac{\omega_{\mathrm{p}}^2}{\omega\left(\omega + j\Gamma\right)}. \tag{2.16}$$

ε_∞ is the value of the permittivity at higher frequencies (IR, visible), which expresses the residual contribution of bound electrons. Γ is the damping pulsation that is inversely proportional to the collision time $\tau(\Gamma = 1/\tau)$. ω_p is the plasma pulsation, which depends upon the free carrier density N:

$$\omega_p^2 = \frac{Ne^2}{m_{\text{eff}}\varepsilon_o}, \tag{2.17}$$

where e is the charge of an electron and m_{eff} is the effective mass of the free charges (electrons or holes) in the material. The concept of effective mass allows one to treat the charge as freely moving in the material: the effect of all the forces induced by the periodic net of positive immobile ions is included in the effective mass, which can be either measured, or determined from quantum mechanics calculation. For metals, ω_p lies within the UV range—i.e., it is larger than 1000 THz. τ is typically smaller than 100 fs, i.e., $\Gamma > 10$ THz. Therefore, the permittivity simplifies into

$$\tilde{\varepsilon} \approx -\frac{\omega_p^2}{\omega\left(\omega + j\Gamma\right)} = -\frac{\omega_p^2}{\omega}\frac{\omega - j\Gamma}{\omega^2 + \Gamma^2}. \tag{2.18}$$

Thus, in the THz range, both the real and imaginary parts of $\tilde{\varepsilon}$ are huge, and metals behave almost as perfect conductors. When one deals with electromagnetic effects at metal surfaces, involving for example surface plasmons (a plasmon is a quantum of plasma oscillation), the bulk Drude model is less valid. Nonlocality should be considered: the response of the free electron gas at a given location in the metal depends on the response of the gas in the neighborhood of this location. Simply speaking, when a collision of a free electron against the surface occurs during a period of the THz wave, it dramatically changes the metal response. This phenomenon is well known as the anomalous conductivity in microwaves [9].

When the doping concentration is well adjusted, the plasma frequency can lie in the THz domain. We then define ω_{lim} as the pulsation for which the real part of the permittivity equals 0:

$$\omega_{\text{lim}} = \sqrt{\frac{\omega_p^2}{\varepsilon_\infty} - \Gamma^2}. \tag{2.19}$$

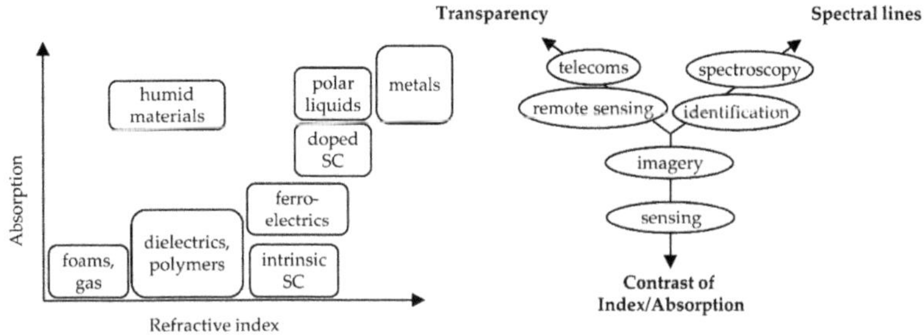

Figure 2.3 Left: Classification of the materials versus their THz properties. Right: Applications of the materials/items versus their main THz properties.

Below ω_{lim}, the semiconductor behaves as a metal, and as a dielectric above ω_{lim}. At ω_{lim}, the free electron gas oscillates as a whole: No wave propagation is allowed in the semiconductor.

From the preceding information on the interactions between light and matter, we can draw guidelines in view of applications of the THz waves (Fig. 2.3):

- Most of dielectric materials are more or less transparent to THz waves. In pure and perfect crystals, phonon lines lead to strong absorption peaks. In more common materials, and especially in amorphous materials, localized vibrations and absorption by impurities reduce the material transparency in the THz range. In this latter case, the absorption spectra do not exhibit peaks, but wide bumps.
- Molecular gases exhibit extremely strong absorption regions made of narrow absorption lines that could overlap. In between absorption regions or absorption peaks, gases are transparent.
- Materials that contain free charges reflect the THz waves. Thus metals, especially noble metals (Ag, Cu, Au, etc.), behave almost as perfect mirrors.
- Liquids made of polar molecules, like water, strongly absorb THz waves, while nonpolar liquids, such as some oils, are rather transparent.

2.3 Present and Foreseen Applications of THz Waves

It follows from the preceding paragraph that THz waves may interact with matter very differently than other parts of the electromagnetic spectrum do, like X-rays, visible or infrared. This opens the route toward novel applications, some of which are described below.

2.3.1 Security

Fabrics are almost transparent to THz waves, at least below 1 THz, making possible the detection of items, like weapons hidden under clothing. The first application in the security domain of the THz technology is to obtain pictures of the observed scene, often a person, in view of detecting prohibited hidden items. Thanks to the spectral signature of many materials, a remote identification of matter is possible in the THz range. This requires spectroscopy data to be obtained, or at least to record a picture at several selected THz frequencies. This hyperspectral imaging would help identify the hidden items, which could be dangerous materials such as explosives, or prohibited items like drugs. Spectroscopy may also be used to detect lethal gases on the battlefield or in public places.

2.3.1.1 Passive and active imaging

To get the image of the hidden items, two techniques are studied, namely passive and active imaging. In passive imaging, the image is formed from the thermal radiation emitted by the body of the inspected people or items. As any matter at a non-zero temperature T, the body radiates a broad spectrum of incoherent electromagnetic waves, which is described by a modified Planck's law:

$$L(f) = \varepsilon_{\text{therm}}(f)\, L_0(f) = \varepsilon_{\text{therm}}(f)\, \frac{2hc}{e^{\frac{hf}{k_B T}} - 1} \left(\frac{f}{c}\right)^3, \qquad (2.20)$$

where $L(f)$ and $L_0(f)$ are the luminances of the body and a blackbody, respectively; i.e., the power radiated by a surface unity of the body per solid angle and per unit of frequency. $\varepsilon_{\text{therm}}(f)$ is the emissivity of the body with regard to the blackbody emission. In the far infrared ($hf \ll k_B T$), Planck's law tends toward the Rayleigh–Jeans law:

$$L(f) = \varepsilon_{\text{therm}}(f)\, L_0(f) \approx 2k_B T\, \varepsilon_{\text{therm}}(f)\, f^2. \qquad (2.21)$$

If a part of the body, for example a metallic weapon hidden under the clothing of a terrorist, is at a smaller temperature $T' = T - \Delta T$ than the body, and assuming $\Delta T \ll T$ and that emissivity varies weakly with temperature, the relative change of far infrared emitted power is:

$$\frac{\Delta P}{P} = \frac{\varepsilon_{\text{therm}}}{L_0} \frac{\partial L_0}{\partial T} \Delta T. \qquad (2.22)$$

The power variation depends linearly on the temperature variation of the body, and also on the emissivity of the body. The THz image gives a map of temperature of the scene. When the scene is observed with an optical system working at the limit of diffraction, i.e., the image of the smallest resolved area at the scene corresponds to the Airy diffraction-limited spot at the camera image plane (for optimized cameras, the size of a pixel equals the Airy spot), a simple photometry calculation leads to the power received by each pixel in a Δf bandwidth:

$$P \approx 2k_B T\, \Delta f. \qquad (2.23)$$

Let us assume that the observed prohibited item shows a temperature difference $\Delta T = 0.1$ K with the human body, which is a realistic value, and that the detector bandwidth is 100 GHz. We get $\Delta P \approx$ 0.3 pW. If the image should be recorded at a video rate (let's say 10 Hz), the power normalized to the recording frequency is $\Delta P \approx$ 100 fW/$\sqrt{\text{Hz}}$ = 10^{-13} W/$\sqrt{\text{Hz}}$. Consequently, the detector must exhibit a noise equivalent power (NEP) better than this value. Today, in 2016, only cryogenic detectors, like hot electron bolometers made with superconductors or single electron quantum dot transistor, reach this requirement.

For security applications, devices operating at a room temperature are preferable from a practical point of view. The most promising are microbolometers and nanotransistors that can be integrated in large arrays. Nevertheless, their NEP is of the order of $10^{-10} \sim 10^{-11}$ W/$\sqrt{\text{Hz}}$, i.e., much larger than the one required for passive imaging (see above). Therefore, these detectors need the scene to be illuminated by an artificial THz source, like a quantum cascade laser. In this active imaging technique, the people to be inspected are thus illuminated by a THz beam (in many countries, this is rejected by the population due to the safety concerns regarding the long-term effects of exposure to electromagnetic radiation). Let us determine the power of a THz source that is required to get an image using room temperature detectors. The scene is at a distance L from the camera, and the power of the THz beam at the object is P_o. We suppose that the beam illuminates an area at the scene corresponding to the spatial resolution of the camera—i.e., that the light scattered by this area is focused by the optics onto a single pixel of the receiving matrix. Assuming a Lambertian scattering, we obtain the power P reaching the pixel:

$$P = \frac{1}{8} P_o \left(\frac{\phi_D}{L} \right)^2 . \tag{2.24}$$

Here ϕ_D is the diameter of the camera optics. For a $\phi_D = 30$ cm optics and a scene located at $L = 10$ m from the camera, we get $P \approx 10^{-4} \times P_o$. For a 10 Hz video rate and a 300-pixel camera, the THz beam illuminates each point of the scene during 0.3 ms. Thus $P/300$ must be larger than the NEP ($10^{-10} \sim 10^{-11}$ W/$\sqrt{\text{Hz}}$), i.e., P_o must be at least of the order of a few mW. In this simple estimation, we have not taken into account the detector efficiency, the transmission of the optics, or the absorption by the atmosphere. This last one is a very limiting factor, mainly due to water vapor, especially for frequencies above 1.1 THz [10]. Therefore, for security imaging, useful frequencies lie between 300 and 800 GHz. Today, most of the systems installed in airports operate at 30 or 90 GHz, i.e., in the microwave regime, because this technology is simple, reliable, and cheap. Also, the properties in the microwave range of most materials of interest are rather similar to their properties at THz frequencies. The main difference is the achievable spatial resolution, which is limited by diffraction to approximately $\lambda/2$. At 0.5 $\sim$ 1 THz, this resolution is typically 150 $\sim$ 300 μm, while it increases to 1.6 $\sim$ 5 mm for microwave cameras. For many applications related to security, this millimeter-scale resolution is sufficient, but it requires very large optics, whose sizes are at least 10 times larger than in the THz range.

2.3.1.2 Hyper-spectral imaging

The main advantage of THz imaging over microwaves techniques is the possibility to benefit from specific spectral signatures of many interesting materials, like explosives. Recording the same image at a few (typically between 4 and 6) different THz frequencies allows one to identify the chemical composition of suspicious items. Nice demonstrations of remote identifications of several chemical compounds have already been published [11–15]. However, most of the relevant absorption peaks of interesting molecules are above 1 THz, and generally between 2 and 5 THz. At these frequencies, the absorption of THz by water vapor in the atmosphere forbids propagation of THz beams over more than a few meters. A smart solution to this question is to carry the THz signature from the scene to the receiver by optical means. Air can be ionized with intense ultrafast laser pulses; thus a plasma is created at the focus of the laser beam. Electrons being emitted from the focus radiate an intense and broadband THz signal [16]. The plasma also emits fluorescence in the visible part of the spectrum. The team led by X.-C. Zhang at RPI Troy has proposed many technological advances in view of remote THz detection. The most promising one is to measure the fluorescence from a plasma generated at the scene [17]. This fluorescence emission is enhanced by a THz field that is also generated at the position of interest by photoionization. Because the phenomenon is instantaneous, the fluorescence spectrum carries the typical features of the THz signal. Finally, we should note that the identification procedure requires us to solve an inverse problem, i.e., to know which molecules or materials give rise to the recorded spectra. As for any inverse problem, a unique solution is very difficult to obtain. However, as the signal is recorded over a wide spectral range, smart signal processing methods can improve the results for identification purposes [18].

2.3.2 Biology

THz waves can be used as a noninvasive probe to test biological materials. They can interrogate vibrational modes of biological molecules, which occur over large portions of the molecular structure. Data from many relevant substances have been published from sugars, amino acids, etc., all with specific signatures in the THz rage—for example, spectra from isomers of benzoic acid may be found in Walther et al. [19]. As these molecular vibrations depend on the shape and modification of the molecule, their THz response could give information on both their geometry and composition. The most famous example is certainly DNA, whose hybridization state (denatured or hybridized) is clearly revealed in the THz refractive index spectrum of DNA films [20]. Also, the dynamics of protein folding in a water solution has been reported using THz spectroscopy [21], and this technique may be used to detect diseased tissues [22]. The THz photon energy and the picosecond timescale resolution of THz techniques are of the same order of magnitude as the energy levels and response times of a variety of interesting biological processes, like proton exchange during enzyme catalysis, vibrations of microtubules [23], vibration of the cell membrane [24], etc. THz sensitivity to water is relevant to reveal drought stress in plants [25].

However, THz investigations in biology suffers some drawbacks: the THz wavelength is much longer than the size of the cells, the strong absorption by water does not permit deep inspection of tissues, and finally signals may be rather difficult to interpret.

2.3.3 Medicine

The applications of THz technology in medicine are rather scarce. Indeed, because the human body, on average, is made at 70% of water, THz waves do not penetrate tissues over more than a few hundreds of microns. Thus THz tools in medicine are restricted both to inspect the surface of extracted organs, and to detect illness of the skin (melanoma, carcinoma, psoriasis, etc.), or to examine thin slices of excised tissues. Skin studies generally reveal an increase in reflectivity of cancerous tissue compared with healthy tissue [26]. It is probable that these diseased tissues contain a higher fraction of water, which results in contrast between the two revealed in THz imaging. Breast tissue exhibits a lower content of water due to the presence of adipose (fatty) tissue, which leads to a smaller THz absorption, thus having the potential of increased penetration depth. However, Ashworth et al. have reported high absorption for breast tumors and for fibrous tissue of the breast where the cancers normally occur [27]. This means that, even if THz imaging is a very promising tool for cancer detection, there remain a number of challenges and additional studies are required for it to become an acceptable medical technique [28]. Other relevant studies concern the observation of skin burns [29] and the hydration of cornea [30]. Another important topic is the biological effect of THz radiation on the illuminated living tissues. At higher powers, thermal effects are dominant [31], and therefore it is advised not to expose people to a THz beam power density higher than 10 mW/cm^2, which is the same as for the millimeter range [32]. Also, Titova et al. [33] observed that exposure of artificial skin to intense THz pulses may cause DNA damage due to phosphorylation of the H2AX molecule. At low power, the weak energy of the THz photons make them non-ionizing. However, the THz radiation can interact with cells and their constituents, and also with macromolecules like proteins by exciting low-frequency collective vibrational modes. In cells, the incoming THz field could be even higher than the electrical field in the membrane, whose related potential difference governs chemical exchange through the membrane. This could perturb the cells bio-operation, a good example of which is the modulation of the firing rate of neurons when exposed to THz radiation, as demonstrated by Siegel and Pikov [34]. Also, genomic instability induced by THz illumination (0.031 mW/cm^2, CW 0.1 THz beam) has been indirectly reported, as explaining an abnormal chromosome number in lymphocytes [35]. Interpretation of these results, and of some others not cited here, is still subject to controversy and merits definitively additional and complementary works [36].

2.3.4 Sensing in Industrial Applications

A large part of THz sensors for industry are based on the imaging of items/defects hidden in a hosting material that is opaque in the visible and infrared domains, but transparent for THz waves. The image can be acquired following different procedures:

- *Contrast of amplitude in transmission.* Strong absorption regions could be humid or electrically charged areas.

Also, regions with different chemical compositions can be inspected. One of the first THz images ever recorded was of a leaf, in which the veins, as expected, contained more water than the other parts of the leaf. For several years, the Nikon Company sold a THz apparatus capable of mapping the doping level in semiconductor wafers. More recent applications [37] include, e.g., in situ real-time monitoring of the fabrication of plastic films, of sheets of paper. Also, THz technology is used for detecting structural defects (hidden cracks, foreign bodies, bad adhesive joints, etc.) in components made of modern materials (glass or carbon fiber reinforced plastics, ceramics, foams, etc.). Promising applications are foreseen in the aircraft industry.

- *Contrast of amplitude in reflection*. This method is used when there is no transmitted signal, i.e., the substrate strongly absorbs the THz waves or reflects them, like metals. Rusty parts of metallic pieces, especially hidden under a coat of paint, are easily detectable via THz inspection.
- Images based on the phase (in the frequency domain) or on the time-of-flight (in the time domain, if dispersion in the sample is not large) of the transmitted signal. The phase change results from the delay of the THz signal propagation due to the reduction of the wave velocity in the regions of high refractive index. Thus the transmitted phase at a given point of the image is proportional to the integral through the sample thickness of the product index of refraction and distance, which is sometimes named optical thickness. This technique is complementary to amplitude contrast. It is applied to test samples transparent in the THz range, for which absorption contrast does not apply, and also samples in which vacuum or air pockets are included. This is the case of items in which several layers of different materials are glued together while the bonding is not perfect. The most famous example was the discovery of such air pockets between thermal tiles and the cabin of the space shuttle Columbia by the team of X.-C. Zhang [38], which explains

why the shuttle was destroyed when re-entering the Earth atmosphere.

2.3.5 Environment

The use of electromagnetic signals, from UV to the far infrared, is the unique remote technique to detect or monitor gases in the atmosphere. Today these techniques are mostly applied to detect pollution in air, and to observe the composition of the high altitude layers of the atmosphere [39], like the ozone concentration in the stratosphere [40]. Many techniques, like LIDAR for the detection of sulfur and nitrogen sulfurs, are very efficient, and thus the sub-THz/THz technology is essential only for the study of molecules that do not show any useful spectral signatures in other frequency ranges, i.e., from UV to infrared. This is, for example, the case of volatile organic compounds (VOCs) [41]. Gases, made of polyatomic molecules with a non-null dipolar transition moment, have far infrared spectra with dense and numerous absorption lines, which originate in the rotational movement of these molecules. The strength of the interaction of these molecules with THz radiation leads to extremely high absorption coefficients, which make it possible to detect very low concentrations of molecules, to a few ppm [42]. This is of the order of magnitude of threshold limit values (TLV) published by some health agencies. Typically, TLVs of cyanhydric acid, trimethyamine, ammoniac are of the order of 10 ppm, while they increase to 100–200 ppm for trichloroethylene and dichloromethane. The advantage of THz techniques over other spectroscopies is the excellent spectral selectivity, thanks to a less pronounced Doppler effect than in the infrared, which permits to record very clear signatures of the molecules. Moreover, THz measurements are possible even in challenging environments (dust, fog, smoke, etc.) [43].

2.3.6 THz Telecoms [44]

Optical telecommunications based on fiber networks already permit the transmission of terabits of data per second. This occurs through wavelength multiplexing. Some tens of channels at different wavelengths, separated by typically 100 GHz, carry ~10 Gbits/s each. At the consumer home, the huge amount of data flux has to be delivered from the fiber to the multimedia equipment. Future super-high-definition TV channels and some professional applications, like telemedicine, will require a very large data bandwidth: In a sense, it follows a Moore's-type law for data rates (40–60 Gbits/s or even higher rates are expected at the consumer electronics). Because of the Nyquist theorem, the necessary carrier frequency of such high data rates is within the range 0.3–1 THz. Thus free-space THz communications (WLAN, WPAN) from the fiber output to the equipments are foreseen. The advantages of the THz technology are the available spectrum and bandwidth, the compactness of components and antennas compared with smaller frequencies, and the transparency of atmosphere in some spectral windows. Shorter-distance applications of telecoms in the THz range could be found on board-to-board and intramachine in computer technology, while p2p links, telecom backhaul, and last-mile access constitute areas of development in longer-distance THz-based telecoms.

2.3.7 Scientific Instrumentation

2.3.7.1 In laboratories

THz spectroscopy is an useful tool in physics and chemistry research. In physics, it allows one to work on low-energy excitations of matter and on physical phenomena occurring on a picosecond or even shorter timescale. This includes phonons in crystals and solid-state matter, Cooper pair dynamics in superconductors, Rydberg states of highly excited atoms, boson peaks in amorphous materials like glasses and polymers, relaxation in ferroelectrics, electrical conductivity in semiconductors, plasmonics in metallic structures and metamaterials, carrier dynamics in quantum dots and nanowires, and so on. Today, the picosecond timescale corresponds to the duration of rapid events in ultrafast electronics components, and thus THz tools find several applications in this area. In chemistry, THz spectroscopy is used for measuring the energy states related to low-frequency molecular oscillation (roto-vibration of the whole molecule skeleton, also called conformational flexibility). It also allows identifying pharmaceutical compounds, and especially polymorphs, which is the ability of a solid material to exist in more than one form or crystal structure. THz spectroscopy is sensitive to the polymorphic state of the crystal because the vibrations in this region are due either to large-amplitude torsional vibrations (intramolecular) or intermolecular interactions between nearest neighboring molecules [45]. Moreover, many chemical reactions occur with a delay of the order of picoseconds [46]. Currently, a very promising topic is the THz study of intermolecular interactions in liquids, especially in water. When excited by a short electromagnetic pulse, the water molecule behaves first like a free molecule in gas (rotation, vibration), then it interacts with its nearest neighbors (what is called the first solvation shell) [47]. THz absorption is related to the second shell, corresponding roughly speaking to two layers of neighboring water molecules. These molecules are bound to the central one by hydrogen bonds. Thus it is extremely sensitive to the hydrogen network and easily perturbed by the presence of a solute. Hydration of small solute molecules or even bigger ones as proteins in water is of crucial importance to understand chemical reactions, especially in biochemistry [48].

2.3.7.2 Radioastronomy

Since the first observation of the cosmic background radiation by A. Penzias and R. Wilson in 1964, radioastronomy has been a very active research field. Roughly speaking, radio-astronomy includes two domains of research activities.

The first one is related to the observation of the background radiation. In the Big Bang theory, our universe was born about 13×10^9 years ago. After about 350,000 years, the density of the initial plasma was sufficiently decreased to allow light and matter to be separated, i.e., light could no more ionize atoms and thus started to propagate in between matter particles. At that time, this initial plasma was at thermal equilibrium, with a temperature T corresponding to the ionization of atoms, i.e., about 3000 K. Thus it radiates a thermal light like an almost perfect blackbody. As cosmology theories describe a universe whose volume varies as the square of time t at constant entropy, it follows from the blackbody model that the (mean) temperature of the universe depends on t as $T(t) \propto t^{-2/3}$. Today this temperature is equal to 2.73 K, which corresponds to a blackbody radiation spectrum that peaks at 160 GHz. The measured spectrum, recorded with several instruments flown on satellites, balloons, and rockets, or located in high-altitude telescopes, fits Planck's law well. Currently, astronomers are looking at possible tiny differences with Planck's law, and mostly at the anisotropy of this radiation (in intensity and polarization), which could be the signature of inhomogeneities in the initial plasma explaining the formation of galaxies. The extremely high sensitivity needed, in terms of equivalent temperature ($\Delta T / T \approx 10^{-6}$), requires the development of amazing THz instruments and detectors. The images of the sky in the millimeter and submillimeter ranges (from 100 to 857 GHz) provided first by the COBE and WMAP satellites, and more recently by the PLANCK satellite, are superb, revealing temperature fluctuations smaller than 1 μK with an angular resolution of a few arc minutes (www.esa.int/our_activities/space_science/planck).

The second THz topic within radioastronomy is the spectroscopy of different astronomical objects:

- Molecular clouds at the origin of stars formation. The evolution of theses clouds is studied by mapping some molecular distribution, through the far infrared emission (e.g., the CO molecule).
- Detection and identification of molecules in interstellar space (what is called nowadays astrochemistry [49]), by recording their far infrared spectra. A large part of these studies is related to big organic molecules [50], which could be at the origin of life on Earth.
- Study of the late evolution stages of stars, by determining their composition from far infrared spectroscopy.
- Observation of the planets of the solar system, mostly to study their atmosphere and to detect and map relevant molecules, like water vapor in the Mars atmosphere.

All these radioastronomy studies in the millimeter and submillimeter range require state-of-the-art instrumentation, which is installed or in construction in ground-based observatories (ALMA, LMT, NOEMA, etc.) or on satellites (HERSCHEL/HIFI, etc.).

2.4 The Need for a Precise Characterization of Materials and Devices

In all the studies and applications described in the preceding paragraphs, the knowledge of the precise complex permittivity of a material, or of its complex index of refraction, is either compulsory or certainly helpful. In the case of the study of materials, it helps to determine the physical origin of the electromagnetic and electric response of the materials, and even more to validate models that fit with experimental data. In the case of imaging, the index of refraction is necessary to determine the expected contrast of the image of a scene. For example, the emissivity $\varepsilon_{\text{therm}}(f)$ (2.19) of a body plays a major role in passive imaging. Thanks to Kirchhoff's law, the emissivity is equal to the normalized amount of light absorbed by a sample, which can be deduced from transmitted and reflected signals. These signals, for simple samples, are easily calculated from the complex value of the index of refraction. The knowledge of this index of refraction is also needed to design optical or quasi-optical components and systems for THz applications. For instance, the focal length of any lens strongly depends on the real part of the refractive index, as well as its transmission (through the Fresnel coefficients), which is even more limited by the absorption losses in the lens material. In fact, the knowledge of the properties of materials is absolutely required to conceive and fabricate any device used to manipulate THz beams or THz fields. Nowadays, a very active domain of THz research concerns metamaterials, waveguides, near-field devices, thin films (i.e., artificial materials with dimensions of the order of or even smaller than the THz wavelength), and so on. The mapping of the THz fields that propagate in such devices depends strongly on the electromagnetic response of each material that constitutes the device. In such examples, the small size of some structures brings an additional difficulty in the sense that the response of nanometric materials can be very different from the bulk one. This difference originates in additional surface effects (structural defects, roughness, scattering of free carriers at the surface, etc.) and/or in the quantum confinement of the carriers. Measuring the THz properties of such shaped materials is again obligatory, at least to know if they differ from the bulk values.

Classical techniques, derived either from optics or from microwaves, give the refractive index or the coefficient of absorption, but rarely both simultaneously.

2.5 Classical Optical Techniques to Determine the Refractive Index of a Material

Generally, in optics, i.e., from the UV to the IR domains, the real part of the index of refraction of a material is obtained by measuring the refraction of a beam at the interface between the material and the surrounding medium. This includes the methods that are summarized in Fig. 2.4. Let us notice that these methods could also be implemented in the microwave domain.

2.5.1 Refractometry: Angle of Deviation

The material to be characterized is a well-defined shape like a rectangular block or, for example, as a prism as shown in Fig. 2.4., then one measures the angle of deviation of an exciting ray. The relation between the angle of deviation D_p, the refractive index n, the angle of incidence θ_o, and the angle α_p of the prism is

$$\sin\left(D_\mathrm{p} - (\theta_\mathrm{o} + \alpha_\mathrm{p})\right) = n \sin\left(\alpha_\mathrm{p} - \sin^{-1}\left(\frac{\sin\theta_\mathrm{o}}{n}\right)\right). \quad (2.25)$$

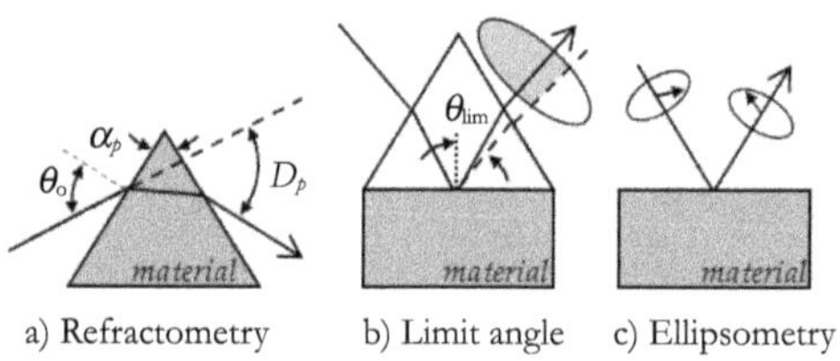

Figure 2.4 Methods for measuring the index of refraction of a material.

In the THz domain, this technique presents two major drawbacks: (i) the requested big amount of material to manufacture a prism whose dimensions are larger than the THz beam diameter (i.e., a few centimeters) and (ii) the difficulty of measuring the angles θ_o and D_p with good accuracy. Typically, an uncertainty of 1° on both θ_o and D_p leads to an error Δn of ~ 0.015 for $n = 1.5$, $\alpha_p = 45^\circ$, and $\theta_o = 30^\circ$.

2.5.2 Refractometry: Limit Angle

The material to be characterized is put against a material of higher index n_p (n_p is known), which is shaped as a prism. From the Snell–Descartes law, it can easily be shown that light impinging from the high-index material side encounters a total reflection at the interface for incident angles larger that the limit angle θ_{lim}:

$$\theta_{lim} = \sin^{-1}\left(\frac{n}{n_p}\right). \tag{2.26}$$

Thus, when looking at the reflected beam (versus the reflection angle at a fixed frequency, or versus frequency at a fixed given angle), one sees a dark zone and a bright zone whose limit appears at θ_{lim}. The measurement of this angle leads to n. Several refractometers (Abbe, Pulfrich, etc.), which differ only by their layout, are based on this principle. The technique, in optics, is more precise than measuring the angle of deviation, because the dark and bright zones are sharply separated.

In the THz domain, these refractometry methods are not employed because of the lack of matrix detectors, which would have made the measurement easy. In addition, many materials cannot be shaped as a prism.

2.5.3 Ellipsometry

This technique is based on the change of the polarization state of an incident beam during reflection at the surface of the material to be characterized. As any incident field can be projected in the plane of incidence (p-polarization) and perpendicularly to this plane (s-polarization), the change can be written as

$$\tan\psi\, e^{j\Delta} = \frac{\tilde{r}_p}{\tilde{r}_s} = \frac{\tilde{r}_{TM}}{\tilde{r}_{TE}}. \tag{2.27}$$

$\tilde{r}_s$ and $\tilde{r}_p$ are the coefficients of reflection for s and p polarized fields (see annex; let us notice that $\tilde{r}_s$ and $\tilde{r}_p$ depend on the complex refractive index $\tilde{n}$ of the material, so they are complex). $\tan\psi$ is the amplitude of this ratio, and Δ its phase. ψ and Δ are measured using rotating polarizers located in the incident and reflected beams. Solving 2.28 with these measured data leads to the value of $\tilde{n}$:

$$\tilde{n} = \tan\theta\sqrt{1 - \frac{2\sin^2\theta}{\left(1 + \tan\psi\, e^{j\Delta}\right)^2}}. \tag{2.28}$$

Ellipsometry gives both real and imaginary parts of $\tilde{n}$; it does not require any reference measurement and it is extremely sensitive as both the amplitude and phase differences are measured. Also it can successfully be applied to detect and characterize substrates covered by thin films, to study anisotropic materials, and so on. Unfortunately, ellipsometry is not accurate for low absorption materials.

Ellipsometry in the far infrared was developed to study the electromagnetic response of materials, like metals, superconductors, semiconductors, etc. First ellipsometers were built with molecular lasers as sources, and bolometers as detectors [51]. More recently, ellipsometers based on THz-TDS were proposed [52]. They will be further described in Chapter 5.

2.5.4 Interferometric Methods

In amplitude splitting interferometers, like Michelson or Mach–Zehnder interferometers, the beam provided by a source is divided in two beams that propagate along two arms and recombine at the detector. Thanks to the principle of superposition, the resulting electromagnetic field is the sum of the two impinging fields. As the fields are trigonometric functions, the resulting field varies as the cosine of the phase difference $\Delta\phi$ between the two arms, which is written as

$$\Delta\phi = \frac{\omega}{2c}\Delta\left(nx\right). \tag{2.29}$$

Here $\Delta(nx)$ is the difference of the product length times refractive index, which is often called optical length. Thus the intensity of the recorded pattern is proportional to $\cos^2\Delta\phi$. The sample to be characterized is installed in one arm of the interferometer, and one looks at the shift of the interference pattern induced by the sample. By inverting Eq. 2.30, one easily gets n if the sample thickness is known. The major difficulty appears when $\Delta\phi$ is larger than π, because of the periodicity of the cosine function (to the square). Also, Fabry–Perot interference inside the sample could make the analysis more difficult. Therefore, the method is not really applicable to thick samples, for which the induced phase variation $\Delta\phi$ is almost impossible to determine. It gives good results for thin films (either stand-alone or deposited on a substrate) and for very low index materials, like gas.

In the far infrared, a Martin–Puplett interferometer is often used [53], which is nothing but a Michelson interferometer in which the beam splitter is a grid polarizer. This configuration does not suffer from interferences occurring in the beam splitter. In addition, the detected signal is less sensitive, compared with the Michelson interferometer, to any fluctuation in the optical path or of the source and detector efficiencies [54]. Such Michelson or Martin–Puplett interferometers are used to perform Fourier transform far-infrared (FTIR) spectroscopy, which will be further explained in Chapter 3.

2.6 Classical Optical Techniques Used to Determine the Absorption of a Material

Absorption is easier to determine than the refraction index. Besides ellipsometry, already described, a simple technique is based on measuring both incident and transmitted signals using a slab with parallel faces of material to be characterized. The coefficient of field transmission of such a slab is given by relation A.33 of the annex:

$$\tilde{t} = \frac{4\tilde{n}}{(\tilde{n}+1)^2} \times \frac{e^{j\frac{\omega}{c}(\tilde{n}-1)d}}{1-\left(\frac{\tilde{n}-1}{\tilde{n}+1}e^{-j\frac{\omega}{c}\tilde{n}d}\right)^2} \equiv \tilde{t}_{\mathrm{Fresnel}} \times \frac{e^{j\frac{\omega}{c}(\tilde{n}-1)d}}{1-FP(\tilde{n},d,\omega)}. \tag{2.30}$$

Expression 2.31 has been established for a sample of thickness d and of index of refraction $\tilde{n}$, which is located in air and illuminated at normal incidence. If the Fabry–Perot term FP in Eq. 2.31 can be neglected (incoherent light illumination, faces not rigorously parallel ...), the expression of $\tilde{t}$ includes only the Fresnel coefficient $\tilde{t}_{\text{Fresnel}}$ and a phase term due to propagation in the sample. The best way of getting the absorption coefficient $\alpha = 2\kappa\omega/c = 2\text{Im}(\tilde{n})\omega/c$ is to measure the transmitted intensity for two samples of different thicknesses d_1 and d_2, and then to take the ratio of both signals:

$$\frac{T_2}{T_1} = \frac{|t_2|^2}{|t_1|^2} = e^{\frac{\omega}{c}\kappa(d_1-d_2)} \quad \Rightarrow \quad \alpha = -2\frac{\ln T_1 - \ln T_2}{d_1 - d_2}. \tag{2.31}$$

Sometimes, when samples of different thicknesses are not available, the transmission (2.30) of a single slab is measured and the Fresnel coefficient t_{Fresnel} is supposed to be equal to 1. If absorption is small ($\kappa \ll n$), the Fresnel coefficient for intensity is

$$T_{\text{Fresnel}} = |t_{\text{Fresnel}}|^2 \approx \frac{16\,n^2}{(n+1)^4}. \tag{2.32}$$

The error made when neglecting t_{Fresnel} is less than 10% for $n < 1.6$. For larger values, this simplification leads to a bad estimation of α. It should be noted that the above also assumes negligible scattering in the material. If it is not, then the calculated value is the attenuation coefficient which is the sum of the absorption and scattering coefficients (see Chapter 10).

2.7 Microwaves Techniques

In the microwave domain, optical techniques (or free-space techniques) are still usable, but generally measurements are performed in a guided wave configuration. The experiments are in fact quite easy to install, because there exist commercial equipment, namely vector network analyzers (VNAs), which measure the magnitude and phase of signals transmitted and reflected by a device. This is a great advantage compared with the optical techniques that are described above. A VNA includes a tunable source and a heterodyne receiver. Signals are carried to and from the device by coaxial cables or, at higher frequencies, by microwaves waveguides.

VNAs do not deliver the coefficients of transmission or of reflection of a device, but the elements s_{ij} of the scattering matrix S (or S matrix). Let us consider a simple electrical quadrupole

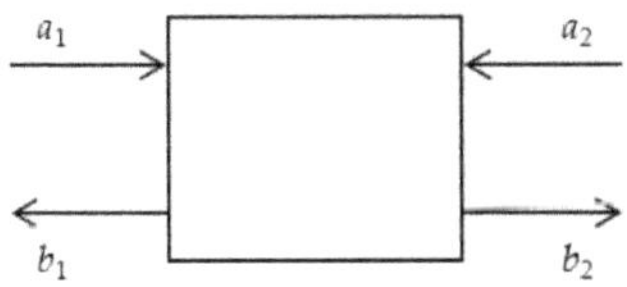

Figure 2.5 Electrical quadrupole (2-port network).

(Fig. 2.5). The current could be injected to the device from the left side (i_1) or from the right side (i_2). These currents are both reflected and transmitted by the quadrupole. One defines the incident and reflected power waves at port i by

$$a_i = \frac{V_i + Z_o i_i}{2\sqrt{|Re(Z_o)|}} = \frac{V_i + Z_o i_i}{2\sqrt{Z_o}}, \quad b_i = \frac{V_i - Z_o^* i_i}{2\sqrt{|Re(Z_o)|}} = \frac{V_i - Z_o i_i}{2\sqrt{Z_o}}. \tag{2.33}$$

Z_o is the impendence of the line or waveguide that feeds the quadrupole (practically, Z_o is real, leading to the second terms of the relations in Eq. 2.33). V_i is the voltage at port i. From Eq. 2.33, one easily gets

$$\begin{pmatrix} b_1 \\ b_2 \end{pmatrix} = S \begin{pmatrix} a_1 \\ a_2 \end{pmatrix} = \begin{pmatrix} S_{11} & S_{12} \\ S_{21} & S_{22} \end{pmatrix} \begin{pmatrix} a_1 \\ a_2 \end{pmatrix}, \tag{2.34}$$

with

$$S_{11} = \frac{V_1^-}{V_1^+}, \quad S_{21} = \frac{V_2^-}{V_1^+}, \quad S_{12} = \frac{V_1^-}{V_2^+}, \quad S_{22} = \frac{V_2^-}{V_2^+}. \tag{2.35}$$

V_i^+ and V_i^- are respectively the incident and reflected voltages at port i. Therefore, the elements of the S matrix correspond to the transmission and reflection coefficients, defined in terms of voltage ratio.

The characterization of a sample is realized by inserting the sample in the waveguide and by measuring the S-matrix elements with and without the sample; the material parameters are extracted from the difference. Generally, this extraction procedure is complicated, especially if several modes of the waveguide are excited. Thus it is advised to run the waveguide in a single-mode operation, if possible the lowest mode. Even in a single-mode operation, the presence of the sample in the waveguide perturbs strongly the wave propagation if the sample is rather big. In this case,

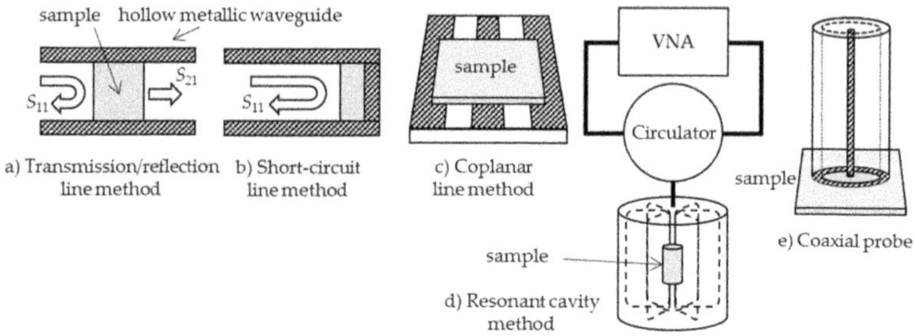

Figure 2.6 Devices for waveguide microwave spectroscopy: (a) hollow metallic waveguide in transmission, (b) hollow metallic waveguide in reflection, (c) coplanar waveguide, (d) resonant microwave cavity (here a cylindrical cavity in which the magnetic field lines of the TE01 mode are shown [dotted lines]), and (e) coaxial probe.

an accurate estimation of the material parameters requires efficient computation codes. When the sample is small or thin, perturbation theories, which lead to analytical expressions, may be used to extract the material parameters.

The most popular microwaves techniques are as follows [55]:

- *Transmission line methods.* The sample is placed inside a hollow waveguide (Figs. 2.6a and 2.6b) or over a 2D waveguide (coplanar waveguide, microstrip line, etc.) (Fig. 2.6c). The sample induces a change of the propagation properties of the transmission line, namely its complex impendance, from which the material parameters could be determined. The parameters extraction is easier if a single mode of the waveguide is excited, and also if the sample induces only a perturbative change in the waveguiding properties. The impendance of the empty waveguide must be well matched to the VNA, in order to avoid standing wave effects. As well, the sample should not induce too strong reflections that may interfere with the signal. Therefore, transmission line methods are well adapted to high-loss materials.
- *Resonant cavity techniques.* The sample is inserted in a resonant cavity, resulting in a shift of the resonance frequency. It is well adapted to rather transparent samples, for

which the technique is precise. Tuning the frequency over a wide range is difficult [56]. If the sample is thin, analytical expressions give both the real and imaginary parts of the dielectric constant, from the measured resonance shift and modification of the cavity quality factor [57]. However, the effect of a thin transparent sample is quite weak and thus the method is not so precise (a few percents) for low-absorbtion materials.

- *Reflection at the open end of coaxial cables.* The cable (probe) is immersed in the material to be characterized, which changes the reflection of the guided microwave signal. The accuracy of the technique is moderate, but it can be applied over rather wide spectral ranges [58].

These microwave techniques can be extended to the THz range. VNAs working at THz frequencies are commercially available (AB Millimètre, Rhode & Schwartz, etc.), but they remain expensive and sweeping a large frequency band requires several sources/detectors.

2.8 The Amazing Time-Domain Technique

In the early 1970s, a revolutionary time-domain technique, not based on refractive or resonance effects, but on the propagation of short electromagnetic pulse, was proposed by Nicolson and Ross [54]. The basic idea was to detect the changes of the temporal shape of the short pulse, induced by propagation through the studied sample. In the original paper by Nicolson and Ross [54], the sample was inserted in a coaxial line, and the pulse rise time was sub-nanosecond. The waveforms reflected and transmitted by the region where the sample is inserted, were measured, and then were Fourier-transformed. Their ratio gave the reflection and transmission scattering parameters S of the sample region, from which the complex permeability and permittivity of the sample material were deduced. In this way, Nicolson and Ross determined the complex permeability and permittivity of Teflon and nylon from 0.1 to 100 GHz. A few years later, the method was applied to higher

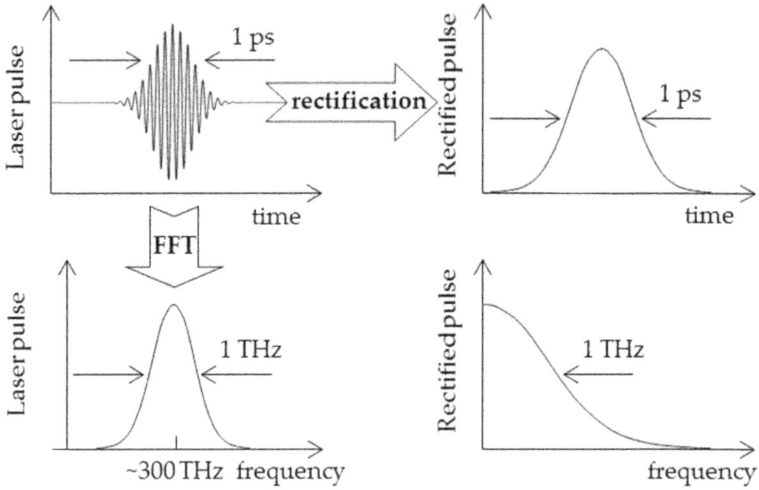

Figure 2.7 Principle of THz broadband generation with ultrashort optical pulses.

frequencies by the team of D. Auston at the Bell Laboratories [60]. To reach the THz range, Auston used a mode-locked laser to produce picosecond electrical pulses and the signals were detected using a photo-conducting switch. Let us notice that the picosecond pulse must not be a wave packet, i.e., a carrier frequency time-limited by an envelope, as it is for ultrashort optical pulses (see Fig. 2.7). The pulse should correspond only to the envelope of the laser pulse or, in other words, the pulse duration should be shorter than the period of the far-infrared frequencies. In such a way, the spectrum of the pulse, as obtained by a Fourier transform of the temporal waveform, is maximum for the zero frequency, and not centered around the carrier frequency (again, for femtosecond laser pulses, the spectrum can be as broad as several tens of THz, but it is centered around the main optical frequency, i.e., around 300 $\sim$ 600 THz for visible or IR pulses).

An easy way to obtain such pulses is to rectify femtosecond laser pulses, i.e., to shift their carrier frequency from the optical range to zero. No electronics is fast enough, even today, to record the shape of the THz waveform and to resolve sub-picosecond signal variations. Time-equivalent sampling, as proposed initially by

Auston, is necessary. This requires a train of laser pulses, coherent detection, and synchronization between emission and detection. Mode-locked lasers deliver such trains of femtosecond pulses with a very stable repetition rate, and even more with a time jitter between two consecutive pulses that is typically of the order or below 1 fs for Ti:Sa Kerr mode-locked lasers, i.e., 10^{-6} of the repetition rate [61]. These performances ensure a high-quality sampling process. By coherent detection, one means that the receiver does not integrate the energy of the incoming pulse, but on the contrary is able to record the signal over a time gate much smaller than any interesting detail of the THz waveform. This was first performed with photo-conducting switches made of semiconductors that exhibit an ultrashort photo-carrier lifetime, due to a high density of traps (structural or stoichiometric defects). An alternative method is electro-optic detection in a nonlinear crystal that mixes THz and optical signals; it is also a very competitive technique showing a very large bandwidth. Finally, the synchronism between emission and detection is achieved by using the same laser beam to drive both emitter and receiver: the beam delivered by the mode-locked laser is separated into two beams by a beam splitter. Thus, identical but delayed pulses are incident onto emitter and detector.

Of course, the main advantage of this optoelectronic sampling technique is the possibility to generate picosecond rectified pulses and to sample them with a great temporal precision. A second remarkable advantage arises from the detection principle: the receiver is sensitive ("on") during a very short period, typically sub-picosecond (Fig. 2.8). Outside this window, it is "off," which

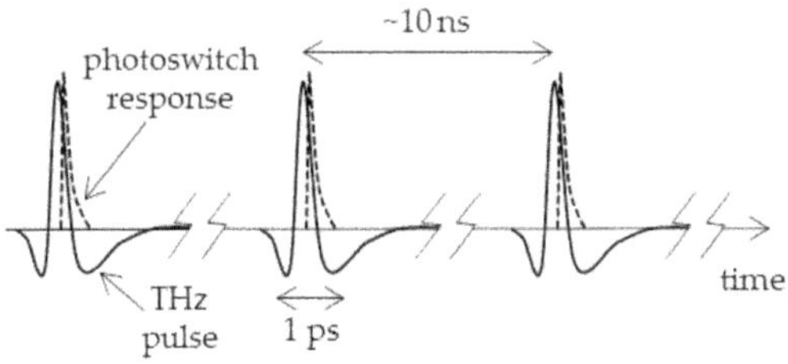

Figure 2.8 Chronology of THz pulses (continuous line) and of the receiver response (dashed line).

Table 2.3 List of most manufacturers of THz spectrometers

Accurion	Germany	www.accurion.com	THz-TDS systems in reflexion.
Advantest	Japan	www.advantest.com	THz-TDS systems (transmission, reflection, ATR...)
Aispec	Japan	www.aispec.com	THz-TDS systems
ARP	USA	http://arphotonics.net/	THz spectrometers based on dendrimer emitters
Bakman Techn.	USA		CW-optoelectronic THz systems
Batop Optoelectronics	Germany	www.batop.de	THz-TDS and fibered THz-TDS systems
Del Mar	USA	www.delmarphotonics.com	THz-TDS systems
Laser Quantum (GigaOptics)	Germany	www.laserquantum.com	ASSP-THz TDS systems
Menlo	Germany	www.menlosystems.com	THz-TDS systems
MicroTech	UK	www.mtinstruments.com	Spectrometers including a BWO source
Picometrix	USA	www.picometrix.com	Fibered THz-TDS systems
Teraview	UK	www.teraview.com	THz-TDS and THz-CW systems working in transmission and reflexion, with imaging options
Teravil	Lithuania	www.teravil.lt	THz-TDS systems
TeTechS	Canada	www.TeTechS.com	THz-TDS kit and components.
Toptica	Germany	www.toptica.com	Fibered THz-TDS and CW systems

means it records only a weak dark current contribution, but no noise originating from the ambient thermal radiation or from the electronic instrumentation. Typically, for photo-conducting receivers, the detection time-window is shorter than 1 ps, while the delay between two consecutive pulses is 10 ns. Thus, compared with a detector of energy, which is not so chopped (on/off), the noise contribution here is 10^4 times smaller! With classical photo-conducting receivers that exhibit an excellent on/off ratio, this leads to impressive dynamic range of about 60 dB! At the end of the 1980s, the initial work by Chueng and Auston [60] was improved by two teams of the IBM company, in the millimeter domain [62] and in the THz domain [63], which designed setups [62, 64] and proposed methodology [62, 63, 65] which are still employed nowadays.

Today, the number of publications related to THz-TDS is huge, and most of the materials and devices have been characterized accordingly. For common THz-TDS setups, the achievable frequency range spreads typically from 0.1 THz up to 5 THz, allowing researchers to obtain the response of materials and devices typically up to $3 \sim 4$ THz. Broader spectral ranges are possible using ultrashort laser pulses of duration shorter than 20 fs that can produce some tens of THz [66], and spectra of materials are published up to $15 \sim 20$ THz [67], i.e., they spread from the millimeter domain (0.1 THz $\leftrightarrow \lambda = 3$ mm) to the infrared (20 THz $\leftrightarrow \lambda = 15$ μm). Moreover, new systems based on THz generation and detection by photoionization of air give similar ultra-broadband performance [68–70]. Nowadays several companies are selling THz-TDS apparatus. The table below lists the suppliers as of January 2013. Some of these systems are very compact and are powered with batteries.

References

1. Houghton, J. T., and Smith, S. D. (1966). *Infra-red Physics*, Clarendon Press, Oxford.
2. Nolt, I. G., Radostitz, J. V., Dilonardo, G., Evenson, K. M., Jennings, D. A., Leofold, K. R., Vanek, M. D., Zink, L. R., Hinz, A., and Chance, K. V. (1987). Accurate rotational constants of CO, HCI, and HF: spectral standards for

the 0.3- to 6-THz (10- to 200-cm^{-1}) region, *J. Mol. Spectrosc.*, **125**, pp. 274–287.

3. Huber, K. P., and Herzberg, G. (1979). *Molecular Spectra and Molecular Structure*, Vol. IV *Constants of Diatomic Molecules*, Van Nostrand, New York.
4. Nakayama, T. (2002). Boson peak and terahertz frequency dynamics of vitreous silica, *Rep. Prog. Phys.*, **65**, pp. 1195–1242.
5. Laib, J. P., Nickel, D. V., and Mittleman, D. M. (2010). Terahertz vibrational modes induced by heterogeneous nucleation in n-alkanes, *Chem. Phys. Lett.*, **493**, pp. 279–282.
6. Schall, M., Walther, M., and Uhd Jepsen, P. (2001). Fundamental and second-order phonon processes in CdTe and ZnTe, *Phys. Rev. B*, **64**, pp. 094301.
7. Wan, F., Han, J., and Zhu, Z. (2008). Dielectric response in ferroelectric $BaTiO_3$, *Phys. Lett. A*, **372**, pp. 2137–2140.
8. Walther, M., Jensby, K., Keiding, S. R., Takahashi, H., and Ito, H. (2000). Far-infrared properties of DAST, *Opt. Lett.*, **25**, pp. 911–913.
9. Reuter, G. E. H., and Sondheimer, E. H. (1948). The theory of the anomalous skin effect in metals, *Proc. R. Soc. London, Ser. A*, **195**, pp. 336–364.
10. Yang, Y., Shutler, A., and Grischkowsky, D. (2011). Measurement of the transmission of the atmosphere from 0.2 to 2 THz, *Opt. Express*, **19**, pp. 8830–8838.
11. Watanabe, Y., Kawase, K., Ikari, T., Ito, H., Ishikawa, Y., and Minamide, H. (2003). Component spatial pattern analysis of chemicals using terahertz spectroscopic imaging, *Appl. Phys. Lett.*, **83**, pp. 800–802.
12. Shen, Y. C., Taday, P. F., Newnham, D. A., Kemp, M. C., and Pepper, M. (2005). 3-D chemical mapping using terahertz pulsed imaging, *SPIE*, **5727**, pp. 24–31.
13. Kemp, M. C., Glauser, A., and Baker, C. (2006). Recent developments in people screening using terahertz technology - seeing the world through terahertz eyes, *Proc. SPIE*, **6212**, pp. 62120T.
14. Luukanen, A., Miller, A. J., and Grossman, E. N. (2005). Passive hyper-spectral terahertz imagery for security screening using a cryogenic microbolometer, *Proc. SPIE*, **5789**, http://dx.doi.org/10.1117/12.608838
15. Meilhan, J., Dupont, B., Goudon, V., Lasfargues, G., Lalanne Dera, J., Nguyen, D. T., Ouvrier-Buffet, J. L., Pocas, S., Maillou, T., Cathabard, O.,

Barbieri, S., and Simoens, F. (2011). Active THz imaging and explosive detection with uncooled antenna-coupled microbolometer arrays, *Proc. SPIE*, **8023**, pp. 80230E.

16. Hamster, H., Sullivan, A., Gordon S., and Falcone, R. W. (1994). Short-pulse terahertz radiation from high-intensity-laser-produced plasmas, *Phys. Rev. E*, **49**, pp. 671–677.
17. Liu, J., Dai, J., and Zhang, X.-C. (2010). Broadband terahertz wave remote sensing using coherent manipulation of fluorescence from asymmetrically ionized gases, *Nat. Photonics*, **4**, pp. 627–631.
18. Trofimov, V. A., Varentsova, S. A., and Chen, J. (2010). Method of THz spectrum dynamics analysis for identification of compound medium, *Proc. SPIE*, **7671**, pp. 76710G.
19. Walther, M., Plochocka, P., Fischer, B., Helm, H., and Uhd Jepsen, P. (2002). Collective vibrational modes in biological molecules investigated by terahertz time-domain spectroscopy, *Biopolymers*, **67**, pp. 310–313.
20. Brucherseifer, M., Nagel, M., Haring Bolivar, P., Kurz, H., Bosserhoff, A., and Buttner, R. (2000). Label-free probing of the binding state of DNA by time-domain terahertz sensing, *Appl. Phys. Lett.*, **77**, pp. 4049–4051.
21. Kim, S. J., Born B., Havenith, M., and Gruebele, M. (2008). Real-time detection of protein–water dynamics upon protein folding by terahertz absorption spectroscopy, *Angew. Chem. Int. Ed.*, **47**, pp. 6486–6489.
22. Png, G. M., Flook, R., Ng, B. W.-H., and Abbott, D. (2009). Terahertz spectroscopy of misfolded proteins in bio-tissue, *Conference IRMMW-THz 2009*, Busan, Korea.
23. Tuszynski, J., Brown, J. A., Crawford, E., Carpenter, E. J., Nip, M. L. A., Dixon, J. M., and Satari, M. V. (2005). Molecular dynamics simulations of tubulin structure and calculations of electrostatic properties of microtubules, *Math. Comput. Modell.*, **41**, pp. 1055–1070.
24. Vos, M. H., Rappaport, F., Lambry, J. C., Breton, J., and Martin, J.-L. (1993). Visualization of coherent nuclear motion in a membrane protein by femtosecond spectroscopy, *Nature*, **363**, pp. 320–325.
25. Jördens, C., Scheller, M., Breitenstein, B., Selmar, D., and Koch, M. (2009). Evaluation of leaf water status by means of permittivity at terahertz frequencies, *J. Biol. Phys.*, **35**, pp. 255–264.
26. Pickwell, E., Cole, B. E., Fitzgerald, A. J., Pepper, M., and Wallace, V. P. (2004). In vivo study of human skin using pulsed terahertz radiation, *Phys. Med. Biol.*, **49**, pp. 1595–1607.

27. Ashworth, P. C., Pickwell-MacPherson, E., Provenzano, E., Pinder, S. E., Purushotham, A. D., Pepper, M., and Wallace, V. P. (2009). Terahertz pulsed spectroscopy of freshly excised human breast cancer, *Opt. Express*, **17**, pp. 12444–12454.
28. Taylor, Z. D., Singh, R. S., Bennett, D. B., Tewari, P., Kealey, C. P., Bajwa, N., Culjat, M. O., Stojadinovic, A., Hubschman, J.-P., Brown, E. R., and Grundfest, W. S. (2011). THz medical imaging: in vivo hydration sensing, *IEEE Trans. THz Sci. Technol.*, **1**, pp. 201–219.
29. Arbab, M. H., Dickey, T. C., Winebrenner, D. P., Chen, A., and Mourad, P. D. (2011). Characterization of burn injuries using terahertz time domain spectroscopy, *SPIE Advanced Biomedical and Clinical Diagnostic Systems IX*, San Francisco, California, USA, pp. 78900Q-7.
30. Singh, R. S., Tewari, P., Bourges, J. L., Hubschman, J. P., Bennett, D. B., Taylor, Z. D., Lee, H., Brown, E. R., Grundfest, W. S., and Culjat, M. O. (2010). Terahertz sensing of corneal hydration, *Conf. Proc. IEEE Eng. Med. Biol. Soc.*, **2010**, pp. 3021–3024.
31. Wilmink, G. J., Rivest, B. D., Ibey, B. L., Roth, C. L., Bernhard, J., and Roach, W. P. (2010), Quantitative investigation of the bioeffects associated with terahertz radiation, *Proc. SPIE*, **7562**, pp. 75620L.
32. Siegel, P. H. (2004). Terahertz technology in biology and medicine, *IEEE Trans. Microwave Theory Tech.*, **52**, pp. 2438–2447.
33. Titova, L. V., Ayesheshim, A. K., Golubov, A., Fogen, D., Rodriguez-Juarez, R., Hegmann, F. A., and Kovalchuk, O. (2013). Intense THz pulses cause H2AX phosphorylation and activate DNA damage response in human skin tissue, *Biomed. Opt. Express*, **4**, pp. 559–568.
34. Siegel, P. H., and Pikov, V. (2010). Impact of low intensity millimetre waves on cell functions, *Electron. Lett.*, **46**, pp. S70–S72.
35. Korestein-Ilan, A., Barbul, A., Hasin, P., Eliran, A., Gover, A., and Korestein, R. (2008). Terahertz radiation increases genomic instability in human lymphocytes, *Radiat. Res.*, **170**, pp. 224–234.
36. Orlando, A. R., and Gallerano, G. P. (2009). Terahertz radiation effects and biological applications, *J. Infrared Millimeter Waves*, **30**, pp. 1308–1318.
37. Jansen, C., Wietzke, S., Peters, O., Scheller, M., Vieweg, N., Salhi, M., Krumbholz, N., Jördens, C., Hochrein, T., and Koch, M. (2010). Terahertz imaging: applications and perspectives, *Appl. Opt.*, **49**, pp. E48–E57.
38. Zhong, H., Karpowicz, N., Jingzhou, X., Yanqing, D., Ussery, W., Shur, M., and Zhang, X.-C. (2004), Inspection of space shuttle insulation foam defects using a 0.2 THz Gunn diode oscillator, *Conference Digest of*

the 2004 Joint 29th International Conference on Infrared and Millimeter Waves and 12th International Conference on Terahertz Electronics, pp. 753–754.

39. Paine, S., and Blundell, R. (2004). The Terahertz Atmosphere, *Proceedings of the 15th International Symposium on Space Terahertz Technology*, pp. 418–425.
40. Jucks, K. W., Johnson, D. G., Chance, K. V., Traub, W. A., Margitan, J. J., Osterman, G. B., Salawitch, R. J., and Sasano, Y. (1998). Observation of OH, HO_2, H_2O, and O_3 in the upper stratosphere: implications for HOx photochemistry, *Geophys. Res. Lett.*, **21**, pp. 3935–3938.
41. Hindle, F., Yang, C., Mouret, G., Cuisset, A., Bocquet, R., Lampin, J.-F., and Ducournau, G. (2009). Recent developments of an opto-electronic THz spectrometer for high-resolution spectroscopy, *Sensors* (Basel, Switzerland), **9**, pp. 9039–9057.
42. Melinger, J. S., Yang, Y., Mandehgar, M., and Grischkowsky, D. (2012). THz detection of small molecule vapors in the atmospheric transmission windows, *Opt. Express*, **20**, pp. 6788–6807.
43. Su, K., Moeller, M., Barat, R. B., and Federici, J. E. (2012). Experimental comparison of terahertz and infrared data signal attenuation in dust clouds, *J. Opt. Soc. Am. A*, **29**, pp. 2360–2367.
44. Nagatsuma, T., Horiguchi, S., Minamikata, Y., Yoshimizu, Y., Hisatake, S., Kuwano, S., Yoshimoto, N., Terada, J., and Takahashi, T. (2013). Terahertz wireless communications based on photonics technologies, *Opt. Express*, **21**, pp. 23736–23747.
45. Taday, P. F., Bradley, I. V., Arnone, D. D., and Pepper M. (2003). Using terahertz pulse spectroscopy to study the crystalline structure of a drug: a case study of the polymorphs of ranitidine hydrochloride, *J. Pharm. Sci.*, **92**, pp. 831–838.
46. Zewail, A. H. (1994). *Femtochemistry: Ultrafast Dynamics of the Chemical Bond*, Vols. 1 and 2, World Scientific, Singapore.
47. Ebbinghaus, S., Kim, S. J., Heyden, M., Yu, X., Heugen, U., Gruebele, M., Leitner, D. M., and Havenith, M. (2007). An extended dynamical hydration shell around proteins, *Proc. Natl. Acad. Sci. U. S. A.*, **104**, pp. 20749–20752.
48. Heyden, M., Sun, J., Forbert, H., Mathias, G., Havenith, M., and Marx, D. (2012). Understanding the origins of dipolar couplings and correlated motion in the vibrational spectrum of water, *J. Phys. Chem. Lett.*, **3**, pp. 2135–2140.

49. Shaw, A. M. (2006). *Astrochemistry: From Astronomy to Astrobiology*, Wiley, Chichester, UK.
50. Jorgensen, J. K., Favre, C., Bisschop, S. E., Bourke, T. L., van Dishoeck, E. F., and Schmalzl, M. (2012). Detection of the simplest sugar, glycolaldehyde, in a solar-type protostar with ALMA, *Astrophys. J. Lett.*, **757**, pp. L4–L9.
51. Barth, K.-L., and Keilmann, F. (1993). Far-infrared ellipsometer, *Rev. Sci. Instrum.*, **64**, pp. 870.
52. Nagashima, T., and Hangyo, M. (2001). Measurement of complex optical constants of a highly doped Si wafer using terahertz ellipsometry, *Appl. Phys. Lett.*, **79**, pp. 3917–3919.
53. Martin, D. H., and Puplett, E. (1969). Polarized interferometric spectroscopy for the millimeter and submillimeter spectrum, *Infrared Phys.*, **10**, pp. 105–109.
54. Lambert, D. K., and Richards, P. L. (1978). Martin-Puplett interferometer: an analysis, *Appl. Opt.*, **17**, pp. 1595–1602.
55. Komarov, V., Wang, S., and Tang., J. (2005). Permittivity and measurements, in *Encyclopedia of RF and Microwave Engineering*, Chang, K. (ed.), John Wiley & Sons, New York.
56. Sheen, J. (2005). Study of microwave dielectric properties measurements by various resonance techniques, *Measurement*, **37**, pp. 123–130.
57. Dube, D. C., Lanagan, L. T., Kim, J. H., and Jang, S. J. (1988). Dielectric measurements on substrates materials at microwave frequencies using cavity perturbation technique, *J. Appl. Phys.*, **63**, pp. 2466–2468.
58. Mosig, J. R., Besson, J. C. E., Gex-Fabry, M., and Gardiol, F. E. (1981). Reflection of an open-ended coaxial line and application to nondestructive measurement of materials, *IEEE Trans. Instrum. Meas.*, **30**, pp. 46–51.
59. Nicolson, A. M., and Ross, G. F. (1970). Measurement of the intrinsic properties of materials by time-domain techniques, *IEEE Trans. Instrum. Meas.*, **19**, pp. 377–382.
60. Cheung, K. P., and Auston, D. H. (1986). A novel technique for measuring far-infrared absorption and dispersion, *Infrared Phys.*, **26**, pp. 23–27.
61. Kim, J., Chen, J., Cox, J., and Kärtner, F. X. (2007). Attosecond-resolution timing jitter characterization of free-running mode-locked lasers, *Opt. Lett.*, **32**, pp. 3519–3521.
62. Pastol, Y., Arjavalingam, G., Halbout, J.-M., and Kopcsay, G. V. (1989). Coherent broadband microwave spectroscopy using picosecond optoelectronic antennas, *Appl. Phys. Lett.*, **54**, pp. 307–309; idem, Dielectric

properties of uniaxial crystals measured with optoelectronically generated microwave transient radiation, *Appl. Phys. Lett.*, **55**, pp. 2277–2279.

63. van Exter, M., Fattinger, Ch., and Grischkowsky, D. (1989). Terahertz time-domain spectroscopy of water vapour, *Opt. Lett.*, **14**, pp. 1128–1130.
64. Fattinger, Ch., and Grischkowsky, D. (1988). Point source teraHz optics, *Appl. Phys. Lett.*, **53**, 1480–1482; idem (1989). TeraHz beams, *Appl. Phys. Lett.*, **54**, pp. 490–492.
65. van Exter, M., and Grischkowsky, D. (1990). Optical and electronic properties of doped silicon from 0.1 to 2 THz, *Appl. Phys. Lett.*, **56**, pp. 1694–1696.
66. Han, P. Y., and Zhang, X.-C. (1998). Coherent, broadband midinfrared terahertz beam sensors, *Appl. Phys. Lett.*, **73**, pp. 3049–3051.
67. Shen, Y. C., Upadhya, P., Linfield, E. H., Beere, H. E., and Davies, A. G. (2003). Ultrabroadband terahertz radiation from low-temperature-grown GaAs photoconductive emitters, *Appl. Phys. Lett.*, **83**, pp. 3117–3119.
68. Vieweg, N., Fischer, B. M., Reuter, M., Kula, P., Dabrowski, R., Celik, M. A., Frenking, G., Koch, M., and Jepsen, P. U. (2012). Ultrabroadband terahertz spectroscopy of a liquid crystal, *Opt. Express*, **20**, pp. 28249–28256.
69. Karpowicz, N., Dai, J., Lu, X., Chen, Y., Yamaguchi, M., Zhao, H., Zhang, X.-C., Zhang, L., Zhang, C., Price-Gallagher, M., Fletcher, C., Mamer, O., Lesimple, A., and Johnson, K. (2008). Coherent heterodyne time-domain spectrometry covering the entire "terahertz gap", *Appl. Phys. Lett.*, **92**, pp. 011131.
70. Ho, I.-C., Guo, X., and Zhang, X.-C. (2010). Design and performance of reflective terahertz air-biased-coherent-detection for time-domain spectroscopy, *Opt. Express*, **18**, pp. 2872–2883.

Chapter 3

Classical Far-Infrared Spectroscopy Techniques

3.1 Introduction

Spectroscopy is the science that makes use of waves, usually over a range of wavelengths or energies, to study matter, in the form of either gas, liquids, solids, or artificial structures. Matter can transmit, scatter, reflect, or radiate waves. The study of the spectrum of these waves gives much information about the matter it interacts with, including composition, density, homogeneity at the wavelength scale, and so forth. Spectroscopy employing electromagnetic waves is actually the most widespread technique, which offers the advantage of remote sensing capability. Spectroscopy, which can be based on radioactivity, sub-atomic particles flux, electrical currents, or even acoustic waves, is rarely used, or applied in limited specialized areas. In this book, we are interested in the far-infrared region of the electromagnetic spectrum, and mostly with spectroscopic measurements that are performed in laboratories using a spectrometer. Spectrometer is the generic name of any spectroscopic instrument which performs measurements over a range of wavelengths. A monochromator is a type of spectrometer

Principles of Terahertz Time-Domain Spectroscopy
Jean-Louis Coutaz, Frédéric Garet, and Vincent Wallace

ISBN 978-981-4774-56-7 (Hardcover), 978-0-203-70969-6 (eBook)
www.panstanford.com

that typically uses a narrowband input source that delivers a monochromatic output beam onto the detector; recording a whole spectrum is achieved by rotating a dispersive element (e.g., a prism) of the spectrometer or scanning the detector position. A spectrograph is a spectrometer that images the studied spectrum onto an array of detectors, so that the spectrum is recorded all at once.

In the far infrared, specific narrow features in the spectrum of the light are fingerprints of the vibrations of entire molecules, of the excitation of phonons, of transitions between shallow energy levels, etc. Broadened spectral tendencies render for the nonresonant polarizability of matter, the influence of free charges, scattering by defects or heterogeneities, and so on. Complementary to the study of these basic phenomena, which permits better understanding of the properties and behaviors of the matter under study, absorption coefficient and refractive index can be extracted from the spectra, which are important input values to design devices for the THz domain, or components and systems running at sub-THz/THz frequencies.

Spectrometers can be divided in two categories (Fig. 3.1), either for emission spectroscopy or for absorption/impedance spectroscopy. In emission spectroscopy, as matter naturally radiates electromagnetic waves, an optical system collects and focuses a part of this emitted radiation onto the spectrometer. The spectrometer is a dispersive system that separates the different wavelengths of the incoming spectrum and directs a very narrow bandwidth beam onto the receiver. The output wavelength may be tuned in order to record the whole spectrum of the incident radiation. In absorption/impedance spectroscopy, matter is inactive at the wavelengths of interest and thus must be illuminated by a source of

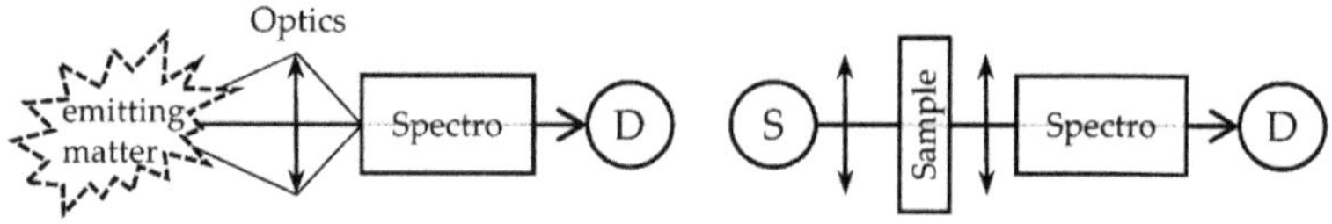

Figure 3.1 Scheme of a spectrometer (left: emission spectroscopy; right: absorption spectroscopy). Spectro is for spectrometer, D for receiver, and S for source.

radiation. Optical components collect and direct the beam delivered by the source onto the sample. The transmitted, reflected, or scattered beam is focused into the spectrometer, and then the spectrally filtered beam impinges the receiver. Note that, strictly speaking, attenuation and not absorption will be recorded for scattering samples (attenuation includes losses by absorption and scattering). The sample could be located before, after, or even inside the spectrometer.

Here we will treat only absorption/impedance spectroscopy, as THz-TDS addresses these parameters. However, emission spectroscopy in the far infrared is widely used, especially in radioastronomy where the remote sources of radiation are planets, stars, interstellar, galactic, and intergalactic clouds, etc., and also the very famous cosmic microwave background (CMB) radiation (see Chapter 2). Cutting-edge techniques, including heterodyne detection and ultra-low-noise cryogenic detectors, permit to record extremely small signals, at almost the quantum limit of incident power [1, 2].

Absorption/impedance spectroscopy gives the transmittance/reflectance properties of a sample, i.e., the percentage of far-infrared light transmitted/reflected by a sample (absorption spectroscopy) or the variation of the phase of this light induced by transmission/reflection through/on the sample (impedance spectroscopy). This kind of spectrometry requires a source and a detector of FIR radiation, some optics to manipulate the FIR beam, and a spectrometer to disperse and select the wavelengths emitted by the source. Depending on the purpose of the spectroscopic study, e.g., performing high-resolution spectroscopy of molecular resonances, monitoring a narrow absorption line, recording broad features over a large spectral range, etc., broadband or narrowband spectra are necessary, and thus the source and receiver must be broadband or narrowband as well. The first common sources of FIR radiation were blackbody-like, which included mercury lamps, and globars (a SiC rod heated up to 2000 K) for higher frequencies ($>15 \sim 20$ THz). Then, in the 1960s, molecular lasers, microwave vacuum tubes, and large instruments were conceived or optimized to produce FIR waves. More recently, electronic circuits succeeded in efficiently delivering sub-millimeter wavelengths. Simultaneously, for higher frequencies, typically over 2 THz, quantum cascade lasers produce powerful monochromatic FIR beams, but require cryogenic operating temperatures. The most popular detectors were cryogenic bolometers and, for operation at room temperature, Golay cells. Other detectors, such as electronic components like Schottky and UTC (uni-traveling carrier) diodes, are rarely employed in FIR spectrometers, because of their limited bandwidth and lower sensitivity.

3.2 Basic Definitions in Spectroscopy

Here we only treat absorption spectroscopy. The spectral response of a sample is obtained by measuring the power spectrum of the beam transmitted or reflected by the sample, and by normalizing it to the incident power spectrum. One defines:

$$\text{transmittance:} \quad T(\omega) = \frac{P_{\mathrm{t}}(\omega)}{P_{\mathrm{i}}(\omega)}, \tag{3.1}$$

$$\text{reflectance:} \quad R(\omega) = \frac{P_{\mathrm{r}}(\omega)}{P_{\mathrm{i}}(\omega)}, \tag{3.2}$$

where $P_{\mathrm{t}}(\omega)$, $P_{\mathrm{r}}(\omega)$, and $P_{\mathrm{i}}(\omega)$ are respectively the transmitted, reflected, and incident powers at angular frequency or pulsation ω. In the case of transmission, a useful quantity is the optical density or absorbance, defined as:

$$\text{absorbance:} \quad A = -\log T(\omega) = -\log \frac{P_{\mathrm{t}}(\omega)}{P_{\mathrm{i}}(\omega)}. \tag{3.3}$$

Absorbance is widely used by chemists, because of (i) its linear variation with the absorbing substance concentration when dealing with a solution (namely Beer's law that is valid at low concentrations), and (ii) its additive when several absorbing substances are in solution:

$$A = \varepsilon_{\mathrm{abs}} Cd, \qquad A_{\mathrm{mixture}} = \sum_i A_{\mathrm{i}}, \tag{3.4}$$

where d is the sample thickness, C is the molar concentration, and $\varepsilon_{\mathrm{abs}}$ is the molar absorptivity. In the case of a mixture, A_{i} is the absorbance of each component i. In fact, for a flat homogeneous and isotropic sample located in air, an analytical value of the transmittance can be established from the Maxwell equations. Let us write again the complex coefficient of transmission (for the THz electrical field) of the sample (see annex, relation A.35):

$$\tilde{t} = \frac{4\tilde{n}}{(\tilde{n}+1)^2} \times \frac{e^{-j\frac{\omega}{c}(\tilde{n}-1)d}}{1-\left(\frac{\tilde{n}-1}{\tilde{n}+1}e^{-j\frac{\omega}{c}\tilde{n}d}\right)^2}. \tag{3.5}$$

In Eq. 3.5, the first term of the right side corresponds to the Fresnel coefficients at the sample interfaces, while the numerator of the ratio (second term) denotes the wave propagation through the sample material, and the denominator renders for the back-and-forth reflections inside the sample (Fabry–Pérot oscillations). For low-index materials, like solutions in chemistry studied in the optical domain, Fresnel coefficients are close to 1, and the Fabry–Pérot effect may be neglected. Expression 3.5 can be approximated by the following simple relation:

$$\tilde{t} \approx e^{j\frac{\omega}{c}(\tilde{n}-1)d} = e^{j\frac{\omega}{c}(n-1)d}e^{-\frac{\alpha}{2}d} \quad \Rightarrow \quad T(\omega) = \left|\tilde{t}\right|^2 \approx e^{-\alpha(\omega)d}. \tag{3.6}$$

This is another expression of Beer's law (3.4), known also as Beer–Lambert law or Beer–Lambert–Bouguer law. Therefore, within this hypothesis, the coefficient of absorption is directly proportional to the absorbance: $A(\omega) = \alpha(\omega) \times d$. However, for most of materials, this crude simplification leads to an erroneous estimation of $\alpha(\omega)$. Figure 3.2 shows the relative error on $\alpha(\omega$ versus the value of the refractive index n, when the Fresnel term is omitted, for $f = 1$ THz and $d = 1$ mm, and for 3 different values of $\alpha(\omega)$. For transparent materials ($\alpha \leq 0.1$ cm^{-1}), the error is larger than 1 % as soon as $n \geq 1.1$, while for absorbing materials (typically $\alpha \geq 10$ cm^{-1}), it remains smaller than 1 % whatever the value of n. A precise determination of $\alpha(\omega)$ requires to rigorously solve Eq. 3.5, as will be described in the next chapter.

The transmittance/reflectance of a sample is measured with a spectrometer. Three main parameters govern the performance of a spectrometer [3, 4], namely the resolving spectral power, the spectrometer throughput, and the signal-to-noise ratio.

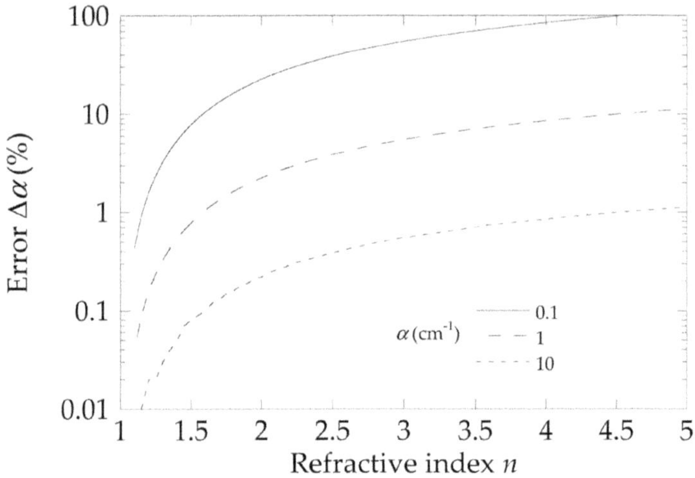

Figure 3.2 Error on the estimation of $\alpha(\omega)$ when the Fresnel term is omitted. The curves are plotted for 3 different values of $\alpha(\omega)$; $f = 1$ THz and $d = 1$ mm.

3.2.1 Resolving Spectral Power

The resolving spectral power R_{res} is the ability of the instrument to separate adjacent spectral lines:

$$R_{\text{res}} = \frac{\lambda}{\delta\lambda} = \frac{\omega}{\delta\omega} = \frac{f}{\delta f}, \tag{3.7}$$

where $\delta\lambda$ is the minimum wavelength interval between two spectral lines that can be separately observed. When dealing with dispersive spectrometers (the output beam direction depends upon the wavelength), the input (D_1) and output (D_2) dispersions of the spectrometer are

$$D_1 = \frac{\partial i_1}{\partial\lambda}, \quad D_2 = \frac{\partial i_2}{\partial\lambda} = \Gamma D_1.$$

Γ is the angular magnification of the spectrometer and i_1 and i_2 the input and output angles. The best efficiency of the spectrometers is obtained when $D_1 = D_2 \equiv D$. With input/output slits of width w, and taking into account the diffraction by the slits, this leads to

$$R_{\text{res}} \approx \frac{\lambda}{w} LD. \tag{3.8}$$

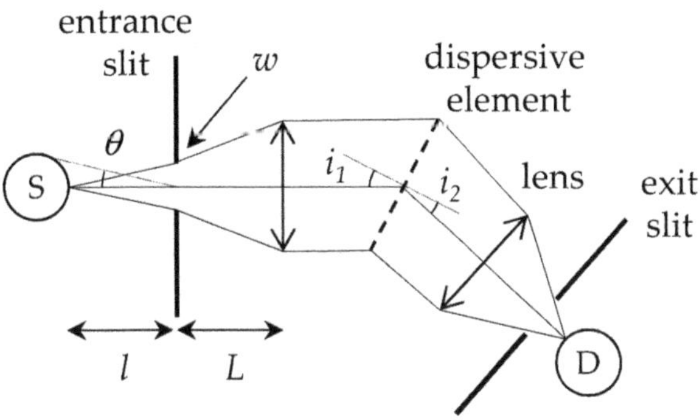

Figure 3.3 Scheme of a spectrometer. S is the source and D the detector.

L is the distance between the slit and the dispersive element (Fig. 3.3).

The resolution is inversely proportional to the slit width. However, decreasing the slit width, i.e., increasing the resolution, reduces as well the amount of light transmitted by the slit, making its detection impossible. Thus one defines the theoretical resolving power $R_{\mathrm{res,o}}$ by setting $w = \lambda$, which is a limit imposed by diffraction:

$$R_{\mathrm{res,o}} = LD. \tag{3.9}$$

Unfortunately, spectrometers are not perfect instruments, and the output exhibits a spectral broadening, which makes the actual resolving power R lower than the theoretical one. The output spectrum of a spectrometer excited by a monochromatic beam is known as the instrumental line profile $A(\lambda)$. It depends on the actual properties of the dispersive element, and on the quality of the system alignment and of its optical components. The FWHM of this line is defined as the bandpass BP(λ) of the spectrometer. The actual power at the detector is the convolution of the impinging collected power and of the profile line:

$$P_{\mathrm{t}}(\lambda) = A(\lambda) \otimes P_{\mathrm{coll}}(\lambda). \tag{3.10}$$

The Rayleigh criterion that defines the minimum resolvable detail in an image may be approximately written here as follows: two lines are resolved if their line profiles cross at their half maximum. Thus $\delta\lambda \approx$ BP(λ), assuming the line profile exhibits a Gaussian or Lorentzian shape.

3.2.2 Spectrometer Throughput

The spectrometer throughput is the amount of light emitted by the source that is gathered and transmitted to the detector. The power of light emitted by the source in a solid angle Ω is $P_{\text{source}}(\lambda) = L(\lambda) S\, \Omega$, where S is area of the source emitting in the normal direction (Fig. 3.3). $L(\lambda)$ is the radiance of the source. The collected power is obtained by setting $\Omega = S'/l^2$, where S' is the entrance slit area and l is the distance between the source and the spectrometer. We obtain

$$P_{\text{coll}}(\lambda) = L(\lambda)\, S\, \frac{S'}{l^2} = L(\lambda)\, S'\, \frac{S}{l^2} \approx L(\lambda)\, S' \sin^2\theta.$$

The numerical aperture NA of the spectrometer is simply $\text{NA} = \sin\theta$ (θ is the half angle of the light cone radiated by the source and collected by the input lens; see Fig. 3.3), when measurements are performed in air. The collected power is given by

$$P_{\text{coll}}(\lambda) = L(\lambda)\, S'\, (\text{NA})^2\,. \tag{3.11}$$

It is often useful to employ the notion of geometric *étendue* G of a beam, which is defined by $G = \iint dS d\Omega$, where an elementary surface dS radiates into an elementary solid angle $d\Omega$. In most spectrometers, this surface is a slit. Hence,

$$P_{\text{coll}}(\lambda) = L(\lambda)\, G_{\text{slit}}. \tag{3.12}$$

Here we have supposed that the source is Lambertian ($L(\lambda)$ does not depend on the angle of emission) and that source and spectrometer are well aligned. For common spectrometers, NA is usually small (typically 0.1 or less), which justifies the paraxial approximation made throughout this paragraph. Finally, the power at the detector is

$$P_{\text{t}}(\lambda) = \int A(\lambda' - \lambda)\, P_{\text{coll}}(\lambda')\, d\lambda' = G_{\text{slit}} \int A(\lambda' - \lambda)\, L(\lambda')\, d\lambda'. \tag{3.13}$$

3.2.3 Signal-to-Noise Ratio

The signal-to-noise ratio (SNR) limits the sensitivity of the spectrometer; it depends on the detector, but also on the power $P_{\text{t}}(\lambda)$ incident on the detector (3.13). It has to be noticed that $P_{\text{t}}(\lambda)$ increases with w, the slit width, but simultaneously the resolving power R_{res} (Eq. 3.8) decreases. The electrical signal $S(\lambda)$ delivered by the detector is usually proportional to the power $P_d(\lambda)$ at the detector:

$$S(\lambda) = \eta\, P_d(\lambda) + S_{dark}, \tag{3.14}$$

where η is the detector sensitivity and S_{dark} is the dark current. $P_d(\lambda)$ includes the useful power $P_t(\lambda)$ together with the power $P_{stray}(\lambda)$ of stray radiations that reach the detector. Moreover, the signal $S(\lambda)$ is affected by noise due to the particle aspect of light (shot noise) and of electrical charges (Johnson noise), and to the electronics, either in the emitter, i.e., the source, or the detector.

3.3 Continuous or Quasi-Continuous Sources

3.3.1 Blackbody Emitters

Historically, the first far-infrared sources were based on the radiation emitted by a discharge between two electrodes. As early as 1896, J. Bose improved the first radiating dipoles by G. Hertz to generate centimeter wavelengths and to characterized materials at these wavelengths [5]. Later on, blackbodies became the most employed sources of far-infrared radiation [6]. A blackbody is a body that radiates thermal light as efficiently as possible.

At a non-null temperature, i.e., >0 K, the basic microscopic constituents of matter, i.e., atoms, molecules or free charges, are agitated by a random movement around their equilibrium position. In gases, for which the theory is the simplest, the temperature is thermodynamically defined as

$$T = \frac{2}{3}\frac{\overline{E_{kin}}}{k_B}, \tag{3.15}$$

where k_B is Boltzmann's constant and $\overline{E_{kin}}$ (cin $=$ KE?) is the averaged value of the kinetic energy of the gas molecules. Any charge that accelerates or slows down radiates an electromagnetic field, and the thermal agitation of the elementary particles (atoms, molecules) in matter results in the emission of thermal radiation by this matter. The theory of the blackbody radiation, definitively established by M. Planck, gives the luminance of the blackbody:

$$L_o(f) = \frac{2hf^3}{c^2}\frac{1}{\exp(hf/k_B T) - 1}. \tag{3.16}$$

Here f is the frequency of the thermal light. The luminance $L_o(f)$ is the power radiated by unit of frequency in a unit solid angle centered perpendicularly to the body surface by a unit area at this surface. Thus its unit is $W/m^2/sr/Hz$.

As we have already written, the blackbody is the most efficient thermal light emitter. This could be more precisely defined using Kirchhoff's law. The incident I_i, reflected I_r, transmitted I_t, and absorbed I_a intensities on/through/by a sample are related by

$$I_i = I_r + I_t + I_a. \tag{3.17}$$

The relation between the relative proportions r, t, and a of respectively the reflected, transmitted, and absorbed intensities writes

$$1 = r + t + a. \tag{3.18}$$

For a transparent body, $a = 0$; for a perfect mirror, $r = 1$, thus $t = a = 0$. A blackbody is an object that absorbs all the incoming light, thus $a = 1$ and $r = t = 0$. For an ideal blackbody, this is verified at any wavelength. For common materials, $r, a \neq 1$, and for colored bodies, they obviously depend on the wavelength.

Kirchhoff has made the observation that the ability of a body to radiate thermal light is proportional to a. Thus the most efficient thermal emitter is the blackbody, since $a = 1$ is maximum. Therefore the luminance $L(f)$ of a common object can be defined versus the one $L_o(f)$ of the blackbody:

$$L(f) = \varepsilon_{\text{therm}}(f) L_o(f), \tag{3.19}$$

where $\varepsilon_{\text{therm}}(f)$ is the emissivity of the body, and Kirchhoff's law imposes:

$$\varepsilon_{\text{therm}}(f) = a. \tag{3.20}$$

This can be simply understood by considering an illuminated sample, which is thick enough to transmit no light. Thus we have, from Eq. 3.18, $1 = r + a$. On the other hand, when reaching the surface of this hot matter sample, the radiation emitted by a restless

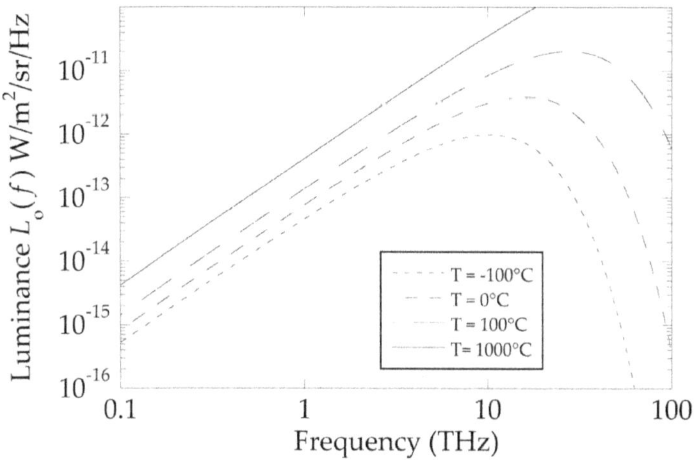

Figure 3.4 Luminance of a blackbody at −100, 0, 100, and 1000°C.

particle (atom, molecule) of the matter is in part (r) reflected back to the bulk and in part transmitted $(1 - r)$ to the upper medium. This light emerging from the surface is the thermal radiation. Thus $\varepsilon_{\text{therm}} = 1 - r$ and consequently $\varepsilon_{\text{therm}} = a$.

The luminance spectrum of the blackbody is shown in Fig. 3.4 for different temperatures. The frequency f_{max}, at which the maximum of $L_o(f)$ occurs, is obtained by taking the derivative of Eq. 3.16:

$$f_{\text{max}}(\text{THz}) = 0.04864 \times T\ (\text{K}) \approx \frac{T\ (\text{K})}{20}. \qquad (3.21)$$

This is known as Wien's law. The spectrum is very broadband around f_{max}. Typically, the −3 dB bandwidth exceeds 30 THz for a body at room temperature. When the body temperature is larger than the room temperature (~300 K), the far-infrared part of the spectrum obeys the Rayleigh–Jeans law, which is the limit of Eq. 3.16 for large values of T:

$$L_o(f) = 2 \times 1.38 \frac{f^2 \times 10}{9E16} T \approx 3 \times 10^{-16} f_{\text{THz}}^2 T. \qquad (3.22)$$

The luminance varies as T and as the square power of the frequency. For an object of surface S emitting a blackbody-like radiation in a given frequency bandwidth Δf in the THz range, the integrated

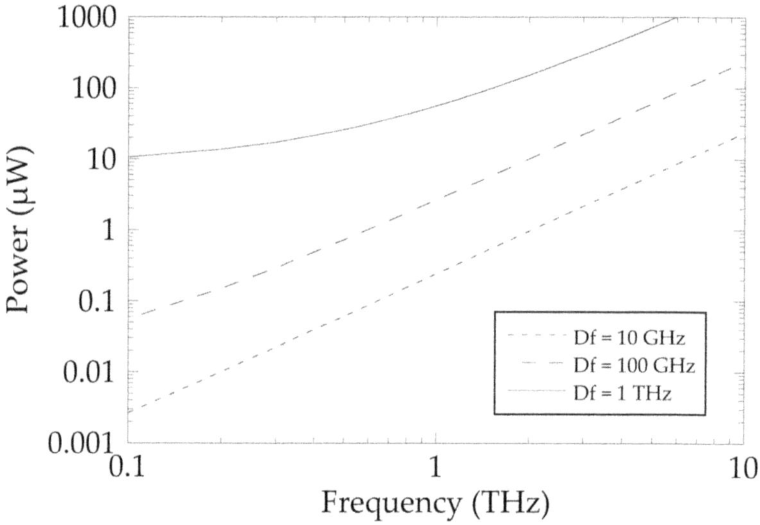

Figure 3.5 Power radiated by a 1 cm^2 blackbody at 100°C in a bandwidth of respectively 10 GHz, 100 GHz, and 1 THz.

power that is radiated in the 2π solid angle (see Fig. 3.5), is expressed as

$$P = 2\pi S \int_{f}^{f+\Delta f} L_{\mathrm{o}}(f)\,df = \frac{4\pi}{3} S \frac{k_{\mathrm{B}} T}{c^2} \left(f^2 + f\Delta f + \frac{\Delta f^2}{3} \right) \Delta f. \tag{3.23}$$

Thus a 1 cm^2 blackbody, in a 100 GHz bandwidth around 1 THz, radiates towards the upper space a power P (nW) $\approx 7 \times T$ (K). At room temperature, $P \approx 2$ μW. Over the whole THz band, generally defined from 0.1 to 10 THz, this power is $P \approx 6$ mW. For practical measurements, mercury (Hg) arc lamps are widely employed as a source of blackbody radiation. The thermal light is emitted by the discharge ($T \approx 5000$ K) but also by the quartz envelop of the lamp ($T \approx 1000$ K). Typically, a 75 ∼ 100 W lamp radiates in a very wide band (0.1 ∼ 20 THz) with a spectral power density of several tens of μW/THz. One important feature of the blackbody emission is that the radiated light is incoherent.

3.3.2 Diodes with a Negative Differential Resistance (NDR)

Devices like Gunn, RTC, or IMPATT diodes could exhibit a negative differential resistance (NDR), which makes them very useful to design high-frequency oscillators and amplifiers. In a standard/simple resistor circuit, the resistance R is constant with applied voltage V (ohmic response). However, in diodes, due to specific basic carrier transport phenomena in the semiconductor materials and/or in the device structure, the $I(V)$ curve of the diode is not linear (non-ohmic behavior), but shows a negative slope for a given voltage range (Fig. 3.6 left). The resistance of the device is not constant with V, and thus must be defined as $R_{\text{dev}} = dV/dI$, namely the differential resistance. The negative slope of the $I(V)$ curve induces a negative value of R_{dev}.

Let us consider a resonant RLC circuit, whose resistance could be negative (Fig. 3.6, right). The differential equation for the current I flowing in the circuit is

$$L\frac{d^2I}{dt^2} + R_{\text{dev}}\frac{dI}{dt} + \frac{I}{C} = 0. \tag{3.24}$$

The solution of Eq. 3.24 writes $I = I_o e^{at}$, with the constant a defined as

$$a = \frac{1}{2L}\left(-R_{\text{dev}} \pm \sqrt{R_{\text{dev}}^2 - 4L/C}\right). \tag{3.25}$$

If $R_{\text{dev}}^2 < 4L/C$, the square root in Eq. 3.25 is imaginary and the current I becomes oscillating at a pulsation ω:

$$I(t) = I_o\, e^{-\frac{R_{\text{dev}}}{2L}t}\, e^{j\omega t}, \quad \omega \equiv \frac{\sqrt{4L/C - R_{\text{dev}}^2}}{2L}. \tag{3.26}$$

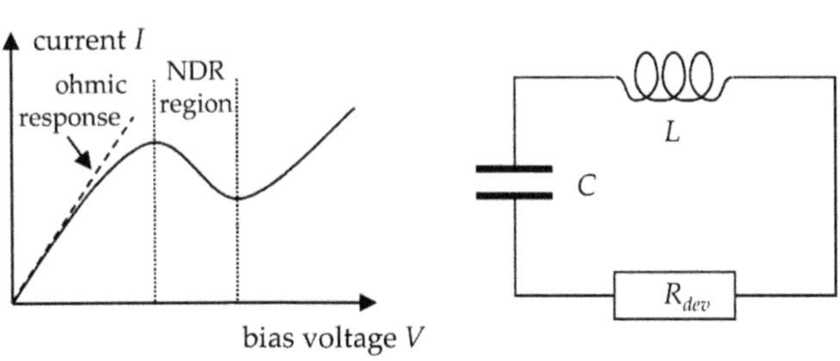

Figure 3.6 Left: $I(V)$ curve with an NDR region. Right: Resonant circuit.

If $R_{dev} > 0$, the oscillation is damped at a rate equal to $2L/R_{dev}$. On the other hand, if $R_{dev} < 0$, the oscillating signal increases with time at a rate $2L/|R_{dev}|$. Therefore, a resonant circuit including a negative resistor can work as an oscillator provided $0 > R_{dev} > -2\sqrt{L/C}$. In actual systems, the NDR diode is located in a millimeter waveguide cavity, which can be modeled as an *LC* circuit.

In Gunn and IMPATT diodes, the NDR regime originates in the carrier velocity overshoot. In semiconductors like GaAs, the conduction band shows a main valley centered around $k = 0$, namely the Γ valley, and satellite valleys (L, X ...), with a less incurved shape. Let us recall that the carrier velocity in the conduction band is $\vec{v} = \vec{\nabla}_{\vec{k}} E/\hbar$, where $\vec{\nabla}_{\vec{k}}$ is the gradient operator in the k space and E is the energy. Thus, due its pronounced curvature, the velocity in the Γ valley is larger than in the L valley. When applying an electric field $\vec{E}$ to the semiconductor, the electrons acquire kinetic energy and thus their wavevectors increase according to the relation $d\vec{k} = q\vec{E}dt/\hbar$. Thus they climb up in the Γ valley and then they are transferred to the L valley. At low electric field, they have no time to reach the L valley because of collisional processes, and thus their mobility is high (Fig. 3.7). At higher field strengths, the scattering to the L valley decreases the mobility and the velocity to saturate [7]. Figure 3.7

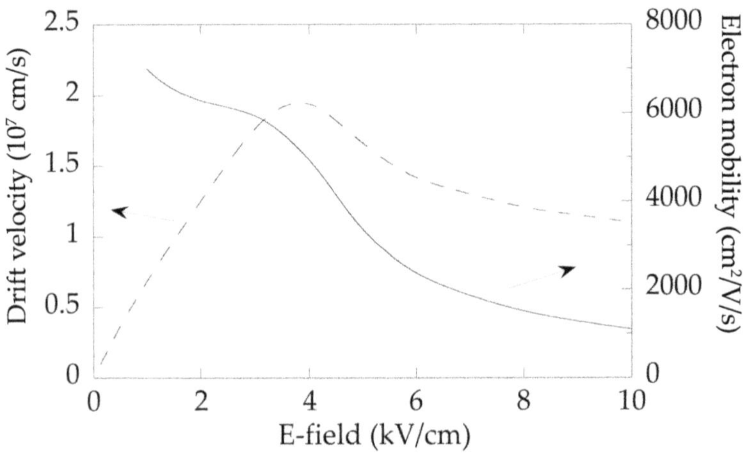

Figure 3.7 Drift velocity and electron mobility of GaAs. Data from the Ioffe Institute (St. Petersburg, Russia) database (www.ioffe.ru).

shows the free electron drift velocity (i.e., the mean velocity induced by an electrical field, in opposition to the random velocity due to the thermal agitation) and the related mobility versus the applied field in GaAs, plotted using data extracted from the Ioffe Institute Database (www.ioffe.ru).

In Gunn diodes, biased in the NDR region, small fluctuations of the electron density near the cathode induce a local perturbation of the electrical field. Due to the NDR regime, the electrons in the perturbed region are gathered together to form a pulse, leading to an increase of the field perturbation. The effect grows up while the electrons move to the anode. Reduction of the electrical field strength in the rest of the sample prevents the simultaneous formation of other pulses. Thus, the repetition rate of the pulses is roughly equal to the carrier time-of-flight, i.e., to the length of the sample divided by the carrier velocity. Typically, a 5-μm thick gallium arsenide (GaAs) diode biased with 3 V delivers a signal at 25 GHz. The best diodes made with indium phosphide (InP) deliver 1 mW at 200 $\sim$ 300 GHz, while the emitted power is higher than 100 mW below 100 GHz. IMPATT diodes are based on the same phenomenon, but an avalanche effect appears during the propagation of the pulse in the NDR region.

Resonant tunneling diodes (RTDs) are formed of a single quantum well (QW) structure surrounded on each side by a very thin layer barrier: this constitutes a double barrier structure. In the QW, the energy of the electrons is quantized. At zero bias, the Fermi level outside the structure is smaller than the first energy level in the resonant tunneling diode, and thus no current flows through the RTD. When increasing the bias voltage, the applied field tilts the energy curve,[a] and the first energy level in the QW equals the Fermi level in the cathode. The current can flow in the diode thanks to a resonant tunnel effect through the first barrier. When increasing the bias further, the first energy level becomes smaller than the Fermi level, so the tunnel effect is no longer resonant and the current decreases. Therefore, the $I(V)$ curve shows a negative slop. RTD-

[a] In semiconductors, the energy of the free carriers is mostly of kinetic origin $E = E_{\text{kin}}$. When a bias voltage V is applied to the material, the energy includes an electrostatic term, namely $E = E_{\text{kin}} + qV$, where q is the free carrier charge. In devices where V varies with the space coordinate x, the energy curve varies as well.

based sources deliver milliwatts at 100 GHz, but only microwatts at $600 \sim 700$ GHz.

3.3.3 Schottky Diode Multiplier

To get THz from electronic devices, we start from a microwave signal of several GHz delivered by a microwave generator, and then the idea is to multiply it in terms of frequency using a nonlinear component. Most of the time, the nonlinear component is a diode. The current/voltage response of a diode is indeed typically nonlinear. The response for an ideal *p-n* diode is given by

$$I = I_S \left(\exp\left(\frac{eV}{k_B T} \right) - 1 \right). \tag{3.27}$$

When an oscillating bias $V = V_o \cos(\omega t)$ is applied to the diode, the current is

$$I = I_S \left(\exp\left(\frac{eV_o \cos(\omega t)}{k_B T} \right) - 1 \right). \tag{3.28}$$

By expanding the exponential term into a series we get

$$\exp\left(\frac{eV_o \cos(\omega t)}{k_B T} \right) - 1 = \sum_{n=1}^{+\infty} \frac{1}{n!} \left(\frac{eV_o \cos(\omega t)}{k_B T} \right)^n. \tag{3.29}$$

And finally, we get:

$$I = \sum_{m=1}^{+\infty} I(m\omega) e^{jm\omega t} = I_S \sum_{m=1}^{+\infty} \left(\sum_{n=m}^{+\infty} \frac{1}{n!} \left(\frac{eV_o}{2k_B T} \right)^n C_n^{\frac{n-m}{2}} \right) e^{jm\omega t}, \tag{3.30}$$

where C_n^k are the binomial coefficients. The harmonic term $I(m\omega)$, at pulsation $m\omega$ (m is an integer), in the current is given by

$$I(m\omega) = I_S \sum_{n=m}^{+\infty} \frac{\left(\frac{eV_o}{2k_B T} \right)^n}{\left(\frac{n-m}{2} \right)! \left(\frac{n+m}{2} \right)!}. \tag{3.31}$$

For realistic values of $eV_o/2k_B T$, the main nonlinear contribution is from the $m = 2$ term of the series:

$$I(2\omega) \approx \frac{I_S}{2} \left(\frac{eV_o}{2k_B T} \right)^2. \tag{3.32}$$

As demonstrated, the diodes show a strong nonlinear behavior that leads to the production of harmonics when biased by an alternating voltage. To reach the THz range, the diodes must also be able to react to high frequencies. The response time of diodes is limited by both the transit time of carriers in the depletion region and the *RC* time constant of the whole electronic circuit. In fast p–n or p–i–n diodes, the resistance of the diode is a few ohms, smaller than the usual 50 Ω circuit impedance. The capacitance of the diode is related to the charges accumulated at the borders of the depletion zone, which are directly proportional to the minority carrier lifetime in the semiconductor. Also the capacitance includes a parasitic contribution from packaging. Typically such diodes can be used up to several hundreds of MHz but reach rarely the THz domain.

To reach higher frequencies, the solution is to employ another type of diodes, namely planar Schottky diodes, which are today the most popular device for building frequency multipliers in the THz range [8]. The size of the active area of the diode can be sub-micronic, resulting in a sub-fF junction capacitance. Parasitic capacitances as well as series resistance can be strongly reduced thanks to smart technical tricks that are permitted by the planar technology, like air-bridge connections. Schottky diode frequency doublers' efficiency is typically 5 ~ 10% at 100 GHz. It decreases to 3 ~ 5% over 500 GHz. Frequency triplers are 2 ~ 3 times less efficient. Such devices are able to deliver hundreds of mW at 100 GHz, and still 0.1 ~ 1 mW at 1 THz. Nevertheless, they exhibit two drawbacks: a limited number of commercial suppliers, and a rather limited bandwidth, which forces one to use several devices to cover the whole THz range. Also, they are inefficient over 2 ~ 3 THz. A THz setup based on the Schottky diode multiplier includes a primary RF source at low frequency (10 ~ 20 GHz) and several multipliers in series.

3.3.4 Electronic Tubes

Backward wave oscillators (BWOs) are electronic tubes whose emission is based on the Smith–Purcell effect.

When an electron moves over a conducting surface, it creates a charge image inside the conductor. If the surface shows a periodic relief of period *d*, i.e., it resembles a diffraction grating, the distance between the electron and its charge image varies periodically. Thus

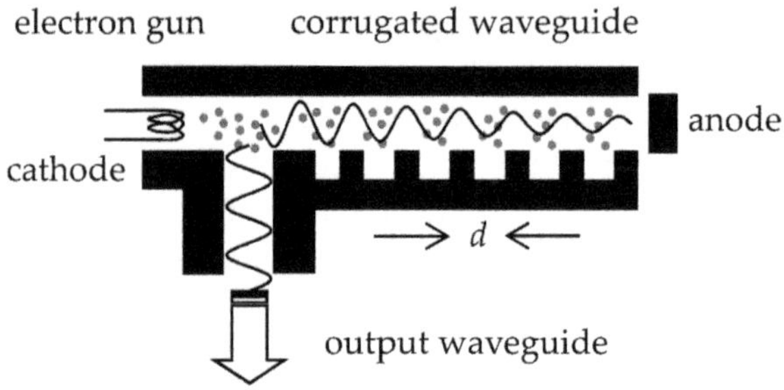

Figure 3.8 Principle scheme of BWO. The period of the corrugation is d.

the electron encounters a periodic potential induced by its charge image, and hence it is subjected to a periodic force. When moving at a velocity v, the spatial period transforms into a temporal period $T = d/v$. It is known from relativity that a charge accelerated by a periodic force radiates an electromagnetic wave of same period. This constitutes the basis of the Smith–Purcell effect: a stream of electrons moving over a periodically engraved metallic surface radiates an electromagnetic wave at frequency $f = 1/T = v/d$.

In BWO and other high-frequency tube sources, the periodic surface constitutes one of the four inner faces of a waveguide. Thus the radiated wave is strongly coupled to a guided mode of the waveguide. Let us call k_r the wavevector component along the waveguide of the excited guided mode. As this mode propagates in a periodic structure, it is diffracted and a part of its power is transferred to a wave of wavevector k_m according to the grating diffraction theory:

$$k_\mathrm{r} + m\frac{2\pi}{d} = k_\mathrm{m}. \tag{3.33}$$

m (integer) is the diffraction order. It follows that the grating period should be

$$d = m\frac{2\pi}{k_\mathrm{m} - k_\mathrm{r}}. \tag{3.34}$$

If the waveguide is monomode, $|k_\mathrm{r}|$ and $|k_\mathrm{m}|$ are equal. Therefore Eq. 3.33 has a solution only if $k_\mathrm{m} = -k_\mathrm{r}$ and m is negative: this means that the wave is scattered backwards. The maximum diffraction

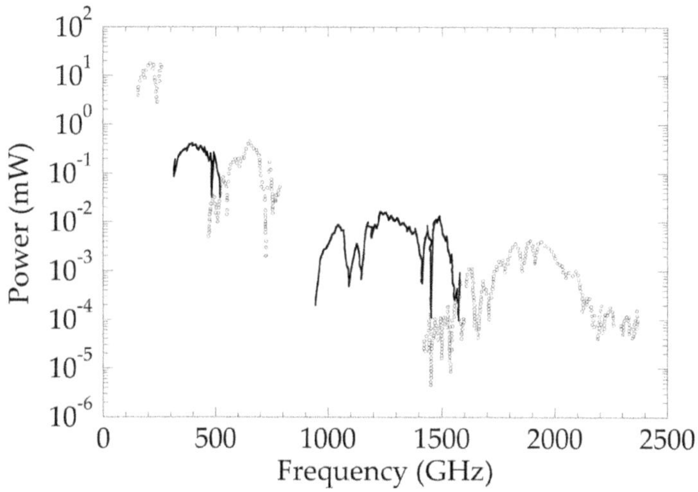

Figure 3.9 Example of spectra delivered by BWO (model QS1-260-2100 BWO-based system, with Schottky multipliers, from Microtech Instruments Inc.).

efficiency is usually obtained in the first order ($m = -1$), so

$$d = \frac{c}{2f\,n_{\text{eff}}}. \tag{3.35}$$

n_{eff} is the effective index of the guided mode. For hollow metallic waveguide, $n_{\text{eff}} \approx 1$, and thus a 1 THz BWO requires $d = 0.15$ mm. The electron velocity must be $v = c/2\,n_{\text{eff}}$. This is achieved with a voltage of several kV between cathode and anode. A magnetic field maintains the electron flow as a spatially limited beam in the waveguide. Typically, BWOs deliver tens of mW up to $0.8 \sim 1$ THz, and this power decreases strongly at higher frequencies (10 mW at 1 THz, 2 mW at 1.2 THz). They can be frequency-tuned over 100–200 GHz by varying the bias voltage or by the magnetic field. This means that several BWOs should be used to cover the whole THz range (Fig. 3.9).

The radiated spectrum shows a very confused shape due to resonances in the corrugated waveguide, but it is highly reproducible and stable, and the spectral width of the emitted signal is excellent ($\Delta f/f < 10^{-4}$), which means BWO emission is quasi-monochromatic.

3.3.5 Optically Driven UTC Diodes

Uni-traveling carrier (UTC) diodes are promising devices to generate THz CW waves [9]. In such devices, incoming photons are absorbed in a highly p-doped zone. The number of photo-generated holes remains small compared with the holes introduced by doping. They do not change the electrical equilibrium in the p zone. Thus only photoelectrons contribute to the photocurrent, but not the holes that move slower. The photoelectrons move towards the positive contact because of diffusion in the absorbing layer and then because of an accelerating field in an intrinsic non-absorbing layer (collecting layer). The response of the UTC diode is thus only limited by the free electron transit time in the absorbing and collecting layers. When excited by the beating of two laser beams, UTC diodes deliver THz signals ranging up to 1 $\sim$ 2 THz, with powers from 100 μW at 100 GHz, decreasing strongly with frequency down to 1 μW at 1 THz.

3.3.6 Other Sources and Future Sources

Among all other sources of THz radiation, we can cite:

(i) Molecular lasers:
The laser emission occurs between rotational levels of a gas of molecules such as methanol, difluoro-methane, etc. The gases are pumped by a CO_2 laser, making the system bulky. The emission is monochromatic and powerful ($\sim$100 mW).

(ii) Synchrotron radiation:
Under the influence of the centripetal acceleration, the electrons rotating in the synchrotron emit electromagnetic waves. The radiated spectrum spreads from low frequencies (<GHz) with a quite constant power up to about 10 THz where it decreases abruptly. The spectral power density is huge ($\sim$10 W/cm^{-1}) and thus the total THz power reaches $\sim$1 kW. Of course, all these values depend on the kinetic energy of the electrons in the synchrotron ring.

(iii) Free electron lasers (FELs):
In a part of the path of an accelerator of electrons (it could be a synchrotron), a magnetic field is applied periodically, by a series of magnets with alternatively inverted magnetic fields

(spatial period d). The electromagnetic radiation emitted by the electrons submitted to this varying Lorentz force is directed along the electron beam. The frequency of this radiation is:

$$f_{\mathrm{FEL}} = 2\frac{c}{d}\left(\frac{E_{\mathrm{kin}}}{mc^2}\right)^2 \Bigg/ \left(1+\frac{1}{2}\left(\frac{eBd}{2\pi mc^2}\right)^2\right), \tag{3.36}$$

where E_{kin} and m are the kinetic energy and the resting mass of the electrons, and B is the magnetic field. This system is located between two mirrors (far-infrared), in order to form a laser cavity. By changing the electron energy, the radiation frequency can be tuned over a very broad band (e.g., $1 \sim 150$ THz at the FELIX facility in the Netherlands). Usually, the emission is pulsed (pulse duration from μs down to ps) and the peak power is large (kW ~ MW). Both synchrotron and FELs deliver superb performance, but they are very bulky and expensive.

(iv) Optical parametric oscillators (OPOs):
They are based on a nonlinear optical phenomenon in a nonlinear crystal (see Annex, paragraph 2): an incoming pump photon at angular frequency ω_{pump} gives rise to two photons respectively at frequencies ω_{idler} (the idler beam) and ω_{THz}. The conservation of energy imposes $\omega_{\mathrm{pump}} = \omega_{\mathrm{idler}} + \omega_{\mathrm{THz}}$. The crystal orientation as well as the direction and polarization of the optical pumping beam are finely chosen in order to achieve phase matching, i.e., the three beams propagate at the same speed, leading to a constructive contribution of any part of the crystal. Main efficient nonlinear crystals are ferroelectrics, like $LiNbO_3$, and they show a large refractive index in the THz range. Thus, the generated THz beam suffers total reflection at the crystal faces, which can be overcome by adding a diffractive structure (prisms, grating) over one face of the crystal. Moreover, the crystal is located in a Fabry–Pérot cavity, in such a way the idler beam oscillates resonantly in this cavity and the idler signal can be seeded by another laser. Typically, such THz OPOs deliver ns pulses (they are indeed pumped by Q-switched lasers) with energy per pulse as high as $0.1 \sim 1$ nJ.

(v) Quantum cascade lasers (QCLs):
QCLs are made of a stack of ultrathin (nanometer) layers that behave as quantum wells for free charges. In each quantum

well, charges cannot take any energy, but can only occupy discrete energy levels. Because of the energy tilting along the QCL by an applied bias, a lower energy level in a first quantum well is equal to a higher energy level in the neighboring quantum well. Thus, electrons or holes can move from a quantum well to the next one by resonant tunneling through a potential barrier in between the two quantum wells. Then they emit a photon to go down to the lower energy state of the second quantum well. By adding several layers, the phenomenon is repeated several times and thus light emission is enhanced: this constitutes the cascade effect. Thus a single charge can participate in the emission of several photons! The other great feature of QCLs resides in the possibility of choosing the emitted wavelength by adjusting the width of the quantum wells, as the energy levels depend on the inverse of this width. Therefore, the QCL wavelength is a function of the design of the structure, and not of the involved materials as it is for semiconductor laser diodes. To ensure an inversion of population between the higher and lower energy levels, QCLs must operate at cryogenic temperatures (we recall that the thermal energy $k_B T \approx 26$ meV at room temperature is larger than the THz photon energy (0.3 $\sim$ 6 meV)). THz QCLs emit typically over 2 $\sim$ 3 THz, but emission at 1.2 THz has been reported [10]. Basically, the emitted frequency of actual QCLs varies as $f = k_B T/h$. Thus at 4 THz, the operation temperature is close to 200 K, while it is only 70 K at 1.5 THz. The highest operation temperatures reported so far are 117 K and 186 K for the CW and pulsed regimes, respectively [11, 12]. Nevertheless, some recent results break this limit: Kumar et al. [13] has reported 1.8 THz QCL operating at 163 K. The QCL output power can be as high as 250 mW [14]. Note that the THz QCL emission is narrowband (monochromatic): the spectral width is of the order of a few tens of kHz. Nowadays, QCLs can also be mode-locked and modulated, and people are looking at tuning the emitted wavelength, for example by using a microelectromechanical system (MEMS) plunger.

Numerous researchers are currently working to improve existing sources or to develop new ones. For example, GaN Gunn diodes are expected to work up to 1 THz, thanks to a higher peak velocity and a higher saturation velocity than in GaAs. Lasing in p-doped germanium is also widely studied. In such a material, the valence bands for heavy and light holes cross at $k = 0$. Transition of holes from the light band to the heavy band, assisted with crossly-applied electric and magnetic field, generates a THz photon. The emission is tunable in the range $1.5 \sim 4$ THz and the radiated power is quite high (W), but the device runs at cryogenic temperatures. Nanotransistors are also promising THz devices. In field-effect transistors (FETs), the current flows between the source and the drain in the thin layer, namely the channel. An electrode, the gate, located over the channel, monitors the current. When the thickness of the channel is of the order of nanometer, the channel can be described as a 2D electron gas. This 2D electron gas can freely oscillate perpendicularly to the layer: this is the plasma oscillation of the system. When injecting the nanotransistor with a steady current, the nonlinear coupling with the plasma oscillations makes the current unstable and a current component pulsating at the plasma frequency appears. This current may radiate THz electromagnetic waves through the transistor electrodes that serve as an antenna. These FETs run at room temperature, but the radiated power remains still low (nW). Using a graphene [15] or carbon nanotube [16] layer as a channel is also a promising technology.

3.4 Large-Band Detectors

3.4.1 General Definitions

A detector is a device that, when illuminated by an electromagnetic beam, supplies a signal, generally an electrical current or voltage, proportional to the impinging beam power or amplitude.

Two families of power detectors may be defined, depending on whether they are sensitive to the incoming photon (photon detectors) and to the integrated energy (thermal detectors). Generally, photon detectors directly produce a current/voltage proportional to the incident electromagnetic power. In energy detectors, the impinging beam power changes some physical parameter of the detecting material, like temperature, pressure, and so on. Then a converter must transform this change into an electrical signal (either current or voltage).

Detectors sensitive to the beam amplitude, sometimes named field detectors, must show a response time shorter than the period of the electromagnetic wave, otherwise the field is integrated over one or several periods of oscillations, and thus the integrated signal is null. In the case of THz waves, this means that the detector response time should be typically shorter than 1 ps. However, no electronic system is able today to deal with sub-picosecond events. Therefore, only repetitive signals may be treated, through time-equivalent techniques. Some smart single-shot techniques are also employed, but more rarely, to measure waveforms of THz pulses. Field detectors will be described in Chapter 4.

3.4.2 Basic Equations of Detection

Let P be the power of a monochromatic light beam received by the detector. The power is equal to the number of impinging photons per second multiplied by the energy hf of each photon. Each photon is absorbed and transformed into a photoelectron in the detector with a probability (or quantum efficiency) η. Thus the induced short-circuit current i delivered by the detector is

$$i = \eta \frac{P}{hf} e = \Re P. \tag{3.37}$$

$\Re = \eta e / hf$ is the responsivity of the detector, which depends on the frequency f of the impinging photons. In fact, we have to take into account the random arrival of the photons onto the receiver. If the emission of each photon by the source is independent from the other ones, the number N of photons arriving during a short time slot Δt obeys the statistics of independent random events, i.e., a Poisson distribution. The photon–electron conversion η is also random, so the number n of electrons photo-generated during Δt obeys the Poisson distribution, assuming N and η are independent. Therefore, any photocurrent i shows a noisy behavior due to the particle character of both photons and electrons. We will now determine this photon noise, which is also called the shot noise.

Let us call $\overline{N}$ the mean value of N and σ_{N} its standard deviation. Mathematically, they are defined by

$$\overline{N} = \frac{1}{p_{\max}} \sum_{p=0}^{p_{\max}} N_{\mathrm{p}}, \quad \sigma_{\mathrm{N}}^2 = \frac{1}{p_{\max}} \sum_{p=0}^{p_{\max}} \left(N_{\mathrm{p}} - \overline{N}\right)^2, \quad p_{\max} = \frac{T}{\Delta t} \gg 1. \tag{3.38}$$

T is the duration over which mean value and standard deviation are calculated. Rigorously, from a mathematical point of view, T must tend towards infinity. The relation between $\overline{N}$ and P is simply given by:

$$\overline{N} = \frac{P}{hf} \Delta t. \tag{3.39}$$

The standard deviation σ_{N} of the Poisson distribution N is $\sigma_{\mathrm{N}}^2 = \overline{N}$. The signal-over-noise ratio SNR is the square of the ratio of the mean value to the standard deviation:

$$\mathrm{SNR}\,(N) = \left(\frac{\overline{N}}{\sigma_{\mathrm{N}}}\right)^2 = \overline{N}. \tag{3.40}$$

The rate of photo-generating charges and its standard deviation obey also the Poisson statistics:

$$\bar{n} = \frac{1}{p_{\max}} \sum_{p=0}^{p_{\max}} n_{\mathrm{p}} = \eta\, \bar{N}, \quad \sigma_{\mathrm{n}}^2 = \frac{1}{p_{\max}} \sum_{p=0}^{p_{\max}} \left(n_{\mathrm{p}} - \bar{n}\right)^2 = \bar{n}, \tag{3.41}$$

$$\Rightarrow \quad \mathrm{SNR}\,(n) = \frac{\bar{n}^2}{\sigma_{\mathrm{n}}^2} = \bar{n}. \tag{3.42}$$

n_{p} is the number of charges photo-generated during the time slot p of duration Δt. The current intensity and the optical power appearing in Eq. 3.37 are nothing but mean values, as usually the detector response time is longer than the short random temporal events, and thus than Δt. Thus,

$$\begin{cases} i \equiv \bar{i} = \frac{\bar{n}}{\Delta t} e, \\ \sigma_{\mathrm{i}}^2 = \frac{1}{p_{\max}} \sum\limits_{p=0}^{p_{\max}} \left(i_{\mathrm{p}} - \bar{i}\right)^2 = \frac{1}{p_{\max}} \left(\frac{e}{\Delta t}\right)^2 \sum\limits_{p=0}^{p_{\max}} \left(n_{\mathrm{p}} - \bar{n}\right)^2 = \left(\frac{e}{\Delta t}\right)^2 \sigma_{\mathrm{n}}^2. \end{cases} \tag{3.43}$$

It follows that the current SNR is:

$$\mathrm{SNR}\,(i) = \frac{\bar{i}^2}{\left(\frac{e}{\Delta t}\right)^2 \bar{n}} = \frac{\Delta t}{e}\bar{i} = \bar{n}. \tag{3.44}$$

If the detected light beam is not monochromatic, Eq. 3.37 becomes:

$$i = \int_0^{\infty} \Re \frac{dP}{d\omega} d\omega = \int_0^{\infty} \Re \frac{dP}{df} df. \tag{3.45}$$

$dP/d\omega$ (or dP/df) is the spectral power density, i.e., the power of the beam in the spectral interval $(\omega, \omega + d\omega)$ or $(f, f + df)$.

If the spectral bandwidth of the detector, including the electronic readout, is BW, the shortest time interval the system can resolved is the width of its impulse response $\Re\,(t)$. The impulse response is the Fourier transform of the responsivity spectrum $\Re\,(\omega)$. Let us assume the response of the detector is equal to 1 between frequencies $f_{\min}$ and $f_{\max} = f_{\min} + \mathrm{BW}$. We can determine $\Re\,(t)$:

$$\begin{aligned}\Re\,(t) &= \frac{1}{\sqrt{2\pi}} \int_{-\infty}^{+\infty} \Re\,(\omega)\, e^{j\omega t} d\omega = \frac{1}{\sqrt{2\pi}} \int_{2\pi f_{\min}}^{2\pi(f_{\min}+\mathrm{BW})} e^{j\omega t} d\omega \\ &= \frac{1}{\sqrt{2\pi}} \left[\frac{e^{j2\pi(f_{\min}+\mathrm{BW})t} - e^{j2\pi f_{\min} t}}{jt}\right] \\ &= \sqrt{2\pi}\,\mathrm{BW}\, e^{j2\pi(f_{\min}+\mathrm{BW}/2)t} \frac{\sin(\pi\,\mathrm{BW}\,t)}{\pi\,\mathrm{BW}\,t}.\end{aligned}$$

The modulus of $\Re\,(t)$ is:

$$|\Re\,(t)| = \sqrt{2\pi}\,\mathrm{BW}\,\mathrm{sinc}\,(\pi\,\mathrm{BW}\,t). \tag{3.46}$$

The time at which the impulse response has decreased by $\frac{1}{2}$ (−3 dB) is defined as the temporal resolution Δt of the detecting system:

$$|\Re\,(t)| = \frac{1}{2} \;\rightarrow\; \mathrm{sinc}\,(\pi\,\mathrm{BW}\,\Delta t) = \frac{1}{2} \;\Rightarrow\; \Delta t \approx \frac{1}{2\mathrm{BW}}. \tag{3.47}$$

We then substitute Δt by $\frac{1}{2}\mathrm{BW}$ in (3.43) and we get:

$$i = 2\mathrm{BW}\, e\, \bar{n} = e\eta\frac{P}{hf}, \quad \sigma_{\mathrm{i}} = 2\mathrm{BW}\, e\, \sigma_{\mathrm{n}}. \tag{3.48}$$

For a Poisson distribution, $\sigma_{\mathrm{n}}^2 = \bar{n}$ and thus:

$$i = e\eta\frac{P}{hf}, \quad \sigma_{\mathrm{i}}^2 = (2\mathrm{BW}\, e)^2\, \sigma_{\mathrm{n}}^2 = (2\mathrm{BW}\, e)^2\, \bar{n} = 2\mathrm{BW}\, e\, i. \tag{3.49}$$

This latter expression is the current noise caused by the photon fluctuation of the impinging light beam, namely the shot noise. The corresponding SNR is:

$$\mathrm{SNR}\,(i) = \frac{i^2}{\sigma_{\mathrm{i}}^2} = \frac{i^2}{2\mathrm{BW}\,e\,i} = \frac{i}{2\mathrm{BW}\,e} = \bar{n}. \tag{3.50}$$

The flow of the charges in the detection circuit, i.e., the photocurrent, induces an additional noise, namely the Johnson noise. In a conductor, the resistance originates from the random collision of the charge carriers with defects in the material (structural and stoichiometric defects or heterogeneities) or, mostly at room temperature, with phonons (i.e., the thermal vibrations in the conductor). If the average displacement of the carriers (let us say electrons) is towards the direction of the anode, their instantaneous movement is random and they can even move for a short while towards the cathode. Thus the contribution of a charge to the current is fluctuating around a mean value. Due to this scattering phenomenon, the current is noisy. Let us call τ the mean time between two collisions (in most of metals at room temperature, $\tau < 100$ fs). We determine the autocorrelation function of the electrons velocity as follows:

$$\begin{aligned} \overline{v\,(t)\,v\,(t+\delta t)} &= \overline{v^2\,(t)}\,P\,[v(t) = v(t+\delta t)] \\ &+ \underbrace{\overline{v(t)v(t+\delta t)}}_{v(t)\neq v(t+\delta t)\,\rightarrow\,=0}\;P\,[v(t) \neq v(t+\delta t)] \quad . \end{aligned} \tag{3.51}$$

In Eq. 3.51, $P\,[v\,(t) = v\,(t+\delta t)]$ is the probability that the velocity at time $t + \delta t$ is the same than at time t. Because the collision phenomenon is random, the mean value $\overline{v\,(t)\,v\,(t+\delta t)}$ is null for $v\,(t) \neq v\,(t+\delta t)$. The statistic of the random collision phenomenon is given by the Poisson law:

$$p\,(k) = e^{-a|\delta t|}\frac{(a\delta t)^k}{k!}, \tag{3.52}$$

where $p\,(k)$ is the probability that k collisions occur during δt. The velocities are the same if there is no collision, i.e., $k = 0$. Therefore,

$$P\,[v\,(t) = v\,(t+\delta t)] = p\,(0) = e^{-a|\delta t|}, \tag{3.53}$$

where $a = 1/\tau$ is the inverse of the electron collision time τ. The time interval must be taken to be positive ($|\delta t|$) in order to have the probability always bound by 1. Thus we obtain

$$\overline{v(t)\,v(t+\delta t)} = \overline{v^2(t)}\,e^{-a|\delta t|} = \sigma_v^2\,e^{-a|\delta t|}. \tag{3.54}$$

The spectral density of the autocorrelation function is

$$\mathrm{TF}\left[\overline{v(t)\,v(t+\delta t)}\right] = \mathrm{TF}\left[\sigma_v^2\,e^{-a|\delta t|}\right] = 2\frac{\sigma_v^2}{\sqrt{2\pi}}\frac{2a}{a^2+(2\pi f)^2}. \tag{3.55}$$

The additional factor of 2 appearing in the third term in Eq. 3.55 comes from the fact that we take the Fourier transform only for positive frequencies, so it compensates for the forgotten energy carried by the negative frequencies. The pulsation $2\pi f$ of the current flowing in the receiving circuit is always much smaller than a, so

$$\mathrm{TF}\left[\overline{v(t)\,v(t+\delta t)}\right] \approx \frac{4\sigma_v^2}{\sqrt{2\pi}\,a}. \tag{3.56}$$

Expression 3.56 does not depend on frequency: the corresponding current noise is a white noise. For a charge moving in one direction (one-dimension model), the standard deviation of the velocity versus temperature T can easily be derived from the basic law of statistical physics:

$$\overline{E_{\mathrm{kin}}} = \frac{1}{2}m\overline{v^2} = \frac{1}{2}k_{\mathrm{B}}T \quad \Rightarrow \quad \mathrm{TF}\left[\overline{v(t)\,v(t+\tau)}\right] \approx 4\frac{k_{\mathrm{B}}T}{\sqrt{2\pi}\,m\,a}. \tag{3.57}$$

The current i_e, carried by one electron moving at velocity v, and thus traveling over a distance L during the time Δt, is

$$i_e = \frac{-e}{\Delta t} = \frac{-e}{L}v \quad \Rightarrow \quad \mathrm{TF}\left[\overline{i(t)\,i(t+\tau)}\right] = 4\frac{k_{\mathrm{B}}T}{\sqrt{2\pi}\,m\,a}\frac{e^2}{L^2}. \tag{3.58}$$

Then, we introduce the resistance R of the conductor whose length and section are L and S, and in which the free electron density is n:

$$R = \frac{m\,a}{n\,e^2}\frac{L}{S} \quad \Rightarrow \quad \mathrm{TF}\left[\overline{i(t)\,i(t+\tau)}\right] = \frac{4\,k_{\mathrm{B}}T}{\sqrt{2\pi}\,R}. \tag{3.59}$$

Finally, we derive the standard deviation of the current by integrating over the whole reachable frequency range:

$$\sigma_i^2 = \sqrt{2\pi} \int_0^{+\infty} \mathrm{TF}\left[\overline{i(t)\, i(t+\tau)}\right] df$$
$$= \sqrt{2\pi} \int_0^{B} \mathrm{TF}\left[\overline{i(t)\, i(t+\tau)}\right] df = \frac{4\, k_B T\, B}{R}. \quad (3.60)$$

This is the expression of the Johnson noise.

Additional sources of noise are usually observed in the detected signal: noise from the power supply of the source, noise in the electronics of the detection circuit (mostly amplification noise), thermal THz signal radiated by the room and/or by the items in the experimental setup, which are often at room temperature (optics, holders, etc.). All this additional noise leads to a noisy current, which is written as

$$i(t) = \bar{i} + \Delta i(t). \quad (3.61)$$

$\Delta i(t)$ is the current noise, which includes the shot and Johnson noise, and whose average value is $\overline{\Delta i(t)} = 0$. The power dissipated by the detected current in the detection circuit is:

$$P(t) = R\, i(t)^2 = R\left(\bar{i} + \Delta i(t)\right)^2. \quad (3.62)$$

The time-averaged value of $P(t)$ is:

$$\overline{P(t)} = R\left(\overline{\bar{i}^2 + 2\Delta i(t)\,\bar{i} + \Delta i(t)^2}\right) = R\bar{i}^2 + R\overline{\Delta i(t)^2}. \quad (3.63)$$

The detected power includes the actual absolute power of the signal, $P = R\bar{i}^2$, and the noise power $R\overline{\Delta i(t)^2}$. The signal-over-noise SNR ratio is defined by the ratio of these two powers:

$$\mathrm{SNR} = \frac{R\bar{i}^2}{R\overline{\Delta i(t)^2}} = \frac{\bar{i}^2}{\overline{\Delta i(t)^2}}. \quad (3.64)$$

The noise-equivalent power (NEP) corresponds to an incoming electromagnetic power that gives rise to SNR = 1. The sensitivity of the detector (defined by 3.37) relates this impinging power and the delivered signal. Here, for a detector supplying a current, relation 3.37 becomes $i = \Re P_{\text{beam}}$. Hence,

$$\text{SNR} = 1 = \frac{R\bar{i}^2}{R\overline{\Delta i\,(t)^2}} = \frac{\Re^2 R P_{\text{beam}}^2}{R\overline{\Delta i\,(t)^2}} \Rightarrow \text{NEP} = P_{\text{beam}}$$

$$= \frac{\sqrt{\overline{\Delta i\,(t)^2}}}{\Re} = \frac{\sigma_{\text{i}}}{\Re}. \quad (3.65)$$

The inverse of NEP is the detectivity D:

$$D = \frac{1}{\text{NEP}}. \quad (3.66)$$

Because NEP is power, its unit is watt. However, in the far infrared, the NEP of many detectors is proportional to $\sqrt{S \times \Delta f}$, where S is their active area and Δf their bandwidth. In order to normalize and compare the experimental NEP values, one is accustomed to measure the NEP value in a given bandwidth and to normalize it to a $\Delta f = 1$ Hz bandwidth. The unity of this normalized NEP is thus $\text{W}/\sqrt{\text{Hz}}$. As well, the normalized detectivity D^* is

$$D^* = \frac{\sqrt{S \times \Delta f}}{\text{NEP}}. \quad (3.67)$$

Its unit is the jones $(\text{cm}\sqrt{\text{Hz}}/\text{W})$.

When observing the thermal emission of a scene, the spectral power density reaching the detector is

$$dP = L(f) S\,\Omega\,df \approx \varepsilon_{\text{therm}}(f) 2k_{\text{B}} T \frac{f^2}{c^2} S\,\Omega\,df. \quad (3.68)$$

Ω is the detection solid angle (i.e. how large the detector appears to an observer looking from the scene), S is the area of the imaged region, T is its temperature, and $\varepsilon_{\text{therm}}(f)$ its emissivity. Therefore, the detected current is

$$i = \int_0^\infty \Re(f) dP = A \times T, \text{ with } A = \frac{2k_{\text{B}}}{c^2} S\,\Omega \int_0^\infty \Re(f)\varepsilon_{\text{therm}}(f) f^2\,df. \quad (3.69)$$

The current is proportional to the temperature T of the scene. To derive Eq. 3.69, we have introduced the Rayleigh–Jeans approximation (3.22), which is valid in the THz range. The noise-equivalent differential temperature (NEDT) is defined as the variation ΔT of temperature of the scene that induces a change of the recorded signal equal to the noise of the detector. In other words, assuming the quantum efficiency η does not vary with frequency, we get

$$\text{NEDT} = \frac{\sigma_\text{i}}{A} = \frac{\Re}{A}\text{NEP} = \frac{\Re \times \text{NEP}}{2e\,k_\text{B}S\,\Omega \int\limits_0^\infty \varepsilon(f) f\,df}. \tag{3.70}$$

3.4.3 The Different Types of Detectors

Classical large-band THz detectors (optoelectronic field detectors will be described in Chapter 4) are based on absorbing the incoming THz energy in a material, leading to an increase in the temperature of this material. This THz-induced variation of temperature modifies some physical properties of the material, which can be measured with common equipment. In pneumatic detectors like a Golay cell, a gas in an expandable cell is heated up by the THz signal and consequently its volume increases. The volume change is measured by optical means. In pyroelectric detectors, a voltage is induced by the temperature change in a pyroelectric crystal. In bolometers, the electrical resistance of the heated material depends on temperature. In a transition-edge sensor (TES), an absorbing layer is attached to a superconducting layer. The temperature of the device is close to the critical temperature T_c of the superconductor, i.e., the temperature at which the transition from the normal state to the superconducting one occurs: any variation, even a weak one, of the superconductor temperature dramatically changes its resistance, which is monitored by a read-out circuit. Superconducting hot-electron bolometers (HEBs) are based on a different principle. A very short bridge fabricated in a thin superconducting film is illuminated by the incoming THz radiation. Let us remind that, in a superconductor, a given percentage of free electrons form pairs, namely the Cooper pairs, through phonon-assisted interaction with the crystal lattice. In the framework of the Bardeen–Cooper–Schrieffer (BCS) theory, the Cooper pair energy is $E_\text{Cooper} = 3.5\,k_\text{B}T_\text{c}$. For incoming photons with an energy larger than E_Cooper, i.e., with a frequency larger than $f\,(\text{THz}) = 3.5\,k_\text{B}T_\text{c}/h \approx 0.073 \times T_\text{c}$, Cooper pairs are broken and supply single "hot" electrons, resulting in an increase of the superconductor bridge resistance. Semiconductor HEBs operate

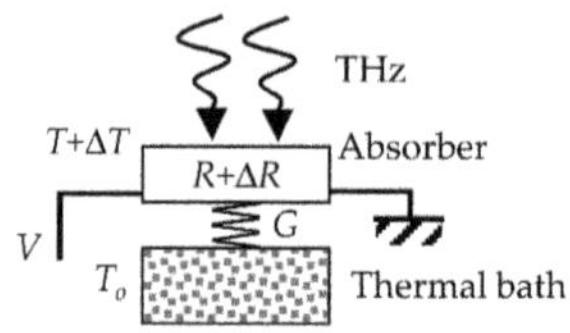

Figure 3.10 Scheme of a bolometer based on a thermally induced variation of the electrical resistance.

also at cryogenic temperatures. In a doped semiconductor, usually InSb, the THz energy is absorbed by the free carriers in the conduction band. Thanks to the conservation of momentum, the excited free electrons climb up along the conduction band in the k space. This changes their velocity and therefore the resistance of the semiconductor, whose variation is proportional to the THz power. As the electron–electron interaction time is much shorter than that of the electron–phonon interaction, the excited electrons come into equilibrium at a temperature higher than the lattice (the transfer of their additional energy to the lattice is done through an electron–phonon process, which takes a longer time). Such photo-excited electrons are named hot electrons.

Let us treat the general equations for bolometers in which the electrical resistance of the heated absorber is measured (Fig. 3.10). The heat capacity of the absorber is C. The absorber is linked to a thermal bath at temperature T_0 via a thermal conductance G. The variation of the absorber temperature T under illumination is given by

$$C\frac{\partial T\left(t\right)}{\partial t}=P_{\mathrm{THz}}\left(t\right)-G\left(T\left(t\right)-T_{0}\right). \tag{3.71}$$

P_{THz} is the absorbed THz power at time t. At equilibrium, $P_{\mathrm{THz}} = G\left(T-T_0\right) = G\,\Delta T$. This change ΔT of temperature induces a variation ΔR of the resistance R of the absorber. ΔR is measured by applying a bias voltage V at the absorber and then reading the current flowing through the absorber. But this current increases also the temperature because of the Joule effect: $G\,\Delta T = P_{\mathrm{THz}} + V^2/R$. Any variation in P_{THz} changes R, which is calculated by differentiating the latter relation: $G\,dT = dP_{\mathrm{THz}} - V^2 dR/R^2$. From this equation, we easily derive the relative change of resistance induced by the absorption of the THz beam:

$$\frac{\Delta R}{R} = \frac{P_{\mathrm{THz}}}{\frac{G}{\alpha_{\mathrm{th}}} + \frac{V^2}{R}}, \qquad \alpha_{\mathrm{th}} = \frac{1}{R}\frac{dR}{dT} = \frac{1}{\rho}\frac{d\rho}{dT}. \tag{3.72}$$

α_{th} is the thermal coefficient of the resistance, and ρ is the resistivity of the material used to make the resistance. Efficient bolometers are obtained by reducing the denominator of Eq. 3.72. This is achieved with materials exhibiting a large α_{th} (superconductors) and/or by minimizing G. This latter condition requires bolometers operating in vacuum (to avoid thermal loss by convection), and/or at low temperature (to reduce thermal radiation), and/or made with membranes on which the absorber is deposited (to reduce thermal loss by conduction).

3.5 Spectrometers

In the optical and infrared domain, most common spectrometers (Fig. 3.11) are based on

(1) a dispersive element (prism or grating),
(2) beam splitting and recombination of the two beams (Michelson and Mach–Zehnder interferometers), and
(3) multiple beam splitting and recombination of the multiple beams (Fabry–Pérot *étalon*).

In the THz domain, similar spectrometers have been designed. The most popular systems are Fourier-transform spectrometers

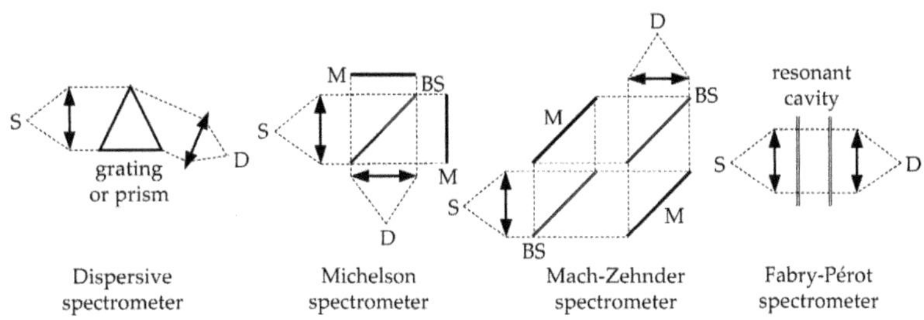

Figure 3.11 Different types of common spectrometers. S is for source, D for detector, M for mirror, and BS for beam splitter.

(see next paragraph) or time-domain spectrometers (see the next chapters). In fact, because the power delivered by the first sources of THz radiation (blackbodies) was rather weak, spectrometers had to exhibit high luminosity and to collect the maximum of emitted light, at the expenses of long recording times. This constraint leads to quite huge systems. Time-domain spectrometers are more compact: some commercial systems have the same size as a notebook. They show also amazing dynamic and SNR, which lead to short recording times. FTIR spectrometers are by definition recording a more intense signal than other spectrometers (Felgett advantage; see next paragraph). Therefore, grating spectrometers and Fabry–Pérot resonators are employed today only for specific THz spectroscopy studies.

Let us first describe dispersive spectrometers based on diffraction grating. Due to the longer wavelengths being used, diffraction gratings for THz waves are easy to manufacture. Typically, the period of the grating should be of the order of the studied wavelengths, i.e., $0.1 \sim 1$ mm for THz waves. The conservation of the tangential component $k_{||}$ of the beam wavevectors at the periodic grating surface leads the following dispersion equation:

$$k_{\mathrm{i}||} + m\,k_{\mathrm{G}} = k_{\mathrm{m}||} \quad \Leftrightarrow \quad \sin\theta_{\mathrm{i}} + m\frac{\lambda}{d} = \sin\theta_{\mathrm{m}}, \tag{3.73}$$

where m is an integer (the diffraction order), θ_{i} is the angle of incidence, θ_{m} is the diffraction angle, d is the grating period, and λ is the wavelength. The angular dispersion of the grating is

$$\frac{\Delta\theta_{\mathrm{m}}}{\Delta\lambda} = -\frac{m}{d\cos\theta_{\mathrm{m}}} = \frac{1}{\lambda}\frac{\sin\theta_{\mathrm{i}} - \sin\theta_{\mathrm{m}}}{\cos\theta_{\mathrm{m}}}. \tag{3.74}$$

It is much bigger than the dispersion of any prism. The effective bandwidth of a grating spectrometer is limited by the free spectral range (FSR). For given angles of incidence and dispersion, Eq. 3.73 exhibits several solutions corresponding to $m\lambda$ remaining constant. Thus two adjacent orders m and $m+1$ diffract two different wavelengths λ_1 and $\lambda_2 (m\lambda_1 = (m+1)\lambda_2)$ in the same directions, which can no more be separated. FSR is the largest spectral range for which wavelengths can be separated by the spectrometer:

$$\mathrm{FSR} = \frac{m}{\lambda}. \tag{3.75}$$

Grating for the far infrared are mostly either engraved at the surface of a metallic slab, or made of metallic strips (mesh gratings). The first ones only work in reflection, but the grooves may be easily tailored, for example to fabricate blaze gratings with optimized efficiency (used in the Littrow mounting, i.e., the diffracted order is backward reflected). Also, their shape can be concave, in order to focus the diffracted light or to compensate for geometrical aberrations in the spectrometer. Moreover, because of the rather large period (sub-millimeter), they can be mechanically manufactured with milling machines. Such spectrometers are very often based on a Czerny–Turner scheme, i.e., a parabolic mirror makes parallel the beam passing through the entrance slit. Thus the grating is wholly enlightened with a parallel beam. A second parabolic mirror focuses the diffracted beam onto the exit slit. These spectrometers show a good spectral resolution, typically $\Delta f/f = 10^{-3}$.

Prism, Mach–Zehnder, and Michelson (in its classical use) interferometers are rarely employed in the THz range, as their performances are inferior to those of FTIR spectrometers (see next paragraph). Fabry–Pérot spectrometers of high finesse exhibit a very high resolution (10^{-4}) and are very compact, so they are suitable for airborne/balloon space observations. The difficulty is to manufacture good mirrors with a specific transmissivity/reflectivity, which are either mesh filters or even multilayer coatings. The spectral tunability is achieved by progressively shifting the gap between the two mirrors of the *étalon*.

3.6 Fourier-Transform Infrared Spectroscopy

Fourier-transform infrared spectroscopy (FTIR) [17, 18] was originally suggested by A. A. Michelson at the end of the 19th century [17], after he invented the Michelson interferometer, dedicated to the observation and measure of the possible effect of Earth's rotation on the speed of light. He clearly understood that the output of the interferometer gives the Fourier transform of the spectrum of the incoming light source. At those times, the interference fringes were measured by Michelson with his eyes as detectors, no computer existed, and performing the Fourier transform of the recorded interferograms was not possible. However, Michelson was smartly able to resolve some doublets, like the doublet of sodium emission [20]. Independently, P. B. Fellgett in the UK in 1949 and P. Jacquinot in France in 1954 realized [21] that the Michelson interferometer exhibits two strong advantages compared with dispersive interferometers, regarding the throughput and the signal-over-noise ratio (today known as Fellgett and Jacquinot advantages; see below). This was the beginning of FTIR spectroscopy: the first apparatus was fabricated and the first spectra produced later in the 1950s by J. D. Strong and colleagues [22, 23]. The last pioneering stage was done by the Digilab Company in the USA, which commercialized the first FTIR system in the late 1960s, namely model FTS 14, based on a Block Engineering interferometer.

As most of commercial equipments are based on the Michelson interferometer scheme, we describe here the FTIR system operation related to the Michelson geometry (Fig. 3.12). The ideal interferometer includes a source that delivers a parallel beam with a large spectral band, an ideal beam splitter at 45° that splits and directs half of the beam power towards each of the two mirrors at normal incidence. The mirror located in arm B of the interferometer is fixed, while the mirror in arm A can be moved along the beam direction. Position $x = 0$ of mirror A corresponds to an equal distance between the two mirrors and the beam splitter (the "lengths" of the two arms

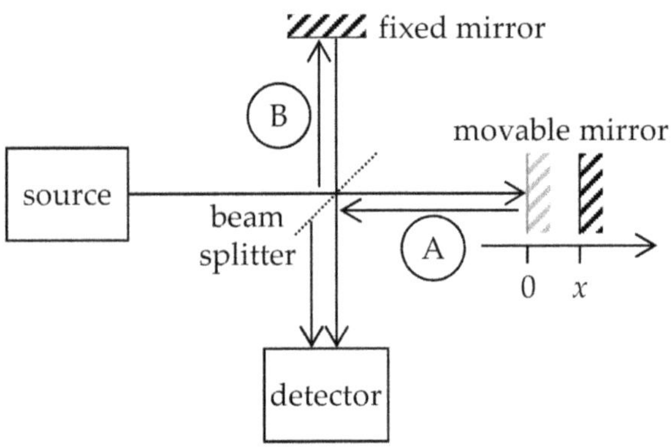

Figure 3.12 Schematic of a Michelson interferometer.

are identical). In other words, when the mirror in A is located at $x = 0$, it corresponds to the symmetry image of mirror B as regards to the beam splitter. Therefore $2x$ is the difference of path encountered by the light beam between the two arms. The beams reflected by the mirrors are recombined at the beam splitter, i.e., they are partly directed towards the detector. Let us first consider a single spectral component of the field delivered by the source, whose pulsation is ω.

The field at the receiver coming from arm B is defined as

$$E_{\mathrm{B}}(t) = E_{\mathrm{o}}(\omega)\cos(\omega t). \tag{3.76}$$

The field from arm A is phase-shifted compared with the one from B, because of the $2x$ additional path:

$$E_{\mathrm{A}}(t) = \eta\, E_{\mathrm{o}}(\omega)\cos(\omega t + \phi). \tag{3.77}$$

A possible imbalance between the arms is accounted by the η coefficient. We suppose here that the two beams exhibit the same polarization. Thus the intensity received by the detector is proportional to the time-averaged square of the sum of the two fields:

$$I(\varphi) \propto \overline{(E_{\mathrm{A}}(t) + E_{\mathrm{B}}(t))^2} = E_{\mathrm{o}}^2(\omega)\,\overline{(\cos(\omega t) + \eta\cos(\omega t + \varphi))^2},$$

$$\Rightarrow I(\varphi) \propto E_{\mathrm{o}}^2(\omega)\,\overline{(\cos^2(\omega t) + 2\eta\cos(\omega t + \varphi)\cos(\omega t) + \eta^2\cos^2(\omega t + \varphi))},$$

$$\Rightarrow I(\varphi) = I_{\mathrm{o}}(\omega)\,\frac{1 + 2\eta\cos\varphi + \eta^2}{2}.$$

If the ambient medium is vacuum or air, the phase difference φ is

$$\varphi = 2\frac{\omega}{c}x \quad \Rightarrow \quad I(x,\omega) = \frac{I_{\mathrm{o}}(\omega)}{2}\left(1 + \eta^2 + 2\eta\cos\left(2\frac{\omega}{c}x\right)\right). \tag{3.78}$$

The intensity is the sum of a constant term and a term varying as the cosine of the phase. Note that $I_{\mathrm{o}}(\omega) = I_{\mathrm{source}}(\omega)\,T_{\mathrm{FTIR}}(\omega)$ is the intensity $I_{\mathrm{source}}(\omega)$ supplied by the source and multiplied by the transmission $T_{\mathrm{FTIR}}(\omega)$ of the interferometer (this transmission includes the spectral responses of the mirrors and the beam splitter). If the source is not monochromatic, but exhibits a spectral density (power per unit of frequency), and if the receiver is broadband, i.e., it collects the energy of all the spectral components, the recorded signal becomes

$$S(x) = \int_0^{+\infty} \Re(\omega)\, I(x,\omega)\, d\omega,$$

$$\Rightarrow S(x) = \int_0^{+\infty} I_{\text{source}}(\omega)\, T_{\text{FTIR}}(\omega)\, \Re(\omega)\, \frac{1+\eta^2+2\eta\cos\left(2\frac{\omega}{c}x\right)}{2}\, d\omega. \tag{3.79}$$

If $\Re(\omega)$ is the detector responsivity (3.37), $S(x)$ is called the interferogram (Fig. 3.12 left) and is proportional to the intensity received by the detector at the location x of the detector. We transform relation 3.79 by writing the cosine as a complex function. As the product $I_{\text{source}}(\omega)\, T_{\text{FTIR}}(\omega)\, \Re(\omega)$ is real, $I_{\text{source}}(-\omega)\, T_{\text{FTIR}}(-\omega)\, \Re(-\omega) = I_{\text{source}}(\omega)\, T_{\text{FTIR}}(\omega)\, \Re(\omega)$. We get:

$$S(x) = S_o + \frac{1}{2}\eta \int_{-\infty}^{+\infty} I_{\text{source}}(\omega)\, T_{\text{FTIR}}(\omega)\, \Re(\omega)\; e^{2j\frac{\omega}{c}x} d\omega. \tag{3.80}$$

S_o is a constant, while the second term is nothing but the inverse Fourier transform of $I_{\text{source}}(\omega)\, T_{\text{FTIR}}(\omega)\, \Re(\omega)$, and where the conjugated variable of ω is $2x/c$, i.e., the time for the light to propagate over the distance $2x$:

$$S(x) = S_o + \eta\sqrt{\frac{\pi}{2}}\, \text{TF}^{-1}\left[I_{\text{source}}(\omega)\, T_{\text{FTIR}}(\omega)\, \Re(\omega)\right]. \tag{3.81}$$

Then, taking the Fourier transform of Eq. 3.81 leads to:

$$I_{\text{source}}(\omega)\, T_{\text{FTIR}}(\omega)\, \Re(\omega) = \delta(\omega) + \frac{1}{\eta}\sqrt{\frac{2}{\pi}}\, \text{TF}\left[S(x)\right]. \tag{3.82}$$

$\delta(\omega)$ is the delta Dirac function, which is the Fourier transform of the constant S_o. Thus, for $\omega \neq 0$, the Fourier transform (see Fig. 3.13, right) of the recorded interferogram is simply the product of the spectra of the source and the receiver sensitivity. If the frequency response $T_{\text{FTIR}}(\omega)\, \Re(\omega)$ of the spectrometer is known, the spectrum $I_{\text{source}}(\omega)$ of the source may be determined by recording the interferogram and then computing its Fourier transform. This latter task was a stumbling block when the first apparatuses were

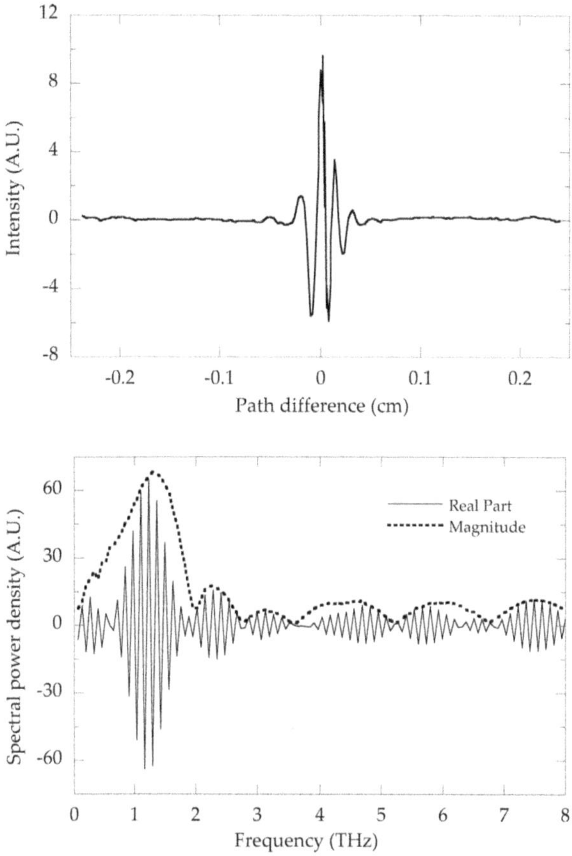

Figure 3.13 Typical FTIR interferogram (left) and its spectrum (right). Adapted with permission from Burton and Akimoto [24].

fabricated in the 1960s. Thanks to the fast Fourier transform (FFT) algorithm proposed by J. W. Cooley and J. Tukey, nowadays the computation of the Fourier transform is almost instantaneous, even with laptop computers.

Actual experimental spectra differ from expression 3.82, because the interferogram is recorded over a finite maximum displacement x_{max} of mirror A. Thus we have to multiply in Eq. 3.82 the actual interferogram term $S(x)$ by a window function $T_{\text{W}}(x)$ that is equal to 1 between x_1 and $x_2 = x_1 + x_{\text{max}}$ (x_1 and x_2 are respectively the shortest and longest values of x), and equal to 0 elsewhere:

$$I_{\text{source}}(\omega)\, T_{\text{FTIR}}(\omega)\, \Re(\omega) = \frac{1}{\eta}\sqrt{\frac{2}{\pi}}\,\text{TF}\left[S(x)\; T_{\text{W}}(x)\right], \quad \omega \neq 0. \tag{3.83}$$

The Fourier transform of a product of two real functions is the correlation of the Fourier transforms of each function. We remind the reader that the Fourier transform of a rectangle function is a sinc function (see Annex A.106). Therefore, for $\omega \neq 0$:

$$I_{\text{source}}(\omega)\, T_{\text{FTIR}}(\omega)\, \Re(\omega) = \frac{1}{\eta}\sqrt{\frac{2}{\pi}}\,\text{TF}\left[S(x)\right] \otimes \text{sinc}\left(\frac{\omega}{c}x_{\text{max}}\right). \tag{3.84}$$

In the case of ideal monochromatic sources of pulsation ω, the interferogram $S(x)$ is thus a sinusoidal function and its Fourier transform is constant. The Fourier transform of the recorded interferogram simply varies as sinc $(x_{\text{max}}\omega/c)$. According to the Rayleigh criterion, the spectral width of this sinc function is the spectral resolution Δf of the interferometer:

$$\Delta f = \frac{c}{2\, x_{\text{max}}}. \tag{3.85}$$

Typically, the best commercial instruments offer a spectral resolution of a few hundreds of MHz, i.e., comparable with grating spectrometers. The advantages of FTIR spectrometers, compared with dispersive spectrometers, are principally twofold. First, in FTIR, all the frequency components of the light are recorded at the same time. Thus the recorded signal is larger than in dispersive spectrometers, for which each individual spectral component is recorded separately. Let us compare a dispersive spectrometer and a FTIR spectrometer under similar experimental conditions (same sources, detectors, etc.). Let us assume that the noise of the recorded current is only the shot noise, $\sigma_{\text{i}} = \sqrt{2\text{BW}e\,i} = \sqrt{e\,i/T}$ (3.49), where T is the recording duration. In FTIR spectrometers, all the spectral components of the light are measured simultaneously during the time T. Therefore, the noise σ_{FTIR} is given by the preceding expression, i.e., $\sigma_{\text{FTIR}} = \sigma_{\text{i}}$. In dispersive spectrometers, if M spectral components are observed, the signal for each of them is recorded during the time interval T/M. Consequently, the shot noise writes $\sigma_{\text{disp}} = \sqrt{eiM/T} = \sqrt{M}\sigma_{\text{FTIR}}$. Thus, the noise in FTIR spectrometer is reduced by a factor $\sqrt{M}$ compared with the one in dispersive spectrometers: this the Fellgett advantage (or multiplex advantage). Secondly, the spectral resolution of dispersive spectrometers is inversely proportional to the width of the entrance/exit slits (3.8), while the spectral resolution of the FTIR spectrometers is only inversely proportional to the maximum displacement of the movable mirror (3.86). Therefore, the throughput of an FTIR instrument is higher than the one of a dispersive instrument; this is called the Jacquinot advantage. In reality, the throughput of FTIR spectrometers is limited by the size of the optical components and the aberrations of the optical systems, but it remains much larger than for classical dispersive spectrometers.

FTIR spectrometers are mostly employed to measure the absorbance of a sample located in front of the detector, after the interferometer. Let $T(\omega)$ be the transmittance of the sample (3.1). The spectrum ($\omega \neq 0$) of the signal reaching the detector is simply expression (3.82) multiplied by the modulus of $T(\omega)$:

$$I_{\text{source}}(\omega)\, T_{\text{FTIR}}(\omega)\, \Re(\omega)\, |T(\omega)| = \frac{1}{\eta}\sqrt{\frac{2}{\pi}}\, \text{TF}\left[S_{\text{sample}}(x)\right]. \quad (3.86)$$

Then one records an interferogram $S_{\text{ref}}(x)$ without any sample and we get

$$I_{\text{source}}(\omega)\, T_{\text{FTIR}}(\omega)\, \Re(\omega) = \frac{1}{\eta}\sqrt{\frac{2}{\pi}}\, \text{TF}\left[S_{\text{ref}}(x)\right]. \quad (3.87)$$

The ratio of these two spectra gives directly $|T(\omega)|$:

$$|T(\omega)| = \frac{\text{TF}\left[S_{\text{sample}}(x)\right]}{\text{TF}\left[S_{\text{ref}}(x)\right]}. \quad (3.88)$$

An excellent estimation of $|T(\omega)|$ is achieved by entering the recorded signals in Eq. 3.88, if one takes care to avoid interferences between the sample and the detector, as well as any variation of the power of the source between the two measurements. Once $|T(\omega)|$ is determined, the Beer–Lambert law (3.6) permits to deduce the coefficient of absorption of the sample. A similar procedure can be used for measurements performed in reflection, in which the sample directs the output light from the spectrometer onto the detector. In this case, a perfect mirror (metal) must serve to record the reference signal.

The refractive index of the sample may also be estimated from $|T(\omega)|$, which shows an oscillating behavior [25]. Let us treat the case of a transparent material ($\kappa = 0$), which leads to simple analytical expressions. The modulus of the transmission by a slab is calculated from Eq. 3.5:

$$|T(\omega)|^2 = \frac{16n^2}{(n+1)^4 + (n-1)^4 - 2\left(n^2-1\right)^2 \cos\left(2\frac{\omega}{c}nd\right)}. \quad (3.89)$$

This function oscillates as $\cos(2\omega nd/c)$, i.e., with a period $\Delta f = c/2nd$, with the assumption that n does not vary so much over the interval Δf. Therefore, the refractive index is obtained from the measured Δf value:

$$n = \frac{c}{2d\Delta f}. \quad (3.90)$$

Another way of characterizing samples with an FTIR interferometer is to place the sample in one arm of the interferometer [26]. This is called dispersive FTIR spectroscopy or dispersive Fourier transform spectroscopy (DFTS) [27, 28]. This technique delivers not only the modulus of $T(\omega)$, like in absorbance spectroscopy described herein before, but also its phase. Let us suppose that the sample is located in arm B of the Michelson interferometer. The modulus of the beam supplied by arm B is multiplied by $|T(\omega)|^2$ (we have to take the square of $|T(\omega)|$ because the beam crosses twice the sample), while there is an additional phase φ_T induced by the propagation through the sample. Therefore, Eq. 3.78 becomes:

$$\begin{aligned} I(\varphi) &\propto \overline{(E_A(t) + E_B(t))^2} \\ &= E_o^2(\omega)\,\overline{\left(\cos(\omega t) + \eta\,|T(\omega)|^2 \cos(\omega t + \varphi + 2\varphi_T)\right)^2}, \\ \Rightarrow I(x,\omega) & \\ &= I_o(\omega)\,\frac{1 + 2\eta\,|T(\omega)|^2 \cos\left(2\frac{\omega}{c}x + 2\varphi_T\right) + \eta^2\,|T(\omega)|^4}{2}. \end{aligned} \quad (3.91)$$

Then, we proceed as previously and we obtain

$$I_{\text{source}}(\omega)\, T_{\text{FTIR}}(\omega)\, \Re(\omega)\, |T(\omega)|^2\, e^{2j\phi_T} = \frac{1}{\eta}\sqrt{\frac{2}{\pi}}\,\text{TF}\left[S_{\text{sample}}(x)\right]. \quad (3.92)$$

The modulus and the phase of $T(\omega)$ are determined by measuring the interferogram with and without (reference) the sample in arm B of the spectrometer:

$$|T(\omega)|\, e^{j\phi_T} = \left(\frac{\mathrm{TF}\left[S_{\mathrm{sample}}(x)\right]}{\mathrm{TF}\left[S_{\mathrm{ref}}(x)\right]} \right)^{1/2}. \tag{3.93}$$

Therefore, the DFTS techniques allow one to determine the modulus and phase of the coefficient of transmission of a sample, and then to get the complex refractive index of the material. In fact, relation 3.93 is the same as the one for THz-TDS, as we will see in the next chapter! The reachable frequency domain is limited by the spectral emission of the source (namely $I_{\mathrm{source}}(\omega)$ in (3.83)) and the sensitivity of the detector ($\Re(\omega)$ in (3.82)). The sub-THz/THz range can be explored using, for example, mercury blackbody-like lamp and cryogenic bolometers [29–31]. However, because the recording times are rather long and the adjustment is quite tricky, DFTS is nowadays not very often employed, and has been supplanted by THz-TDS.

One of the technical difficulties encountered when designing and assembling an FTIR spectrometer is related to the beam splitter of the Michelson interferometer. Because the interferogram amplitude linearly depends on η (3.81), the beam splitter efficiency must be as close as possible to 50–50%, i.e., half of the beam amplitude is reflected, while the resting half is transmitted. Therefore, the beam splitter must be as transparent as possible. For convenient geometrical design, a beam splitter at 45° from the direction of the incoming beam is preferable, as it points the reflected and transmitted beams towards perpendicular directions. Such beam splitters are mostly either a dielectric transparent slab with plane and parallel faces, or a metallic mesh. When dealing with transparent slabs of thickness d located in air, their coefficients of transmission and reflection are given by expression A.32 (see Annex) and for a 45° incidence are given by

$$|t|^2 = \frac{(1-A)^2}{1+A^2-2A\cos(2kd)}, \quad |r|^2 = \frac{2A\,(1-\cos(2kd))}{1+A^2-2A\cos(2kd)}, \tag{3.94}$$

with $k = \frac{2\pi f}{cn}\sqrt{\frac{2n^2-1}{2}}$. A is a coefficient that depends on the THz beam polarization:

$$A_{\mathrm{TE}} = \left(\frac{1-\sqrt{2n^2-1}}{1+\sqrt{2n^2-1}}\right)^2, \quad A_{\mathrm{TM}} = \left(\frac{n^2-\sqrt{2n^2-1}}{n^2+\sqrt{2n^2-1}}\right)^2.$$

Because of the Fabry–Pérot effect in the slab, transmission and reflection exhibit an oscillating behavior with frequency. The period Δf of the oscillations is deduced from $\Delta\,(2kd) = \frac{4\pi d}{cn}\sqrt{\frac{2n^2-1}{2}}\Delta f = 2\pi$, yielding

$$\Delta f = \frac{cn}{d\sqrt{2\,(2n^2-1)}}\;. \tag{3.95}$$

The relative amplitude A_{fringe} and the contrast C of the fringes are

$$A_{\mathrm{fringe}} = |t|^2_{\max} - |t|^2_{\min} = |r|^2_{\max} - |r|^2_{\min} = \frac{4A}{(1+A)^2},$$
$$C_{|t|^2} = \frac{|t|^2_{\max} - |t|^2_{\min}}{|t|^2_{\max} + |t|^2_{\min}} = \frac{2A}{1+A^2}, \quad C_{|r|^2} = \frac{|r|^2_{\max} - |r|^2_{\min}}{|r|^2_{\max} + |r|^2_{\min}} = 1. \tag{3.96}$$

For the FTIR spectra not perturbed by beam-splitter fringes, either Δf must be smaller or much bigger than the spectral resolution of the spectrometer, or the amplitude of the fringes must be small, if possible of the order of the noise. Δf smaller than the resolution is almost impossible. Indeed, a common $\Delta f = 1$ GHz requires a $d \approx 170$ mm thick slab of index $n = 1.5$. A better solution is to employ a film, i.e., d small: the period Δf is large, and thus the reflectivity and transmittivity of the beam splitter is almost constant over a rather wide band. A common material is biaxially oriented polyethylene terephthalate (boPET), commercially known as Mylar®, whose refractive index in the THz range is about 1.85. Films 3-μm thick are commercially available, whose corresponding Δf exceeds 50 THz. The reduction of the fringe amplitude requires a small index of refraction. Typically, in TM polarization, the fringe amplitude is smaller than 5% for $n < 1.5$. Above this value, it increases quite strongly to reach, e.g., 50% for $n \approx 3.4$ (refractive index of Si or GaAs). In addition, beam splitters should obey a second condition, i.e., $|t|^2 \approx |r|^2 \approx 50\%$. This condition is difficult to establish for

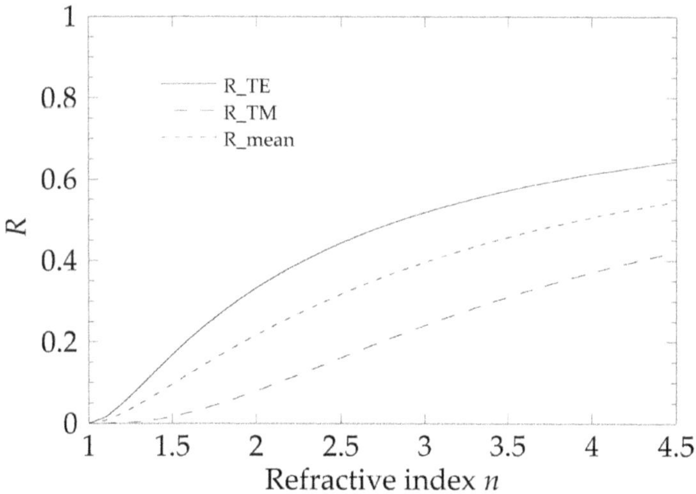

Figure 3.14 Approximated reflection coefficient R (3.97) by a slab of index n.

an oscillating response (3.94). To get around this problem [15], one considers only the first beam transmitted/reflected by the slab and the first Fabry–Pérot rebound, instead of using the rigorous relation (3.94). In addition, one supposes that the condition $\cos(2k_{1z}d) = 1$ (constructive interferences) is fulfilled. The transmitted/reflected beam intensities are the sum of these two terms:

$$R = |r|^2 = R_{\text{air/slab}} + R_{\text{air/slab}} T^2_{\text{air/slab}},$$
$$T = |t|^2 = T^2_{\text{air/slab}} + T^2_{\text{air/slab}} R^2_{\text{air/slab}}, \tag{3.97}$$

with

$$R_{\text{air/slab}} = \left(\frac{\alpha \cos\theta_{\text{i}} - n\cos\theta_{\text{t}}}{\alpha \cos\theta_{\text{i}} + n\cos\theta_{\text{t}}} \right)^2, \quad T_{\text{air/slab}} = 1 - R_{\text{air/slab}},$$
$$\alpha_{\text{TE}} = 1, \quad \alpha_{\text{TM}} = n^2. \tag{3.98}$$

Figure 3.14 presents the reflectivity R (3.97) of a slab for TE, TM, and unpolarized beam versus the refractive index of the slab. $R \approx 50\%$ is achieved for $n = 2.861$ (TE), $n = 5.497$ (TM), and $n = 3.928$ (unpolarized). This shows that high-index materials permit to fulfill the condition $|t|^2 \approx |r|^2 \approx 50\%$. Intrinsic, i.e., undoped, semiconductors exhibit a fairly constant and high-value refractive

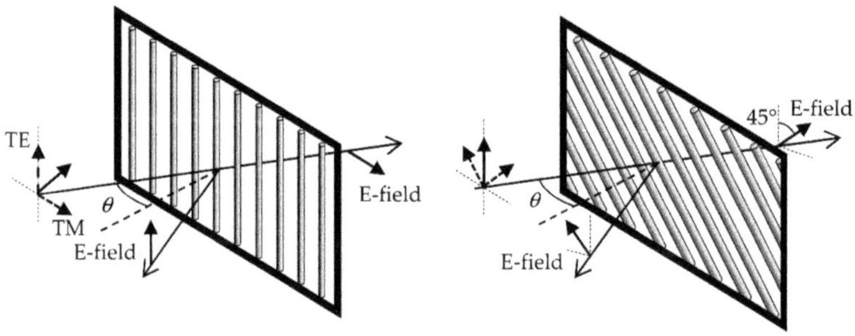

Figure 3.15 Transmission/reflection by a wire grid. Left: Wires perpendicular to the plane of incidence. Right: Wires aligned at 45° to the plane of incidence.

index over the whole THz range. For example, $n \approx 3.6$ for GaAs and $n \approx 3.4$ for Si. Because it is widespread and cheap, Si is a preferable material for THz beam splitters. However, this analysis was done within the assumption that there are no interference effects ($\cos(2kd) = 1$), which is difficult to achieve practically.

To avoid the slab problem, an elegant solution has been proposed by Martin and Puplett [32, 33]. The idea is to replace the slab with a device made of parallel metallic wires (wire grid) tightly wound on a frame so they are taut and periodically arranged [34]. The diameter and periodicity d of the wires are sub-wavelength. Because of the sub-wavelength character, the grid behaves as an effective medium and does not exhibit any grating-like diffraction. The advantages of wire grids are that they produce no Fabry–Pérot effect (the grid could be considered as a 2D device, whose thickness is null) and their splitting efficiency is almost constant over a very large band of frequencies.

However, the transmission by a wire grid is polarization-sensitive. Simply speaking, if the E field of the incoming beam is directed perpendicularly to the wires (i.e., the E field is in the plane of incidence; this is the p polarization or TM polarization; see Fig. 3.15), it moves the free electrons of the metal in the transversal direction along the wires. Because of the small diameter, this movement is almost forbidden. Thus the interaction between the EM wave and the grid is very weak, and this TM-polarized beam is nicely transmitted by the grid. Because of energy conservation, the

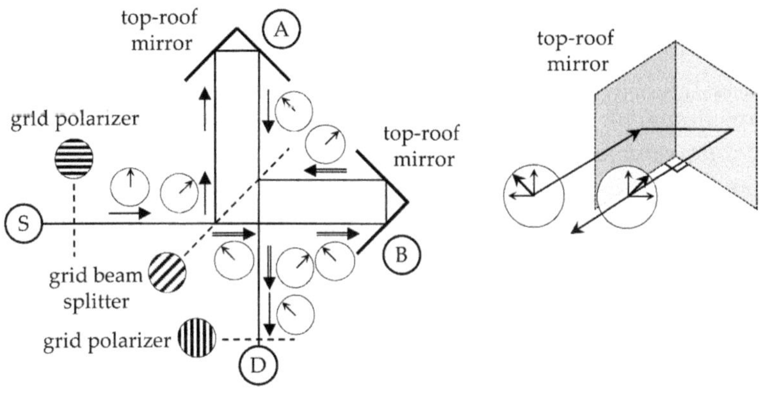

Figure 3.16 Left: Scheme of a Martin–Puplett interferometer. Right: 90° rotation of polarization with a right-angle roof top reflector.

TM beam is not reflected. For TE beams, the E field and consequently the movement of electrons are directed along the wire. The resulting electrical current is not restricted by the geometry: it is almost similar to the one induced when illuminating a continuous metal sheet. Because of this similarity and of the sub-wavelength period, the grid behaves as a metal sheet. Thus the TE E field is largely reflected and not transmitted. Typically, more than 98–99% of the power of the TM beam is transmitted by actual devices, while less than 1% is reflected. The requested sub-wavelength geometry of the grid imposes d smaller than $\lambda/10$. For larger values of d, transmission decreases while reflection increases. Modern wire grid polarizers are made of free standing metal wire grids (tungsten or gold-coated tungsten, minimum diameter 3–5 μm, minimum period 2–10 μm) or of metal strips deposited over a very thin dielectric film (PET, SiC, etc.).

Let us examine the polarization of the beams inside the Martin–Puplett interferometer. We consider for example the beam reflected by the wire grid and propagating in arm A (see Fig. 3.16, left). Its polarization is aligned along the wires of the grid. If reflected in arm A by a planar mirror, it would impinge again on the grid with the same polarization. Therefore, it would be fully reflected by the grid towards the source, and would never reach the detector D. A similar phenomenon would occur in arm B. Therefore, the beam polarization in each arm has to be rotated by 90°, in order for the beams reflected from arms A and B to reach the detector. Unfortunately, efficient broadband half-wave plates that are commonly used to rotate the polarization of an optical beam do not exist for THz waves. However, rooftop reflectors rotate the polarization, but at the expense of a geometrical shift of the reflected beam. The rotation of polarization, when the beam is polarized at 45°, is depicted in Fig. 3.16 (right). A 90° angle of the top-roof reflector yields a backward reflection, and thus the device acts as a mirror and can be installed in the interferometer. Nevertheless, the two beams, from arms A and B, which reach the detector, are orthogonally polarized and thus cannot produce interferences. This is overcome by inserting a polarizer at 45° in front of the detector to match the polarizations.

Compared with a Michelson interferometer, 50% of the incoming light is lost in the Martin–Puplett instrument, because of the polarizer in front of the detector. But its great advantage is the use of a polarizing beam splitter, namely a wire grid which improves performance over a wide spectral range and suppresses background noise.

References

1. Sizov, F. (2010). Terahertz radiation sensors, *Opto-Electron. Rev.*, **18**, pp. 10–36.
2. Rogalski, A., and Sizov, F. (2011). Terahertz detectors and focal plane arrays, *Opto-Electron. Rev.*, **19**, pp. 346–404.
3. Bousquet, P., and Greenland, K. M. (1971). *Spectroscopy and Its Instrumentation*, Institute of Physics Publishing, London.
4. Harvey, D. (2012). Chapter 10 - Spectroscopic methods, in *Analytical Chemistry 2.0*, Saylor Academy, pp. 543–666. Available at: www.saylor.org/site/wpcontent/uploads/2012/07/Chapter1011.pdf.
5. Emerson, D. T. (1997). The work of Jagadis Chandra Bose: 100 years of mm-wave research by, *IEEE Trans. Microwave Theory Tech.*, **45**, pp. 2267–2273.
6. Palik, E. D. (1977). History of far-infrared research. I. The Rubens era, *J. Opt. Soc. Am.*, **67**, pp. 857–865.

7. Son, J., Sha, W., Kim, J., Norris, T. B., Whitaker, J. F., and Mourou, G. A. (1993). Transient velocity overshoot dynamics in GaAs for electric fields <200 W/cm, *Appl. Phys. Lett.*, **63**, pp. 923–925.
8. Maestrini, A., Thomas, B., Wang, H., Jung, C., Treuttel, J., Jin, Y., Chattopadhyay, G., Mehdi, I., and Beaudin, G. (2010). Schottky diode-based terahertz frequency multipliers and mixers, *C. R. Phys.*, **11**, pp. 480–495.
9. Nagatsuma, T., and Ito, H. (2011). Chapter 14 - High-power RF uni-traveling-carrier photodiodes (UTC-PDs) and their applications, in *Advances in Photodiodes*, Dalla Betta, G. F. (ed.), InTech, Rijeka, Croatia. Available from: http://www.intechopen.com/books/advances-in-photodiodes/high-power-rf-uni-traveling-carrier-photodiodes-utc-pds-and-their-applications.
10. Walther, C., Fischer, M., Scalari, G., et al. (2007). Quantum cascade lasers operating from 1.2 to 1.6 THz, *Appl. Phys. Lett.*, **91**, pp. 131122 (1–3).
11. Williams, B. S., Kumar, S., Hu, Q., et al. (2005). Operation of terahertz quantum-cascade lasers at 164 K in pulsed mode and at 117 K in continuous-wave mode, *Opt. Express*, **13**, pp. 3331–3339.
12. Kumar, S., Hu, Q., and Reno, J. L. (2009). 186 K operation of terahertz quantum cascade lasers based on a diagonal design, *Appl. Phys. Lett.*, **94**, pp. 131105 (1–3).
13. Kumar, S., Chan, C. W. I., Hu, Q., et al. (2010). A 1.8-THz quantum cascade laser operating significantly above the temperature of $\hbar\omega$/kB, *Nat. Phys.*, **7**, pp. 166–171.
14. Williams, B., Kumar, S., Hu, Q., and Reno, J. (2006). High-power terahertz quantum-cascade lasers, *Electron. Lett.*, **42**, pp. 89–90.
15. Koppens, F. H. L., Mueller, T., Avouris, Ph., Ferrari, A. C., Vitiello, M. S., and Polini, M. (2014). Photodetectors based on graphene, other two-dimensional materials and hybrid systems, *Nat. Nanotechnol.*, **9**, pp. 780–793.
16. He, X., Fujimura, N., Meagan Lloyd, J., Erickson, K. J., Alec Talin, A., Zhang, Q., Gao, W., Jiang, Q., Kawano, Y., Hauge, R. H., Léonard, F., and Kono, J. (2014). Carbon nanotube terahertz detector, *Nano Lett.*, **14**, pp. 3953–3958.
17. Bell, R. J. (1972). *Introductory Fourier Transform Spectroscopy*, Academic Press, New York.
18. Genzel, L. (1998). Far-infrared fourier transform spectroscopy, in *Millimeter and Submillimeter Wave Spectroscopy of Solids*, **74**, Grüner, G. (ed.), Topics in Applied Physics, Springer-Verlag, Berlin, pp. 169–220.

19. Michelson, A. A. (1891). On the application of interference methods to spectroscopy measurements - I, *Phil. Mag.*, **31**, pp. 338–346; idem (1892). On the application of interference methods to spectroscopy measurements - II, *Phil. Mag.*, **34**, pp. 280–299.
20. Michelson, A. A. (1902). Excerpt from *Light Waves and their Uses*, University of Chicago Press, reproduced in *Appl. Opt.*, **6**, pp. 692–693 (1967).
21. Loewenstein, E. V. (1966). The history and current status of fourier transform spectroscopy, *Appl. Opt.*, **5**, pp. 845–854.
22. Gebbie, H. A., and Vanasse, G. A. (1956). Interferometric spectroscopy in the far infra-red, *Nature*, **178**, p. 432.
23. Strong, J., and Vanasse, G. A. (1959). Interferometric spectroscopy in the far infrared, *J. Opt. Soc. Am.*, **49**, pp. 844–849.
24. Burton, C. H., and Akimoto, Y. (1980). A polarizing Michelson interferometer for the far-infrared and millimetre regions, *Infrared Phys.*, **20**, pp. 115–120.
25. Sakai, K. (1972). Michelson-type Fourier spectrometer for the far infrared, *Appl. Opt.*, **11**, pp. 2894–2901.
26. Bell, E. E. (1966). Measurement of the far infrared optical properties of solids with a Michelson interferometer used in the asymmetric mode: Part I, mathematical formulation, *Infrared Phys.*, **6**, pp. 57–74.
27. Birch, J. R., and Parker, T. J. (1979). A bibliography on dispersive Fourier transform spectrometry, *Infrared Phys.*, **19**, pp. 201–215.
28. Parker, T. J. (2009). Fourier transform spectroscopy of solids at terahertz frequencies, *THz Sci. Technol.*, **2**, pp. 75–89.
29. Birch, J. R. (1983). Optical constants of some commercial microwave materials between 90 and 1200 GHz, *IEE Proc. on Microwaves, Optics and Antennas*, **130**, pp. 327–330.
30. Afsar, M. N., Moonshiram, A., and Wang, Y. (2003). Measurement of transmittance and permittivity of dielectric materials using dispersive Fourier transform spectroscopy, *Microw. Opt. Technol. Lett.*, **38**, pp. 27–30.
31. Afsar, M. N. (1986). Precision millimeter-wave measurements of complex refractive index, complex dielectric permittivity, and loss tangent of common polymers, *IEEE Trans. Instrum. Meas.*, **36**, pp. 530–536.
32. Martin, D. H., and Puplett, E. (1970). Polarised interferometric spectrometry for the millimetre and submillimetre spectrum, *Infrared Phys.*, **10**, pp. 105–109.

33. Martin, D. H. (1982). Martin-Puplett interferometric spectrometers, in *Infrared and Millimeter Waves*, **6**, Button, K. J. (ed.), Academic Press, New York, pp. 65–148.

34. Sentz, A., Pyee, M., Gastaud, C., Auvray, J., and Letur, J.-P. (1978). Construction of parallel grids acting as semitransparent flat mirrors in the far infrared, *Rev. Sci. Instrum.*, **49**, pp. 926–927.

Chapter 4

THz-TDS Instrumentation

4.1 Introduction

As explained in Chapter 2, the principle of THz-TDS is based on measuring the impulse response of a sample. This response must be determined at the picosecond scale in order to obtain the frequency response of the sample up to the THz frequency range. Thus, the procedure requires picosecond or even shorter pulses, and a detector with an equivalent temporal resolution. These ultrafast signals and responses are mostly produced by optoelectronic means, i.e., by exciting both emitter and receiver with sub-picosecond laser pulses. Compression of electrical pulses along a nonlinear microwave line produces picosecond pulses [1] as well, which may be used to perform THz spectroscopy [2]. We will focus here only on optoelectronic techniques as they are by far the most popular. First, we will describe the most common optoelectronic emitters and receivers, namely photo-conducting antennas and electro-optic crystals. Then we will present the time-equivalent sampling technique that is needed to measure the temporal shapes of short pulses. Thereafter we will introduce THz-TDS and the extraction of the sample parameters from the THz-TDS data.

Principles of Terahertz Time-Domain Spectroscopy
Jean-Louis Coutaz, Frédéric Garet, and Vincent Wallace

ISBN 978-981-4774-56-7 (Hardcover), 978-0-203-70969-6 (eBook)
www.panstanford.com

4.2 THz Pulse Generation by Optoelectronic Means

4.2.1 Photo-conducting Antenna

This is certainly the most widespread optoelectronic generator of THz pulses. Basically, the device includes two electrodes deposited on a semiconductor substrate in which free carriers exhibit an ultrashort lifetime (i.e., the average time for charge carriers to recombine is very short). A bias voltage is applied to the electrodes that are separated by a small distance, a. In the absence of illumination, the semiconductor is highly resistive (intrinsic). Thus, no current flows between the electrodes that hold opposite charges, $\pm Q$. The device behaves as an electrical dipole, whose dipolar moment p is defined as $p = Qa$ (here we are only interested in the magnitude of the signal, so we write values as scalars, even if rigorously they are vectors). Then the gap between the two electrodes is illuminated by a short laser pulse with a photon energy above the semiconductor bandgap: this generates free carriers (electrons and holes) by photo-absorption. This photo-induced carrier population first changes the permittivity of the semiconductor material, then the carriers are accelerated by the applied field, and finally some of them reach the electrodes. All these phenomena induce a variation of the dipolar moment, which produces the dipole emitting electromagnetic waves. In the far-field region, the radiated field is proportional to the second time-derivative of the dipole moment [3]:

$$E\,(t) \propto \frac{\partial^2 p\,(t)}{\partial t^2} \propto a\frac{\partial^2 Q\,(t)}{\partial t^2} \propto a\frac{\partial i\,(t)}{\partial t}. \tag{4.1}$$

To derive Eq. 4.1, we considered that the distance a between the electrodes is fixed, and thus only the dipole charge Q varies. Moreover, the time derivative of the charge is nothing but the current $i(t)$ that flows in the device. This current originates from the polarization effect, i.e., the variation of the semiconductor permittivity that leads to the so-called displacement current, and in the movement of the free charges (conduction current). The establishment of the displacement current is almost instantaneous, and thus it follows the variation of the laser pulse intensity. The conduction current is governed by the response time of the electrical circuit: practically, only conduction in the gap and in the electrode wires close to the gap contributes to the THz radiation. When the carrier lifetime in the semiconductor is ultrashort, carriers recombine quickly, even before reaching the electrodes. The pulse of current $i(t)$ is thus very sharp and short, and its time-derivative is maximized. Consequently, the THz far field is quite intense, as deduced from relation 4.1. The spectrum of the far field is given by

$$E(\omega) = FT\,[E(t)] \propto \int_{-\infty}^{+\infty} \frac{\partial i(t)}{\partial t} e^{-j\omega t} dt \propto j\omega\, FT\,[i(t)]. \quad (4.2)$$

It depends directly on the Fourier transform (FT) of the current pulse. Let us estimate the order of magnitude of $i(t)$ in a simple static model (the rigorous calculation requires us to solve time differential equations, which are described, for example, in [4]). We consider only the contribution of the conduction current of electrons in the ohmic regime. Thus, the photo-conduction current is

$$i = n\,e\,v\,S = \eta \frac{e}{h\,f} P_{\text{laser}}\, \mu \frac{V_{\text{bias}}}{a^2}. \quad (4.3)$$

n is the density of photo-generated carriers, V_{bias} is the applied voltage, e and μ are respectively the charge and mobility of the electrons. Relation 4.3, which is similar to relation 3.37 written in the case of photo-conduction, is obtained assuming the incoming laser power is focused over the whole switch area, where η is the photo-electric conversion efficiency. It follows that

$$E(\omega) \propto \omega\mu\, V_{\text{bias}} P_{\text{laser}}. \quad (4.4)$$

The THz electric field amplitude is thus proportional to the bias voltage, to the carrier mobility and to the exciting laser power. Therefore, not only must the semiconductor material exhibit a short free carrier lifetime, but also a high carrier mobility and a high breakdown field (in order to bias the antenna as much as possible). Also, a high dark resistivity is necessary to get $i(t)$ a short as possible (i.e., no current before and after the laser pulse excitation) and thus to maximize the THz emission (see Eq. 4.2).

Let us treat two extreme cases. If the current pulse is Gaussian-shaped with a duration of τ_{curr}, we obtain

$$i(t) = i_o\, e^{-\left(2\sqrt{\ln 2}\frac{t}{\tau_{\text{curr}}}\right)^2} \quad \Rightarrow \quad E(\omega) \propto i_o\, \omega\tau_{\text{curr}}\, e^{-\left(\frac{\omega\tau_{\text{curr}}}{4\sqrt{\ln 2}}\right)^2}. \quad (4.5)$$

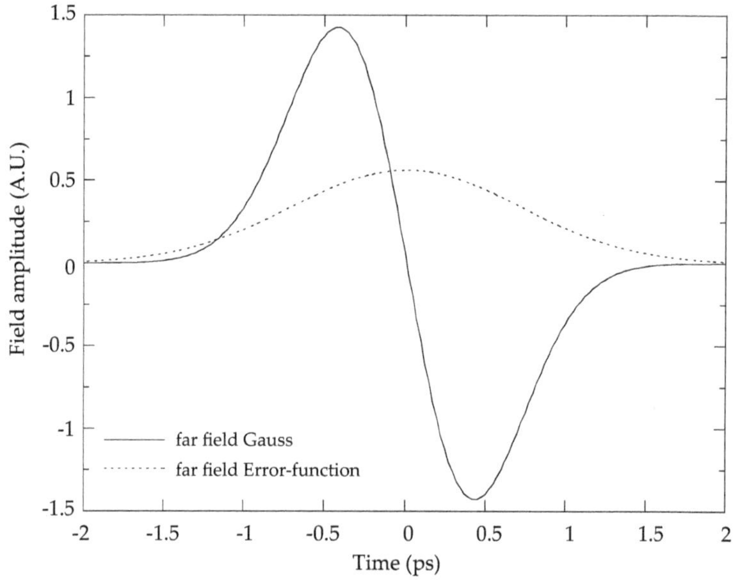

Figure 4.1 Far-field waveform generated by photoconduction with a Gaussian current pulse (1 ps pulse width) and an error-function-like current (1 ps rise time).

This is typically (see Figs. 4.1 and 4.2) what is obtained with semiconductors, like low-temperature (LT) grown GaAs, where the carrier lifetime is smaller than 1 ps. The −60 dB energy bandwidth spreads up to $f_{\max} \approx \pi/(2\,\tau_{\text{curr}})$, while the maximum of the THz field is obtained at $f_{\text{opt}} = \sqrt{2\ln 2}/(\pi\,\tau_{\text{curr}}) \approx 0.375/\tau_{\text{curr}} \approx f_{\max}/4.19$. Therefore, the shorter the current pulse duration τ_{curr}, the wider the bandwidth.

On the other hand, if the free carrier lifetime in the semiconductor is very long (as in silicon, which can be on the order of milliseconds), the current obeys an error-function-like behavior when excited by a Gaussian laser pulse:

$$i\,(t) = i_0 \int_{-\infty}^{t} e^{-\left(\frac{t'}{\tau}\right)^2} dt' = i_0 \tau \frac{\sqrt{\pi}}{2}\left(1 + erf\left(\frac{t}{\tau}\right)\right) \qquad (4.6)$$

The waveform (Fig. 4.1) is still very short, but it does not show the dipolar behavior obtained with a Gaussian pulse (i.e., a positive peak followed by a negative one). In addition, the signal is rather weak. The Fourier transform of this current is obtained from its derivative:

$$\frac{d}{dt}i\,(t) = i_0\,\tau\frac{\sqrt{\pi}}{2}e^{-\left(\frac{t}{\tau}\right)^2} \Rightarrow FT\left[\frac{d}{dt}i\,(t)\right] = j\omega\,FT\,[i\,(t)]$$
$$= i_0\frac{\tau^2}{4}e^{-\left(\frac{\omega\tau}{2}\right)^2},$$
$$\Rightarrow FT\,[i\,(t)] = i_0\frac{\tau^2}{4j\omega}e^{-\left(\frac{\omega\tau}{2}\right)^2}.$$

Therefore, the far field is

$$E\,(\omega) \propto j\omega\,FT\,[I\,(t)] \propto \frac{I_0\tau}{2\sqrt{\pi}}e^{-\left(\frac{\omega\tau}{2}\right)^2}. \qquad (4.7)$$

Figure 4.2 shows the amplitude spectra of the far field generated by a Gaussian current pulse (Eq. 4.5) and a step current (Eq. 4.7), both with a 0.5 ps pulse width or rise time. The step current leads to a signal about 10 times smaller than the Gaussian pulse, and

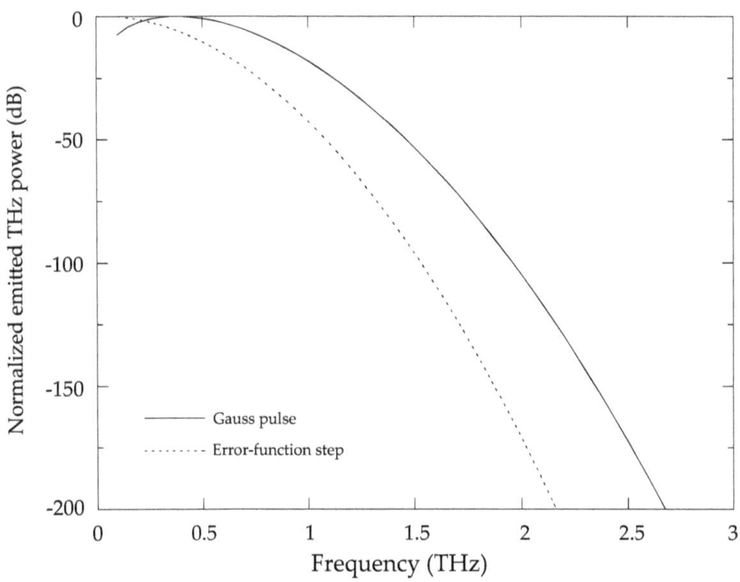

Figure 4.2 Far-field amplitude generated by photoconduction with a Gaussian current pulse (1 ps pulse width) and an error-function-like current (1 ps rise time).

thus its bandwidth is narrower, but it does not show a decrease at lower frequencies. For actual materials, the free carrier population exhibits a tail, mostly due to the longer lifetime and slower mobility of holes. Thus, experimental records lie between the two curves of Fig. 4.2. Therefore, the gain in using an ultrafast material for emission, in terms of power or bandwidth, is not so large and cheaper semiconductors like intrinsic GaAs are good materials for manufacturing THz photo-conducting emitters. Moreover, the higher mobility in good-quality bulk wafers, like GaAs, could compensate for the effect of a longer lifetime compared with ultrafast semiconductor layers.

As already explained, the THz far field is generated by the variation of the current in the gap in between the electrodes, but also in the electrodes nearby the switch. At longer distances from the switch, high-frequency components of the current do not propagate in the metallic wires. Thus, the electrical pulse becomes longer and does not contribute efficiently to the THz generation. The first switch for microwave optoelectronic generation was proposed by G. Mourou et al. [5]: a simple gap fabricated in a microstrip waveguide deposited over GaAs fed an antenna through a coaxial cable. Great improvement was brought by the team of D. Auston [6], who located the switch at the center of a dipolar antenna biased through a coplanar waveguide (CPW). The scheme was even upgraded by D. Grischkowsky et al. [7], who designed a narrower gap region at the center of the CPW. This design allows to increase the bias field by shortening the gap width, while keeping unchanged the CPW impedance. A slightly higher THz power is delivered when electrodes exhibit a sharp shape (bow tie) [8]. Also, plasmonic electrodes, made of parallel strips, permit to reduce the distance between the two electrodes without increasing the capacity of the switch. The carrier moves fast under a ballistic transport regime, and thus it leads to an enhanced emission [9].

Ultrashort carrier lifetime in semiconductors [10, 11] is achieved by introducing in the material a high density of traps that promptly capture the free carriers. These traps are structural or stoichiometric defects. The first ones are made by damaging the semiconductors with a beam of high-energy ions, by implanting the semiconductors with ions, or by growing polycrystalline films.

Figure 4.3 Different geometries of the electrodes of a photo-conducting switch. From left to right: sliding switch, Grischkowsky switch, switch with bowtie, and plasmonic electrodes.

Stoichiometric defects are mostly obtained by an epitaxial growth at a low temperature (200 ~ 300°C) of III–V or ternary/quaternary compounds. One of the species is in excess, leading to antisite, interstitial, or vacancy defects. The bad quality of such grown films is improved by a thermal annealing post treatment at a higher temperature (500 ~ 700°C). For an operation at a laser wavelength of 0.8 μm (Ti:Sa lasers), low-temperature-grown GaAs (LT-GaAs) is definitively the optimal material. The lifetime of free electrons could be adjusted from below 100 fs up to several tens of picoseconds, while the carrier mobility remains high. In LT-GaAs, the traps are antisite arsenic ions in excess, which form an EL2-like defect level at the center of the bandgap. The ionized trap density may be adjusted by compensating with beryllium doping. Similarly, the ternary LT-GaBiAs alloy is used for operation around 1 μm (Nd-YAG or Yb:YVO_4 lasers, Yb fiber lasers, etc.). At the telecoms wavelength (~1.5 μm ↔ ~0.8 eV), a good material is still required. Typical semiconductors with a 0.8 eV bandgap are InGaAs or GaInAlAs. LT growth procedures are no more suitable, because the EL2 defect is too close to the conduction band, as the bandgap is about twice as narrow than in LT-GaAs, which makes the semiconductors almost conductive. A smart solution is to fabricate sandwiches of alternative layers of photoconductive LT-InGaAs and highly resistive LT-InAlAs layers [12].

4.2.2 Emission at the Surface of a Semiconductor Wafer

Another possibility of optically generating THz pulses with a semiconductor is based on the surface effect [11]. Because of the decrease in the charge density at the surface of a semiconductor, the energy bands are bent, which results in a native electrical field perpendicular to the surface. When the semiconductor is illuminated by

Table 4.1 Relative THz surface generation efficiency of some semiconductors

	p-InAs	n-InAs	InGaAs	GaBiAs	GaAs	InSb	HgCdTe	Ge
@ 800 nm	1	0.50	0.50	0.15	0.08	0.06	0.05	0.03
@ 1030 nm	1	0.30	0.15	0.10	0.07	0.15	0.15	0.09

a laser pulse, photo-generated carriers are accelerated by this field. The equivalent dipole is oriented normal to the surface, and thus the dipolar radiation is maximum along the surface. This problem is overcome by illuminating the semiconductor under an angle with a rather wide beam. In this latter case, the THz beam direction is governed by a Snell–Descartes law involving both optical and THz photons. Therefore, the direction of the emitted THz radiation can be finely adjusted.

An additional effect may enhance the radiated THz field magnitude. In the photo-generated plasma, particles with similar charges repel and consequently they diffuse into the material. Electron and hole mobilities are different: electrons generally move faster than holes. This difference leads to a transient dipole at the surface, which contributes to the THz radiation. This is called the Dember effect. Also, in non-centrosymmetric semiconductors, optical rectification by bound electrons contributes to the THz generation.

Generation at semiconductor surfaces is very easy to set up, and cheap. It produces signals almost similar to the ones obtained with photo-conducting antennas. Table 4.1, adapted from [11], gives the efficiency of THz surface emission from different semiconductors normalized to the one from doped (p) InAs. Nevertheless, the quality of the THz beam shape is more difficult to control, and this beam propagates with an angle compared with the incident laser beam, which complicates the setup.

4.2.3 Optical Rectification in Electro-optic Crystals

This is a purely nonlinear optical effect occurring in non-centrosymmetric crystal (see Annex). Although called optical rectification ($\omega_{\text{laser}} - \omega_{\text{laser}} = 0$), the involved nonlinear phenomenon is, strictly speaking, a frequency difference ($\omega_1 - \omega_2 = \omega$) between two spectral components $\omega_1 = \omega_{\text{opt}} + \omega$ and $\omega_2 = \omega_{\text{opt}}$ of the pump laser pulse. The propagation equation of the THz field at frequency ω is given by relation A.58 of the annex:

$$\Delta\vec{E}_\omega(\vec{r}) + \varepsilon(\omega)\frac{\omega^2}{c^2}\vec{E}_\omega(\vec{r})$$
$$= -\left(\frac{\omega}{c}\right)^2 \overleftrightarrow{\chi}^{(2)}\left(\omega = \omega_{\text{opt}} + \omega - \omega_{\text{opt}}\right) : \vec{E}_{\omega_{\text{opt}}+\omega}(\vec{r})\,\vec{E}_{\omega_{\text{opt}}}(\vec{r}). \quad (4.8)$$

In fact, the nonlinear term (on the right side) of Eq. 4.8 must include all the spectral components of the pumping laser pulse that contribute to the THz generation [13]:

$$\Delta\vec{E}_\omega(\vec{r}) + \varepsilon(\omega)\frac{\omega^2}{c^2}\vec{E}_\omega(\vec{r})$$
$$= -\left(\frac{\omega}{c}\right)^2 \int_{\text{pulse}} \overleftrightarrow{\chi}^{(2)}\left(\omega = \omega_{\text{opt}} + \omega - \omega_{\text{opt}}\right) : \vec{E}_{\omega_{\text{opt}}+\omega}(\vec{r})\,\vec{E}_{\omega_{\text{opt}}}(\vec{r})\,d\omega_{\text{opt}}. \quad (4.9)$$

Assuming the nonlinear process occurs in spectral regions far from optical resonances in the electro-optic crystal, the nonlinear susceptibility remains almost constant over the integration spectral range. Moreover, for sake of clarity, we treat the case in which only one component $\chi^{(2)}$ of the tensor contributes, which allows us to write simple analytical scalar expressions. Consequently, the right-hand term of Eq. 4.9 simplifies into

$$-\frac{1}{\varepsilon_0 n_{\omega_{\text{opt}}}}\left(\frac{\omega}{c}\right)^2 \chi^{(2)}\left(\omega, \omega_{\text{opt}}\right) I_\omega(\vec{r}), \quad (4.10)$$

in which the value of $\chi^{(2)}\left(\omega, \omega_{\text{opt}}\right)$ is taken at the central frequency ω_{opt} of the laser pulse and $I_\omega(\vec{r}) = n_{\omega_{\text{opt}}}\varepsilon_0 c \int_{\text{pulse}} \vec{E}_{\omega_{\text{opt}}+\omega}(\vec{r})\,\vec{E}_{\omega_{\text{opt}}}(\vec{r})\,d\omega_{\text{opt}}$ is proportional to the laser pulse intensity. For sake of simplicity, let us suppose that all the beams propagate along the direction z. Under all the preceding assumptions, we get

$$\frac{d^2}{dz^2}E_\omega(z) + \varepsilon(\omega)\frac{\omega^2}{c^2}E_\omega(z) = -\frac{\mu_0\omega^2}{n_{\omega_{\text{opt}}}c}\chi^{(2)}\left(\omega, \omega_{\text{opt}}\right) I_\omega(z). \quad (4.11)$$

The laser pulse shape, at time t and location z, is given by

$$I(t, z) = I_0\left(t - \frac{z}{v_G}\right)e^{-\alpha_{\omega_{\text{opt}}}z}. \quad (4.12)$$

$v_G = \partial\omega/\partial k_{\omega_{opt}}$ is the group velocity of the light, i.e., the velocity at which the pulse maximum runs. By taking the Fourier transform of Eq. 4.12, we obtain the spectral components of the laser pulse:

$$I_\omega(z) = I_o(\omega)\, e^{-j\frac{\omega}{c}n_G z} e^{-\alpha_{\omega_{opt}} z}, \tag{4.13}$$

where $n_G = c/v_G$ is the group index at optical frequencies. Thus, the solution of Eq. 4.11 is

$$E_\omega(z) = \frac{\mu_o\omega^2}{n_{\omega_{opt}}c} \frac{\chi^{(2)}(\omega, \omega_{opt})\, I_o(\omega)}{j\frac{\omega}{c}(n_G + n_\omega) + \frac{\alpha_\omega}{2} + \alpha_{\omega_{opt}}}$$

$$\times \frac{e^{-\left(j\frac{\omega}{c}n_G + \alpha_{\omega_{opt}}\right)z} - e^{-\left(j\frac{\omega}{c}n_\omega + \frac{\alpha_\omega}{2}\right)z}}{j\frac{\omega}{c}(n_\omega - n_G) + \frac{\alpha_\omega}{2} - \alpha_{\omega_{opt}}}. \tag{4.14}$$

The last fraction in relation 4.14 has the unit of a length. Its modulus corresponds to the effective length $L(z)$ of the crystal that contributes to the THz generation:

$$L(z) = \sqrt{\frac{e^{-2\alpha_{\omega_{opt}}z} + e^{-\alpha_\omega z} - 2e^{-\left(\frac{\alpha_\omega}{2} + \alpha_{\omega_{opt}}\right)z}\cos\left(\frac{\omega}{c}(n_\omega - n_G)z\right)}{\frac{\omega^2}{c^2}(n_\omega - n_G)^2 + \left(\frac{\alpha_\omega}{2} - \alpha_{\omega_{opt}}\right)^2}}. \tag{4.15}$$

Thus, the modulus of THz field (Eq. 4.14) varies like $L(z)$:

$$|E_\omega(z)| = \frac{\mu_o\omega^2}{n_{\omega_{opt}}c} \frac{\chi^{(2)}(\omega, \omega_{opt})\, I_o(\omega)}{\sqrt{\frac{\omega^2}{c^2}(n_G + n_\omega)^2 + \left(\frac{\alpha_\omega}{2} + \alpha_{\omega_{opt}}\right)^2}} L(z). \tag{4.16}$$

This electric field is plotted versus z in Fig. 4.4. In the general case, it shows an oscillatory behavior whose amplitude decreases with the distance because of attenuation. The oscillation period becomes larger when THz and optical waves propagate with the same n and α parameters, i.e., when $L(z)$ varies proportionally with z. For crystals absorbing both in the optical and THz domains, this maximum generation efficiency is achieved when

$$n_\omega = n_G, \qquad \frac{\alpha_\omega}{2} = \alpha_{\omega_{opt}}. \tag{4.17}$$

This is the phase-matching condition (see annex A.62). Let us notice that the equality of the indices in Eq. 4.17 can also be written as

$$n_{THz} - n_{opt} = \frac{\partial\, n_{opt}}{\partial\, \omega_{opt}}\omega_{opt} \tag{4.18}$$

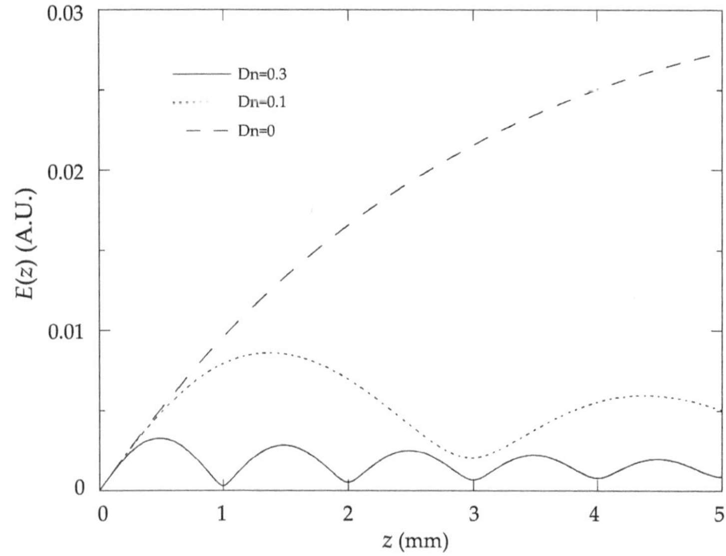

Figure 4.4 Amplitude of the THz field (Eq. 4.16) generated at 1 THz in an absorbing EO crystal versus the distance z for different values of $\Delta n = n_\omega - n_G$, ($\Delta n = 0$, 0.1, and 0.3, marked "Dn" in the figure). The parameters are $n_{\omega_{opt}} = n_G = 2$, $\alpha_{\omega_{opt}} = 0.5\ \text{cm}^{-1}$, and $\alpha_\omega = 5\ \text{cm}^{-1}$.

Under this condition, optical pulse and THz waves propagate at the same speed, and the contribution from any part of the crystal adds constructively to the contributions of other parts. Thus, the field strength varies linearly with the propagation length z in the crystal:

$$|E_\omega(z)| = \frac{\mu_0\omega^2\chi^{(2)}(\omega,\omega_{opt})\,I_0(\omega)}{2n_{\omega_{opt}}c\sqrt{\frac{\omega^2}{c^2}n_\omega^2+\alpha_{\omega_{opt}}^2}}\,e^{-\alpha_{\omega_{opt}}z}z$$

$$\Rightarrow |E_\omega(z)| \approx \mu_0\omega\frac{\chi^{(2)}(\omega,\omega_{opt})\,I_0(\omega)}{2n_{\omega_{opt}}n_\omega}\,e^{-\alpha_{\omega_{opt}}z}z. \quad (4.19)$$

The latter expression is obtained with the realistic simplification $n_\omega\omega/c \gg \alpha_{\omega_{opt}}$.

Let us calculate the temporal waveform of the generated THz signal by taking the inverse Fourier transform of Eq. 4.19. To get simple expressions, we neglect the absorption at optical frequencies, and the dispersion of the refractive indices and the nonlinear coefficient in Eq. 4.19. We get

$$E\left(t,z\right)=-\frac{\mu_o}{2n_\omega n_{\omega_{\text{opt}}}}\chi^{(2)}\left(\omega,\omega_{\text{opt}}\right)z\frac{dI_o\left(t-\frac{n_G z}{c}\right)}{dt}. \tag{4.20}$$

The time shift is related to the delay due to propagation in the crystal. Therefore, the THz waveform is proportional to the time derivative of the laser pulse, which is comparable to the case of the photo-conducting antenna. In this latter case, the THz waveform is rigorously proportional to the time derivative of the photo-induced current, which is longer than the laser pulse. Thus, optical rectification in EO crystals is expected to supply shorter THz pulses than photo-conducting antennas, and thus broader spectra. In reality, dispersion effects and the lack of a perfect phase matching makes the two techniques comparable in terms of bandwidth. This is shown in Fig. 4.5, derived from Fig. 5 of a recent paper by E. Matsubara et al. [14], in which experimental THz bandwidths, generated either with a PC antenna, an EO crystal or by photo-induced air plasma, are plotted versus the laser probe pulse duration. The advantage, in terms of bandwidth, of any one of these techniques is not obvious; as expected from Eq. 4.19, the product bandwidth–laser pulse duration $\Delta f \times \tau$ is of the order of 1 (straight line in Fig. 4.5).

The generated THz power can be estimated as follows. $I_o(t')$ is the power density of the laser beam, and therefore $I_o(t') = P_{\text{laser}}(t')/S$, where S is the section of the laser beam. From Eq. 4.20, we deduce the THz power $P_{\text{THz}}(t)$:

$$\begin{aligned}P_{\text{THz}}\left(t\right)&=\frac{n_\omega}{2Z_o}E^2\left(t\right)S\\&=\frac{n_\omega}{2Z_o}\left(\frac{\mu_o}{\sqrt{2}n_{\omega_{\text{opt}}}n_\omega}\chi^{(2)}\left(\omega,\omega_{\text{opt}}\right)z\frac{1}{S}\frac{d\,P_{\text{laser}}\left(t'\right)}{dt'}\right)^2\end{aligned} \tag{4.21}$$

Assuming a Gaussian temporal shape of the laser pulse, we get

$$P_{\text{THz}}\left(t\right)=\frac{4\mu_o}{cn_{\omega_{\text{opt}}}^2n_\omega S}\left(d^{(2)}z\frac{t}{\tau^2}P_{\text{laser}}e^{-\left(\frac{t}{\tau}\right)^2}\right)^2 \tag{4.22}$$

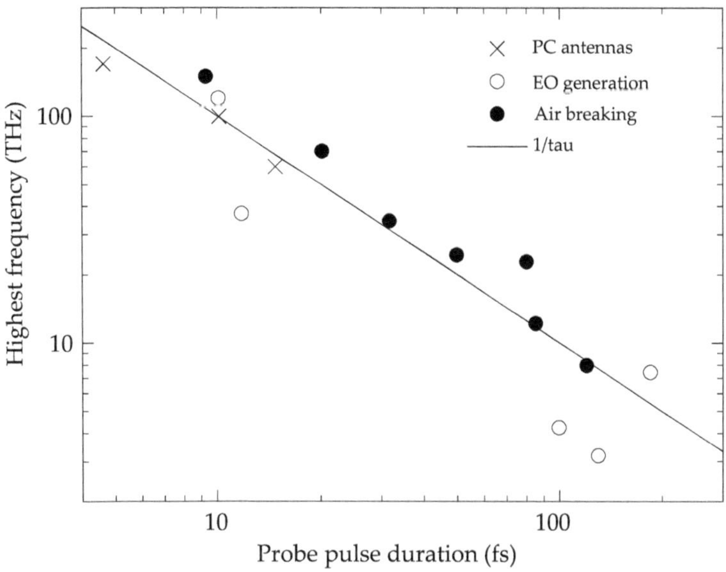

Figure 4.5 Highest experimental THz frequency generated and measured in a TDS system versus the laser probe pulse duration. Full circles, open squares, and crosses correspond to data achieved by different teams respectively with air-breaking, EO sampling, and a PC antenna. The continuous line represents the relation $\Delta f \times \tau$. Adapted from Matsubara et al. [14].

The nonlinear coefficient $d^{(2)}$ is equal to $\chi^{(2)}(\omega, \omega_{\text{opt}})/2$. Then we derive the energies of each THz and laser pulses:

$$\begin{cases} E_{\text{THz}}^{\text{pulse}} = \int\limits_{\text{pulse}} P_{\text{THz}}(t)\,dt = \sqrt{\dfrac{\pi}{2}}\dfrac{\mu_o \left(d^{(2)} z P_{\text{laser}}\right)^2}{4cn_{\omega_{\text{opt}}}^2 n_\omega S\,\tau}, \\ E_{\text{laser}}^{\text{pulse}} = \int\limits_{\text{pulse}} P_{\text{laser}}(t)\,dt = \tau P_{\text{laser}}\sqrt{\pi}. \end{cases} \tag{4.23}$$

Finally, the efficiency of the rectification process is

$$\eta = \frac{E_{\text{THz}}^{\text{pulse}}}{E_{\text{laser}}^{\text{pulse}}} = \frac{\mu_o \left(d^{(2)} z\right)^2}{4\sqrt{2}\,c\,n_{\omega_{\text{opt}}}^2 n_\omega S\,\tau^2} P_{\text{laser}}. \tag{4.24}$$

Using the following typical values ($d^{(2)} = 10$ pm/V, $n_{\omega_{\text{opt}}} = n_\omega = 2$, $\tau = 50$ fs) and a $z = 1$ mm long crystal, $\eta \approx 3.5 \times 10^{-18} P_{\text{laser}}/S$. P_{laser}/S is the peak power density of the laser pulses, which should not exceed the damaged threshold in the nonlinear crystals, typically

20 GW/cm^2 for femtosecond laser pulses (i.e., a 1 mJ pulse, of duration 50 fs, which illuminates the crystal over 1 cm^2). Using this upper limit power density, we obtain 0.7 % of efficiency. In actual experiments, one should consider absorption, as well as the effect of other limiting phenomena, like Cerenkov radiation or two-photon absorption. Efficiencies of a small percentage have been reported using a smart geometry scheme (tilted wave front) [15].

For Ti:Sa lasers ($\lambda \approx 0.8$ μm), ZnTe exhibits good performance: even if its nonlinear coefficient is not so large ($r_{41} \approx 4$ pm/V), the phase matching condition is almost satisfied ($c/(n_{\text{THz}} - n_{\text{opt}}) = 0.4$ ps/mm) [16]. At telecoms wavelength ($\lambda \approx 1.5$ μm), organic crystals like DAST ($r_{11} = 47$ pm/V) are very efficient [13].

Let us recall that optical rectification is a nonlinear process, and therefore the generated THz power, i.e., modulus of expression Eq. 4.23 to the square, varies as the square of the optical pump power. For non-amplified laser systems, photo-conducting is more efficient, while optical rectification is preferable at higher optical powers, for example with amplified systems. Also, as the nonlinear phenomenon is almost instantaneous (the deformation of the atomic cloud takes a few femtoseconds), the related THz bandwidth is wide and limited only by the bandwidth of the phase matching conditions and by the absorption by optical phonons in the crystal.

4.3 THz Pulse Detection

Time-resolved detection of THz pulsed waves is mostly performed by using a receiver sensitive to the electric field of the waves.

4.3.1 Photo-conducting Detectors

Such detectors work almost as photo-conducting emitters do. Basically, the device is the same: two electrodes are deposited over a semiconductor substrate in which free carriers exhibit an ultrashort lifetime. However, no bias voltage is applied to the electrodes, but these electrodes are connected to an ammeter that measures the current flowing from one electrode to the other one. When a laser pulse illuminates the gap between the electrodes, it creates a photo-induced free charge carrier population. If no bias is applied, the carriers recombine after a time delay equal to the carrier lifetime, and no net current is read by the ammeter. On the other hand, if simultaneously to the laser pulse excitation, a THz pulse illuminates the switch, the electric field of the THz accelerates the photo-induced free-carrier population. Then a current is read by the ammeter.

Let us consider the free photo-electron contribution, and we call $dN(t')$ the number of free electrons generated during the time slot dt' by the laser pulse arriving at time t' in the gap region. Photogeneration in a semiconductor makes this number $dN(t')$ proportional to the laser power when saturation does not take place (see Chapter 3, Section 3.4.2):

$$dN(t') = \eta \frac{P_{\text{laser}}(t')}{hf} dt'. \tag{4.25}$$

η is the quantum efficiency, and f the frequency of the incoming photons. At a later time, t after t', the total number $N(t)$ of free electrons is equal to the summation over all the times t' of the resting number of electrons created at time t'. As the decay law $G(\Delta t)$ of the free electron population depends only on the time difference $\Delta t = t - t'$ between generation and observation, we get:

$$N(t) = \int_{-\infty}^{t} G(t - t')\, dN(t') = \frac{\eta}{hf} \int_{0}^{+\infty} P_{\text{laser}}(t - t'')\, G(t'')\, dt''. \tag{4.26}$$

The current at time t is the product of the number of free electrons, each with charge $-e$ and of their velocity $v(t)$. The velocity is equal to the carrier mobility times the applied electrical field (here the impinging THz field):

$$v(t, \tau) = \mu(t)\, E_{\text{THz}}(t - \tau). \tag{4.27}$$

Here τ is the time delay between emission and detection. The current density is thus given by

$$\begin{aligned} J(t, \tau) &= -e\, N(t)\, v(t, \tau) \\ &= -\eta\, e\, \mu\, E_{\text{THz}}(t - \tau) \int_{0}^{+\infty} P_{\text{laser}}(t - t'')\, G(t'')\, dt''. \end{aligned} \tag{4.28}$$

Because of causality, i.e., the effect appears after its cause, $G(t) = 0$ for $t < 0$. Thus, the integral over t'' can be spread from $-\infty$ to $+\infty$. The electronics that reads the current is slow compared with the duration of the short electrical events: it delivers a signal S that is time-integrated, and that depends only on τ:

$$S(\tau) = \int_{t=-\infty}^{+\infty} J(t,\tau)\,dt$$
$$= -\frac{\eta\, e\, \mu}{h\, f} \int_{t=-\infty}^{+\infty} \int_{t''=-\infty}^{\infty} E_{\mathrm{THz}}(t-\tau)\, I_{\mathrm{laser}}\left(t-t''\right) G\left(t''\right) dt''\, dt. \tag{4.29}$$

We take the Fourier transform of $S(\tau)$ (see annex):

$$S(\omega) = FT\,[S(\tau)] = -\frac{\eta\, e\mu}{h\, f} G^{*}(\omega)\, I_{\mathrm{laser}}(\omega)\, E^{*}_{\mathrm{THz}}(\omega). \tag{4.30}$$

This gives the spectral components of the recorded signal, which are proportional to the spectral components of the incoming THz field. It is also proportional to the response function of the switch $G(\omega)$, which should be as flat and as broad as possible, i.e., it should be as fast as possible in the time domain. Let us remember than $G(t)$ describes the decay law of the photo-carriers in the switch, which is governed by both the carrier lifetime in the semiconductor and the response time of the detection circuit. Therefore, contrary to photo-conducting emitters, receivers must be fabricated from semiconductors with fast response times, i.e., short carrier lifetimes.

4.3.2 Electro-optic Detection

Electro-optic detection is based on the Pockels effect in a nonlinear crystal (see Annex), which is the reverse process to optical rectification. The incident THz field causes the crystal to be birefringent, which means the refractive index in one axis of the crystal is different to the others. This modifies the equation of the index ellipsoid as demonstrated in the Annex. We start with a static approach, i.e., both THz and laser fields are continuous. To help the reader, we reproduce here relation A.71:

$$x^2\left(\frac{1}{n_{\mathrm{x}}^2}+\sum_k r_{1k}E_{\mathrm{THz,k}}\right)+y^2\left(\frac{1}{n_{\mathrm{y}}^2}+\sum_k r_{2k}E_{\mathrm{THz,k}}\right)$$
$$+z^2\left(\frac{1}{n_{\mathrm{z}}^2}+\sum_k r_{3k}E_{\mathrm{THz,k}}\right)+2yz\sum_k r_{4k}E_{\mathrm{THz,k}} \qquad (4.31)$$
$$+2xz\sum_k r_{5k}E_{\mathrm{THz,k}}+2xy\sum_k r_{6k}E_{\mathrm{THz,k}}=\frac{1}{n^2}$$

As shown in the Annex (see expressions A.76), the refractive index values along y and z are changed by the applied THz field.

Let us now consider an optical beam, linearly polarized at the entrance of the crystal, which propagates along direction X in the crystal of length L. At the output of the crystal, the two components of the optical field are

$$E_{\mathrm{Y}}=E_{\mathrm{oY}}\cos\left(\omega_{\mathrm{opt}}t-\frac{\omega_{\mathrm{opt}}}{c}n_{\mathrm{Y}}L\right),$$
$$E_{\mathrm{Z}}=E_{\mathrm{oZ}}\cos\left(\omega_{\mathrm{opt}}t-\frac{\omega_{\mathrm{opt}}}{c}n_{\mathrm{Z}}L\right), \qquad (4.32)$$

where E_{oY} and E_{oZ} are the components at the entrance of the crystal, with reference of the laboratory coordinates X, Y, Z, while the principal axes of the crystal are x, y, z, as defined previously. Calculating n_{Y} and n_{Z} using Eq. A.76 requires one to know the relative orientation of the crystal (x, y, z) with regard to X, Y, Z. We set

$$\phi=\frac{\omega_{\mathrm{opt}}}{c}n_{\mathrm{Y}}L,\quad \Delta=\frac{\omega_{\mathrm{opt}}}{c}\left(n_{\mathrm{Z}}-n_{\mathrm{Y}}\right)L. \qquad (4.33)$$

Hence,

$$E_{\mathrm{Y}}=E_{\mathrm{oY}}\cos\left(\omega_{\mathrm{opt}}t-\phi\right),\quad E_{\mathrm{Z}}=E_{\mathrm{oZ}}\cos\left(\omega_{\mathrm{opt}}t-\phi-\Delta\right). \qquad (4.34)$$

Through a simple trigonometric calculation, we combine both E-field components (4.34) and eliminate the $\omega_{\mathrm{opt}}t-\phi$ term. We get

$$Y'^2+Z'^2-2Y'Z'\cos\Delta=1$$
$$\Leftrightarrow\left(X'\ Y'\right)\begin{pmatrix}1 & -\cos\Delta\\ -\cos\Delta & 1\end{pmatrix}\begin{pmatrix}Y'\\ Z'\end{pmatrix}=1, \qquad (4.35)$$

with $Y'=\frac{E_{\mathrm{Y}}}{E_{\mathrm{oY}}\sin\Delta}$, $Z'=\frac{E_{\mathrm{Z}}}{E_{\mathrm{oZ}}\sin\Delta}$. One might recognize this as the equation of an ellipse, which tells us that the linearly polarized

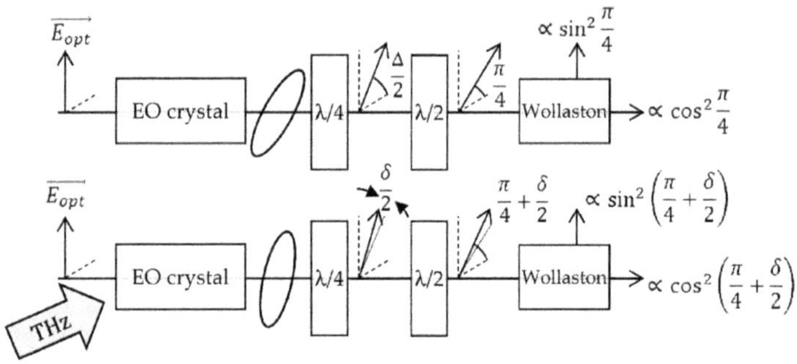

Figure 4.6 Scheme of EO detection, showing the polarization of the optical probe beam. Top without incoming THz field, bottom with incoming THz field. $\lambda/4$ and $\lambda/2$ are quarter- and half-wave plates, respectively.

beam has been transformed into an elliptically polarized one. The lengths of the axes of the ellipse are a' and b', defined by $(a')^{-1} = \sqrt{2}\cos(\Delta/2) = \sqrt{\lambda_a}$ and $(b')^{-1} = \sqrt{2}\sin(\Delta/2) = \sqrt{\lambda_b}$, where λ_a and λ_b are the eigenvalues of the matrix in Eq. 4.35. In the $Y'Z'$ referential, the angle of the (Y', Z') vector as regard to the Y' axis is $\theta_{Y'Z'} = \pi/4$. Thus, the ellipse is oriented with an angle θ_{YZ} in the reference of the laboratory such as $\tan\theta_{XY} = \frac{E_{oZ}}{E_{oY}}$. The ellipticity $\xi = a'/b' = \tan(\Delta/2)$ is measured by converting the transmitted beam into a linearly polarized one thanks to a $\lambda/4$ wave plate located into the transmitted beam and whose neutral axes are along the directions of the principal axes of the ellipse. The orientation of the field after the waveplate makes an angle $\Delta/2$ with the neutral axes of the wave plate.

The induced electro-optic change of the refractive index is measured using a polarimetric technique (see Fig. 4.6). We define ϕ and Δ as in Eq. 4.33, but now the values include the nonlinear contribution:

$$\phi = \frac{\omega_{\text{opt}}}{c} n_Y L,\ \Delta = \frac{\omega_{\text{opt}}}{c} \left(n_Z(0) - n_Y(0)\right) L + \delta = \Delta_0 + \delta, \quad (4.36)$$

with

$$\delta = -\frac{1}{2}\frac{\omega_{\text{opt}}}{c}\left(n_z^3(0)\sum_k r_{3k}E_{\text{THz},k} + n_y^3(0)\sum_k r_{2k}E_{\text{THz},k}\right) L \ll \Delta. \quad (4.37)$$

To derive Eq. 4.37, we used expression A.76 of n_y and n_z. Typically, the r_{ij} coefficients are of the order of pm/V and $n \approx 2$. A 1 μW beam at 1 THz with a diameter of 1 cm corresponds to an E field of 3.1 V/m. Thus $|\delta| \approx 25 \times 10^{-8} L$: For millimeter- or even centimeter-thick crystals, $|\delta|$ is extremely small, justifying the approximation $\delta \ll \Delta$.

After propagation in the crystal, the polarization of the optical beam is elliptic. The ellipticity is now given by

$$\xi = \frac{a'}{b'} = \tan \frac{\Delta_0 + \delta}{2}. \tag{4.38}$$

Therefore, the linearly polarized beam, after the quarter-wave plate, which rotates the phase by $\pi/2$, makes now an angle $(\Delta_0 + \delta)/2$ with the neutral axes of the plate. The angular rotation induced by the electro-optic effect is $\delta/2$.

To measure this angle, the easiest and most popular method is polarimetry. After the quarter-wave plate (Fig. 4.6), the transmitted polarized beam goes through a polarizing prism, like a Wollaston prism, whose axes make a 45° angle with the polarization of the beam without applied THz field. Thus, the prism outputs two beams, whose intensities are respectively $I_1(0) = I_o \cos^2(\pi/4)$ and $I_2(0) = I_o \sin^2(\pi/4)$. As the polarization of the beam rotates by an angle $\delta/2$ when the THz beam illuminates the EO crystal, the 45° angle becomes $(\pi/4) + (\delta/2)$ and thus the intensities of the two beams exiting by the Wollaston prism are changed into

$$I_1\left(E_{\mathrm{THz}}\right) = I_o \cos^2\left(\frac{\pi}{4} + \frac{\delta}{2}\right), \quad \text{and} \quad I_2\left(E_{\mathrm{THz}}\right) = I_o \sin^2\left(\frac{\pi}{4} + \frac{\delta}{2}\right).$$

One records these two values (e.g., with balanced photodiodes), and one calculates their normalized difference:

$$S = \frac{I_1\left(E_{\mathrm{THz}}\right) - I_2\left(E_{\mathrm{THz}}\right)}{I_1\left(E_{\mathrm{THz}}\right) + I_2\left(E_{\mathrm{THz}}\right)} = \frac{\cos^2\left(\frac{\pi}{4} + \frac{\delta}{2}\right) - \sin^2\left(\frac{\pi}{4} + \frac{\delta}{2}\right)}{\cos^2\left(\frac{\pi}{4} + \frac{\delta}{2}\right) + \sin^2\left(\frac{\pi}{4} + \frac{\delta}{2}\right)}$$
$$= \sin \frac{\delta}{2} \approx \frac{\delta}{2}. \tag{4.39}$$

The normalized intensity S gives directly the angle δ, and thus the strength of the THz field through relation 4.37. The properties of common EO crystals are given in Table 4.2. ZnTe tends to be the most popular material to use.

The above analysis is a static one, i.e., the temporal aspects of both THz and laser pulses are ignored. However, this analysis

Table 4.2 Electro-optic materials used in THz optoelectronics

	ZnTe	GaAs	DAST	GaSe	GaP
Crystallographic group	$\bar{4}3m$	$\bar{4}3m$	m	$\bar{6}2m$	$\bar{4}3m$
n_{opt}at (λ_{opt} (µm))	2.85 (0.8)	3.61 (0.9)			3.18 (1)
n_g (λ_{opt} µm)	3.2 (0.8)	4.96 (0.9)			
r_{ij} (pm/V)	$r_{41} = 3.9$	$r_{41} = 1.4$	$r_{11} = 47$	$d_{eo} = 24.3$	
n_{THz} (at f (THz))	3.17 (1)	3.6 (1–2)	2.3 (3)	n_o 3.25 n_e 2.47	3.34
$f_{TO\text{-}phonon}$ (THz)	5.3	8.0		0.584	11

gives the main properties and parameters of the polarimetric EO detection of a THz electric field. Nevertheless, a precise description of the technique should be based first on a spectral decomposition of the pulses, and then on solving the nonlinear equations as done above, in which propagation effects should be accounted. This quite long calculation is very well explained in a paper by Gallot and Grischkowsky [17]. The main result is

$$I_1\left(E_{THz},\tau\right) - I_2\left(E_{THz},\tau\right) \propto \int_{-\infty}^{+\infty} E_{THz}(\omega)\, f(\omega)\, e^{-j\omega\tau}d\omega, \qquad (4.40)$$

with

$$f(\omega) = C_{opt}(\omega)\,\chi^{(2)}_{eff}\left(\omega,\omega_{opt}\right)\frac{e^{j\Delta kz}-1}{j\Delta kz}, \qquad \Delta k = \frac{\omega}{c}\left(n_\omega - n_G\right),$$

$$C_{opt}(\omega) = \int_{-\infty}^{+\infty} E^*_{opt}\left(\omega_{opt}\right) E_{opt}\left(\omega_{opt}-\omega\right) d\omega' \propto P_{laser}.$$

$E_{opt}(\omega_{opt})$ is the envelop of the laser pulse field, and $C_{opt}(\omega)$ is the autocorrelation function of this envelop. All the other notations are the same as for optical rectification (see Eqs. 4.11–4.14). Therefore, the Fourier transform of Eq. 4.40 is

$$FT\left[I_1\left(E_{THz},\tau\right) - I_2\left(E_{THz},\tau\right)\right] \propto E_{THz}(\omega)\, f(\omega). \qquad (4.41)$$

The Fourier transform of the recorded signal is the product of the THz field to be measured and a function that renders for the frequency response of the EO detection system when triggered by the laser pulse.

4.4 The Principle of Equivalent Time Sampling

At the time of writing, no electronic apparatus is capable of recording the ultrashort waveform delivered by a THz-TDS setup. As already written, this signal exhibits features that should be recorded with a precision of tens of femtoseconds. Even if single-shot measurements of the temporal waveform can be performed [18], the most efficient way of getting the waveform is to benefit from the ultrastable repetition of pulses delivered by femtosecond mode-locked lasers. The basic idea is to temporally read the waveform, step by step, with a slow field detector that integrates the signal, by mixing the waveform signal with a probe pulse in a nonlinear device. This device is an electric field detector, either a photo-conducting switch (see the nonlinear Eq. 4.30) or an EO crystal (see Eq. 4.41). To ensure good synchronism between the THz and the probe pulse, both are generated using the same laser. One part of the laser beam, selected by a beam splitter, excites the THz emitter, while the second part of the beam illuminates the receiver. The laser pulses at the receiver are delayed with regard to the ones at the emitter thanks to an optical delay line which is used to change the optical path length. The delay line is typically made with a mirror, or in fact a retro-reflector made of a prism (corner cube) or, better, of perpendicular mirrors, which reflects the impinging beam at normal incidence. The mirror is located on a motorized stage. For each displacement x of the mirror, the time delay is changed by

$$\tau = \frac{2\,x}{c}\,. \tag{4.42}$$

The factor of 2 accounts for the back and forth propagation in the delay line. Let us now derive what signal is delivered by the slow detector. All the temporal functions used here are defined to peak at $t = 0$. The nonlinear device makes the product of both functions, i.e., $S_{\text{probe}}\,(t-\tau) \times S_{\text{THz}}\,(t)$. The THz waveform is depicted by the temporal function $S_{\text{THz}}\,(t)$ and the laser pulse by $S_{\text{probe}}\,(t)$. At the nonlinear device, the laser probe is delayed by τ and thus written as $S_{\text{probe}}\,(t-\tau)$. As the bandwidth of the nonlinear device and/or of the detector is not infinite, the detection response time is not instantaneous and writes $G\,(t)$. Thus, the effective probe signal at time t results from the integration of all the probe signal effects occurring at t'' before t:

$$S_{\text{probe}}^{\text{eff}}(t,\tau) = \int_{-\infty}^{t} S_{\text{probe}}\left(t''-\tau\right) G\left(t-t''\right) dt''. \tag{4.43}$$

Let us set $t' = t - t''$. We simplify Eq. 4.43 into

$$\begin{aligned} S_{\text{NL}}(t,\tau) &= S_{\text{probe}}^{\text{eff}}(t,\tau) \times S_{\text{THz}}(t) \\ &= S_{\text{THz}}(t) \int_{0}^{+\infty} S_{\text{probe}}\left(t-t'-\tau\right) G\left(t'\right) dt'. \end{aligned} \tag{4.44}$$

As already explained, because of causality, the impulse response of the nonlinear detection system $G(t')$ is null for negative t'. Equation 4.44 becomes

$$S_{\text{NL}}(t,\tau) = \int_{-\infty}^{+\infty} S_{\text{probe}}\left(t-t'-\tau\right) G\left(t'\right) S_{\text{THz}}(t)\, dt'. \tag{4.45}$$

Then the signal is time-integrated by slow electronics:

$$\begin{aligned} S(\tau) &= \int_{-\infty}^{+\infty} S_{\text{NL}}(t,\tau)\, dt \\ &= \int_{-\infty}^{+\infty}\int_{-\infty}^{+\infty} S_{\text{probe}}\left(t-t'-\tau\right) G\left(t'\right) S_{\text{THz}}(t)\, dt' dt. \end{aligned} \tag{4.46}$$

The last stage of the calculation is to Fourier-transform $S(\tau)$:

$$S(\omega) = FT\left[S(\tau)\right] = \int_{-\infty}^{+\infty} S(\tau)\, e^{-j\omega\tau}\, d\tau$$

$$= \int_{-\infty}^{+\infty}\left(\int_{-\infty}^{+\infty}\left(\int_{-\infty}^{+\infty} S_{\text{probe}}\left(t-t'-\tau\right) e^{-j\omega\tau}\, d\tau\right) S_{\text{THz}}(t)\, dt\right) G\left(t'\right) dt'.$$

Then we set $t'' = t - t' - \tau$:

$$S(\omega) =$$

$$\int_{-\infty}^{+\infty}\left(\int_{-\infty}^{+\infty}\left(\int_{-\infty}^{+\infty} S_{\text{probe}}\left(t''\right) e^{-j\omega t} e^{j\omega t'} e^{-j\omega t''}\, dt''\right) S_{\text{THz}}(t)\, dt\right) G\left(t'\right) dt'.$$

Finally, by identifying the different terms of this latter relation, we obtain

$$S(\omega) = G^{*}(\omega) \times S_{\text{probe}}(\omega) \times S_{\text{THz}}(\omega), \tag{4.47}$$

with $S_{\text{probe}}(\omega) = FT\left[S_{\text{probe}}(t)\right]$, $S_{\text{THz}}(\omega) = FT\left[S_{\text{THz}}(t)\right]$, and the spectral response of the nonlinear detection is $G(\omega) = FT\left[G(t)\right]$.

Time-domain spectroscopy is based on relation 4.47. The Fourier transform of the recorded signal is proportional to the product of the Fourier transforms (i.e., of the spectra) of the THz signal, the probe signal, and the nonlinear detector response. Any change in the THz spectrum may be recorded and evaluated, assuming the probe and detector spectra remain constant. To obtain the most faithful spectrum of the THz signal, the probe signal should be extremely short, corresponding to a wide spectrum almost flat over the bandwidth of interest. As well, the nonlinear detector should be as fast as possible. In this ideal case of an instantaneous detection response and ultrashort laser pulse (Dirac-like), $S_{\text{probe}}(\omega)$ and $G(\omega)$ are constant and $S(\omega)$ is directly proportional to $S_{\text{THz}}(\omega)$.

In reality, the waveforms are not recorded continuously versus the time delay τ, but at several temporal points $\tau_m = m\,\Delta\tau$, where $\Delta\tau$ is the interval between two records, and m is an integer. The delay line allows one to explore a maximum displacement distance X, i.e., a maximum time delay $\Delta T = 2X/c = M\,\Delta\tau$, in which M is the maximum value of m. Therefore, the waveform is measured over the time window ΔT and at discrete time values τ_m. Its spectrum is thus calculated using the discrete time Fourier transform (see Annex):

$$S_{\text{rec}}(\omega) = \frac{\Delta T}{2\pi}\,\text{sin}\,c\left(\frac{\Delta T}{2}\omega\right) \otimes \sum_{m=0}^{M-1} S(m\,\Delta\tau)\,e^{-jm\omega\Delta\tau}. \tag{4.48}$$

As explained in the Annex, the spectral resolution δf is limited by the sinc function in Eq. 4.48 and its expression is

$$\delta f = \frac{1}{\Delta T} = \frac{c}{2X}. \tag{4.49}$$

Thus, δf is inversely proportional to the maximum displacement distance (X) of the delay line of the setup. Typically, X is of the order of $5 \sim 10$ cm, which gives $\delta f = 1.5 \sim 3$ GHz. This is the main handicap of THz-TDS compared with other spectroscopy

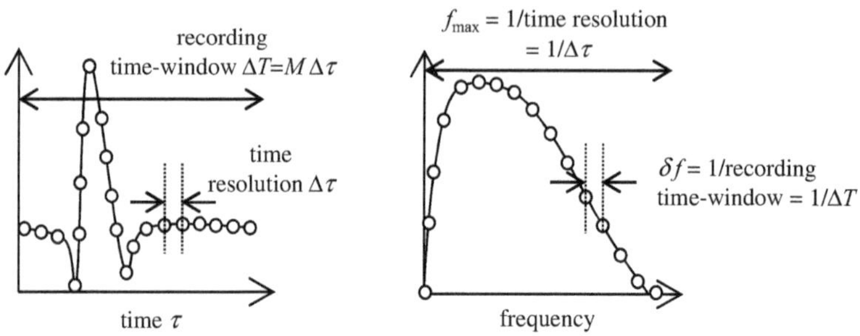

Figure 4.7 Typical THz-TDS waveform (left) and its spectrum (modulus, right).

techniques employing continuous wave signals, like FTIR. Longer distances of X are obviously possible, but difficult to manage. First, the geometrical adjustment of the setup becomes tricky, because the delay line should not induce any geometrical shift of the laser beam, at the micrometer scale (remember, for example, that the laser beam must be focused at the gap of photo-conducting antennas, whose dimension is a few micrometers). Secondly, records become lengthier, and one should check that during this long recording time, there is no variation of the laser intensity. Artificially enlarging the recording time window by adding zeros to the waveform does not improve the frequency resolution (see Annex), but only makes the spectra smoother.

The achievable bandwidth starts from δf up to $f_{\max} = 1/2\,\tau = \delta x c$, where δx is the displacement step of the delay line. Common motorized stages permit incremental movements δx shorter than 1 μm, corresponding to a THz bandwidth larger than some tens of THz. Therefore, the bandwidth of THz-TDS setups is not limited by the Fourier transform procedure, but by the bandwidth of the emitting and receiving antennas.

The relationships between the recorded waveform parameters (time increment and time window) and the computed spectrum parameters (bandwidth, frequency resolution) are sketched in Fig. 4.7.

The speed of the motorized stage of the optical delay line determines the THz waveform acquisition time. If long integration times are required (i.e., to maximize signal-to-noise) at each point, then the stage can be stepped or the delay line can be moved by a servo motor, i.e., moved at a constant speed, and data is collected at regular time intervals. For many applications, a faster acquisition time is needed, and thus an oscillating delay line can be employed. One of the simplest and cheapest types are those designed around a voice coil, which can be driven with a function generator producing a sinusoidal or sawtooth output. These can typically be made to oscillate at say 10 Hz with enough amplitude (distance travelled) to collect an entire THz waveform in 50 ms, if both the upsweep and downsweep are used. Much faster systems have been proposed, e.g., rotating mirrors [19]; however, faster acquisition does mean compromising on signal to noise, and care must be taken to ensure the data is collected over a linear range of motion of the oscillating delay line. This is further discussed in Chapters 11 and 12.

4.5 From Time-Domain to Frequency Domain

Figure 4.8 shows a typical THz pulse recorded with a classical THz-TDS setup. Here, the setup uses LT-GaAs photo-conducting antennas as both emitter and receiver. A high-resistivity silicon hyper-hemispheric lens is attached to the antennas. The pulse is recorded without any sample in between the antennas. The measurement is performed in air with a low humidity of about 5%. The recording equivalent time window is 56.50 ps. The time delay between two consecutive recorded data is 27.6 fs and the number of temporal data points is 2048. Thus, in the frequency domain, the resolution is 17.7 GHz and the useful spectral range is 18.12 THz. After the first ultra-short waveform, i.e., for $t > 5$ ps, the signal shows weak features spreading up to 40 $\sim$ 50 ps. They correspond to the signature of long resting oscillations of residual water molecules along the THz beam in between the THz antennas.

The next stage in the THz spectroscopy procedure is to take the Fourier transform of the waveform. Figure 4.9 presents the modulus (magnitude) of the spectrum of the waveform shown in Fig. 4.8. The spectrum extends up to about 5 THz, which is much smaller than the maximum frequency (18.12 THz) allowed by the Fourier transform. Thus, the spectrum is limited by the bandwidth of the THz antennas.

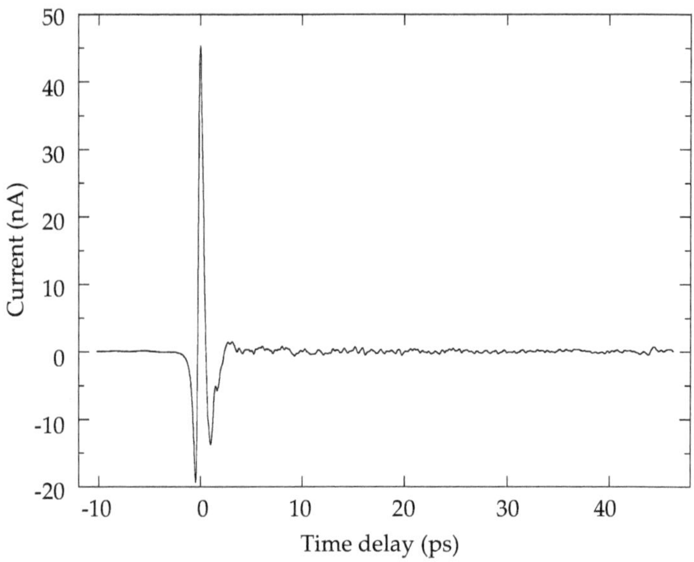

Figure 4.8 Typical temporal waveform recorded in a THz TDS setup.

At higher frequencies, only noise is seen, with an average value equal to 0.6 nA/THz.

The peak value of the spectrum in Fig. 4.9 is 1 μA/THz at 0.415 THz. Thus, the dynamic range, defined as the maximum signal divided by the noise, is around 1700 (~32 dB). When dealing with intensity instead of field amplitude, the dynamic range must be squared and takes the value $\sim 2.9\times\ 10^6$, i.e., 64 dB. At lower frequencies, <0.4 THz, the spectrum decreases abruptly. It is due to two additional reasons. First, because a THz antenna behaves as a dipole, the radiated far field varies as the time derivative of the dipolar current. The Fourier transform of this derivative is the product of the current spectrum and of the frequency (see relation (A.56) in the Annex). Therefore, the far-field spectrum magnitude tends to zero at lower frequencies. The second reason originates in the radiation gain of the antenna, which includes both photoconducting switch and hemispheric lens. As lower frequencies, the antenna size is of the order of the THz wavelength or even smaller. The radiated signal is no more directed in a narrow solid angle, and the signal received at the detector decreases as well.

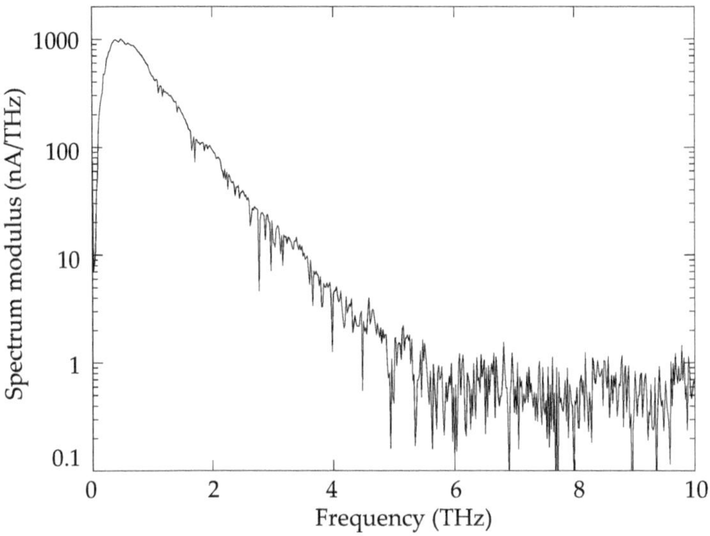

Figure 4.9 Modulus of the Fourier transform of the waveform shown in Fig. 4.8.

The low-amplitude lines observed in the spectrum, e.g., the strong ones near 3 THz, are due to absorption by the vapor molecules along the THz beam. Here, as the measurement is performed in dry air, their strength is rather weak. They correspond, in the time domain, to the long-term ripples observed in Fig. 4.8.

Figure 4.10 presents the phase, in degrees, of the Fourier transform of the waveform. As the phase is obtained through the inverse tangent of the ratio of the imaginary part over real part of the Fourier transform, the phase is given by any Fourier transform code in a π-angular interval.

For example, on Fig. 4.10, the phase is given between -90° and $+90^{\circ}$. This means that each time the phase reaches $\pm 90^{\circ}$, there is an addition or subtraction of π to the actual phase. On a wide frequency scale as in Fig. 4.10, these π-jumps give a random appearance to the phase curve. Nevertheless, if we zoom in on Fig. 4.10 which is presented in Fig. 4.11, the effect of the π-jumps is clearly seen. One says that the extracted phase is wrapped. Removing these artificial π-jumps is known as phase unwrapping. It simply entails comparing two successive extracted phases. If their

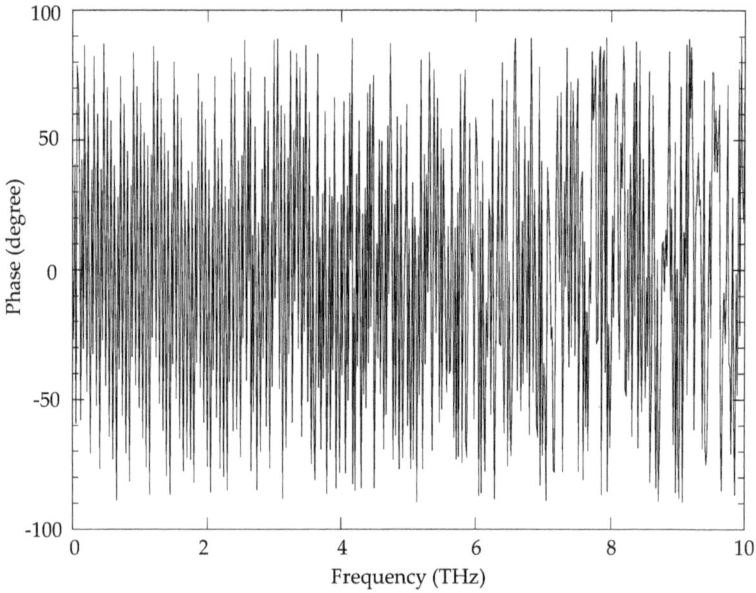

Figure 4.10 Phase of the Fourier transform of the waveform shown in Fig. 4.7.

difference exceeds a given threshold value, the necessary $\pm\pi$ is added (Fig. 4.12). Of course, when dealing with signals transmitted by a sample, one must be sure that abrupt transmission features, like at the border of forbidden bands of metamaterials, do not induce a variation of phase larger than π between two consecutive recorded frequencies. When this is doubtful, the best is to record several temporal signals over a different time-window, in order to vary the frequency step of the extracted spectra. This allows one to resolve abrupt transitions, and not to mix them up with the phase-wrapping effect. A complementary solution could be to add zeros to the waveforms: this just inserts additional spectral points to the phase curve, but it does not improve the spectral resolution as already explained. This procedure may help remove artefacts, i.e., π-jumps due to noise.

Figure 4.12 depicts the unwrapped phase of the spectrum of the waveform shown in Fig. 4.8. Let us notice that it is the phase of the received signal, and not of the signal radiated by the emitter.

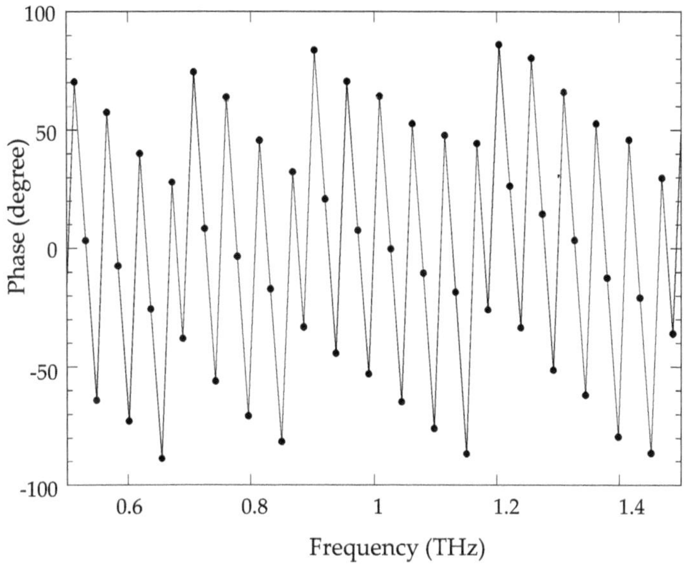

Figure 4.11 Zoom between 0.5 and 1.5 THz of Fig. 4.10. The calculated points are depicted by dots, while the continuous line is a guide to the eyes.

At higher frequencies, i.e., over 2.5 THz, some random jumps, due to the noise, are seen. A zoom of Fig. 4.12, in the low-frequency region (Fig. 4.13), shows that the phase data are affected by noise at the lower frequencies, as explained by the decrease of THz beam intensity at these frequencies. Moreover, below the smallest frequency of the experimental spectrum, the phase is unknown. Therefore, correcting the phase, with the continuity argument explained previously, which was used to remove the π-jumps, is not possible. We can conclude that the absolute phase of the signal, in THz-TDS, is not known. However, if there is no "phase accident" due to the setup (antennas, optics), the phase varies as

$$\phi = \frac{\omega}{c} n(\omega) x, \tag{4.50}$$

where x is the propagation length of the THz beam and $n(\omega)$ is refractive index of the medium in which the beam propagates. Therefore, the tendency of the spectral data follows a linear variation with frequency (continuous line in Fig. 4.13), because this phase behavior is dominated by the ω/c term and consequently, at

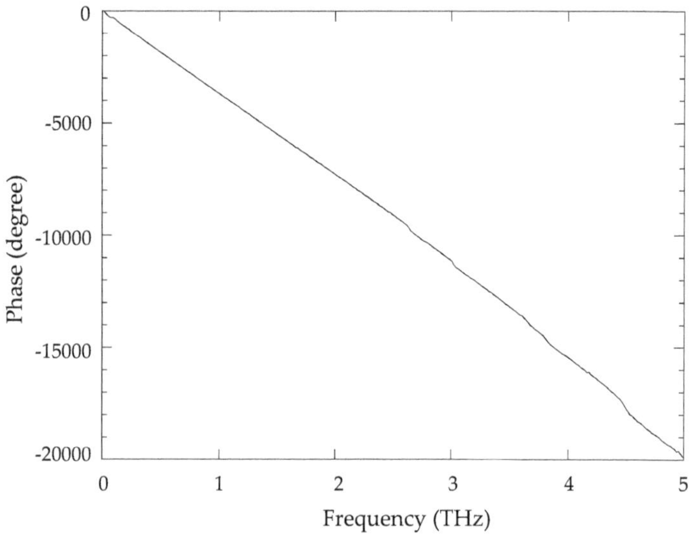

Figure 4.12 Unwrapped phase of the waveform shown in Fig. 4.8.

$\omega = 0, \phi = 0$. Thus, one may force the phase to be null at $\omega = 0$ by adding a constant value in such a way that the linear extrapolation of the corrected phase data crosses 0 at $\omega = 0$ (see Fig. 4.13).

Because all the spectral components of the THz pulse are radiated at the same time in the emitter, and because they propagate mostly in air or vacuum, in which dispersion is negligible, we can ask ourselves about the origin of the large phase plotted in Fig. 4.12 and 4.13 and conclude that it is a simple numerical effect. Indeed, in order to record the whole waveform, the measurement is started before the pulse arrives at the receiver. In Fig. 4.7, the record begins 10 ps before the occurrence of the peak of the THz pulse. When performing the FFT of the waveform, the computer keeps the origin of the time at $t_0 = -10$ ps. This induces an artificial phase shift equal to $\omega\, t_0 = -20\,\pi\, f_{\text{THz}}$ rad $\approx -3600\, f_{\text{THz}}$ deg. This huge linear phase shift hides any small real feature in the phase spectrum.

To clearly emphasize the spectral features of $\phi\,(\omega)$, it is preferable to divide the phase by ω/c. The obtained value has the unity of an optical length. It is usually named the effective length:

$$x_{\text{eff}} = \frac{\phi}{\omega/c}. \tag{4.51}$$

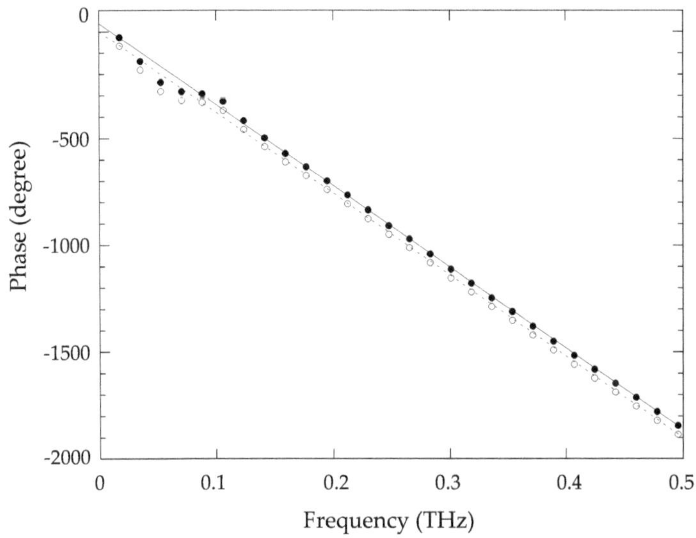

Figure 4.13 Unwrapped phase of the waveform shown in Fig. 4.9 (black dots): the continuous line is a linear fit. Open circles represent the corrected phase data, whose linear fit (dotted line) crosses the origin 0.

The effective length relating to the phase shown in Fig. 4.12 is presented in Fig. 4.14. Both uncorrected and corrected curves are plotted. The overall value of 3 mm corresponds well to the 10 ps shift of the time origin. Weak features are clearly seen, like small peaks and dips from 0.5 to 2.6 THz due to water vapor absorption in the setup, or like dispersion effect in the optics (silicon domes/lenses at the antennas, GaAs switch substrates, optics misalignment, etc.). Moreover, when the THz signal decreases above 2.6 THz, random π-jump phases due to noise appear. The curves are noisy and tend to $\pm\infty$ at low frequencies, because of the division by ω. This phenomenon can be explained differently; when ω tends to zero, one reaches the electrostatic regime, for which no propagation occurs, i.e., all the phenomena are instantaneous. Hence, the wave velocity $x_{\text{eff}}/c = \phi/\omega$ becomes infinite.

Finally, let us point out that all the spectral features in both modulus (Fig. 4.9) and phase (Fig. 4.12) will be removed when dealing with characterization of samples by THz-TDS. As a reference and a sample signal will be recorded, the ratio of their moduli and

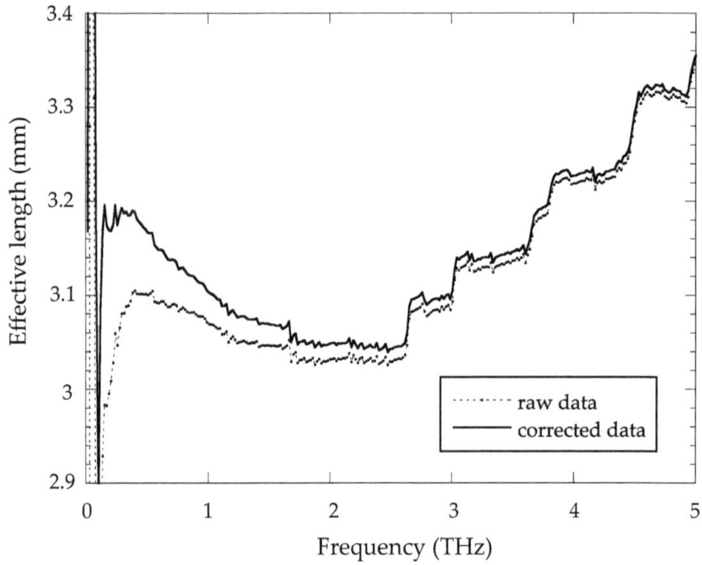

Figure 4.14 Effective length related to the phase shown in Fig. 4.12.

the difference of their phases will remove these features, as far as the signals remain over the noise level.

4.6 THz-TDS Setups

4.6.1 Common Systems Based on Photo-conducting Antennas

A THz-TDS instrument commonly includes a femtosecond pulsed laser, an emitting antenna and a receiving one, some THz optical components that steer the THz beam from the emitter to the receiver via a sample, an optical delay line and an electronics that drives the delay line and simultaneously records the signal. Most of TDS setups are either built with photo-conducting antennas or with electro-optic antennas, or a combination of the two.

A typical THz-TDS scheme based on photo-conducting antennas is shown in Fig. 4.15. The optical beam delivered by a femtosecond mode-locked laser is divided in two beams by a beam splitter.

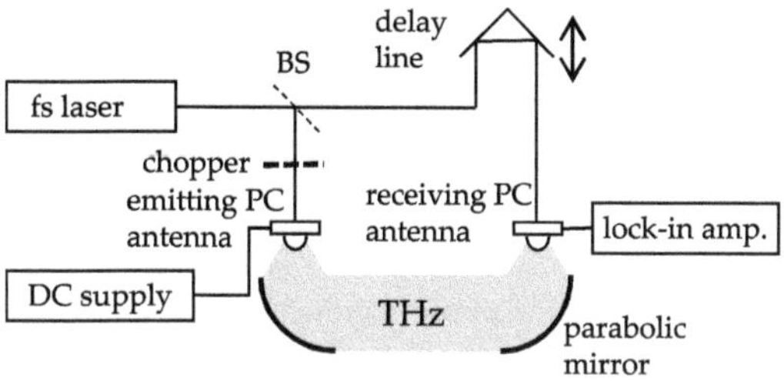

Figure 4.15 Basic THz-TDS setup with photo-conducting (PC) antennas. BS is for beam splitter.

The first one excites the emitting antenna. A lens (not shown here) focuses the laser beam at the gap of the antenna switch. An optimized focusing is important as it ensures a good THz signal. This is achieved by attaching the antenna on a three-axis micrometric stage. Typically, a mechanical chopper is used to modulate the laser beam at a frequency of the order of 1 kHz, in order to perform a lock-in amplification of the detected current. The chopper could be located in the THz beam, but it should be much bigger as the THz beam is wider than the laser beam. Also, instead of using a chopper, one can bias the emitting antenna with an alternative voltage. This permits higher modulation frequencies, but this adds noise to the detected signal, because the electrical connections of the antenna weakly radiate at this modulation frequency, and this radiation is not filtered by the lock-in amplifier. To reduce the experimental noise, the DC voltage supply that biases the antenna, must be as constant as possible. Because the power consumption of photo-conducting switches is low, a battery is well adapted to bias the switch. In most of such setups, the THz radiation goes through the antenna substrate. However, the switch can be considered as a dipole located at the surface of the antenna substrate and oriented parallel to the surface. Thus, it radiates towards both the substrate and the upper medium (Fig. 4.16a). The THz field in the substrate is approximately equal to the one in the upper medium times the ratio of the refractive indices at THz frequencies [20, 21]:

$$\frac{E_{\mathrm{THz}}\,(\text{substrate})}{E_{\mathrm{THz}}\,(\text{upper medium})} \approx \frac{n_{\text{substrate}}}{n_{\text{upper medium}}}. \qquad (4.52)$$

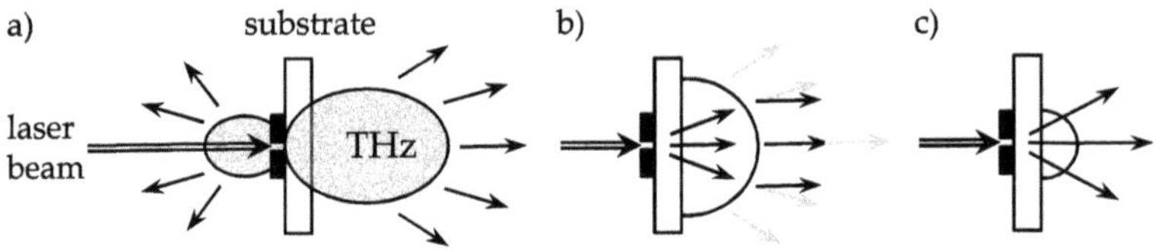

Figure 4.16 (a) Scheme of a photo-conducting switch antenna. The two black rectangles represent the biasing electrodes. THz radiation emerges from the gap in between. (b) Same antenna with a hyper-hemispherical lens. The black arrows sketch the higher frequency THz beam, while dashed grey ones are related to the lower frequency beam. (c) Same antenna with a small hemispherical lens.

When employing semiconductors like GaAs or Si, the THz field propagating through the substrate is ~3.4 times stronger than the one propagating in air (or vacuum). This is why it is preferable to deal with this field. Moreover, as the substrate is opaque to the laser light, the THz beam is free of any laser photon that may perturb the sample or the receiver. However, this scheme presents some inconvenient issues: (i) There is still small dispersion and attenuation in the substrate that truncate the highest THz frequencies. Therefore, when looking a very large THz bandwidth, it is preferable to work with the THz field radiated to the air (reflection scheme) [22]. (ii) The THz beam propagating into the substrate encounters a reflection at the opposite side of the substrate and consequently Fabry–Perot oscillations in the substrate. To avoid or at least to minimize this effect, a hemispherical lens is attached against the substrate, if possible with the same refractive index at THz frequencies. Fabry–Perot oscillations are less pronounced, because any reflection at the spherical surface is defocused. Moreover, because of the bigger thickness of the substrate-prism assembly, the temporal period of the Fabry–Perot rebounds is larger, and these rebounds can be removed from the detected signal by a proper time-windowing. In addition, oblique rays from the switch that were totally reflected at the semiconductor–air interface, are now transmitted into the lens, and by geometry, they do not suffer a total reflection at the lens surface.

First designs of the switch–lens antenna device included quite big hemispherical lenses [23, 24]. The idea was to use the lens as a focusing element, which delivers a parallel THz beam if the gap is located at the focus of the lens. For a spherical interface between a medium of refractive index n and air, the focal point is located at a distance h from the interface:

$$h = R\frac{n}{n-1}, \tag{4.53}$$

in which R is the radius of the spherical interface. In the case of a hemispherical lens pressed against the substrate of thickness d, the lens must be cut at a thickness $R + \Delta$, with

$$\Delta = \frac{R}{n-1} - d, \tag{4.54}$$

For a $d = 500$ μm thick wafer, a $R = 0.5$ cm lens, and with $n = 3.4$, we obtain $\Delta = 1.58$ mm. In this case, the switch is exactly at the focus of the lens. This geometry works rather well for higher frequencies and does not require any additional optics if both emitting and receiving are face to face. At lower frequencies, diffraction cancels the focusing effect of the lens, and the beam is not parallel [24]. For example, a simple diffraction model shows that the beam radiated from the lens is almost Gaussian, with a waist given by $W_o \approx R/\sqrt{2}$ [24]. Therefore, the beam divergence is

$$\phi \approx \frac{c}{\pi f\, W_o} \approx \sqrt{2}\frac{c}{\pi f\, R} \approx \frac{0.135}{f_{\mathrm{THz}}\, R_{\mathrm{mm}}}. \tag{4.55}$$

For a typical commercial Si hemispherical lens, $R = 0.6$ mm, and $\phi \approx 1.3^\circ$ at 1 THz, but 12.9° at 100 GHz.

So, it is better to use a smaller lens in such a way that the center of the lens is at the gap location. The lens plays no more a focusing role, and the emitting device appears still as a point source, but with a more powerful output and fewer interference effects in the substrate. On the other hand, making the THz beam parallel requires additional optics, mostly one or two parabolic mirrors. A setup with four identical parabolic mirrors is shown on Fig. 4.17. In between the first (or last) two mirrors, the beam is parallel, while in between the central mirrors, the beam is focused. Typically, with commercial parabolic mirrors of $F = 100$ mm focal length and 2″ of diameter (51 mm), (i.e., the F number ≈ 2) the THz beam, when it is parallel, is about 50 mm of diameter D (waist ≈ 25 mm). In the focusing region,

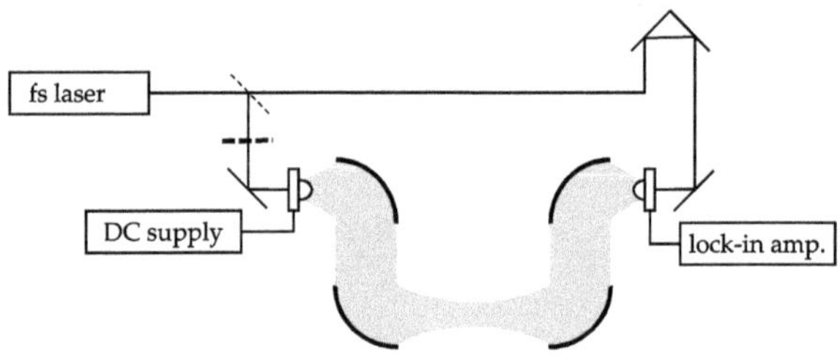

Figure 4.17 THz-TDS setup with photo-conducting (PC) antennas and four parabolic mirrors. In between the first or last two mirrors, the THz beam is parallel. At the center of the THz path, the beam is focused.

a Gaussian calculation gives the beam waist W_o at the focus as well as its Rayleigh length Z_R:

$$2W_o \approx \frac{4c}{\pi f}\frac{F}{D} \approx \frac{4 \times 10^{-4}}{f_{\mathrm{THz}}}\frac{F}{D}, \qquad Z_R \approx \frac{\pi W_o^2 f}{c} \approx \frac{10^{-4}}{f_{\mathrm{THz}}}\left(4\frac{F}{D}\right)^2. \tag{4.56}$$

For $F = 100$ mm and $D = 50$ mm, $2W_o \approx 0.8/f_{\mathrm{THz}}$ mm ($Z_R \approx 6.4/f_{\mathrm{THz}}$ mm). Thus at 100 GHz, $2W_o \approx 8$ mm ($Z_R \approx 64$ mm), while it is 0.8 mm at 1 THz ($Z_R \approx 6.4$ mm). Let us notice that any focused light beam encounters a phase variation of π at the focus, known as the Gouy shift [25, 26]. This is also the case in the THz-TDS setup shown in Fig. 4.17. When performing THz spectroscopy, one must take into account the Gouy shift (see Chapter 12).

At the receiver, both THz and optical beams excite the photo-conducting switch that supplies a current proportional to both the THz field and the laser intensity (Eq. 4.29). The mean value of this current is very low, i.e., in the picoampere to nanoampere range, and it must be read with a lock-in amplifier that is locked to the chopper frequency. We will crudely estimate the detected current, in the case of photo-conducting antennas (the same in both emission and detection). The emitting antenna acts as a varying dipole $p(t)$, whose total power radiated in a half space $P_{\mathrm{THz}}(t)$ is

$$P_{\mathrm{THz}}(t) = \frac{\mu_o}{12\pi c}\left(\frac{d^2p}{dt^2}\right)^2 = \frac{\mu_o}{12\pi c}d^2\left(\frac{di}{dt}\right)^2. \tag{4.57}$$

d is the gap width and $i(t)$ is the photocurrent flowing in the switch. We assume here that the carrier generation is almost instantaneous (short laser pulse duration) and that the carrier population decreases exponentially. The current can be deduced from the number $N(t)$ of photocarriers:

$$N(t) \approx \eta P_{\text{o}} \tau_{\text{laser}} \frac{\lambda_{\text{laser}}}{hc} e^{-\frac{t}{\tau_{\text{current}}}} = \eta \frac{P_{\text{laser}}}{f_{\text{rep}}} \frac{\lambda_{\text{laser}}}{hc} e^{-\frac{t}{\tau_{\text{current}}}},$$

$$\Rightarrow \quad i(t) = \frac{N(t)\, qv}{d} = \frac{N(t)\, q V_{\text{bias}}}{d^2} \mu, \tag{4.58}$$

$$P_{\text{THz}}(t) = \frac{\mu_{\text{o}}}{12\pi c} d^2 \left(\frac{di}{dt}\right)^2 = \frac{\mu_{\text{o}}}{12\pi c} \left(\frac{q V_{\text{bias}} \mu}{d} \frac{dN(t)}{dt}\right)^2. \tag{4.59}$$

P_{laser} is the mean value of the laser power, which is approximated by $P_{\text{laser}} = f_{\text{rep}} \int_{1/f_{\text{rep}}} P_{\text{laser}}(t)\, dt \approx f_{\text{rep}} P_{\text{o}} \tau_{\text{laser}}$, f_{rep} is the laser pulse repetition frequency, and P_{o} is the laser pulse peak power. μ is the photocarrier mobility, η is the photoelectric conversion efficiency, and τ_{current} is the decay time of the photo-induced current. From these relations (4.57–4.59), one deduces the expression of $P_{\text{THz}}(t)$. Assuming all this power is collected by the optics and focused at the detector at the limit of diffraction (spot size $S_{\text{THz}} \approx \lambda_{\text{THz}}^2$), we obtain the THz field $E_{\text{THz}}(t)$ at the detector:

$$E_{\text{THz}}(t) = 2\mu_{\text{o}} c \sqrt{\frac{P_{\text{THz}}(t)}{S_{\text{THz}}}}. \tag{4.60}$$

Finally, the detected current is

$$i_{\text{THz}}(t) = \frac{N(t)\, q\mu}{d} E_{\text{THz}}(t). \tag{4.61}$$

Typically, for LT-GaAs antennas, the gap width of the switch is $d \approx 10$ μm, and the bias voltage is ~10 V. We assume a 2000 $\text{cm}^2/\text{V/s}$ carrier mobility [27] and a 1 ps current decay time. Such antennas are commonly excited by a Ti-S laser ($\lambda = 0.8$ μm) delivering 100 fs pulses at a 100 MHz repetition rate ($P_{\text{laser}} = 10$ mW at the antennas). With $S_{\text{THz}} = 1\ \text{mm}^2$ and $\eta = 1$ we obtain $i_{\text{THz,peak}} \approx 64$ mA, $\overline{i_{\text{THz}}} \approx 32$ μA, $P_{\text{THz,peak}} \approx 1$ mW, and $\overline{P_{\text{THz}}} \approx 50$ nW. The mean values have been evaluated with the $\frac{1}{2}$ factor due to the effect of the chopper that blocks the laser beam for half the time. Actually, one must take into account additional losses in the setup: only ~50% of the THz beam is transmitted at the GaAs–air interface, ~70% of the laser intensity enters the LT-GaAs material, ~30% of the THz beam is well

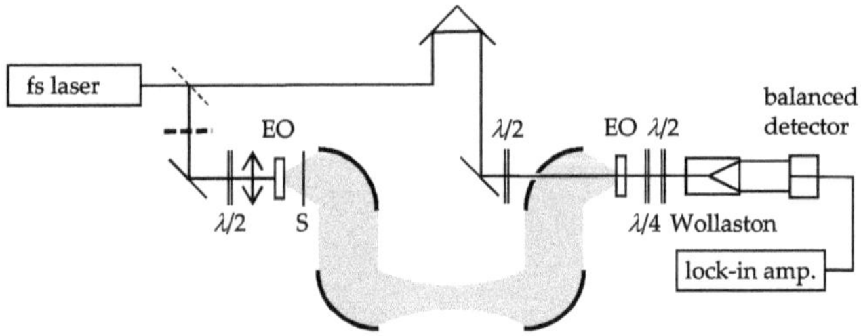

Figure 4.18 Typical electro-optic THz-TDS setup. EO is for electro-optic crystal, S is a filter that blocks laser light, and $\lambda/2$ and $\lambda/4$ are respectively the half- and quarter-wave plates.

collected by the emitting or receiving optics (taking into account total reflection in the lens). Note that these values must be squared to account for both emitter and receiver. Also, if the impedance of the CPW line of the antennas is 50 Ω while the lock-in is 1 kΩ, another factor 20 is lost. Finally, the preceding values must be divided by about 1000, leading to

$$i_{\text{THz,peak}} \approx 50\ \mu\text{A}, \quad \overline{i_{\text{THz}}} = 2.5\ \text{nA},$$
$$P_{\text{THz,peak}} = 1\ \mu\text{W}, \ \overline{P_{\text{THz}}} = 50\ \text{pW}.$$

This is the order of magnitude of what is experimentally achieved.

4.6.2 Common Systems Based on Electro-optic Crystals

Photo-conducting antennas can be substituted by electro-optic antennas, both in emission and in detection [28]. Both systems exhibit comparable performance. A typical electro-optic THz-TDS setup is presented in Fig. 4.18.

The overall scheme is very similar to the photo-conducting setup shown in Fig. 4.17. THz waves are generated by optical rectification in the first electro-optic crystal. With non-amplified lasers, it is compulsory to focus the laser beam onto the crystal, thanks to a lens. Also, a half wave-plate makes easier the adjustment of the direction of the laser polarization regarding the orientation of the crystal. After the crystal, it is preferable to block the laser beam in order to avoid any illumination of a sample located in the THz beam. This is done by inserting a slab of material that is transparent to the THz waves and opaque to the visible or infrared

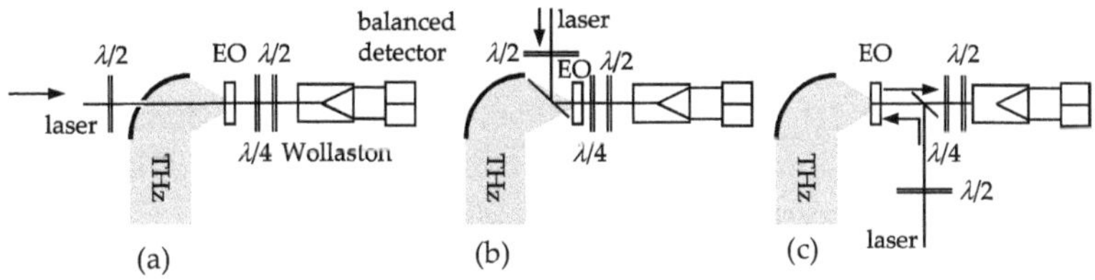

Figure 4.19 Electro-optic detection schemes for THz-TDS setup.

laser beam. Black polyethylene is a relevant material, which in addition shows a low index in the THz domain, minimizing Fresnel reflections, and whose thickness is quite small. One can employ other organic materials like Teflon, in which the laser light is stopped by scattering. High-resistivity (HR) silicon is also transparent in the THz range and absorbing visible and NIR wavelengths. However, if employing a powerful laser, free carrier photo-excitation occurs in the semiconductor, resulting in absorption of the THz waves.

The electro-optic detector is illuminated simultaneously by the THz beam and the laser probe beam. Two configurations are possible: the two beams are co- or contra-propagating. In the first scheme that corresponds to the one shown in Fig. 4.19, the propagation of the beams must be superimposed before the electro-optic crystal. It could be done with a dichroic beam splitter located at 45° in the THz beam (Fig. 4.19b). This splitter is transparent at the THz frequencies, while it reflects more or less totally the optical beam. Pellicles made of micrometer-thick organic films are widely used. A second solution (Fig. 4.19a) is to transmit the optical beam through a small hole at the center of the last parabolic mirror. The part of the THz beam that is not reflected because of the hole is negligible, since the hole is typically of 1 mm diameter, while the THz beam waist at the mirror is of the order of a few centimeters. In the last scheme (Fig. 4.19c), the THz beam and the optical probe beam enter the opposite faces of the electro-optic crystal. The optical beam is reflected backwards towards the polarimetric detector [29]. Finally, the polarization of the probe beam, which is elliptical after crossing the electro-optic crystal, is made linear by a quarter wave-plate, and this linear polarization is oriented at 45° to the axis a polarizing prism (here a Wollaston prism; see

Section 4.2.2 and Fig. 4.5). The two beams emerging from the prism are perpendicularly polarized and impinge on a balanced device with two photodetectors which delivers a signal proportional to the difference of the intensities of these two beams. When no THz signal illuminates the crystal, the signal is null.

4.6.3 ASOPS Systems

The delay line that is necessary to record the time-equivalent THz waveform can be replaced by a system of two synchronized mode-locked lasers whose repetition rates are different [30]. This technique, named asynchronous optical sampling (ASOPS), is derived from optical pump-and-probe spectroscopy [31]. A first laser with a repetition period T_1 excites the emitter. A second one, of repetition period $T_2 = T_1 + \Delta T$, triggers the receiver.

Both are synchronized at time $t = 0$, i.e., the THz emitted pulse and the pulse of the second laser beam are incident on the receiver at the same instant (see Fig. 4.20). The beating period T_{beat}, i.e., the time delay between the first and second coincidence of THz and optical pulses, is simply given by

$$T_{\text{beat}} = n\,T_1 = (n-1)\,T_2 = \frac{T_1\,T_2}{\Delta T}, \quad \text{with } n = \frac{T_2}{\Delta T}. \tag{4.62}$$

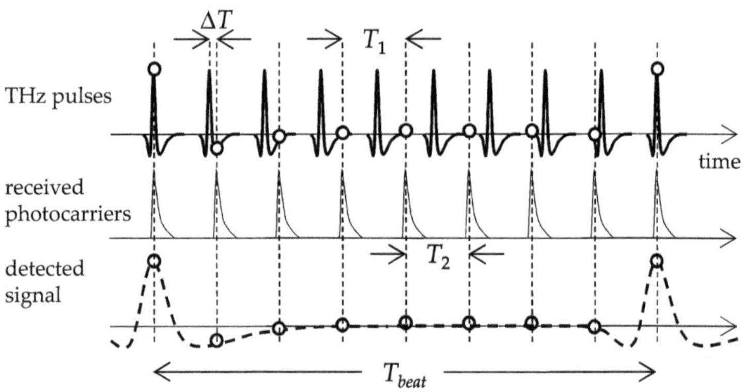

Figure 4.20 ASOPS waveforms. Upper trace: THz pulses at the receiver. Central trace: Number of photocarriers in the receiver. Lower trace: Current delivered by the receiver. The open circles represent the temporal data selected in the THz trace, and the corresponding current data.

As sketched in Fig. 4.20, the THz pulses are generated with a repetition period T_1 and they are sampled over an equivalent repetition period T_{beat}. Thus, the ASOPS technique performs a lengthening of the time scale. The time extension is

$$G = \frac{T_{\text{beat}}}{T_1} = \frac{T_2}{\Delta T} = \frac{T_1 + \Delta T}{\Delta T} \tag{4.63}$$

In the frequency domain, the spectral resolution is inversely proportional to the recording time window (A.120). Here (upper trace of Fig. 4.20), the temporal time data are recorded over an equivalent time window equal to T_{beat}, i.e., corresponding to an actual T_1 window. Therefore, the frequency resolution is

$$\delta f = \frac{1}{T_1} = f_1, \tag{4.64}$$

where f_1 is the repetition frequency of the first laser that excites the THz emitting antenna. On the same trace, one sees that the temporal data are recorded with an equivalent ΔT time shift. The accessible frequency range is therefore (A.117)

$$f_{\text{max}} = \frac{1}{2\,\Delta T} = \frac{f_1 f_2}{2\,(f_1 - f_2)} = \frac{f_1 f_2}{2\,f_{\text{beat}}}. \tag{4.65}$$

The main advantages of ASOPS are that (i) no mechanical delay line is required, so that the recording duration to obtain a whole spectrum is much faster than with classical THz-TDS systems, and (ii) its frequency resolution is excellent. The main drawbacks are that two mode-locked lasers are required, which makes the system quite expensive, and the need for perfect synchronism and stability of the pulse trains delivered by the two lasers. If one of the lasers could operate with a free (passive) mode-locking, the second one must be locked to the first, usually thanks to an active mode-lock operation. This requires good feedback electronics, which can be coupled to some atomic frequency standard, like a rubidium clock, when dealing with high-resolution spectroscopy [32]. Also, the difference $f_{\text{beat}} = f_1 - f_2$ of the repetition frequencies of the two lasers should be tunable (see below). The repetition frequency of a mode-locked laser is $c/2L$, where L is the optical length of its cavity. Changing the repetition frequency requires to vary L, which is achieved by moving a mirror of the laser cavity with a piezoelectric actuator.

Typically, the repetition frequency of common femtosecond mode-locked lasers is of the order of 100 MHz, but dedicated models could reach 1 GHz [33]. The beating frequency is of the order of 1 kHz (it could be decreased down to 5 Hz for high-resolution experiments [32]). Hence, typical temporal magnification is $G \approx f/f_{\text{beat}} \approx 10^5 \sim 10^6$. Thus, a record over 100 ps is spread over an equivalent time of $0.1 \sim 1$ ms. The corresponding trace can easily be viewed and recorded with a fast oscilloscope.

The SNR of the signal is proportional to collecting time window, i.e., to the period T_{beat} (4.62) of the ASOPS signal. It is easy to show, from Eq. 4.62, that $T_{\text{beat}} = 1/f_{\text{beat}}$. In addition, the accessible frequency range (Eq. 4.65) varies also as $T_{\text{beat}} = 1/f_{\text{beat}}$. Thus, it is desirable to increase f_{beat} in view of higher ASOPS performance. But, on the other hand, the temporal resolution increases with f_{beat} ($\Delta T = f_{\text{beat}}/f^2$). A good compromise [34] is $f_{\text{beat}} < \Delta T f F$, where ΔT is the desired temporal resolution and F is the bandwidth of the detection system.

4.6.4 "Poor Man's" (Cost-Effective) Systems

THz-TDS is based on the fast temporal variation of a dipole, either in a photo-conducting or an electro-optical antenna, which leads to a strong far-field EM radiation with a spectrum corresponding to the Fourier transform of the time derivative of the dipolar current. Detection is based on the correlation between this THz field and an optical signal similar to the one that excites the emitter. From these considerations, we were able to write down the time dependence of the detected waveform (4.29):

$$S(\tau) \propto \int_{-\infty}^{+\infty}\int_{-\infty}^{+\infty} E_{\text{THz}}(t-\tau)\, I_{\text{laser}}(t-t')\, G(t')\, dt'dt, \qquad (4.66)$$

as well as its spectrum:

$$S(\omega) \propto E^*_{\text{THz}}(\omega)\, I_{\text{laser}}(\omega)\, G^*(\omega). \qquad (4.67)$$

G is the response of the receiver.

In Eqs. 4.66 and 4.67, the laser intensity should exhibit temporal variations at the picosecond scale, i.e., a spectrum of THz bandwidth. As seen previously, this is nicely achieved with picosecond pulses of

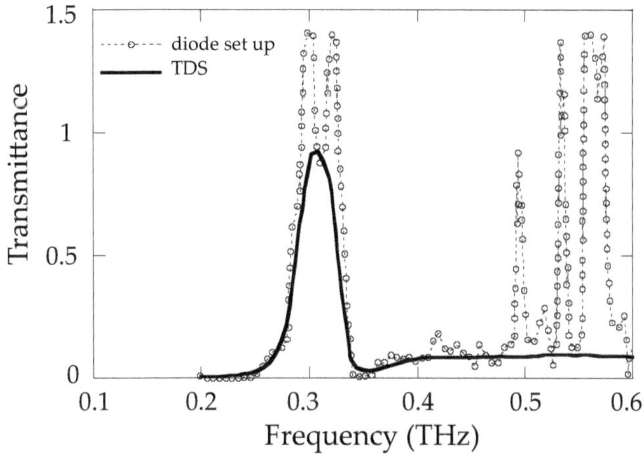

Figure 4.21 THz transmittance of a metal hole array obtained with a cost-effective system based on multimode laser diodes (dots). The large narrow peaks are due to the mixing of the laser modes. For comparison, the curve (continuous line) obtained with a classical THz-TDS setup is plotted. Adapted from Morikawa et al. [37].

mode-locked lasers. But pulses are not compulsory. When not mode-locked, multi-mode lasers show a chaotic output, resulting from the incoherent summation of all the modes:

$$E_{\text{laser}}(t) = \sum_{m} A_{\text{m}} e^{j(\omega_{\text{m}} t + \phi_{\text{m}})}, \quad \omega_{\text{m}} = m\frac{\pi c}{L}. \tag{4.68}$$

L is the length of the laser cavity and A_{m} and ϕ_{m} are respectively the amplitude and phase of the longitudinal mode m. Laser diodes running in a CW multimode regime are well suited to generating and detecting THz waves. This has been proposed and demonstrated by the team of M. Hangyo [35], who used LT-GaAs photo-conducting antennas. Typically, such diodes deliver tens of mW and the spectrum reaches a few hundreds of GHz. However, the spectrum displays strong lines due to the mixing of the laser modes m, whose number is typically $10 \sim 15$. Nevertheless, this cheap system permits one to perform sub-THz spectroscopy [36]. The shape of the spectrum can also be made smoother by a proper filtering of the laser diode beam [37].

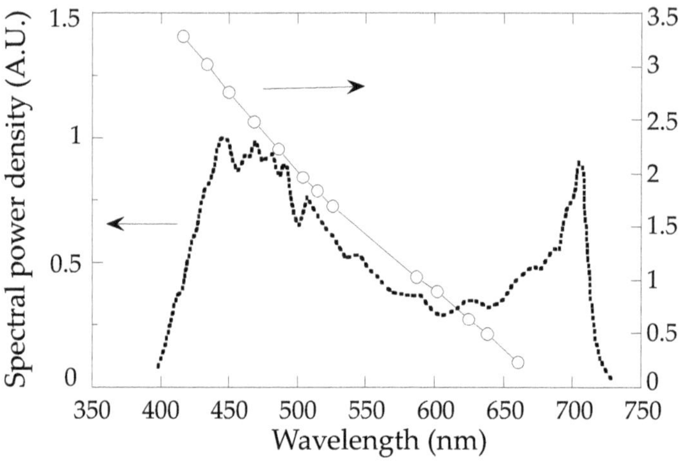

Figure 4.22 Supercontinuum spectral power density (dotted line, left vertical axis) and relative time at which the spectral component is observed (open circles, left vertical axis) versus wavelength. Adapted from Schmidhammer et al. [43].

4.6.5 Single-Shot THz-TDS Systems

In a chirped laser pulse, the different spectral components are time-shifted; the time shift varies monotonously with frequency. It is obtained from a regular laser pulse that propagates in a dispersive setup, or from the pulse generation process itself in some lasers like DBR ones. Thus, there is an unambiguous relation between time and optical frequency, and recording the spectrum of such pulses is equivalent to address the temporal shape of the pulses (see Fig. 4.22 extracted from [43]). The procedure is even simpler if the chirp is linear, i.e., the frequency varies linearly with time along the pulse duration. This has been initially exploited to obtain the temporal shape of sub-nanosecond electrical transients [38, 39]. Then it was applied to recording the temporal shape of THz waveforms in a single-shot procedure [40–43]. The time-equivalent resolution is $\Delta t = \sqrt{T_{\text{las}} T_{\text{ch}}}$, where T_{ch} is the duration of the chirped pulse and T_{las} is the duration of the laser pulse [44]. The duration of the chirped pulse T_{ch} should be at least equal to that T_{THz} of the THz pulse to be measured. Therefore, $\Delta t = \sqrt{T_{\text{las}} T_{\text{THz}}}$. As compared to common

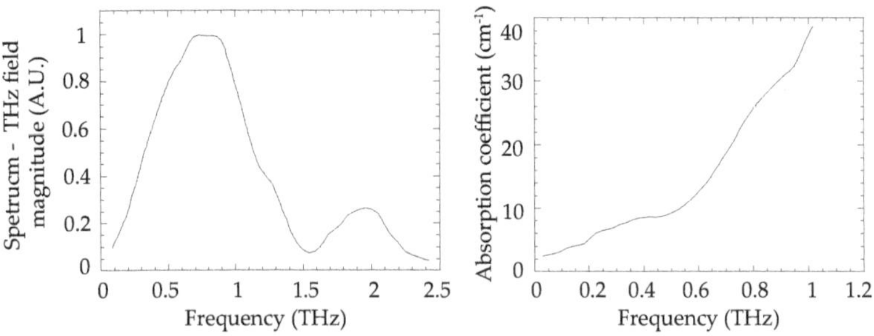

Figure 4.23 Left: Spectrum delivered by the single-shot THz-TDS. Right: Absorption coefficient spectrum of a glass sample determined from a single shot acquisition. Adapted from Jeunesse and Schmidhammer [45].

THz-TDS systems that include a delay line, the temporal resolution decreases by a factor $\sqrt{T_{\mathrm{THz}}/T_{\mathrm{las}}}$ [44]. For example, with a 50 fs laser pulse and a THz waveform that spreads over 20 ps, the single-shot THz-TDS system exhibits a 1 ps time resolution, i.e., about 20 times worse than with classical setups.

More recently, it was applied to THz-TDS by P. Jeunesse and U. Schmidhammer [45]. A broadband supercontinuum is generated by focusing a powerful laser pulse, delivered by an amplified system, in a sapphire crystal. The Stokes side of this supercontinuum, whose chirp is controlled through propagation in a glass slab, serves as probe for the THz signal. The THz pulse is emitted through optical rectification in a ZnTe crystal. The detection part includes another ZnTe crystal and a polarizing beam splitter that separates the two components of the transmitted probe beam and directs them toward two spectrometers. Figure 4.23 (left) shows a spectrum obtained with this single-shot system, which spreads from 0.1 to 2.5 THz. The absorption coefficient of a glass sample, determined form the single-shot measurement, is plotted on Fig. 4.23 (right). Compared to common THz-TDS setups, the *SNR* is worse (in [45], the reported value is $SNR = 40$ for a single-shot acquisition) and the frequency resolution is poorer. Also, an amplified femtosecond laser is requested to produce the super-continuum. But the great advantage of this system is the extremely fast recording length, which permits one to perform a THz characterization of even moving samples [45].

References

1. van der Weide, D. W., Bostak, J. S., Auld, B. A., and Bloom, D. M. (1993). All-electronic generation of 880 fs, 3.5 V shockwaves and their application to a 3 THz free-space signal generation system, *Appl. Phys. Lett.*, **62**, pp. 22–24.
2. Bostak, S., van der Weide, D. W., Bloom, D. M., Auld, B. A., and Özbay, E. (1994). All-electronic terahertz spectroscopy system with terahertz free-space pulses, *J. Opt. Soc. Am. B*, **11**, pp. 2561–2565.
3. Jackson, J. D. (1962). *Classical Electrodynamics*, John Wiley and Sons, New York.
4. Duvillaret, L., Garet, F., Roux, J.-F., and Coutaz, J.-L. (2001). Analytical modeling and optimization of terahertz time-domain spectroscopy experiments using photoswitches as antennas, *IEEE J. Sel. Top. Quantum Electron.*, **7**, pp. 615–623.
5. Mourou, G., Stancampiano, C. V., Antonetti, A., and Orszag, A. (1981). Picosecond microwave pulses generated with a subpicosecond laser-driven semiconductor switch, *Appl. Phys. Lett.*, **39**, pp. 295—296.
6. Smith, P. R., Auston, D. H., and Nuss, M. C. (1988). Subpicosecond photoconducting dipole antennas, *IEEE J. Quantum Electron.*, **24**, pp. 255—260.
7. Katzenellenbogen, N., and Grischkowsky, D. (1991). Efficient generation of 380 fs pulses of THz radiation by ultrafast laser pulse excitation of a biased metal-semiconductor interface, *Appl. Phys. Lett.*, **58**, pp. 222–224.
8. Cai, Y., Brener, I., Lopata, J., Wynn, J., Pfeiffer, L., and Federici, J. (1997). Design and performance of singular electric field terahertz photoconducting antennas, *Appl. Phys. Lett.*, **71**, pp. 2076–2078.
9. Berry, C. W., Wang, N., Hashemi, M. R., Unlu, M., and Jarrahi, M. (2013). Significant performance enhancement in photoconductive terahertz optoelectronics by incorporating plasmonic contact electrodes, *Nat. Commun.*, **4**, p. 1622.
10. Krotkus, A., and Coutaz, J.-L. (2005). Non-stoichiometric semiconductor materials for terahertz optoelectronics applications, *Semicond. Sci. Technol.*, **20**, p. S142.
11. Krotkus, A. (2010). Semiconductors for terahertz photonics applications, *J. Phys. D*, **43**, p. 273001.
12. Sartorius, B., Roehle, H., Künzel, H., Böttcher, J., Schlak, M., Stanze, D., Venghaus, H., and Schell, M. (2009). All-fiber terahertz time-domain

spectrometer operating at 1.5 μm telecom wavelengths, *Opt. Express*, **16**, pp. 9565–9570.

13. Schneider, A., Neis, M., Stillhart, M., Ruiz, B., Khan, R. U. A., and Günter, P. (2006). Generation of terahertz pulses through optical rectification in organic DAST crystals: theory and experiment, *J. Opt. Soc. Am. B*, **23**, pp. 1822–1835.
14. Matsubara, E., Nagai, M., and Ashida, M. (2013). Coherent infrared spectroscopy system from terahertz to near infrared using air plasma produced by 10-fs pulses, *J. Opt. Soc. Am. B*, **30**, pp. 1627–1630.
15. Huang, S. W., Granados, E., Huang, W. R., Hong, K. H., Zapata, L., and Kärtner, F. X. (2013). High conversion efficiency, high energy terahertz pulses by optical rectification, *Opt. Lett.*, **38**, pp. 796–798.
16. Rice, A., Jin, Y., Ma, X. F., Zhang, X.-C., Bliss, D., Larkin, J., and Alexander, M. (1994). Terahertz optical rectification from (110) zincblende crystals, *Appl. Phys. Lett.*, **64**, pp. 1324–1326.
17. Gallot, G., and Grischkowsky, D. (1999). Electro-optic detection of terahertz radiation, *J. Opt. Soc. Am. A*, **16**, pp. 1204–1212.
18. Jiang, Z., and Zhang, X.-C. (1998). Electro-optic measurement of THz field pulses with a chirped optical beam, *Appl. Phys. Lett.*, **72**, pp. 1995–1997.
19. Xu, J., and Zhang, X.-C. (2004). Circular involute stage, *Opt. Lett.*, **29**, pp. 2082–2084.
20. Lukosz, W., and Kunz, R. E. (1977). Light emission by magnetic and electric dipoles close to a plane interface. I. Total radiated power, *J. Opt. Soc. Am.*, **67**, pp. 1607–1615.
21. Lukosz, W., and Kunz, R. E. (1977). Light emission by magnetic and electric dipoles close to a plane interface. II. Radiation patterns of perpendicular oriented dipoles, *J. Opt. Soc. Am.*, **67**, pp. 1615–1619.
22. Shen, Y. C., Upadhya, P. C., Beere, H. E., Linfield, E. H., Davies, A. G., Gregory, I. S., Baker, C., Tribe, W. R., and Evans, M. J. (2004). Generation and detection of ultrabroadband terahertz radiation using photoconductive emitters and receivers, *Appl. Phys. Lett.*, **85**, pp. 164–166.
23. Fattinger, Ch., and Grischkowski, D. (1990). Terahertz beams, *Appl. Phys. Lett.*, **54**, pp. 490–492.
24. Jepsen, P. U., Jacobsen, R. H., and Keiding, S. R. (1996). Generation and detection of terahertz pulses from biased semiconductor antennas, *J. Opt. Soc. Am. B*, **13**, pp. 2424–2436.

25. Gouy, G. (1890). Optique: sur une propriété nouvelle des ondes lumineuses (Optics: on a new property of light waves), *Comptes rendus* (Académie des sciences), **CX24**, pp. 1251–1253.
26. See for example: Saleh, B. E. A., and Teich, M. C. (1991). Chapter 3 - Beam optics, in *Fundamentals of Photonics*, John Wiley and Sons, New York, pp. 80–107.
27. Němec, H., Pashkin, A., Kužel, P., Khazan, M., Schnüll, S., and Wilke, I. (2001). Carrier dynamics in low-temperature grown GaAs studied by terahertz emission spectroscopy, *J. Appl. Phys.*, **90**, p. 1303.
28. Nahata, A., Weling, A. S., and Heinz, T. F. (1996). A wideband coherent terahertz spectroscopy system using optical rectification and electro-optic sampling, *Appl. Phys. Lett.*, **69**, pp. 2321–2323.
29. Qiao, W., Stephan, D., Hasselbeck, M., Liang, Q., and Dekorsy, T. (2012). Low-temperature THz time domain waveguide spectrometer with butt-coupled emitter and detector crystal, *Opt. Express*, **20**, pp. 19769–19777.
30. Yasui, T., Saneyoshi, E., and Araki, T. (2005). Asynchronous optical sampling terahertz time-domain spectroscopy for ultrahigh spectral resolution and rapid data acquisition, *Appl. Phys. Lett.*, **87**, p. 061101.
31. Elzinga, P. A., Kneisler, R. J., Lytle, F. E., Jiang, Y., King, G. B., and Laurendeau, N. M. (1987). Pump/probe method for fast analysis of visible spectral signatures utilizing asynchronous optical sampling, *Appl. Opt.*, **26**, pp. 4303–4309.
32. Yasui, T., Kawamoto, K., Hsieh, Y.-D., Sakaguchi, Y., Jewariya, M., Inaba, H., Minoshima, K., Hindle, F., and Araki, T. (2012). Enhancement of spectral resolution and accuracy in asynchronous-optical-sampling terahertz time-domain spectroscopy for low-pressure gas-phase analysis, *Opt. Express*, **20**, pp. 15071–15078.
33. Krauß, N., Nast, A., Heinecke, D. C., Kölbl, C., Barros, H. G., and Dekorsy, T. (2015). Fiber-coupled high-speed asynchronous optical sampling with sub-50 fs time resolution, *Opt. Express*, **23**, pp. 2145–2156.
34. Takagi, Y., and Adachi, S. (1999). Subpicosecond optical sampling spectrometer using asynchronous tunable mode-locked lasers, *Rev. Sci. Instrum.*, **70**, pp. 2218–2224.
35. Tani, M., Matsuura, S., Sakai, K., and Hangyo, M. (1997). Multiple-frequency generation of sub-terahertz radiation by multimode LD excitation of photoconductive antenna, *IEEE Microwave Guided Wave Lett.*, **7**, pp. 282–284.

36. Morikawa, O., Tonouchi, M., and Hangyo, M. (2000). A cross-correlation spectroscopy in subterahertz region using an incoherent light source, *Appl. Phys. Lett.*, **76**, pp. 1519–1521.

37. Morikawa, O., Fujita, M., Takano, K., and Hangyo, M. (2011). Sub-terahertz spectroscopic system using a continuous-wave broad-area laser diode and a spatial filter, *J. Appl. Phys.*, **110**, p. 063107.

38. Valdmanis, J. A. (1986). In *Ultrafast Phenomena V*, Fleming, G. R., and Siegman, A. E. (eds.), Springer, New York, p. 82.

39. Galvanauskas, A., Tellefsen, Jr., J. A., Krotkus, A., Oberg, M., and Broberg, B. (1992). Real-time picosecond electro-optic oscilloscope technique using a tunable semiconductor laser, *Appl. Phys. Lett.*, **60**, pp. 145–147.

40. Jiang, Z., and Zhang, X.-C. (1998). Single-shot spatiotemporal terahertz field imaging, *Opt. Lett.*, **23**, pp. 1114–1116.

41. Jamison, S., Shen, J., MacLeod, A. M., Gillespie, W. A., and Jaroszynski, D. A. (2003). High-temporal-resolution, single-shot characterization of terahertz pulses, *Opt. Lett.*, **28**, pp. 1710–1712.

42. van Tilborg, J., Schroeder, C. B., Tóth, C., Geddes, C. G. R., Esarey, E., and Leemans, W. P. (2007). Single-shot spatiotemporal measurements of high-field terahertz pulses, *Opt. Lett.*, **32**, pp. 313–315.

43. Schmidhammer, U., DeWaele, V., Marquès, J.-R., Bourgeois, N., and Mostafavi, M. (2009). Single shot linear detection of 0.01–10 THz electromagnetic fields, *Appl. Phys. B*, **94**, pp. 95–101.

44. Sun, F. G., Jiang, Z., and Zhang, X. C. (1998). Analysis of terahertz pulse measurement with a chirped probe beam, *Appl. Phys. Lett.*, **73**, pp. 2233–2235.

45. Jeunesse, P., and Schmidhammer, U. (2013). "On-the-fly" monitoring with a single-shot terahertz time-domain spectrometer, *IEEE Sens. J.*, **13**, pp. 44–49.

Chapter 5

Extracting the Complex Refractive Index from Transmission TDS Data

5.1 Introduction

We have seen in the previous chapter that THz-TDS allows one to acquire the complex transmission $\tilde{t}_{\text{meas}}(\omega)$ or reflection $\tilde{r}_{\text{meas}}(\omega)$ coefficients of a sample over the whole experimental range. The next stage is to extract the complex refractive index $\tilde{n}(\omega) = n(\omega) - j\kappa(\omega)$ of the material of the sample, i.e., to solve what is called an inverse problem. $n(\omega)$ is the refractive index, while $\kappa(\omega)$ is the extinction coefficient of the material. The basic principle of any extraction method is to employ a model for the transmission or reflection coefficients, which is function of $\tilde{n}$, and then to find the right value of $\tilde{n}$ that makes equal the measured and calculated coefficients (up to now, for sake of simplicity, we omit to write the ω dependence of $\tilde{n}$, n and κ). Thus, one must solve one of these two equations:

$$\tilde{t}(n, \kappa, \omega) - \tilde{t}_{\text{meas}}(\omega) = t(n, \kappa, \omega)e^{j\varphi(n,\kappa,\omega)} - t_{\text{meas}}(\omega)e^{j\varphi_{\text{meas}}(\omega)} = 0, \tag{5.1}$$

$$\tilde{r}(n, \kappa, \omega) - \tilde{r}_{\text{meas}}(\omega) = r(n, \kappa, \omega)e^{j\varphi(n,\kappa,\omega)} - r_{\text{meas}}(\omega)e^{j\varphi_{\text{meas}}(\omega)} = 0. \tag{5.2}$$

Principles of Terahertz Time-Domain Spectroscopy
Jean-Louis Coutaz, Frédéric Garet, and Vincent Wallace

ISBN 978-981-4774-56-7 (Hardcover), 978-0-203-70969-6 (eBook)
www.panstanford.com

For flat and homogeneous samples, of thickness d, with parallel faces illuminated by a plane wave, the coefficients of transmission and of reflection at normal incidence are given by expressions A.35 of the Annex:

$$\tilde{t}(n,\kappa,\omega) = \frac{4\tilde{n}}{(\tilde{n}+1)^2} \times \frac{e^{-j\frac{\omega}{c}(\tilde{n}-1)d}}{1-\left(\frac{\tilde{n}-1}{\tilde{n}+1}e^{-j\frac{\omega}{c}\tilde{n}d}\right)^2}, \tag{5.3}$$

$$\tilde{r}(n,\kappa,\omega) = \frac{1-\tilde{n}}{1+\tilde{n}} \times \frac{1-e^{-2j\frac{\omega}{c}\tilde{n}d}}{1-\left(\frac{\tilde{n}-1}{\tilde{n}+1}e^{-j\frac{\omega}{c}\tilde{n}d}\right)^2}. \tag{5.4}$$

As the experimental and theoretical coefficients are complex, the problem consists in finding the zero of a complex function in the complex plane. Because of the exponential complex functions $\exp(-j\frac{\omega}{c}\tilde{n}d)$ in (5.3) and (5.4), $\tilde{t}(n,\kappa,\omega)$ and $\tilde{r}(n,\kappa,\omega)$ exhibit an oscillating behavior versus n, as seen in Fig. 5.1. Moreover, expressions 5.1 and 5.2 cannot be analytically inversed.

Therefore solving Eqs. 5.1 and 5.2 is a delicate task. Actually, one can start from a given value of (n, κ) and calculate the difference $\Delta\tilde{s} = \tilde{s} - \tilde{s}_{\text{meas}}$ ($\tilde{s} = \tilde{r}$ or $\tilde{s} = \tilde{t}$) by increasing step by step the values of n and κ. When a minimum of $\Delta\tilde{s}$ is found, one reduces the incremental values Δn or $\Delta\kappa$ and restarts the sweeping procedure until the desired precision is obtained. This procedure must be applied to each of the minima of $\Delta\tilde{s}$, which are multiple because of the oscillations explained above. Moreover, the calculation must be performed for each frequency of the spectrum. So it is quite long and time-consuming. Commercial software, like Mathematica or Mathlab, are also able to deliver the value of n and κ from $\Delta\tilde{s} = 0$, but the computation is time-consuming. Thus, if one wants a fast extraction, smart methods should be used.

We will describe below the main extraction techniques that have been proposed to invert Eq. 5.3, i.e., to treat the transmission THz-TDS case. THz-TDS reflection will be treated in the next chapter. Equation 5.3 is defined in the frequency domain. However, one can also directly retrieve the complex index of refraction from the waveforms, i.e., in the time domain, as will be described later.

The extraction algorithms not only deliver the spectrum of the complex refractive index, but can also give an accurate estimation

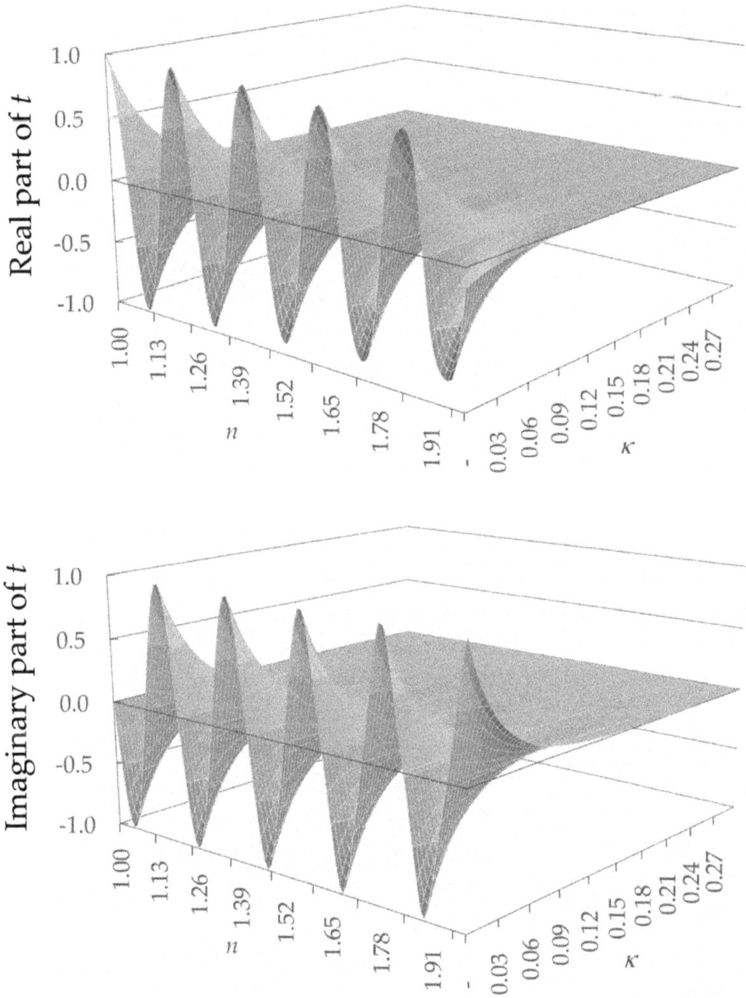

Figure 5.1 Real (top) and imaginary (bottom) parts of t versus n and κ. The maps are plotted for $\omega d/c = 30$ (e.g., a $d = 1.5$ mm thick slab characterized at 1 THz).

of the sample thickness or may reduce the uncertainty due to random errors. This will be addressed in Chapter 12, dedicated to the precision of THz-TDS characterization.

5.2 Method by Duvillaret, Garet, and Coutaz [1]

This method was the first to be published. The basic idea is to compute an approximated value n_a of n by neglecting the Fabry–Pérot term in Eq. 5.3. The corresponding κ_a value is found to obey an equation that is easy to solve. Then (n_a, κ_a) is used to calculate the Fabry–Pérot term in Eq. 5.3. Expression 5.3 is then divided by this Fabry–Pérot term, and thus it shows a much reduced oscillating behavior. Then, the problem of getting the solution of Eq. 5.1 is solved by defining an error function whose zero is obtained after a very few iterations. With modern computers, the determination of (n, κ) over the whole experimental THz range is almost instantaneous. Before describing the procedure in detail, let us explain how it was established. In Fig. 5.2 (top), we plot the phase of the transmission coefficient shown in Fig. 5.1.

The phase map appears almost flat because it is mostly determined by the propagation exponent $-\omega(n-1)d/c$ (the contribution to the phase of the Fresnel coefficients is weaker). This is clearly observed in Fig. 5.2 (bottom), where we plot the phase divided by $n-1$: it oscillates slightly around the value of $-\omega\, d/c$ (here -30). Thus, equalizing the measured phase and this propagation exponent gives a rather good estimation n_a of n. From n_a, the corresponding κ_a can be obtained by taking the modulus of Eq. 5.1 and solving the resulting equation. As the difference between (n, κ) and (n_a, κ_a) is rather weak, calculating the Fabry–Pérot term with (n_a, κ_a) and dividing $\tilde{t}(n, \kappa, \omega)$ by this approximated term attenuates strongly the oscillations.

Let us now describe the method in detail. According to what is explained above, the approximated value n_a is

$$n_a = 1 + \frac{\varphi_{\text{meas}}}{\omega\, c} d. \tag{5.5}$$

Do remember that, for the sake of simplicity, we omit to write that the involved values depend on ω. Then this value n_a is entered into the modulus of Eq. 5.1, in order to find κ_a (let us notice that we already made use of the phase of Eq. 5.1, which leads us to Eq. 5.5).

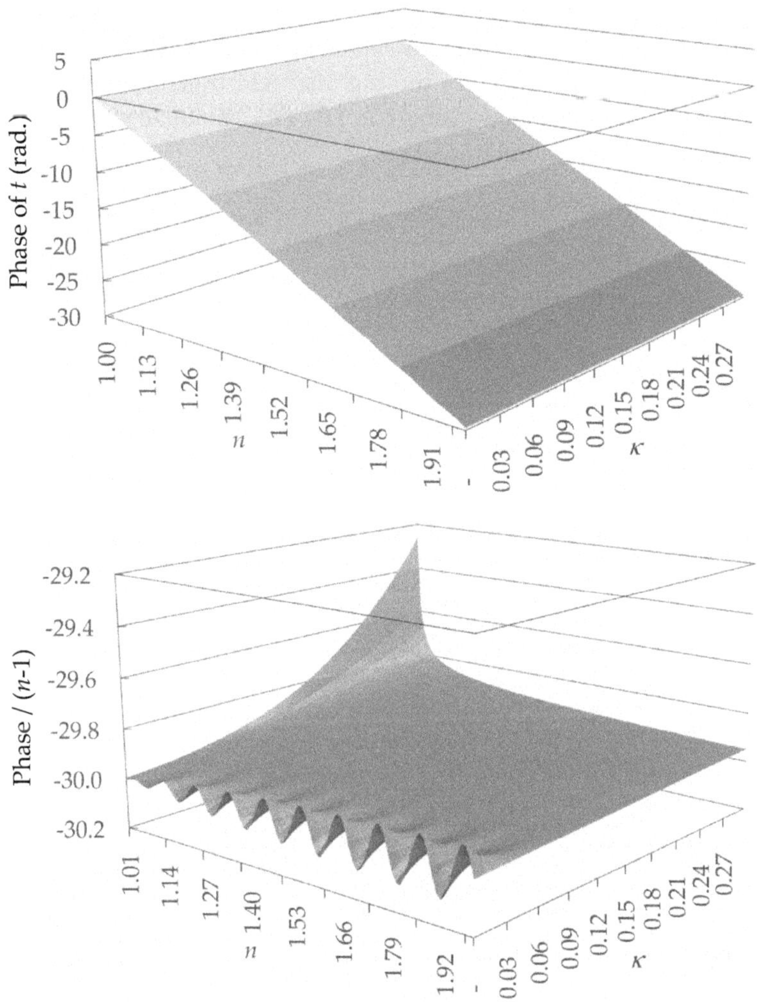

Figure 5.2 Modulus (top) and phase (bottom, normalized by $n - 1$) of t versus n and κ. As for Fig. 5.1, the maps are plotted for $\omega\, d/c = 30$.

We get

$$t_{\text{meas}}(\omega) = \frac{4\,|n_{\text{a}} - j\kappa_{\text{a}}|}{|n_{\text{a}} + 1 - j\kappa_{\text{a}}|^2} \times \frac{e^{-\frac{\omega}{c}\kappa_{\text{a}} d}}{\left|1 - \left(\frac{n_{\text{a}}-1-j\kappa_{\text{a}}}{n_{\text{a}}+1-j\kappa_{\text{a}}} e^{-j\frac{\omega}{c}n_{\text{a}}d} e^{-\frac{\omega}{c}\kappa_{\text{a}}d}\right)^2\right|}. \tag{5.6}$$

Then, we take a development into series of Eq. 5.6, assuming terms in κ_a can be neglected, except the exponential ones ($\exp(-\omega\kappa_a d/c)$). Indeed, if the κ_a values are too big, the field transmitted by the sample is too weak and transmission THz-TDS cannot be performed. Therefore, Eq. 5.6 simplifies into

$$e^{-\frac{\omega}{c}\kappa_a d} + \left(\frac{n_a - 1}{n_a + 1}\right)^2 \cos\left(2\frac{\omega}{c}n_a d\right) e^{-3\frac{\omega}{c}\kappa_a d} = \frac{(n_a + 1)^2}{4n_a} t_{meas}(\omega). \tag{5.7}$$

This equation may be written in a canonical form:

$$X^3 + BX + C = 0, \tag{5.8}$$

with

$$X = e^{-\frac{\omega}{c}\kappa_a d},$$
$$B = \frac{(n_a + 1)^2}{(n_a - 1)^2 \cos\left(2\frac{\omega}{c}n_a d\right)},$$
$$C = \frac{-(n_a + 1)^4\, t_{meas}(\omega)}{4n_a (n_a - 1)^2 \cos\left(2\frac{\omega}{c}n_a d\right)}. \tag{5.9}$$

This is a cubic equation, whose solution is obtained with the Cardano method:

$$\left(\frac{C}{2}\right)^2 + \left(\frac{B}{3}\right)^3 > 0 : X = A - \frac{B}{3A},$$
$$A = \sqrt[3]{\left|\frac{C}{2}\right| + \sqrt{\frac{C^2}{4} + \frac{B^3}{27}}}, \tag{5.10}$$

$$\left(\frac{C}{2}\right)^2 + \left(\frac{B}{3}\right)^3 < 0 : X = -2\sqrt{\frac{-B}{3}} \cos\left(m\frac{\pi}{3} + \frac{1}{3} a\cos\sqrt{-\frac{27C^2}{4B^3}}\right),$$
$$m = -1, 0, 1. \tag{5.11}$$

The values thus achieved (n_a, κ_a) are very close to the actual ones (n, κ). This is clearly observed in Fig. 5.3, where the relative difference between n and n_a, and between κ and κ_a, are plotted. n_a is equal to n within an error less than 1% over the whole range of realistic (n, κ) values. For κ, the difference is within a few percent, but increases with both κ and n.

The next stage is to evaluate more precisely (n, κ), starting from the approximate (n_a, κ_a) values. The principle of the method is to

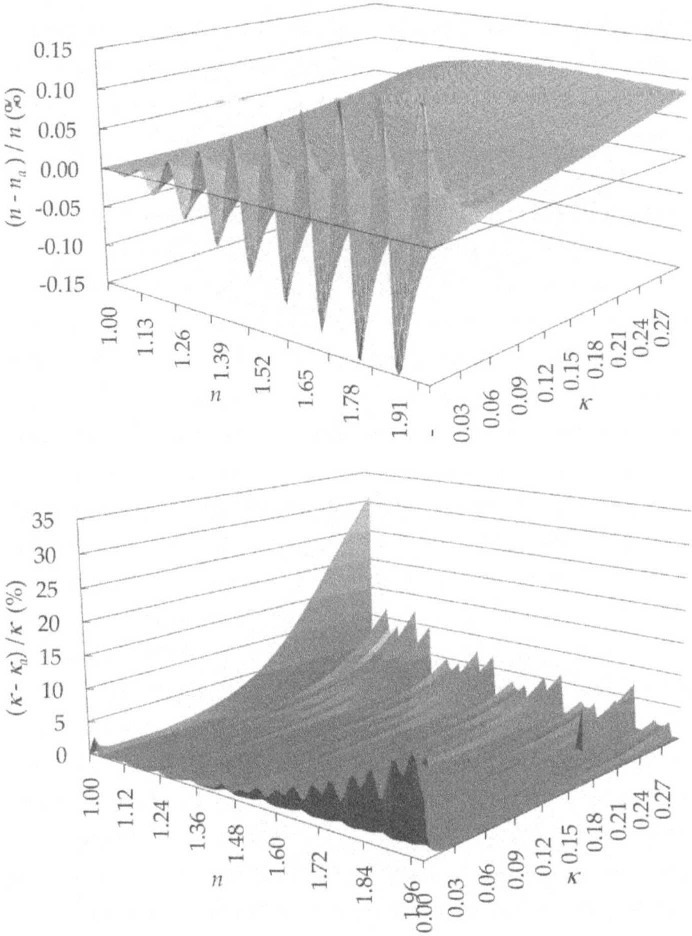

Figure 5.3 Difference between the actual value n and the approximated one n_a (top), and between κ and κ_a (bottom), vs. n and κ. Maps are plotted for $\omega d/c = 30$.

divide $\tilde{t}_{\text{meas}}(\omega)$ by the Fabry–Pérot term calculated with the (n_a, κ_a) values:

$$\tilde{t}'(\omega) = \frac{\tilde{t}_{\text{meas}}(\omega)}{1 - \left(\frac{\tilde{n}_a - 1}{\tilde{n}_a + 1} e^{-j\frac{\omega}{c}\tilde{n}_a d}\right)^2}. \tag{5.12}$$

As $(n_a, \kappa_a) \approx (n, \kappa)$, we almost get rid of the oscillations in $t(n, \kappa, \omega)$ due to the Fabry–Pérot term. Figure 5.4 shows the modulus and

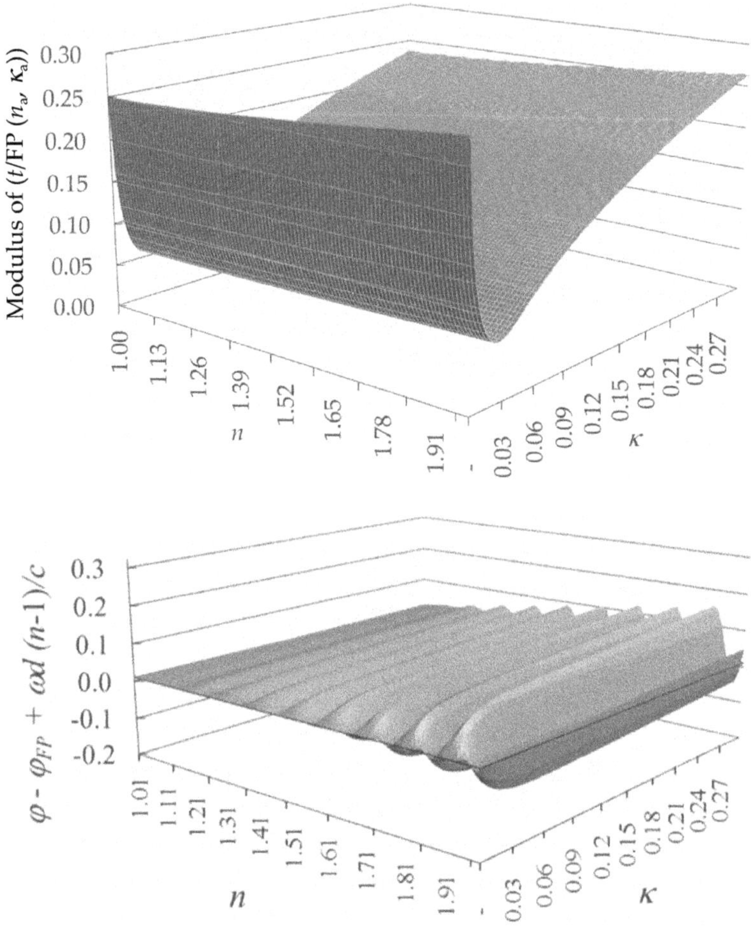

Figure 5.4 Modulus and normalized phase of $t'(\omega)$, with $\omega\, d/c = 30$.

phase of $\tilde{t}'(\omega)$. Therefore, the problem is now to solve

$$\frac{4\tilde{n}}{(\tilde{n}+1)^2} \times e^{-j\frac{\omega}{c}(\tilde{n}-1)d} = \tilde{t}'(\omega). \tag{5.13}$$

There exists no analytical solution of Eq. 5.13. A fast way of deriving the numerical solution of Eq. 5.13 is to define an error function and to minimize it. Here a relevant error function is

$$\delta(n, \kappa) = \delta\rho^2 + \delta\varphi^2, \tag{5.14}$$

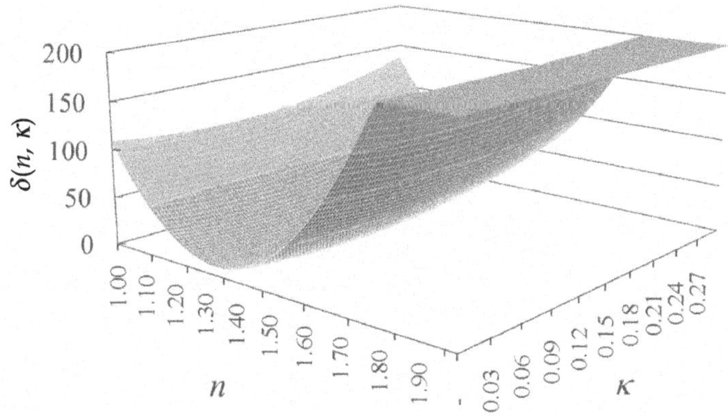

Figure 5.5 Error function $\delta(n, \kappa)$ versus n and κ. The map is plotted for $t_{\text{meas}} = 0.05$, $\varphi_{\text{meas}} = 10$ rad, and $\omega d/c = 30$.

with

$$\delta\rho = \ln(t) - \ln(t_{\text{meas}}), \quad \delta\varphi = \varphi - \varphi_{\text{meas}}. \tag{5.15}$$

This error function exhibits a single zero (Fig. 5.5). Moreover, close to the zero, the curve resembles a paraboloid. Thus, reaching the zero is quite easy, by using, for example, the steepest descent method.

This method is applicable and has been validated over a huge range of values, with n starting from 1 up to 5, and κ from 0 to 10 (i.e., a power absorption varying from 0 to more than 100–1000 cm^{-1} over the 0.1–10 THz range). It can be applied to a large collection of sample thicknesses, from sub-micrometer films to bulky slabs whose thickness exceeds several centimeters. Figure 5.6 shows an example of the precision that can be achieved with this extraction technique. The complex transmission coefficient of a 1 mm thick sample has been calculated versus frequency using relation 5.3, and given values of n and α, plotted in the upper curves of Fig. 5.6. Then the extraction method was applied to these computed spectra to retrieve n and α. The lower curve shows the difference between the so-extracted values and the values input in the calculation. Typically, n and α are obtained with a precision respectively better than 0.01% and 1%. Note that because α tends to zero at lower frequencies, here its precision is limited by rounding errors.

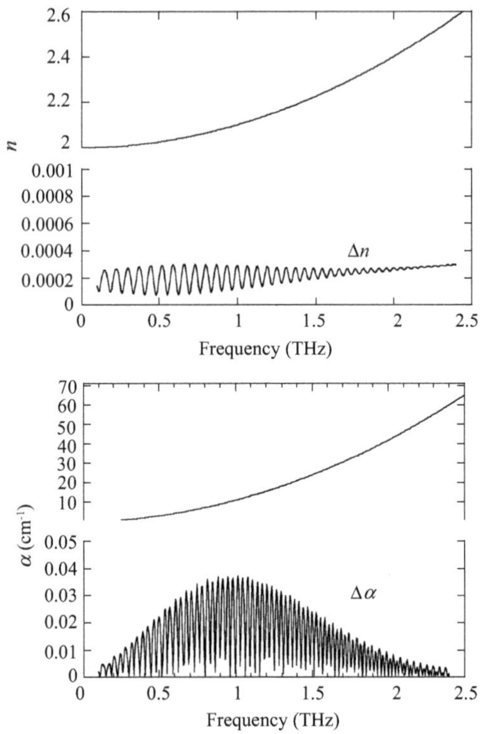

Figure 5.6 Precision of the method, i.e., the errors Δn (top) and $\Delta\alpha$ (bottom) between the actual and retrieved values. The upper traces depict n and α versus frequency.

5.3 Derived Methods

Dorney et al. [2] brought some improvements to the method by Duvillaret et al. [1]. First they took into account a possible oblique incidence. Thus, the transmission coefficient (5.3) becomes

$$\tilde{t}(n,\kappa,\omega) = \frac{4\tilde{n}\cos\theta\ \cos\beta\, e^{-j\frac{\omega}{c}(\tilde{n}-\cos(\theta-\beta))d}}{(\tilde{n}\cos\theta\ +\cos\beta)^2} \times \left(1+\sum_{k=1}^{\infty}\left(\frac{\tilde{n}\cos\theta-\cos\beta}{\tilde{n}\cos\theta+\cos\beta}e^{-j\frac{\omega}{c}\tilde{n}d}\right)^{2k}\right), \tag{5.16}$$

where θ is the angle of incidence and $\sin\theta = n\sin\beta$. Then they define the following error function:

$$\delta(n,\kappa) = \sum_{\omega} \left[|t_{\text{meas}}(\omega) - t(n,\kappa,\omega)| + |\varphi_{\text{meas}}(\omega) - \varphi(n,\kappa,\omega)| \right]. \tag{5.17}$$

(n, κ) is retrieved by minimizing $\delta(n, \kappa)$. The procedure starts from the same n_a as in the preceding method (see Eq. 5.5) and from $\kappa_a = 0$. Then the values of n and κ are updated at each calculation step i as follows:

$$\begin{cases} n_i(\omega) = n_{i-1}(\omega) + \varepsilon \times (\varphi_{\text{meas}}(\omega) - \varphi(\omega)), \\ \kappa_i(\omega) = \kappa_{i-1}(\omega) + \varepsilon \times (t_{\text{meas}}(\omega) - t(\omega)). \end{cases} \tag{5.18}$$

The coefficient $\varepsilon(\sim 0.01)$ quantifies the effect of the magnitude and phase errors on the new value of the complex index of refraction. The procedure is repeated until $\delta(n, \kappa)$ is no longer decreasing.

Pupeza et al. [3] started from the same error function (5.17) as Dorney et al. [2]. But this error function is minimized thanks to a Nelder–Read algorithm, which is a simplex direct search method to minimize nonlinear scalar functions of real multivariables [4]. This algorithm allows one to take into account the actual number of recorded temporal echoes in the transmitted waveform, while theoretical expression (Eq. 5.3) is obtained for an infinite number of echoes. This is of utmost importance, since actual signals are recorded over a limited equivalent time window, and thus only some of all the rebounds in the waveform are recorded. In most of the cases, the rebounds outside the time window are below the noise level and do not contribute to the signal. But in some cases, like thin samples, contribution rebounds may fall outside the recorded time-window. To take into account a given number of rebounds, Eq. 5.3 is rewritten as

$$\tilde{t}(n,\kappa,\omega) = \frac{2\tilde{n}}{(1+\tilde{n})^2} e^{-j\frac{\omega}{c}(\tilde{n}-1)d} \left[1 + \sum_{i=1}^{i_{\max}} \left(\frac{1-\tilde{n}}{1+\tilde{n}} e^{-j\frac{\omega}{c}\tilde{n}d} \right)^{2i} \right]. \tag{5.19}$$

In Eq. 5.19, the index i denotes the rebound i of the pulse in the sample. The maximum value $i_{\max}$ of i is set when the amplitude of pulse i is equal or below the noise floor. The algorithm starts from approximated values of n and κ determined by supposing that first the attenuation of the pulse directly transmitted through the sample is simply due to absorption. Thus, $\kappa = -c\ln\left(t_{\text{meas}}(\max)\right)/(\omega\, d)$ and then the phase change is only due to propagation through the sample, i.e., $n = 1 + c\,\Delta t/d$, where Δt is the time delay, introduced by the sample, of the pulse arrival at the receiver.

More recently, Scheller [5] got rid of the Fabry–Pérot oscillations in the $\tilde{t}_{\text{meas}}(\omega)$ records by a proper spectral filtering. The procedure starts from n_{a} given by Eq. 5.5 and κ_{a} defined by

$$\kappa_{\text{a}}(\omega) = \frac{c}{\omega\, d}\ln\left(\frac{(n+1)^2}{4\, n}t_{\text{meas}}(\omega)\right). \tag{5.20}$$

In fact, relation 5.20 is obtained from 5.6 in which the Fabry–Pérot term is omitted as well as the κ terms in the Fresnel coefficients. Assuming that the refractive index n of the sample is constant over the studied THz range, Fabry–Pérot oscillations occur at frequencies:

$$f_m = m \times \frac{c}{2\, n_{\text{a}}\, d}, \quad m \text{ positive integer.} \tag{5.21}$$

Therefore, the modulus $t_{\text{meas}}(\omega)$ is the product of a smooth spectrum and of an oscillating function. Its Fourier transform shows pronounced peaks at f_m. Figure 5.7 illustrates a typical example, i.e., the THz-TDS signal transmitted by a 1.065 mm BK7 glass sample. Fabry–Pérot oscillations are clearly seen in the modulus of the transmitted spectrum (Fig. 5.7, top). Due to absorption in the glass, both signal and oscillation decrease with frequency, and over 1 THz, due to the low transmission of the sample, the signal is hidden by noise. Figure 5.7 (bottom) shows the Fourier transform of the modulus spectrum. Instead of time t, the coordinate of the x axis is the equivalent length, i.e., $x = t \times c/2n$, where $n \approx 2.5$ is a rough estimation of the average refractive index of BK7. The Fourier transform exhibits a strong peak at 1.07 mm, i.e., at a distance corresponding to the sample thickness. As well, the second-order peak is seen at 2.15 mm. By a proper filtering of these peaks, and an inverse Fourier transform, the transmission spectrum is made free of the oscillations (see dashed curve in Fig. 5.7, top). From this "smooth" spectrum, the initial values n_{a} (5.5) and κ_{a} (5.19) are precisely obtained.

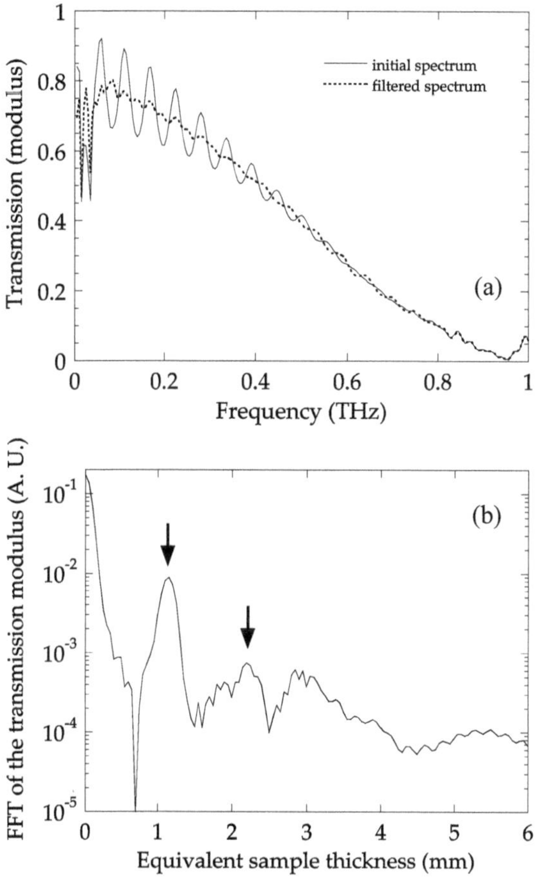

Figure 5.7 Transmission spectrum (initial and filtered data) through a 1.065 mm BK7 glass sample (top), and corresponding Fourier transform of this spectrum versus the equivalent propagation distance in the sample (adapted from Scheller [5]). Arrows point out d (sample thickness) and $2d$.

Then, Scheller defines $n(\omega) = \xi\,(n_a(\omega) - 1) + 1$ and $\kappa(\omega) = \psi\,\kappa_a(\omega)$, where ξ and ψ are scalar parameters to be optimized. These values are entered in $\tilde{t}(n, \kappa, \omega)$ and the error function $\delta(n, \kappa) = \sum_{\omega} \left|\tilde{t}_{meas}(\omega) - \tilde{t}(n, \kappa, \omega)\right|$ is minimized with a Nelder–Read algorithm.

Another method has been introduced recently by Krüger and coworkers [6]. These authors start the computation procedure from approached values n_s and κ_s of n and κ. κ is estimated from the Beer–Lambert law and n from the time delay Δt between the transmitted and reference pulses:

$$\kappa_s(\omega) \approx -\frac{c}{2\omega d} \ln\left(\left|\tilde{t}(\omega)\right|^2\right), \quad n_s \approx 1 + c\frac{\Delta t}{d} \tag{5.22}$$

These values are made closer to the actual values n and κ by adding correcting terms to κ_s in Eq. 5.22 and calculating n_s from the argument of $\tilde{t}(\omega)$. Then the roots of Eq. 5.1 are calculated with $n = n_s$ and $\kappa = \kappa_s$. During the searching procedure, the complex plane is divided in domains in such a way that, after a few iterations based on the Cauchy's principle that gives the number of roots per domain, only one root is located in the domain. The size of the smallest domain gives the precision with which n_s and κ_s are determined. The method has been applied to different organic compounds, like anthracene, EAN, or capric acid. The precision on n is better than 1%.

5.4 Samples of Different Thicknesses Made with the Same Material

This is almost the ideal case! Let us consider two samples of thicknesses d_1 and $d_2 = d_1 + \delta$ made from the same material. One measure will serve as reference for the second one. Both thicknesses are large enough to permit a proper time-windowing of the directly transmitted pulse. We easily get the relative differential transmission:

$$\begin{aligned}\Delta\tilde{t}(\omega) &= \frac{\tilde{t}_2\,(n,\kappa,\omega) - \tilde{t}_1(n,\kappa,\omega)}{\tilde{t}_1\,(n,\kappa,\omega)} \\ &= e^{-j\frac{\omega}{c}(\tilde{n}-1)\delta} - 1 \\ &= e^{-j\frac{\omega}{c}(n-1)\delta}e^{-\frac{\omega}{c}\kappa\delta} - 1,\end{aligned} \tag{5.23}$$

from which we deduce the complex refractive index of the material:

$$\begin{cases} n(\omega) = 1 - \dfrac{c}{\omega\delta} a\tan \dfrac{\mathrm{Im}\Delta\tilde{t}(\omega)}{1 + \mathrm{Re}\Delta\tilde{t}(\omega)}, \\ \kappa(\omega) = -\dfrac{c}{\omega\delta} \ln \dfrac{-\mathrm{Im}\Delta\tilde{t}(\omega)}{\sin\left(\dfrac{\omega}{c}(n(\omega)-1)\delta\right)}. \end{cases} \tag{5.24}$$

Expression 5.23 is useful if $\tilde{t}_1(n, \kappa, \omega) \neq 0$, i.e., if the reference sample does not show a pronounced Fabry–Pérot minimum at the studied frequency. The determination is direct as far as $\omega(n-1)\delta/c < \pi/2$, i.e., $f < c/4(n-1)\delta$. Otherwise, the phase change due to the additional thickness δ is only known modulo π, and one encounters the problem of phase wrapping, as explained in Chapter 4. For a material with $n \approx 2, \delta \approx 10$ μm allows unambiguous characterization of the material up to 7.5 THz.

5.5 Thin Films

Thin films obey the following condition:

$$n\, d\omega/c \ll 1. \tag{5.25}$$

Typically, at 1 THz, $n\, d \ll 10$ μm. Therefore, for common materials, “thin films” means films of thickness of the order or smaller than a micrometer.

5.5.1 Stand-Alone Thin Films

The expression 5.3 of the transmission coefficient simplifies into

$$\tilde{t}(n, \kappa, \omega) \approx 4\,(n - j\kappa)\, \frac{1 - j\frac{\omega}{c}(n-1)d}{(n+1-j\kappa)^2} e^{-\frac{\omega}{c}\kappa d}. \tag{5.26}$$

For most of materials, especially dielectric ones, $\kappa \ll n$ and thus

$$\Rightarrow \begin{cases} n \approx 1 + \dfrac{c}{\omega\, d}\varphi_{\mathrm{meas}}, \\ \kappa \approx \frac{c}{\omega\, d}\left(\ln\left|1 + \dfrac{c}{\omega\, d}\varphi_{\mathrm{meas}}\right| - \ln t_{\mathrm{meas}} - 2\ln\left|1 + \dfrac{c}{2\omega\, d}\varphi_{\mathrm{meas}}\right|\right). \end{cases} \tag{5.27}$$

t_{meas} and φ_{meas} are the modulus and phase of the measured transmission. The thinner films that can be characterized will induce a phase change equal to the phase noise:

$$\varphi_{\mathrm{meas}}(\mathrm{min}) = \frac{\omega}{c}\left[(n-1)d\right]_{\mathrm{min}} = \varphi_{\mathrm{noise}}. \tag{5.28}$$

When expression 5.28 is verified, the error on n is 100%. If one desires to determine n with an error less than 10%, we get

$$\left[(n-1)d\right]_{\mathrm{min}} = 10\frac{c}{\omega}\varphi_{\mathrm{noise}}. \tag{5.29}$$

Typically, for modern THz-TDS setups, $\sigma^2_{\mathrm{noise}} \approx 3 \times 10^{-4}$. At 1 THz, films with $[(n-1)d]_{\mathrm{min}} \approx 10$ μm can be well characterized. They can even be thinner (a few micrometers) if one accepts a larger imprecision.

5.5.2 Thin Films Deposited over a Substrate

Here, transmission through the uncovered substrate will serve as a reference. This substrate should preferably be characterized before the thin film deposition, or after removing the deposited thin film. In both cases, one should check that deposing or removing the film does not alter the substrate properties, for example because of heating or mechanical stress. One can also deposit the film over a limited region of the substrate. Reference and signal will be recorded by moving the sample perpendicularly to the THz beam [7, 8]. This technique is not recommended if one cannot measure the bare substrate independently. Indeed, a variation of the substrate thickness or a homogeneity between the covered and bare regions could induce a change of THz transmission larger than the influence of the thin film. For sub-micronic films, this constraint is difficult to fulfill.

The transmission, at normal incidence, through a substrate covered by a film, is derived from Eq. A.30:

$$\tilde{t}(n, \kappa, \omega) = \frac{1}{M_T(2,2)} = \frac{1}{\left(\prod_{p=0}^{2} M_{p,p+1}\right)(2,2)}, \tag{5.30}$$

with

$$M_{p,\,p+1} = \frac{1}{t_{p,\,p+1}} \begin{pmatrix} e^{j\frac{\omega}{c}\left(\tilde{n}_p - \tilde{n}_{p+1}\right)z_p} & r_{p,\,p+1}e^{j\frac{\omega}{c}\left(\tilde{n}_p + \tilde{n}_{p+1}\right)z_p} \\ r_{p,\,p+1}e^{-j\frac{\omega}{c}\left(\tilde{n}_p + \tilde{n}_{p+1}\right)z_p} & e^{-j\frac{\omega}{c}\left(\tilde{n}_p - \tilde{n}_{p+1}\right)z_p} \end{pmatrix}.$$

In the general case, inverting Eq. 5.29 is very difficult because the signature of the thin film is strongly perturbed by the Fabry–Pérot resonances in the substrate. The problem is simplified by making some realistic hypothesis. The substrate is supposed to be transparent ($\tilde{n}_s = n_s$) and thick enough to permit a time-windowing of only the directly transmitted pulse. The film is supposed to be thin (condition 5.24). A simple expression is obtained for the relative change, as regards the bare substrate, of the THz field transmitted by the covered substrate:

$$\Delta\tilde{t}\left(n_f, \kappa_f, \omega\right) = \frac{\tilde{t}_{\text{covered}}\left(n_f, \kappa_f, \omega\right) - \tilde{t}_{\text{bare}}\left(n_s, \omega\right)}{\tilde{t}_{\text{bare}}\left(n_s, \omega\right)}. \tag{5.31}$$

In Eq. 5.30 and the following equations, subscripts f, s, and a refer respectively to the film, the substrate, and the ambient medium (air or vacuum). We get

$$\tilde{t}_{\text{covered}}\left(n_f, \kappa_f, \omega\right) = t_{as}t_{f,a}t_{s,f}\frac{e^{-j\frac{\omega}{c}(d_f - d_s)}e^{-j\frac{\omega}{c}\tilde{n}_f d_f}}{1 + r_{s,f}r_{f,a}e^{-2j\frac{\omega}{c}\tilde{n}_f d_f}}, \tag{5.32}$$

$$\tilde{t}_{\text{bare}}\left(n_s, \omega\right) = t_{as}t_{s,a}e^{-j\frac{\omega}{c}(d_f - d_s)}e^{-j\frac{\omega}{c}d_f}. \tag{5.33}$$

The Fresnel coefficients t and r are given in the Annex (A.25). Therefore,

$$\Delta\tilde{t}\left(n_f, \kappa_f, \omega\right) = \frac{t_{f,a}t_{s,f}}{t_{s,a}}\frac{e^{-j\frac{\omega}{c}(\tilde{n}_f - 1)d_f}}{1 + r_{s,f}r_{f,a}e^{-2j\frac{\omega}{c}\tilde{n}_f d_f}} - 1. \tag{5.34}$$

As the film is thin, we take a development in series of the exponential terms in Eq. 5.33 and we obtain

$$\Delta\tilde{t}\left(n_f, \kappa_f, \omega\right) \approx -j\frac{\omega}{c}d_f\frac{\tilde{n}_f^2 - 1}{n_s + 1}, \tag{5.35}$$

$$\Rightarrow n_f^2 \approx 1 - \frac{\left(n_s + 1\right)c}{d_f\,\omega}\text{Im}\Delta\tilde{t}(\omega), \quad \kappa_f \approx \frac{-\left(n_s + 1\right)c}{2d_f\,\omega\,n_f}\text{Re}\Delta\tilde{t}(\omega). \tag{5.36}$$

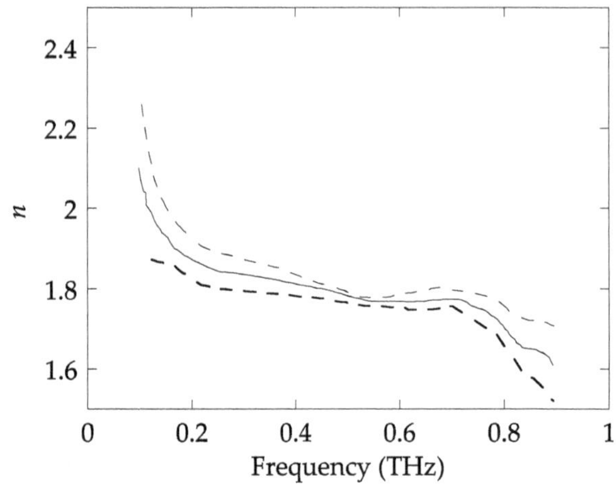

Figure 5.8 Refractive index versus frequency of a 50 μm thick porous silicon film deposited over a 305 μm thick HR silicon wafer (extracted from Brucherseifer et al. [8]). The dashed lines represent the experimental errors.

Figure 5.8 depicts the refractive index of a 50 μm thick porous silicon layer (70% porosity) deposited over a 305 μm thick high resistivity silicon wafer [9]. Because the film is thin, $\Delta\tilde{t}(\omega)$ is small and thus the uncertainties in the determination of n_f and κ_f are large (see Fig. 5.8), mostly originating in the poor signal-to-noise ratio. This has been greatly improved by Brucherseifer et al. [8], who made used of a double modulation of the signal: the THz beam is chopped at a given frequency, and the sample is periodically moved perpendicularly to the beam (one part of the sample is covered, the other one is bare) at another frequency.

The detection with two lock-in amplifiers benefits a 30 dB noise reduction compared to the classical single modulation scheme. The characterization of a 100 nm SiO_2 film deposited over a HR Si wafer was made possible (Fig. 5.9).

5.5.3 Metallic and Highly Conducting Thin Films

Highly conducting thin films include metal layers, but also superconductors, graphene, etc. The electrical conductivity originates from both conduction and displacement currents, through the Maxwell–

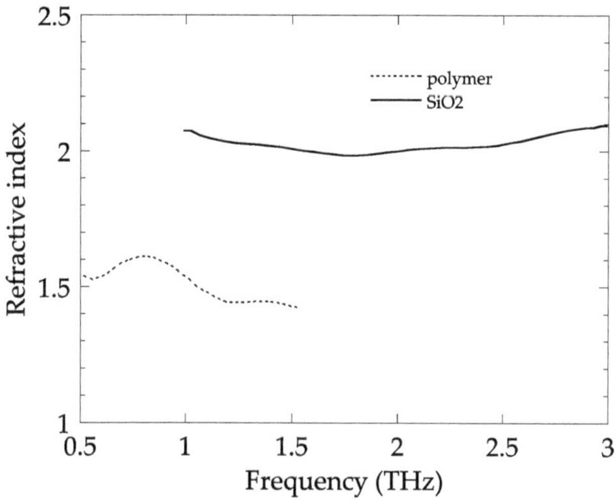

Figure 5.9 Refractive index versus frequency of a 1.2 μm thick polymer film (AZ 5214 photoresist on glass) and a 100 nm SiO_2 film on HR Si. Adapted from Brucherseifer et al. [8].

Faraday equation (A.2):

$$\vec{\nabla} \times \vec{H} = \frac{\partial \vec{D}}{\partial t} + \vec{J} = (j\omega\varepsilon_0\tilde{\varepsilon} + \tilde{\sigma})\,\vec{E} \equiv j\omega\varepsilon_0\tilde{\varepsilon}_{\text{total}}\vec{E}. \tag{5.37}$$

$\tilde{\varepsilon}_{\text{total}}$ represents the total relative permittivity of the material, which includes the contribution of the conduction current. In fact, except the ω dependence in Eq. 5.37, separating the conduction and displacement current contributions is difficult because, for example, the imaginary part of $\tilde{\sigma}$ can be interpreted as the real part of $\omega\tilde{\varepsilon}$. For highly conducting materials, the conductivity $\tilde{\sigma}$ is the dominant term, and thus

$$\tilde{\sigma} = j\omega\varepsilon_0\tilde{\varepsilon}_{\text{total}}. \tag{5.38}$$

When performing a THz-TDS characterization of the highly conducting film deposited over a substrate, relation 5.35 is still valid, providing one substitutes $\tilde{n}_{\text{f}}^2$ by $\tilde{\varepsilon}_{\text{total}}$, which leads to

$$\Delta\tilde{t}\,(n_{\text{f}}, \kappa_{\text{f}}, \omega) \approx \frac{d_{\text{f}}}{c} \times \frac{j\omega - \frac{\tilde{\sigma}}{\varepsilon_0}}{n_{\text{s}} + 1}, \tag{5.39}$$

$$\Rightarrow \quad \sigma_r \approx -\frac{\varepsilon_0 c}{d_{\text{f}}}\,(n_{\text{s}} + 1)\,\text{Re}\left(\Delta\tilde{t}\right), \quad \frac{\sigma_{\text{i}}}{\varepsilon_0} \approx \omega - \frac{c}{d_{\text{f}}}\,(n_{\text{s}} + 1)\,\text{Im}\left(\Delta\tilde{t}\right). \tag{5.40}$$

If $\tilde{\sigma}$ is large, the ω term in Eq. 5.39 can be neglected:

$$\tilde{\sigma} \approx c\varepsilon_0 \frac{n_s + 1}{d_f} \frac{\tilde{t}_{bare} - \tilde{t}_{covered}}{\tilde{t}_{bare}}. \tag{5.41}$$

Another equivalent way of addressing the extraction of the conductivity of highly conducting thin films is to deal with the ratio of the transmissions by the covered sample and the bare sample [10]:

$$\frac{\tilde{t}_{covered}(n_f, \kappa_f, \omega)}{\tilde{t}_{bare}(n_s, \omega)} = \frac{\tilde{t}_{f,a}\tilde{t}_{s,f}}{\tilde{t}_{s,a}} \frac{e^{-j\frac{\omega}{c}(\tilde{n}_f - 1)d_f}}{1 + \tilde{r}_{s,f}\,\tilde{r}_{f,a}\,e^{-2j\frac{\omega}{c}\tilde{n}_f d_f}},$$

$$\Rightarrow \quad \tilde{\sigma} \approx j\omega\varepsilon_0 \left(1 + \frac{n_s + 1}{j\frac{\omega}{c}d_f}\left(1 - \frac{\tilde{t}_{covered}(n_f, \kappa_f, \omega)}{\tilde{t}_{bare}(n_s, \omega)}\right)\right). \tag{5.42}$$

Assuming the conductivity is large, we get an expression similar to Eq. 5.41:

$$\tilde{\sigma} \approx c\varepsilon_0 \frac{n_s + 1}{d_f} \frac{\tilde{t}_{bare} - \tilde{t}_{covered}}{\tilde{t}_{bare}}. \tag{5.43}$$

Figure 5.10 shows the surface resistance R_S of a 90 nm thick YBCO superconducting film deposited over a $LaAlO_3$ wafer. R_S is the real part of the surface impedance $\tilde{Z}_S$ defined by $\tilde{Z}_S = \sqrt{j\mu_o\omega/\tilde{\sigma}}$.

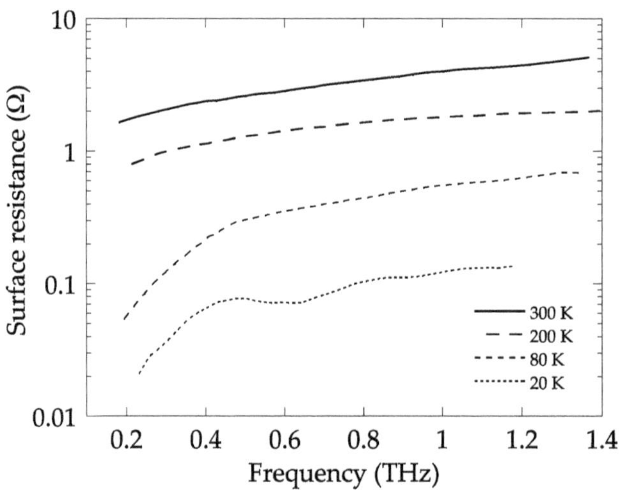

Figure 5.10 Surface resistance, at different temperatures, versus frequency of a YBCO film grown onto a $LaAlO_3$ wafer. Adapted from Gao et al. [10].

The curves are plotted for different temperatures. Over the critical temperature T_c, only "normal" electrons contribute to R_S, while below T_c, mainly Cooper pairs participate to R_S, which explains the change of slope of the recorded curves especially at low temperature [10].

5.6 Thick Samples

In the case of thick samples, rebounds in the waveform are well separated, and one can stop the waveform record before the first rebound appears. Thus, the Fabry–Pérot term does not appear in the expression of the coefficient of transmission (5.3), which is simply written as

$$\tilde{t}(n, \kappa, \omega) = \frac{4\tilde{n}}{(\tilde{n}+1)^2}\, e^{-j\frac{\omega}{c}(\tilde{n}-1)d}. \tag{5.44}$$

The equation above is solved with the techniques previously described. However, as the thick sample transmits a THz signal, it means that the material is quite transparent, i.e., $\kappa \ll 1$. Thus, one can make use of an approximated expression of Eq. 5.44:

$$\tilde{t}(n, \kappa, \omega) \approx \frac{4n}{(n+1)^2}\, e^{-j\frac{\omega}{c}(n-1)d} e^{-\frac{\omega}{c}\kappa d}.$$

This relation leads directly to

$$n = 1 - \frac{c}{\omega d}\varphi,\ \kappa = -\frac{c}{\omega d}\ln\left(t\frac{(n+1)^2}{4n}\right). \tag{5.45}$$

The limited time-window implies a rather weak spectral resolution. If the record is performed over a time slot corresponding to the delay between two consecutive rebounds, we get a spectral resolution equal to

$$\Delta f = \frac{1}{\Delta t_{\text{record}}} = \frac{c}{2n\,d} \quad \Leftrightarrow \quad \Delta f\,(\text{GHz}) = \frac{150}{n\,d\,(\text{mm})}. \tag{5.46}$$

For instance, a 1 cm thick sample of material with $n = 1.5$ would be characterized with a resolution $\Delta f = 10$ GHz. Explosive materials (see Fig. 5.11) have been characterized in this way [11, 12].

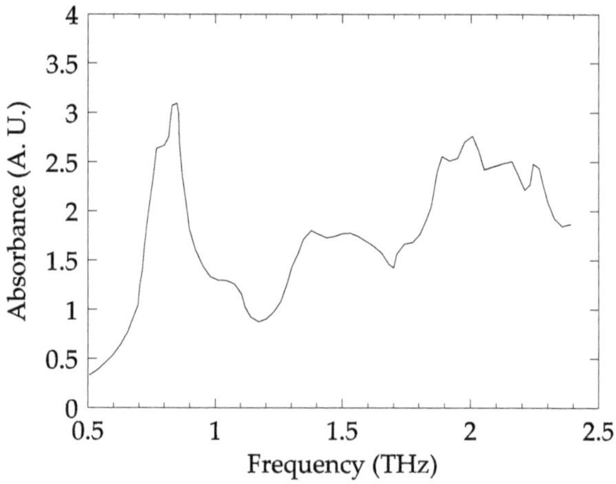

Figure 5.11 Absorbance of RDX. Adapted from Jian et al. [11].

5.7 Layered Samples

The first method to solve the problem of the extraction of both the complex refractive index and the thickness of an unknown layer in a multi-stacked sample was proposed by R. Wilk and coworkers [13]. First, the transfer function of the multilayer structure is written down in such a way that the number of rebounds in the structure is clearly pointed out. Let us examine the beam that encounters only one rebound in a structure with $N = 3$ layers (Fig. 5.12). In [13], the propagation in each layer i of thickness d_i and refractive index $\tilde{n}_i$ is expressed as

$$\tilde{P}_i = \exp\left(-j\frac{\omega}{c}\tilde{n}_i d_i\right). \tag{5.47}$$

As well, the Fresnel transmission and reflection coefficients at the interface between layers i and $i + 1$ are $\tilde{t}_{i,i+1}$ and $\tilde{r}_{i,i+1}$ ($i = 0$ corresponds to the medium [air] before the sample). It follows that the contribution to the transmitted field of the ray that is reflected at the 1/2 interface (first row in Fig. 5.12) is

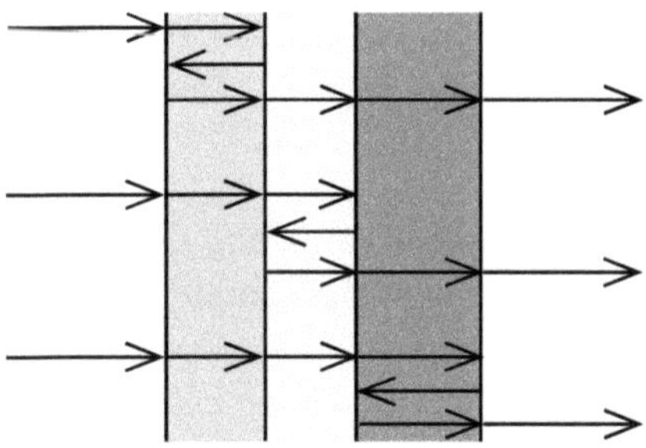

Figure 5.12 Ray-tracing equivalent scheme depicting the 3 solutions for 1 rebound in a 3-layer structure.

$$\tilde{E}_{t,1} = \tilde{E}_o \tilde{t}_{01} \underbrace{\underbrace{\tilde{P}_1}_{+1} \tilde{r}_{12} \underbrace{\tilde{P}_1}_{-1} \tilde{r}_{10} \underbrace{\tilde{P}_1}_{+1} \tilde{t}_{12} \underbrace{\tilde{P}_2}_{+1} \tilde{t}_{23} \underbrace{\tilde{P}_3}_{+1} \tilde{t}_{34}}_{C_1} \exp\left(-j\frac{\omega}{c}(Z-d)\right), \tag{5.48}$$

where Z is the distance between the emitting and receiving antennas, $d = \sum_{i=1}^{N=3} d_i$ is the total thickness of the sample. The $+1$ (-1) subscript indicates that the propagation related to $\tilde{P}_i$ is forward (backward). The whole set of subscripts, namely here ($+1$, -1, $+1$, $+1$, $+1$), forms a row of a matrix $\widehat{M}$. The other rows of this "resonant" matrix are obtained by considering all the possibilities of having $\Delta = 1$ rebound in the structure, as shown on Fig. 5.12. Here we get

$$\widehat{M}(N=3, \Delta=1) = \begin{pmatrix} +1 & -1 & +1 & +1 & +1 \\ +1 & +1 & -1 & +1 & +1 \\ +1 & +1 & +1 & -1 & +1 \end{pmatrix}. \tag{5.49}$$

The transmission coefficient of the sample is then

$$t = P_o(-d)\sum_{r=1}^{R} C_r. \tag{5.50}$$

R is the number of rows r of all the matrices M that are taken into account. C_r is the product of Fresnel coefficients and propagation terms related to row r. For example, C_1 is shown in Eq. 5.48. The computation is performed up to the number of rebounds Δ_{max}, which is calculated as follows:

$$\Delta_{max} \leq \frac{T_{window} - t_{transit}}{2\, t_i^{min}}. \tag{5.51}$$

T_{window} is the total equivalent recording time, $t_{transit}$ is the time of transit of a THz pulse through the sample, and t_i^{min} is the shortest time of transit through the thinner (in terms of optical thickness) layer of the stack. Once Δ_{max} has been evaluated, the extraction of the unknown layer parameters is performed with a code similar to the one proposed by the same team for treating a single slab (method described in Section 5.3) [3].

The method by Wilk et al. [13] has been generalized by Hejase et al. [14, 15]. These authors addressed both the cases of oblique incidence and total multilayer characterization, i.e., contrary to Wilk et al. [13], all the layers of the stack are unknown. This requires one to write expression 5.3 in the case of oblique incidence, and also to enlighten propagation in each layer. Let us call N the number of layers and θ the angle of incidence. The transmission coefficient is written as

$$\tilde{t}\,(\omega, \theta) = \exp\left(-jk_{z,0}d\right) \prod_{i=0}^{N} \tilde{S}_{i,i+1} e^{-j\,\tilde{k}_{z,i} d_i}, \tag{5.52}$$

in which d_i is the thickness of layer i, and

$$\tilde{S}_{i,i+1} = \frac{\tilde{t}_{i,i+1}}{1 - \tilde{r}_{i+1,i}\tilde{r}_{i+1,i+2} e^{-j\,2\tilde{k}_{z,i+1} d_{i+1}}}, \qquad \tilde{k}_{z,i} = \frac{\omega}{c}\sqrt{\tilde{n}_i^2 - \sin^2\theta}. \tag{5.53}$$

This formulation is similar to the one given in the Annex, based on a transmission matrix of the multilayer sample. For each frequency, the number of unknown values here is $3N$, because there are three values to be determined, namely n_i, κ_i, and d_i, for each layer i. Each measurement of $\tilde{t}\,(\omega, \theta)$ gives two quantities (magnitude and phase). Therefore, the problem can be mathematically solved only if M different measurements are performed ($M = 3N/2$ (N even) or $M = 3\,(N+1)/2$ (N odd)). This is achieved by changing the angle of

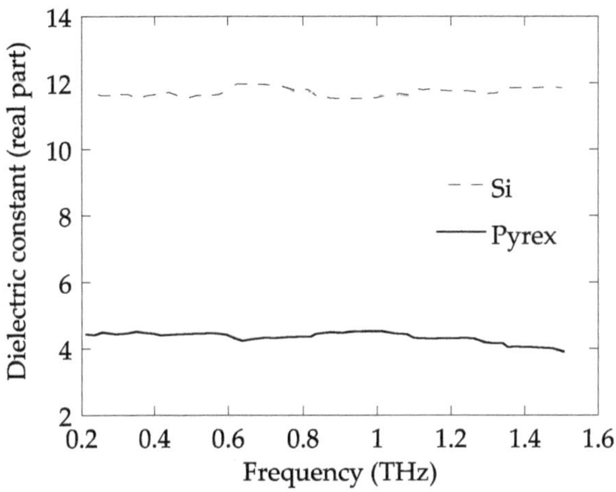

Figure 5.13 Dielectric constant versus frequency of HR-Si and Pyrex extracted from the THz-TDS data measured with a 2-layer sample. Adapted from Hejase [15].

incidence. This leads to the following group set of equations:

$$\begin{cases} \left|\tilde{t}\left(\omega,\theta_1\right)\right| = \left|\tilde{t}_{\text{meas}}\left(\omega,\theta_1\right)\right|, & \arg\left(\tilde{t}\left(\omega,\theta_1\right)\right) = \arg\left(\tilde{t}_{\text{meas}}\left(\omega,\theta_1\right)\right) \\ \quad\cdots & \quad\cdots \\ \left|\tilde{t}\left(\omega,\theta_m\right)\right| = \left|\tilde{t}_{\text{meas}}\left(\omega,\theta_m\right)\right|, & \arg\left(\tilde{t}\left(\omega,\theta_m\right)\right) = \arg\left(\tilde{t}_{\text{meas}}\left(\omega,\theta_m\right)\right) \\ \quad\cdots & \quad\cdots \\ \left|\tilde{t}\left(\omega,\theta_M\right)\right| = \left|\tilde{t}_{\text{meas}}\left(\omega,\theta_M\right)\right|, & \arg\left(\tilde{t}\left(\omega,\theta_M\right)\right) = \arg\left(\tilde{t}_{\text{meas}}\left(\omega,\theta_M\right)\right) \end{cases}$$

Hejase et al. [14] solved these equations with the secant method (the solution of the equation $f(x) = 0$ is obtained for discrete values x_n of x using the recurrent formula $x_{n+1} = x_n - f(x_n)(x_n - x_{n-1})/(f(x_n) - f(x_{n-1}))$. Note that this method is very sensitive to the initial values entered in the code. The method is validated in [14] for single slabs of different dielectric materials (HDPE, Si, Pyrex, etc.). The material parameters are retrieved within the error bars of published data, while the thicknesses differ from the measured ones by a few percent.

Figure 5.13, adapted from [15], shows the dielectric constant of Si and Pyrex determined from the experimental transmission of a 2-stack sample (0.52 mm of Si and 0.52 mm of Pyrex). This method is mostly limited by the experimental uncertainty in the setting of the different requested angles and by the sensitivity of the secant method to initial values entered in the calculation.

5.8 Parametric Methods

Recently, Ahmed and coworkers [16] have proposed to define the error function over the whole achievable spectrum. If m spectral data are recorded, one defines the vectors $\vec{n}$ and $\vec{\kappa}$, and $\overrightarrow{\delta M}$ and $\overrightarrow{\delta\phi}$, whose m components are respectively $n_i = n(\omega_i)$, $\kappa_i = \kappa(\omega_i)$, $\delta M_i = \ln\left(|t(\omega_i)| / \left|t_{\exp}(\omega_i)\right|\right)$, $\delta\phi_i = \arg(t(\omega_i)) - \arg\left(t_{\exp}(\omega_i)\right)$. Then the following expression is minimized:

$$\left.\overrightarrow{\delta M}^T \cdot \overrightarrow{\delta M} + \zeta \overrightarrow{\delta\phi}^T \cdot \overrightarrow{\delta\phi}\right|_{\vec{n},\vec{\kappa}}, \tag{5.54}$$

where ζ is a regularization factor. In this case, the number $2m$ of unknown values (n_i, κ_i) is equal to the number of experimental values M_i and ϕ_i. Consequently the mathematical problem has a solution. If the sample thickness d is unknown, the number of unknown values is $2m + 1$ and thus the problem cannot be solved.

Getting the solution is based on the following hypothesis: one supposes that the electromagnetic response of the sample can be described by a physical model (Debye, Lorentz, etc.). Such models depend on a few parameters, whose number is much smaller than $2m$. Therefore, the problem has a solution, which can be rapidly obtained. Not only the n and k spectra are determined, but also the sample thickness. Figure 5.14, adapted from [16], shows the refractive index of a 10 μm thick fullerene sample, with data (full circles) measured by [17] and extracted with the present parametric method [16]. Here the response of fullerene is modeled with a Debye model. We can see the good agreement between data achieved with classical extraction techniques and with the present method.

This method was improved by Kniffin and Zurk [18], who generalized it to samples showing multiple resonances in the THz range and also to the reflection case. Figure 5.15 presents the

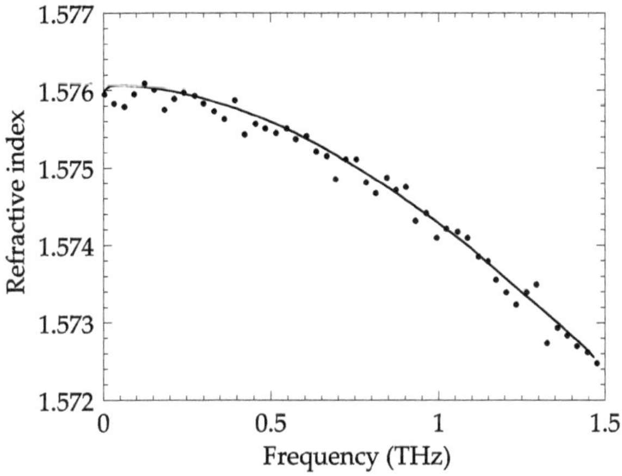

Figure 5.14 Refractive index versus frequency of a 10-μm thick fullerene sample. Adapted from Hejase [15].

refractive index spectrum of a powder mixture of polyethylene and lactose, extracted with the parametric method [18], which exhibits two strong resonances.

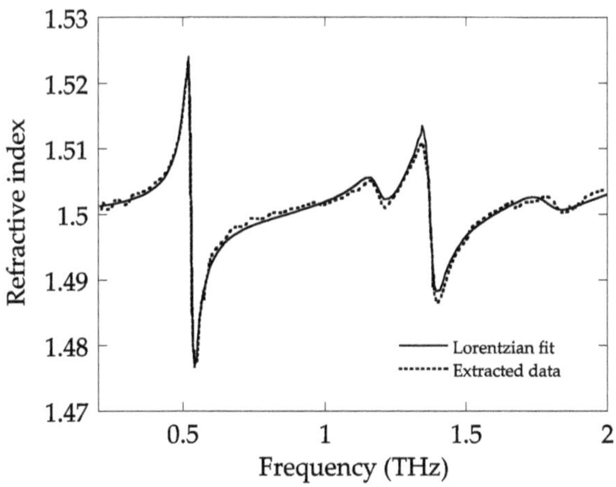

Figure 5.15 Refractive index versus frequency of a 1.428 mm thick pellet of 10% of lactose monohydrate in polyethylene powder. Adapted from Kniffin and Zurk [18].

In conclusion, when the electromagnetic response of the sample material can be pre-empted, this parametric method is very efficient. However, if this response is more complicated than the one described by simple solid-state models, which is the case for most materials, the method cannot be applied or at least is rather imprecise.

5.9 Extraction in the Time Domain

Recently, Li et al. [19] have determined the complex refractive index of materials by rebuilding the THz waveform, i.e., by performing the analysis in the time domain instead of in the frequency domain. The basic idea is to minimize an error function defined in the time domain:

$$\delta(n, \kappa) = \sum_{i=0}^{T/\Delta\tau} \left(t(n, \kappa, i\Delta\tau) - t_{\text{meas}}(i\Delta\tau)\right)^2. \tag{5.55}$$

Here, $\Delta\tau$ is the equivalent time interval between two successive records, T is the equivalent time duration of the record, $t_{\text{meas}}(i\Delta\tau)$ is a measured waveform data and $t(n, \kappa, i\Delta\tau)$ is a calculated waveform data. $t(n, \kappa, i\Delta\tau)$ is obtained by a Fourier transform of $\tilde{t}(n, \kappa, \omega)$ (5.3). The procedure starts from the same initial values $n_a(\omega)$ and $\kappa_a(\omega)$ as in the frequency domain, which were defined in Eqs. 5.5 and 5.20, respectively. The minimization of Eq. 5.14 is performed with the Nelder–Mead algorithm [4]. In [19], Li et al. compare the material parameters obtained with their time domain method and the frequency domain method by Pupeza et al. [3], for a high-index transparent material (e.g., high resistivity silicon) and low-index materials showing spectral resonances (e.g., lactose and glucose). Typically, the agreement between the two methods is within 2%. The main drawback of the time domain method is that a Fourier transform must be calculated for each temporal data, while frequency methods need only one Fourier transform of the waveforms; thus, here the computation time is much longer.

References

1. Duvillaret, L., Garet, F., and Coutaz, J.-L. (1996). A reliable method for extraction of material parameters in terahertz time-domain spectroscopy, *IEEE J. Sel. Top. Quantum Electron.*, **2**, pp. 739–746.
2. Dorney, T. D., Baraniuk, R. G., and Mittleman, D. M. (2001). Material parameter estimation with terahertz time-domain spectroscopy, *J. Opt. Soc. Am. A*, **18**, pp. 1562–1571.
3. Pupeza, I., Wilk, R., and Koch, M. (2007). Highly accurate optical material parameter determination with THz time-domain spectroscopy, *Opt. Express*, **15**, pp. 4335–4350.
4. Lagarias, J. C., Reeds, J. A., Wright, M. H., and Wright, P. E. (1998). Convergence properties of the Nelder-Mead simplex method in low dimensions, *SIAM J. Optim.*, **9**, pp. 112–147.
5. Scheller, M. (2014), Data extraction from terahertz time domain spectroscopy measurements, *J. Infrared Milli. Terahz. Waves*, **35**, pp. 638–648.
6. Kruger, M., Funkner, S., Brundermann, E., and Havenith, M. (2011). Uncertainty and ambiguity in terahertz parameter extraction and data analysis, *J. Infrared Milli. Terahz. Waves*, **32**, pp. 699–715.
7. Jiang, Zh., Li, M., and Zhang, X.-C. (2000). Dielectric constant measurement of thin films by differential time-domain spectroscopy, *Appl. Phys. Lett.*, **76**, pp. 3221–3223.
8. Brucherseifer, M., Haring Bolivar, P., and Kurz, H. (2002). Combined optical and spatial modulation THz-spectroscopy for the analysis of thin-layered systems, *Appl. Phys. Lett.*, **81**, pp. 1791–1793.
9. Labbé-Lavigne, S., Barret, S., Garet, F., Duvillaret, L., and Coutaz, J.-L. (1998). Far-infrared dielectric constant of porous silicon layers measured by terahertz time-domain spectroscopy, *J. Appl. Phys.*, **83**, p. 6007.
10. Gao, F., Whitaker, J. F., Uher, C., Hou, S. Y., and Phillips, J. (1995). High frequency surface impedance and penetration depth of $YBa_2Cu_3O_7$ films: coherent time-domain spectroscopy method, *IEEE Trans. Appl. Supercond.*, **5**, pp. 1970–1974.
11. Chen, J., Chen, Y., Zhao, H., Bastiaans, G. J., and Zhang, X.-C. (2007). Absorption coefficients of selected explosives and related compounds in the range of 0.1–2.8 THz, *Opt. Express*, **15**, pp. 12060–12067.

12. Chen, Y., Liu, H., Deng, Y., Schauki, D., Fitch, M. J., Osiander, R., Dodson, C., Spicer, J. B., Shur, M., and Zhang, X.-C. (2004). THz spectroscopic investigation of 2,4-dinitrotoluene, *Chem. Phys. Lett.*, **400**, pp. 357–361.
13. Wilk, R., Pupeza, I., Cernat, R., and Koch, M. (2008). Highly accurate THz time-domain spectroscopy of multilayer structures, *IEEE J. Sel. Top. Quantum Electron.*, **14**, pp. 392–398.
14. Hejase, J. A., Rothwell, E. J., and Chahal, P. (2013). A multiple angle method for THz time-domain material characterization, *IEEE Trans. Terahertz Sci. Technol.*, **3**, pp. 656–665.
15. Hejase, J. A. (2012). THz time-domain methods for material characterization of layered dielectric media, PhD thesis, Michigan State University.
16. Ahmed, O. S., Swillam, M. A., Bakr, M. H., and Li, X. (2010). Efficient optimization approach for accurate parameter extraction with terahertz time-domain spectroscopy, *J. Lightwave Technol.*, **28**, pp. 1685–1692.
17. Chen, X., Ma, M., Ji, T., Wu, S., and Zhu, Z. (2007). Optical properties of carbon materials in THz region, in *Proc. Joint 32nd Int. Conf. Infrared and Millimeter Waves/15th Int. Conf. Terahertz Electron.*, pp. 550–552.
18. Kniffin, G. P., and Zurk, L. M. (2012). Model-based material parameter estimation for terahertz reflection spectroscopy, *IEEE Trans. Terahertz Sci. Technol.*, **2**, pp. 231–241.
19. Li, X., Hong, Z., He, J., and Chen, Y. (2010). Precisely optical material parameter determination by time domain waveform rebuilding with THz time-domain spectroscopy, *Opt. Commun.*, **283**, pp. 4701–4706.

Chapter 6

Extracting the Refractive Index from Reflection TDS Data

6.1 Introduction

The coefficient of reflection at normal incidence of a flat slab of thickness d located in air is

$$\tilde{r}\,(n,\kappa,\omega)=\frac{1-\tilde{n}}{1+\tilde{n}}\times\frac{1-e^{-2j\frac{\omega}{c}\tilde{n}d}}{1-\left(\frac{\tilde{n}-1}{\tilde{n}+1}e^{-j\frac{\omega}{c}\tilde{n}d}\right)^{2}}. \qquad (6.1)$$

The extraction of n and κ is done by solving the equation

$$\tilde{r}\,(n,\kappa,\omega)=\tilde{r}_{\mathrm{meas}}(\omega). \qquad (6.2)$$

This equation is akin to Eq. 5.1 for the transmission coefficient, for example, it incorporates the same Fabry–Pérot term. Therefore, when taking into account rebounds in the sample, extraction methods similar to the ones described in Chapter 5 can be used to get the complex refractive index (see, e.g., [1]).

It is important to note that performing THz-TDS in reflection at normal incidence requires a semitransparent mirror that directs the reflected beam towards the receiver (this last one cannot be located at the same position as the emitter). Therefore, only a quarter of the incident signal is useful. Several setups run at a small incident angle instead, typically 10° (Fig. 6.1).

Principles of Terahertz Time-Domain Spectroscopy
Jean-Louis Coutaz, Frédéric Garet, and Vincent Wallace

ISBN 978-981-4774-56-7 (Hardcover), 978-0-203-70969-6 (eBook)
www.panstanford.com

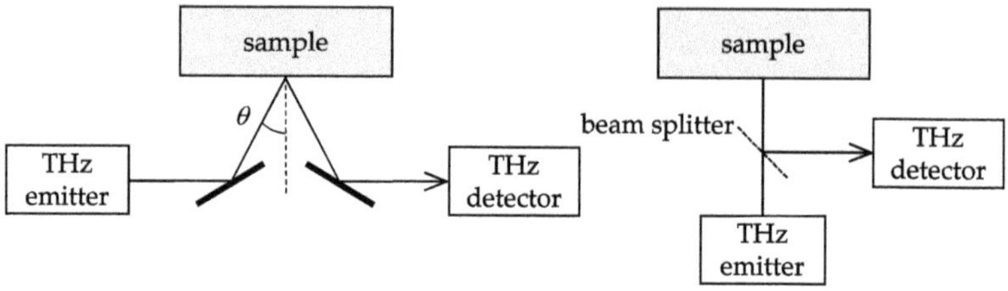

Figure 6.1 Schematic of reflection configurations.

6.2 Parametric Method

The parametric method [2], described in Section 5.8 for the transmission case, is also well adapted to THz-TDS in reflection. It permits to take into account any position difference δ between the sample and the requested reference mirror. Let us treat the simple case of a thick sample (rebounds inside the sample can be time windowed), for which the reference mirror is inaccurately position shifted by δ. The reflected field at the front surface of the layer is

$$\tilde{E}_{\text{sample}}(\omega) = \tilde{r}\,(n,\kappa,\omega)\,\tilde{E}_i(\omega) = \frac{1-\tilde{n}}{1+\tilde{n}}\tilde{E}_i(\omega). \tag{6.3}$$

The reflected field by a perfect metallic mirror that is misplaced by δ as compared to the position of the sample front surface is calculated at the position of the sample surface:

$$\tilde{E}_{\text{ref}}(\omega) = e^{-2j\frac{\omega}{c}\delta}\,\tilde{E}_i(\omega). \tag{6.4}$$

The coefficient of reflection corresponding to this experimental case is

$$\tilde{r}_{\text{meas}}(\omega) = \frac{\tilde{E}_{\text{sample}}(\omega)}{\tilde{E}_{\text{ref}}(\omega)} = \tilde{r}\,(n,\kappa,\omega)\,e^{2j\frac{\omega}{c}\delta}. \tag{6.5}$$

Thus, the phase spectrum (plotted versus f) should exhibit a mean linear slope equal to $4\pi\,\delta/c$. This measured value is used to remove the phase term proportional to δ in Eq. 6.5. Then the extraction method is the same as in transmission (Section 5.8).

Figure 6.2, adapted from Kniffin [3], depicts the refractive index of a pellet of 10% (by weight) α-lactose monohydrate in

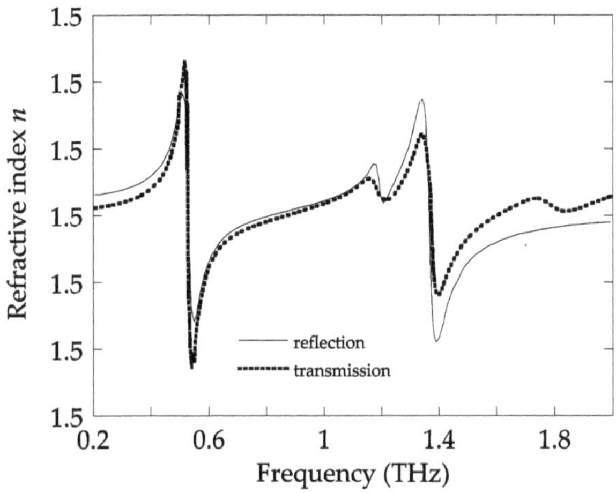

Figure 6.2 Refractive index versus frequency of a powder mixture of 10% α-lactose monohydrate in polyethylene. The curves are calculated with the parametric method from reflection or transmission THz-TDS data. Adapted from Kniffin [3].

polyethylene powder. The curves have been obtained with the parametric method using either reflection (continuous line) or transmission (dotted line) THz-TDS data. Here the parametric method is based on a Lorentzian model for the resonances. The shift, due to a small difference δ between the positions of the sample and the reference mirror, is corrected as explained above. Both curves are roughly similar, with well-evidenced resonances. The differences arise from a lower SNR in reflection.

6.3 Optically Thick Samples

In most practical cases, the problem is simpler than trying to solve Eq. 6.2. Figure 6.3 shows the map of the modulus r and of the phase φ of $\tilde{r}(n, \kappa, \omega)$ versus n and κ, calculated from Eq. 6.1 with $d\,\omega/c = 30$, same sample as in Chapter 5 on transmission THz-TDS. This corresponds, for example, to a 1.5 mm thick slab characterized at 1 THz). At smaller values of κ, Fabry–Pérot oscillations are clearly observed. They disappear, i.e., they contribute to the transmission

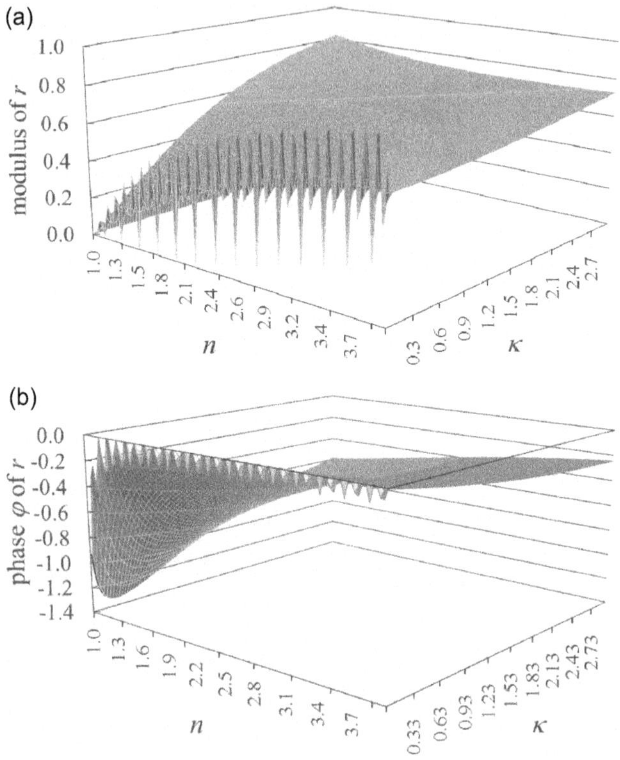

Figure 6.3 Modulus (a) and phase (b) of $\tilde{r}\,(n, \kappa, \omega)$ versus n and κ. The map is calculated for $d\,\omega/c = 30$.

for less than 1 %, when $\alpha = 2\omega\kappa/c$ is larger than α_{lim}:

$$\alpha_{\text{lim}} \approx \frac{12 - 5 \times n + 1.25 \times n^2 - 0.1 \times n^3}{d} \tag{6.6}$$

with α_{lim} in cm^{-1} and d in cm. The value of α_{lim} is plotted in Fig. 6.4 versus n for different values of the sample thickness d. Typically, for a 1 mm thick sample, α should be larger than 60 cm^{-1} to avoid any interference effects inside the sample.

Therefore, for $\alpha > \alpha_{\text{lim}}$, the Fabry–Pérot term in Eq. 6.1 may be neglected and the sample can be treated as semi-infinite. The coefficient of reflection takes a very simple expression:

$$\tilde{r}\,(n, \kappa, \omega) = \frac{\tilde{n} - 1}{\tilde{n} + 1}, \tag{6.7}$$

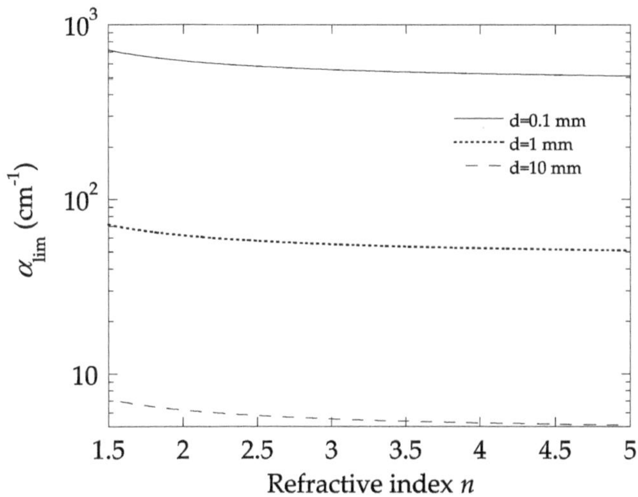

Figure 6.4 α_{lim} versus n for different thicknesses of the sample.

which can be easily inverted:

$$n(\omega) = \frac{r^2(\omega) - 1}{1 + r^2(\omega) - 2r(\omega)\cos\varphi(\omega)},$$

$$\kappa(\omega) = \frac{2r(\omega)\sin\varphi(\omega)}{1 + r^2(\omega) - 2r(\omega)\cos\phi(\omega)}. \tag{6.8}$$

Let us notice that the phase φ of the approached expression 6.7 of $\tilde{r}(n, \kappa, \omega)$ is given by

$$\tan\varphi = \frac{-2\kappa}{n^2 + \kappa^2 - 1}. \tag{6.9}$$

For a semi-infinite-like sample, the phase is not null only if the material exhibits losses, i.e., $\kappa \neq 0$. In other words, if the beam is reflected by a perfectly transparent material, its phase is not changed during the reflection process. Moreover, the phase is smaller than $\pi/2$ (Fig. 6.5), so there is no problem of unwrapping the phase as it is in transmission, except when the refractive index tends towards 1.

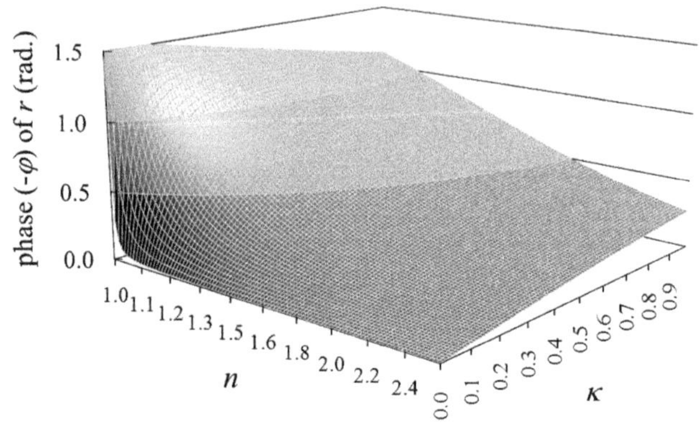

Figure 6.5 The phase φ of the coefficient of reflection versus n and κ.

6.4 Reflection THz-TDS for Characterization of Liquids

As will be shown in Chapter 12 devoted to precision, a difficult problem, when performing THz-TDS in reflection, is to record both signal and reference waveforms without changing the phase origin between the two measurements. This means that the reference mirror and the sample should be placed at the same position, without any angular tilt.

When dealing with liquids, the problem could be solved in a very simple manner [4]. The bottom face of the recipient should be a transparent slab with parallel faces (Fig. 6.6), which is thick enough

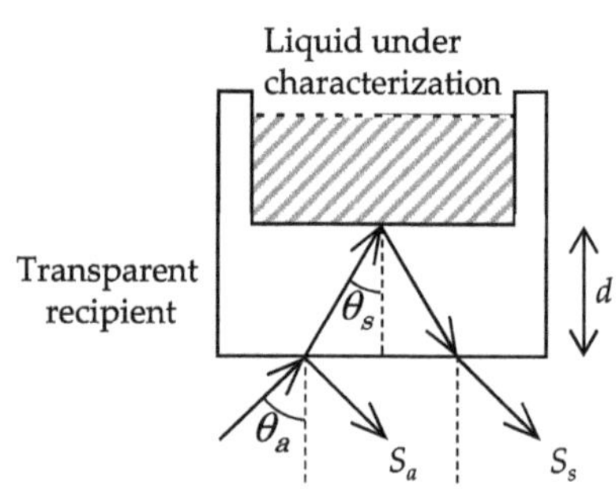

Figure 6.6 THz-TDS characterization in reflection of a liquid.

(thickness d) to allow one to record only the first pulses reflected either at the air/recipient interface or at the recipient/liquid interface, and to reject the rebounds by a proper time-windowing. The coefficients $\tilde{r}_a$ and $\tilde{r}_s$, respectively at the air/slab and slab/liquid interfaces, are

$$\tilde{r}_a = \frac{\tilde{k}_{az} - \tilde{k}_{sz}\,\tilde{\alpha}_{as}}{\tilde{k}_{az} + \tilde{k}_{sz}\,\tilde{\alpha}_{as}},$$
$$\tilde{r}_s = \frac{2\tilde{k}_{az}}{\tilde{k}_{az} + \tilde{k}_{sz}\,\tilde{\alpha}_{as}}\frac{\tilde{k}_{sz} - \tilde{k}_{lz}\,\tilde{\alpha}_{sl}}{\tilde{k}_{sz} + \tilde{k}_{lz}\,\tilde{\alpha}_{sl}}\frac{2\tilde{k}_{sz}}{\tilde{k}_{sz} + \tilde{k}_{az}\,\tilde{\alpha}_{sa}}e^{-2j\tilde{k}_{sz}d}. \quad (6.10)$$

It is easy to show that the complex refractive index $\tilde{n}_l$ of the liquid is a function of the ratio $\tilde{r}_s/\tilde{r}_a$:

$$\tilde{n}_l^2 = \sin^2\theta + \left(n_s^2 - \sin^2\theta\right)\left(\frac{\frac{\tilde{r}_s}{\tilde{r}_a} - \varsigma}{\frac{\tilde{r}_s}{\tilde{r}_a} + \varsigma}\right)^2. \quad (6.11)$$

with $\varsigma = 4\frac{\cos\theta}{1-n_s^2}\sqrt{n_s^2 - \sin^2\theta}\,e^{-j2\frac{\omega}{c}\sqrt{n_s^2-\sin^2\theta}d}$.

Thus, measuring $\tilde{r}_s/\tilde{r}_a$ leads directly to $\tilde{n}_l$ through relation 6.11, assuming n_s, θ, and d are known. Let us notice that the ratio $\tilde{r}_s/\tilde{r}_a$ is free of phase error as far as the thickness of the recipient wall is known. Figure 6.7 exhibits the two pulses reflected at the outer and inner faces of a bottle of PET filled with pure water [4]. This technique is also used for characterizing or imaging biological tissues [5]. Pressing the rather soft biological tissue against a transparent plate, like quartz, makes it flat and facilitates the measurement. When the plate is not thick enough, the temporal tail of the first pulse could be superimposed with the second pulse, and thus it perturbs the signature of the sample to be characterized. This is the baseline effect, which can be removed by first measuring a known sample, and thus calculating the baseline by comparison between simulated and measured waveforms (see, e.g., in Fig. 6.7b [6]).

6.5 Attenuated Total Reflection (ATR) THz-TDS

When dealing with liquids, the ATR scheme (Fig. 6.8) [7] is an elegant solution to keep immobile the interface at which THz reflection occurs. Of course, it could be applied to other materials

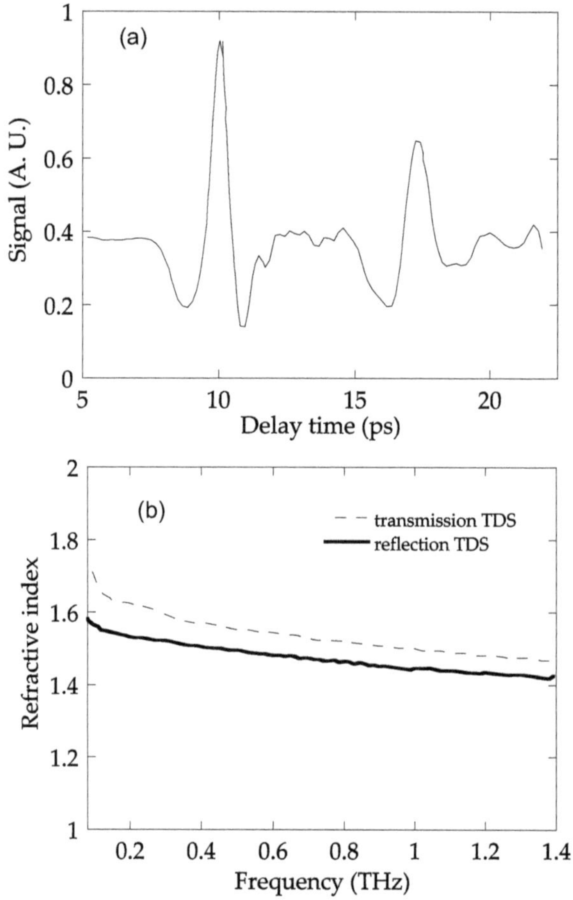

Figure 6.7 (a) Waveform reflected by a PET bottle (0.65 mm wall thickness) filled with pure water. The first pulse is reflected at the air/PET interface, while the second one is reflected at the PET/water interface. Adapted from Parrott et al. [5]. (b) Refractive index of isopropanol measured through a quartz window (continuous line) corrected from the baseline contribution, and measured with a classical THz-TDS transmission technique (dashed line). Adapted from Huang et al. [6].

that are not liquid, like gas, biological tissues, organic materials and so one. In the case of liquids and gas, the ATR scheme allows one to avoid artefacts (interferences) of the classical transmission geometry, which can be caused by the walls of the cell that contains

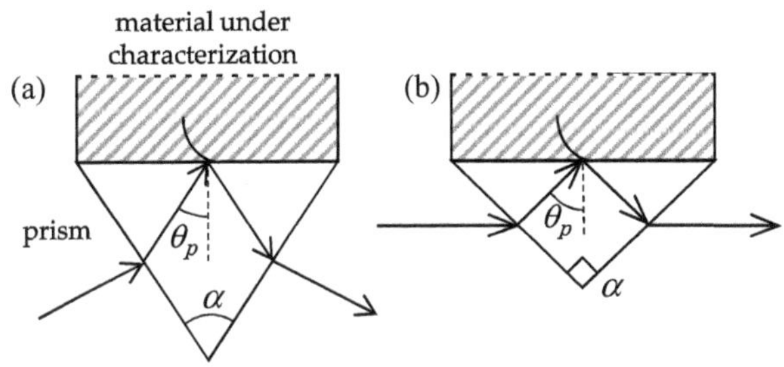

Figure 6.8 ATR schemes: (a) common scheme, (b) Dove prism scheme.

the liquid. As compared to the preceding paragraph, the ATR scheme does not ask for time-windowing the waveforms. Another great advantage of ATR geometry is the evanescent character of the THz field in the material to be characterized, as will be explained below.

The material to be characterized is located (Fig. 6.8a) over a prism of higher index, which serves as entrance wall for the containing cell. Interferences within the prism do not exist. If the angle of incidence θ_p inside the prism is larger than the limit angle θ_l (defined by $\sin\theta_l = n/n_p$, where n_p and n are the refractive indices of respectively the prism and the material. Note that because of the sinus function that is always smaller than 1, the limit angle does exist only if $n_p > n$), the THz beam is totally reflected at the prism/material interface.

From a practical point of view, using a Dove prism (Fig. 6.8b) facilitates the setup building up, because the incident and transmitted beams propagate along the same axis. So, the Dove prism device may be installed in a classical THz-TDS system without changing the geometrical scheme of the setup. The Dove prism is right-angle prism ($\alpha = 90°$). The angle of incidence θ_p at the interface is

$$\theta_p = \frac{\pi}{4} + \text{asin}\left(\frac{1}{n_p\sqrt{2}}\right) \tag{6.12}$$

To avoid any dispersion with the frequency, n_p must be constant over the studied spectral range. This is, fortunately, the case for intrinsic monoatomic semiconductors like high-resistivity silicon, whose refractive index is 3.42 at THz frequencies. However, the angle

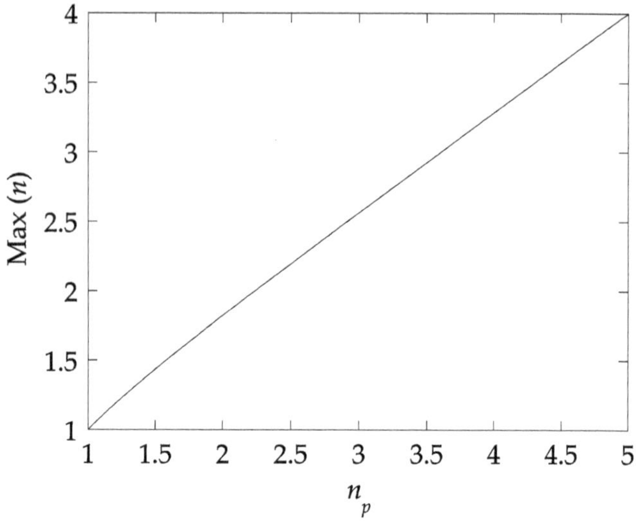

Figure 6.9 The maximum value of the refractive index n of a substance that can by studied by ATR spectroscopy with a Dove prism whose refractive index is n_p.

of incidence must also fulfill the condition $\theta_\mathrm{p} > \theta_1$. Both conditions lead to

$$n < n_\mathrm{p} \sin\left(\frac{\pi}{4} + \mathrm{asin}\left(\frac{1}{n_\mathrm{p}\sqrt{2}}\right)\right). \tag{6.13}$$

ATR THz-TDS with a Dove prism of refractive index n_p can be performed only if the refractive index n of the studied substance obeys inequality 6.13. Figure 6.9 shows the maximum value of n versus n_p. For the usual HR-Si prism, n must be smaller than 2.87. Moreover, the ATR scheme offers two strong advantages:

- The THz field is evanescent in the material, i.e., it shows an exponential decay inside the material. Therefore, if the material sample is thicker than the penetration depth, it could be considered semi-infinite and the THz signal is not perturbed by Fabry–Pérot interferences.
- The variation of phase due to the reflection is much bigger than in the case of simple reflectometry, so the technique is very sensitive.

We will now describe in detail the technique and thus explain more precisely these two advantages. Assuming the sample of material is thick enough to avoid any Fabry–Pérot resonance, the coefficient of reflection, at the interface prism/material, is written as

$$\tilde{r}_{\mathrm{TE}}(n,\kappa,\omega)=\frac{\cos\theta_{\mathrm{p}}-\sqrt{\tilde{N}^2-\sin^2\theta_{\mathrm{p}}}}{\cos\theta_{\mathrm{p}}+\sqrt{\tilde{N}^2-\sin^2\theta_{\mathrm{p}}}},\quad \tilde{N}\equiv\frac{\tilde{n}}{n_{\mathrm{p}}},\tag{6.14}$$

$$\tilde{r}_{\mathrm{TM}}(n,\kappa,\omega)=\frac{\tilde{N}^2\cos\theta_{\mathrm{p}}-\sqrt{\tilde{N}^2-\sin^2\theta_{\mathrm{p}}}}{\tilde{N}^2\cos\theta_{\mathrm{p}}+\sqrt{\tilde{N}^2-\sin^2\theta_{\mathrm{p}}}}.\tag{6.15}$$

Both expressions 6.14 and 6.15 include the same square root term, namely $\sqrt{\tilde{N}^2-\sin^2\theta_{\mathrm{p}}}$. When the real part of $\tilde{N}^2-\sin^2\theta_{\mathrm{p}}$ becomes negative, the square root is mostly imaginary. The reflection changes from a refracting behavior to a total reflection one. The limit is given by

$$\sin\theta_{\mathrm{p}}<\sin\theta_{\mathrm{l}}\equiv\frac{n}{n_{\mathrm{p}}}\sqrt{1-\frac{\kappa^2}{n^2}}.\tag{6.16}$$

The limit angle θ_{l} is the same as defined earlier, but the effect of losses in the studied material is now taken into account. Introducing θ_{l} simplifies the expressions of the reflection coefficients:

$$\tilde{r}_{\mathrm{TE}}(n,\kappa,\omega)=\frac{\cos\theta_{\mathrm{p}}-\sqrt{X^2-2jNK}}{\cos\theta_{\mathrm{p}}+\sqrt{X^2-2jNK}},$$

$$X^2=\sin^2\theta_{\mathrm{l}}-\sin^2\theta_{\mathrm{p}},\tag{6.17}$$

$$\tilde{r}_{\mathrm{TM}}(n,\kappa,\omega)=\frac{(N^2-K^2-2jNK)\cos\theta_{\mathrm{p}}-\sqrt{X^2-2jNK}}{(N^2-K^2-2jNK)\cos\theta_{\mathrm{p}}+\sqrt{X^2-2jNK}},\tag{6.18}$$

with $\tilde{N}=N-jK$. The behavior of reflectivity at the prism/material interface is governed by the X coefficient, which is real if $\theta_{\mathrm{p}}<\theta_{\mathrm{l}}$ and imaginary if $\theta_{\mathrm{p}}>\theta_{\mathrm{l}}$. This can be clearly shown in the simplified case of low absorption ($K\ll N$), which leads to

$$\tilde{r}_{\mathrm{TE}}(n,\kappa,\omega)\approx\frac{\cos\theta_{\mathrm{p}}-X}{\cos\theta_{\mathrm{p}}+X}\left(1+j\frac{NK}{X}\frac{2\cos\theta_{\mathrm{p}}}{\cos^2\theta_{\mathrm{p}}-X^2}\right),\tag{6.19}$$

$$\tilde{r}_{\mathrm{TM}}(n,\kappa,\omega)\approx\frac{N^2\cos\theta_{\mathrm{p}}-X}{N^2\cos\theta_{\mathrm{p}}+X}\left(1+j2\frac{NK}{X}\cos\theta_{\mathrm{p}}\frac{N^2-2X^2}{N^4\cos^2\theta_{\mathrm{p}}-X^2}\right).\tag{6.20}$$

In the case of refraction (X real), we get

$$\begin{cases} r_{\mathrm{TE}} \approx \dfrac{\cos\theta_{\mathrm{p}} - \sqrt{N^2 - \sin^2\theta_{\mathrm{p}}}}{\cos\theta_{\mathrm{p}} + \sqrt{N^2 - \sin^2\theta_{\mathrm{p}}}}, \tan\varphi_{\mathrm{TE}} \approx \dfrac{2NK\cos\theta_{\mathrm{p}}}{\left(1 - N^2\right)\sqrt{N^2 - \sin^2\theta_{\mathrm{p}}}}, \\ r_{\mathrm{TM}} \approx \dfrac{N^2\cos\theta_{\mathrm{p}} - \sqrt{N^2 - \sin^2\theta_{\mathrm{p}}}}{N^2\cos\theta_{\mathrm{p}} + \sqrt{N^2 - \sin^2\theta_{\mathrm{p}}}}, \tan\varphi_{\mathrm{TM}} \approx \dfrac{\tan\varphi_{\mathrm{TE}}}{1 - \left(N^2 + 1\right)\cos^2\theta_{\mathrm{p}}}. \end{cases}$$

In the case of total reflection (X imaginary), the expressions are

$$r_{\mathrm{TE}} \approx 1 - \frac{2NK\cos\theta_{\mathrm{p}}}{\left(N^2-1\right)\sqrt{\sin^2\theta_{\mathrm{p}}-N^2}}, \qquad \tan\varphi_{\mathrm{TE}} \approx -2\frac{\sqrt{\sin^2\theta_{\mathrm{p}}-N^2}}{\cos\theta_{\mathrm{p}}},$$

$$r_{\mathrm{TM}} \approx 1 - \frac{2NK\cos\theta_{\mathrm{p}}}{\left(N^2-1\right)\sqrt{\sin^2\theta_{\mathrm{p}}-N^2}}\frac{2\sin^2\theta_{\mathrm{p}}-N^2}{1-\cos^2\theta_{\mathrm{p}}\left(1+N^2\right)}, \qquad \tan\varphi_{\mathrm{TM}} = \frac{\tan\varphi_{\mathrm{TE}}}{N^2}.$$

Thus, when total internal reflection occurs, the phase of the coefficient of reflection depends mostly on n, while its modulus is also governed by κ. This is opposite to what happens in the regime of refraction.

The component normal to the interface of the wavevector in the material specifies the propagating or evanescent behavior of the field in the material. It is given by

$$\tilde{k}_{\mathrm{z}} = \frac{\omega}{c} n_{\mathrm{p}} \sqrt{\tilde{N}^2 - \sin^2\theta_{\mathrm{p}}}. \tag{6.21}$$

For angles below the limit angle ($\theta_{\mathrm{p}} > \theta_{\mathrm{l}}$), the square root term in Eq. 6.21 is mostly positive and real, and thus $\tilde{k}_{\mathrm{z}}$ is mostly real: the field propagates in the material. For angles over the limit angle, $\tilde{k}_{\mathrm{z}}$ is mostly imaginary. The propagation exponent $-j\,\tilde{k}_{\mathrm{z}} z$ is mostly real and negative, which means that the field decays exponentially in the material. Therefore, the field is no more propagating into the material: it is evanescent. The penetration depth δ of the THz field in the sample is the inverse of the imaginary part of $\tilde{k}_{\mathrm{z}}$:

$$\delta = \frac{1}{\mathrm{Im}\left(\tilde{k}_{\mathrm{z}}\right)}. \tag{6.22}$$

Figure 6.10a depicts the map of δ versus n and κ, for a silicon prism at $f = 1$ THz and $\theta_{\mathrm{p}} = 60^\circ$, in TE polarization.

Typically, δ is smaller than 200 μm, except when n becomes close to n_{p} and κ remains small. For κ large and n small, which is quite unrealistic as materials exhibit commonly a large value

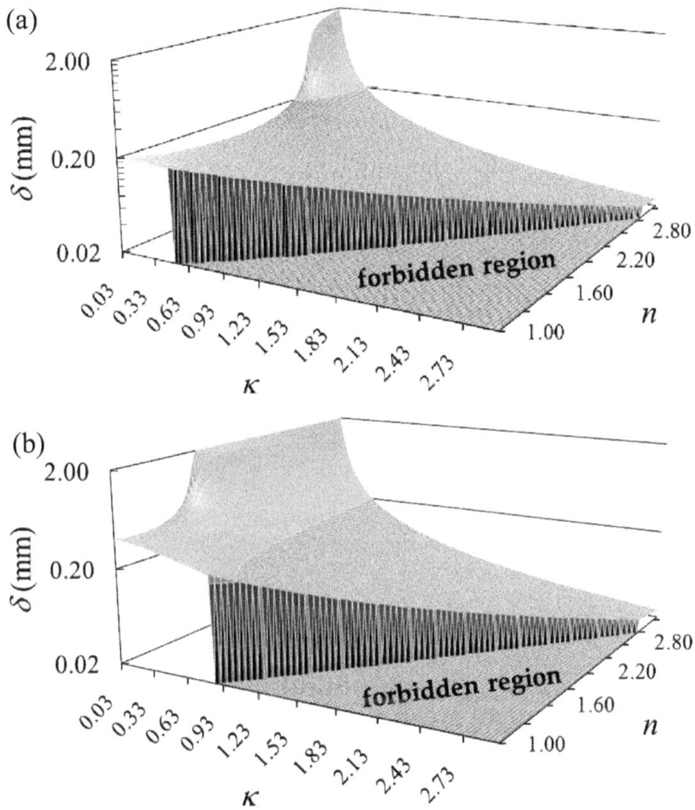

Figure 6.10 δ versus n and κ for a Si prism ($n_p = 3.4$) at $f = 1$ THz. (a) $\theta_p = 60°$, (b) $\theta_p = 30°$.

of n when κ is large, the relation (6.16) is not met, and thus, ATR spectroscopy cannot be performed: corresponding data are not plotted on Fig. 6.10. Therefore, ATR scheme is very sensitive to the layer of material in contact with the prism. To test a larger volume of material (i.e., increase depth), one should adjust the incidence angle to a smaller value. For example (Fig. 6.10b), a 30° incidence angle permits to study a 400 μm thick layer of material at the prism/substance interface. The great advantage of the field confinement is that, generally, the signal does not undergo multiple reflections and, therefore, is not troubled by Fabry–Pérot effects within the sample.

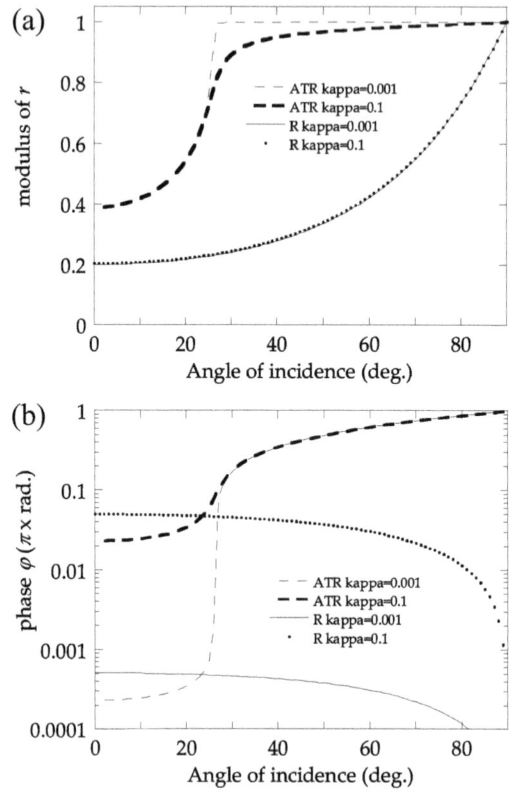

Figure 6.11 Modulus (a) and phase (b) of $\tilde{r}_{\mathrm{TE}}$ versus the angle of incidence, in an ATR scheme (dashed lines) and a classical reflection scheme (continuous/dotted lines). The sample material parameters are $n = 1.5$ and $\kappa = 10^{-3}$ or $\kappa = 10^{-2}$. In the ATR scheme, $n_{\mathrm{p}} = 3.4$ and the phase is multiplied by -1 (thus the phase is positive and could be plotted on a log scale).

Figure 6.11 shows the modulus r (a) and phase ϕ (b) of the ATR reflection signal versus the incident angle, when the prism is made of intrinsic silicon ($n_{\mathrm{p}} = 3.4$), and the studied material has a refractive index $n = 1.5$, and is either transparent ($\kappa = 10^{-3}$, i.e., $\alpha \approx 0.4$ cm^{-1} at $f = 1$ THz) or absorbing ($\kappa = 0.1$, i.e., $\alpha \approx 40$ cm^{-1} at $f =$ 1 THz). Note that the phase in the ATR scheme is negative: here the phase has been made positive by multiplying it by -1, in order to plot it on a logarithm scale. Together, we draw the related curves of

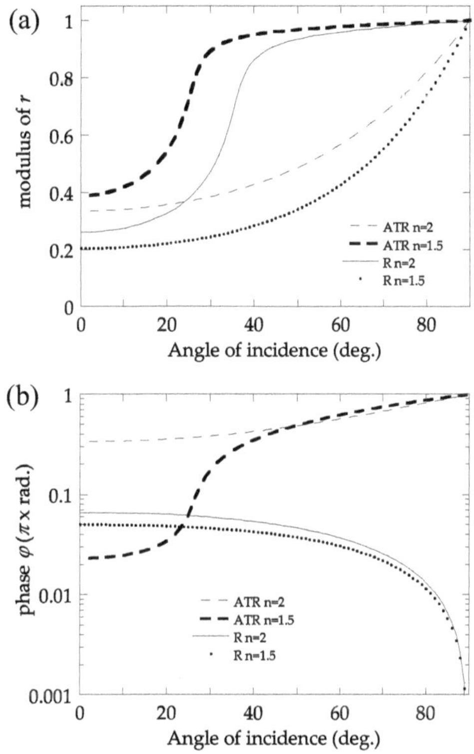

Figure 6.12 Modulus (a) and phase (b) of $\tilde{r}$ versus the angle of incidence, in an ATR scheme (dashed lines) and a classical reflection scheme (continuous/dotted lines). The material parameters are $\kappa = 0.01$, and $n =$ 1.5 or 2. In the ATR scheme, $n_p = 3.4$ and the phase is multiplied by -1 (thus the phase is positive and could be plotted on a log scale).

the signal reflected at an interface air/sample, which corresponds to a classical THz-TDS in reflection. Above the limit angle ($\theta_l = 26.18°$), r tends to 1 (r equals 1 if $\kappa = 0$). Increasing the losses (κ) makes a smoother transition from refraction to total reflection. As discussed above, the modulus r is sensitive to κ in the total reflection regime, while in the refraction regime only the phase depends notably on κ.

This analysis is confirmed by Fig. 6.12, which is similar to Fig. 6.11, but here the refractive index has been changed from $n = 1.5$ to 2, while κ is kept constant ($\kappa = 0.1$). ATR THz-TDS characterization has been mostly applied to liquids (water [8],

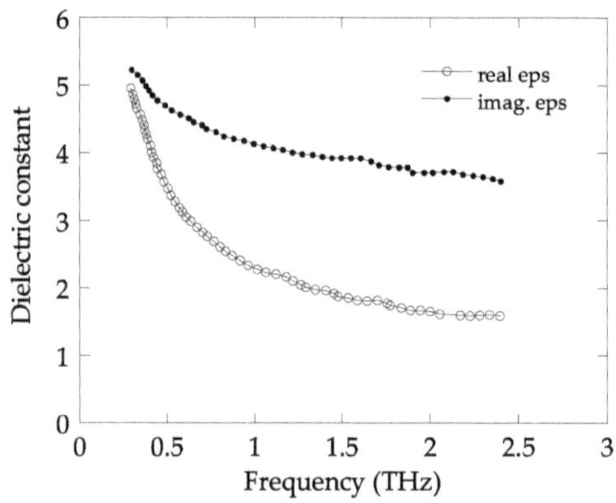

Figure 6.13 Real and imaginary part of the dielectric constant of water measured with ATR THz-TDS. Extracted from Nagai et al. [8].

alcohols [9]), but also to organic and pharmacological materials [9]. Figure 6.13, adapted from [8], shows the THz parameters of water measured with a THz-TDS ATR set up. Today, this technique is also employed to monitor the variation of the chemical properties of liquids [10, 11]. Compact systems have been manufactured [12].

6.6 Ellipsometry

In the optical domain, ellipsometry is an elegant way of characterizing a material, i.e., getting data to calculate its complex refractive index $\tilde{n}$, without performing interferometric measurements that are required to obtain the necessary phase of the signals. Ellipsometry is amazingly sensitive to very thin surface films or small amounts of adsorbed materials. Ellipsometry can be performed in both reflection and transmission modes. However, the latter is limited to transparent samples. Let us restrict here the description of the technique to the reflection scheme and to semi-infinite samples. The coefficient of reflection $\tilde{r}$ at the interface between two media noted 1 and 2 depends on the polarization of the incident beam, as given by relation A.21 of the Annex:

$$\tilde{r} = \frac{\tilde{k}_{1z} - \tilde{k}_{2z}\tilde{\alpha}_{12}}{\tilde{k}_{1z} + \tilde{k}_{2z}\tilde{\alpha}_{12}}. \tag{6.23}$$

z is the direction normal to the interface, and $\tilde{k}_{iz} = \omega\,\tilde{n}_{\text{i}}\cos\theta_{\text{i}}/c$ is the z component of the wave-vector in medium i. For TE polarization, $\tilde{\alpha}_{12} = \tilde{\mu}_1/\tilde{\mu}_2$, while $\tilde{\alpha}_{12} = \tilde{\varepsilon}_1/\tilde{\varepsilon}_2$ in the TM case. Generally, medium 1 is air ($n_1 = 1$, $k_{1z} = \omega\cos\theta/c$, where θ is the angle of incidence) and medium 2 is the sample to be characterized ($\tilde{n}_2 = \tilde{n}$). As $\tilde{n}$ is complex, $\tilde{\theta}_2$ is also complex: it is preferable to employ $k_{2z} = \sqrt{\tilde{k}_2^2 - k_x^2} = \sqrt{\tilde{n}_2^2 - \sin^2\theta} \times \omega/c$. Because $\tilde{\alpha}_{12}$ is not the same for TE and TM components of the E field, a beam polarized in an intermediate direction encounters a rotation of its polarization on reflection. Ellipsometry is based on retrieving the parameters ($\tilde{n} = n - j\kappa$) of the sample from the measured reflection coefficient and the related change of polarization. One measures the modulus r and phase φ of $\tilde{r}$ in both TE and TM polarization, and then one takes the ratio of both $\tilde{r}_{\text{TM}}$ and $\tilde{r}_{\text{TE}}$:

$$\tilde{\rho} = \tan\psi\, e^{j\Delta} = \frac{\tilde{r}_{\text{TM}}}{\tilde{r}_{\text{TE}}} = \frac{r_{\text{TM}}}{r_{\text{TE}}}\, e^{j(\varphi_{\text{TM}} - \varphi_{\text{TE}})}. \tag{6.24}$$

Entering in Eq. 6.24 the expression 6.23 of the coefficients of reflection leads easily to

$$\tilde{\rho} = \frac{\sin^2\theta - \sqrt{\tilde{n}^2 - \sin^2\theta}\cos\theta}{\sin^2\theta + \sqrt{\tilde{n}^2 - \sin^2\theta}\cos\theta}. \tag{6.25}$$

To obtain Eq. 6.25, we have supposed that the sample is not magnetic. Expression 6.25 is then inverted to determine $\tilde{n} = n - j\kappa$:

$$\tilde{n} = \sin\theta\sqrt{1 + \tan^2\theta\left(\frac{\tilde{\rho} - 1}{\tilde{\rho} + 1}\right)^2}. \tag{6.26}$$

Introducing in Eq. 6.26 the ellipsometric parameters ψ and Δ leads to

$$\begin{cases} n^2 - \kappa^2 = \sin^2\theta\left(1 + \tan^2\theta\left(\dfrac{\cos^2 2\psi - \sin^2\Delta\sin^2 2\psi}{(1 + \cos\Delta\sin 2\psi)^2}\right)\right), \\ 2n\kappa = \dfrac{\sin^2\theta\tan^2\theta\sin\Delta\sin 4\psi}{(1 + \cos\Delta\sin 2\psi)^2}. \end{cases} \tag{6.27}$$

The main advantages of ellipsometry are that no reference measure is necessary, and the phase value is very sensitive to the sample surface state, especially to any very thin layer on top of the sample. Ellipsometry in the far infrared was established as early as the

1960s [13, 14]. Most systems [15] were using a Fourier transform spectrometer in which the sample, under oblique incidence, as well as the rotating polarizers, were located at the output of the spectrometer (see Chapter 3, Section 3.6). Studies were also performed with systems built with molecular lasers [16], or even using synchrotron radiation [17]. Most of these studies were devoted to solid state physics, for example for determining the gap of superconductors [18], phonon modes in crystals [19], etc. More recently, CW THz ellipsometry, using a BWO as a source, was applied to the identification of liquids, using an ATR Dove prism scheme [20].

6.6.1 THz-TDS Ellipsometry

While, in the optical and IR domains, determining ψ and Δ requires one to perform polarimetry, i.e., measuring the reflected beam intensity, whereas rotating both the incident and detected polarizations, it is done directly in the THz range, thanks to the ability of THz-TDS to deliver both amplitude and phase of the signals. Therefore, using THz-TDS, ellipsometry needs only two measurements with the same sample, i.e., in the TE and TM polarization cases. This is the great advantage over classical reflectometry, in which a reference mirror should be located at the exactly same position than the sample, leading to strong phase errors. The drawback is the lack of polarizers that are efficient over the whole THz range, as well as a poor sensitivity (see below). However, monitoring the polarization of the THz beam could be done by acting on the polarization of the pump laser beam, when dealing with THz-TDS systems in which emitter and receiver are EO antennas [21–23].

THz-TDS ellipsometry has been first proposed and demonstrated by the team of M. Hangyo at the University of Osaka [24], who extracted the complex refractive index of doped Si. Later on, the ellipsometer was used for solid state studies (phonons in $SrTiO_3$) [25], thin-film analysis [26], and electrical transport effects [27]. Their THz-TDS setup is shown in Fig. 6.14. The emitted THz beam is polarized at 45° from the setup plane. A grid polarizer selects either

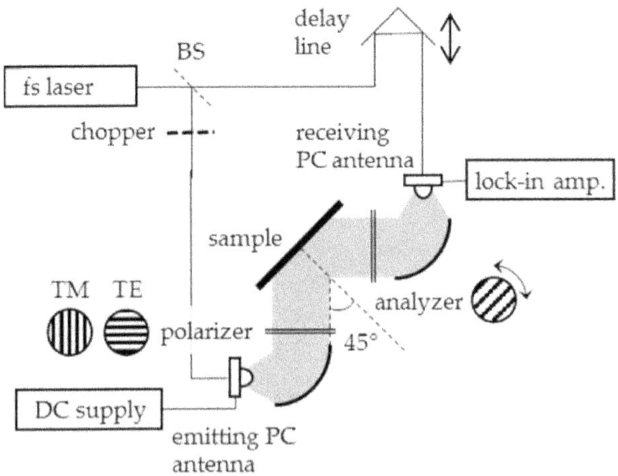

Figure 6.14 THz-TDS ellipsometry setup. Adapted from Nagashima and Hangyo [24].

the TE polarized (in the setup plane) or TM polarized (perpendicular to the setup plane) component of the THz field. The sample is illuminated at 45° and reflects the beam towards the receiver. In between, a second polarizer aligned in the direction of the first one permits the receiver to detect the right polarization. Both polarizers are tungsten wire grids.

Figure 6.15, adapted from Nagashima and Hangyo [24], shows the refractive index n of a p-doped Si wafer (DC resistivity 0.136 Ω/cm). The continuous curve is calculated using the Drude model. At low frequencies, the calculated n values increase, and the theoretical curve is no longer in good agreement with the experimental data.

In fact, this illustrates the main drawback of the technique. When the material exhibits large n or κ, $\tan\psi$ tends to 1 and Δ tends to 0. The relative signal-over-noise ratios become smaller, making the uncertainties on the extracted parameters larger.

As a consequence, any small error made on the experimental values of the angle of incidence θ, the phase difference Δ or the moduli ratio ψ, leads to a rather large error on the extracted n and

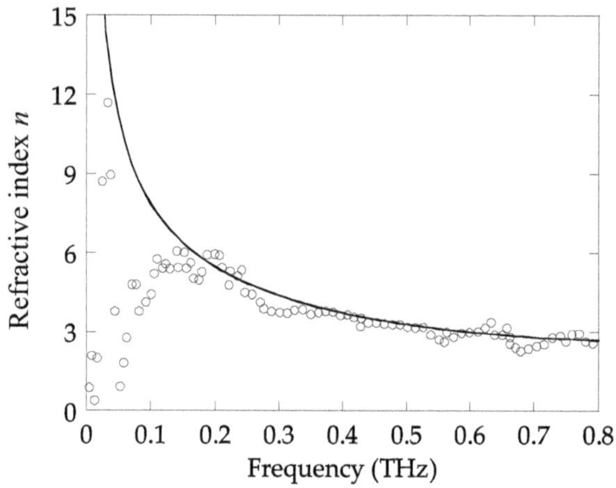

Figure 6.15 Plot of n versus frequency of a p-doped Si wafer measured by THz-TDS ellipsometry (open circles: experimental data; continuous curve: Drude model). Adapted from Nagashima and Hangyo [24].

κ. This is depicted in Fig. 6.16, where we plot the calculated map of the error Δn and $\Delta\kappa$ induced by an error $\delta\theta = 1°$. Similarly, an experimental error $\delta\Delta = 1°$ on the phase Δ result in the error Δn and $\Delta\kappa$ plotted in Fig. 6.17. Typically, these realistic errors $\delta\theta$ and $\delta\Delta$ induce large errors on n and κ. Especially when the phase Δ is badly measured, the error made on the determination of κ is huge. The discrepancy observed in Fig. 6.15 is explained Nagashima and Hangyo [24] by a $0.7°$ error of the angle of the polarizer.

To reduce the experimental errors, Neshat and Armitage [28] have built a THz-TDS ellipsometer with an $8\times$ F confocal scheme (F is the focal length of the THz optics) and fed by fiber-coupled antennas: thus one can vary the angle of incidence θ without readjusting the optical path in the setup. Moreover, the polarization response of the antennas is determined through the calibration method explained in Neshat and Armitage [29]. This permits the authors to get rid of the effect of some azimuthal offset angles due to the uncertainty in the polarizer angle, or misalignment of the components, especially a surface tilt $\delta\theta$ of the sample. Typically, the method corrects the angle ψ and Δ by several degrees in the

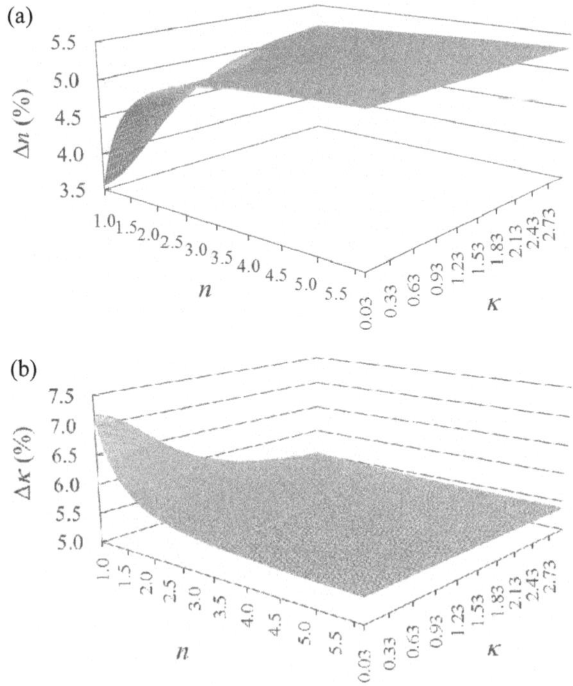

Figure 6.16 Δn (a) and $\Delta\kappa$ (b) versus n and κ deduced from ellipsometric data when the angle of incidence θ (here 45°) is wrong by an error $\delta\theta = 1^\circ$.

case of HR silicon, which leads to an improved, by more than 10%, determination of the refractive index n.

6.6.2 Thin-Film Ellipsometry

In optics, ellipsometry is widely used to characterize very thin films (nanometer thickness) deposited over a substrate, especially in microelectronics, for which surfaces of semiconductor wafers suffer from oxidation or other contamination effects. Also, ellipsometry can give information about the diffusion of dopants at the semiconductor surface. In the THz domain, ellipsometry can also be employed to characterize thin films. Let us suppose that the covered substrate is studied in reflection and that it is thick enough to omit Fabry–Pérot resonances in the substrate. But we will

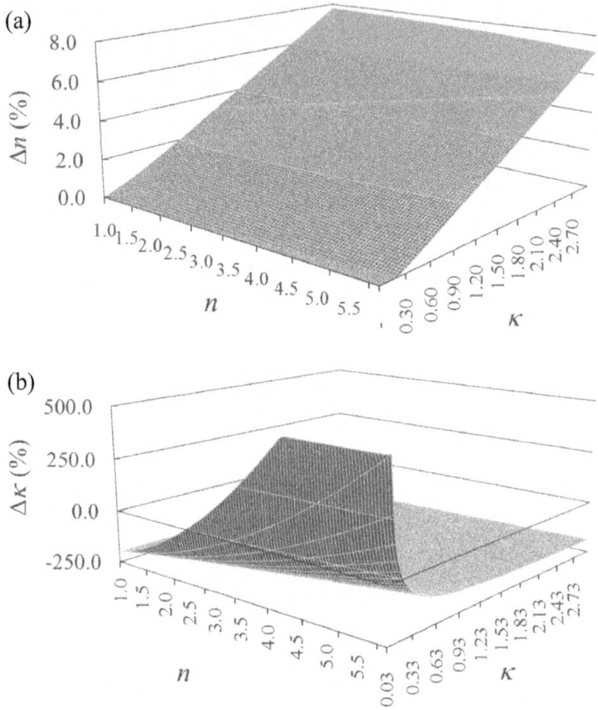

Figure 6.17 Δn (a) and $\Delta\kappa$ (b) versus n and κ deduced from ellipsometric data when the phase Δ is measured with an error $\delta\Delta = 1°$.

take into account Fabry–Pérot resonances in the film. Within these assumptions, the coefficient of reflection (Eq. A.32) is written as

$$\tilde{r} = \frac{\tilde{r}_{a,f} + \tilde{r}_{f,s} e^{-2j\,\tilde{k}_{fz}\,d_f}}{1 + \tilde{r}_{a,f}\tilde{r}_{f,s} e^{-2j\,\tilde{k}_{fz} d_f}}. \tag{6.28}$$

In Eq. 6.28, the subscripts a, f, and s are respectively related to the surrounding medium (air), the film and the substrate, and z is the direction normal to the surface. The coefficients of reflection at the interfaces are given by

$$\tilde{r}_{a,f} = \frac{\tilde{k}_{az} - \tilde{k}_{fz}\,\tilde{\alpha}_f}{\tilde{k}_{az} + \tilde{k}_{fz}\,\tilde{\alpha}_f},\ \tilde{r}_{f,s} = \frac{\tilde{k}_{fz} - \tilde{k}_{sz}\,\tilde{\alpha}_s}{\tilde{k}_{fz} + \tilde{k}_{sz}\,\tilde{\alpha}_s}, \tag{6.29}$$

where $\tilde{\alpha}_i\,(TM) = \tilde{\varepsilon}_{i-1}/\tilde{\varepsilon}_i = \tilde{n}_{i-1}^2/\tilde{n}_i^2$ ($i = a,\ f$ or s) and $\alpha_i\,(TE) = 1$. The component k_z of the wavevector is

$$\tilde{k}_{iz} = \frac{\omega}{c}\sqrt{\tilde{n}_i^2 - \sin^2\theta}. \tag{6.30}$$

For very thin films, Eq. 6.28 is simplified by taking the first-order development of the exponential terms:

$$\tilde{r} \approx \frac{\tilde{r}_{a,f} + \tilde{r}_{f,s}}{1 + \tilde{r}_{a,f}\,\tilde{r}_{f,s}} - 2j\,\tilde{k}_{fz}\,d_f \tilde{r}_{f,s} \frac{1 - \tilde{r}^2_{a,f}}{\left(1 + \tilde{r}_{a,f}\,\tilde{r}_{f,s}\right)^2} = \tilde{r}_o + \Delta\tilde{r}. \qquad (6.31)$$

The first term in Eq. 6.31 is the coefficient of reflection $\tilde{r}_{\mathrm{o}}$ of the bare substrate, while the second one corresponds to the variation of reflection $\Delta\tilde{r}$ induced by the thin film.

Figure 6.18 shows the ellipsometric parameters ψ and Δ for a conducting Si wafer ($\tilde{\varepsilon}_{\mathrm{S}} = 11.6 - j\,0.2$, i.e., $n_{\mathrm{s}} \approx 3.4$ and $\alpha_{\mathrm{s}} \approx 12.5$ cm^{-1} at 1 THz) covered by a thin dielectric film ($\tilde{\varepsilon}_{\mathrm{S}} = \varepsilon_{\mathrm{S}} = 2.25$), versus the thickness d_f of the film. The curves show the first Fabry–

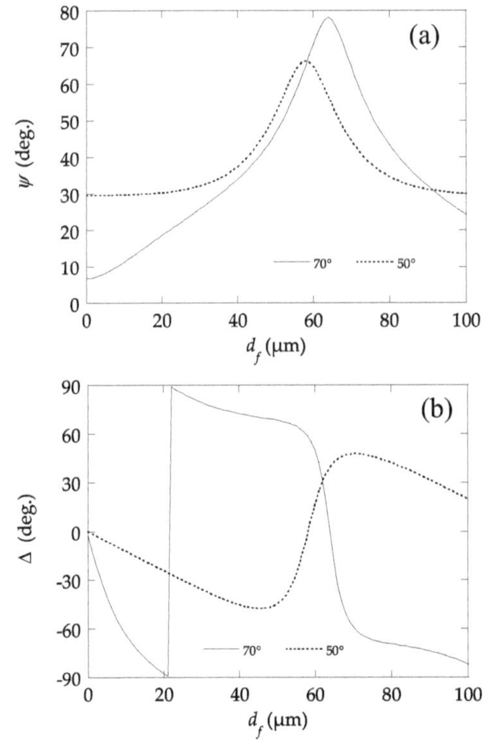

Figure 6.18 Ellipsometric parameters ψ (a) and Δ (b) versus d_{f} for $\theta = 50^\circ$ and 70° at $f = 1$ THz. The sample is a doped-Si substrate ($\tilde{\varepsilon}_{\mathrm{s}} = 11.6 - j\,0.2$) covered by a film ($\tilde{\varepsilon}_{\mathrm{f}} = 2.25$) of thickness d_{f}.

Pérot resonance in the film around $d_f \approx 60$ μm. The resonance is more pronounced when θ is bigger. The resonance makes both ψ and Δ varying with d_f and thus it determines the sensitivity of ellipsometry to detect and characterize thin films. For example, if the ellipsometric THz-TDS system allows measuring ψ with a resolution of 1°, the influence of a 4 μm film will be clearly observed as $\psi \approx$ 6.75° for the bare substrate, while $\psi \approx 7.75°$ with the 4 μm cover film. Looking at the phase angle Δ, a similar angular resolution will lead to a 0.12 μm sensitivity. To retrieve the refractive index of the film, a few times thicker film is necessary (roughly speaking, the precision on $\tilde{n}_f$ will be equal to the ratio film thickness over minimum detectable film thickness). This was reported by Neshat

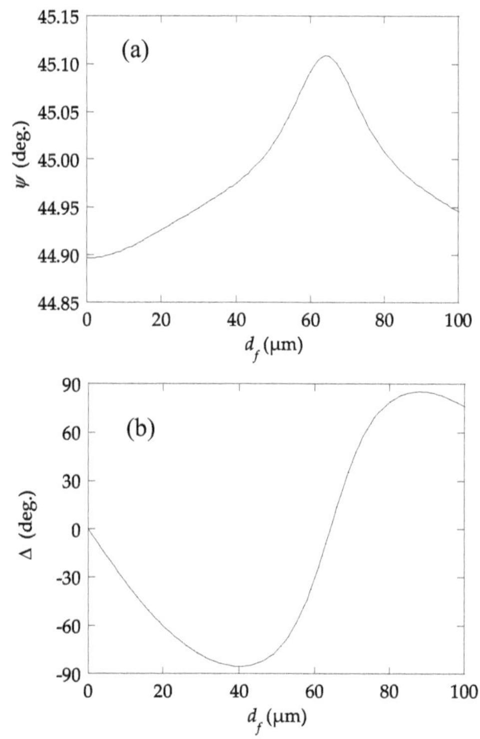

Figure 6.19 Ellipsometric parameters ψ (a) and Δ (b) versus d_f for $\theta = 70°$. The sample is an aluminum substrate ($\tilde{\varepsilon}_s = 3 \times 10^4 - j\,1 \times 10^6$) covered by a film ($\tilde{\varepsilon}_f = 2.25$) of thickness d_f.

and Armitage [28], who determined both d_f and n_f of a 1.95 μm thick oxide layer over a highly doped Si wafer (resistivity 0.011 Ω·cm).

When the film is deposited over a metal, i.e., a highly reflecting material, the effect on ψ is even weaker, while the variation $\Delta = 1^\circ$ permits to observe a 0.3 μm thick film (Fig. 6.19).

6.6.3 Generalized Ellipsometry

Generalized ellipsometry [30] is the extension of classical ellipsometry to the study of anisotropic materials. Due to anisotropy, the TE and TM polarizations are no more eigenmodes of the reflection scheme, i.e., a TE or TM incident beam is transformed, by reflection on the anisotropic sample, into an elliptically polarized one, with both TE and TM components. The easiest way to deal with the modeling of generalized ellipsometry is to employ the Jones matrix formalism [31]. This paragraph is inspired by the clear and synthetic presentation of generalized ellipsometry given by M. Schubert in [32]. We write the reflected field $\vec{E}_r$ as a tensor function of the incident field $\vec{E}_i$:

$$\vec{E}_r = \begin{pmatrix} B_p \\ B_s \end{pmatrix} = \overleftrightarrow{J} : \vec{E}_i = \begin{pmatrix} \tilde{j}_{pp} & \tilde{j}_{sp} \\ \tilde{j}_{ps} & \tilde{j}_{ss} \end{pmatrix} \begin{pmatrix} A_p \\ A_s \end{pmatrix}$$

$$= \overleftrightarrow{r} : \vec{E}_i = \begin{pmatrix} \tilde{r}_{pp} & \tilde{r}_{sp} \\ \tilde{r}_{ps} & \tilde{r}_{ss} \end{pmatrix} \begin{pmatrix} A_p \\ A_s \end{pmatrix}. \tag{6.32}$$

Here, for the sake of simplicity, the indices p and s are associated respectively to the TM and TE polarizations, A_i and B_i ($i = s$ or p) are the components of the incident or reflected fields, $\overleftrightarrow{J}$ is the Jones matrix. It follows immediately that the Jones matrix is the reflection matrix $\overleftrightarrow{r}$, so that $\tilde{j}_{ik} = \tilde{r}_{ik}$ ($i, k = s$ or p). The ellipsometric angles ψ and Δ are defined as previously (Eq. 6.24):

$$\tilde{\rho} = \tan\psi\, e^{j\Delta} = \frac{B_p}{A_p}\frac{A_s}{B_s}$$

$$= \frac{\tilde{r}_{pp}A_p + \tilde{r}_{sp}A_s}{A_p}\frac{A_s}{\tilde{r}_{ps}A_p + \tilde{r}_{ss}A_s}. \tag{6.33}$$

By setting the ratio of the incident amplitudes $\chi = A_p/A_s$, we get

$$\tilde{\rho} = \tan\psi\, e^{j\Delta} = \frac{\frac{\tilde{r}_{pp}}{\tilde{r}_{ss}} + \chi^{-1}\frac{\tilde{r}_{sp}}{\tilde{r}_{ss}}}{1 + \chi\frac{\tilde{r}_{ps}}{\tilde{r}_{ss}}} \equiv \frac{\tilde{R}_{pp} + \chi^{-1}\tilde{R}_{sp}}{1 + \chi\tilde{R}_{ps}} \tag{6.34}$$

$\tilde{R}_{ik} \equiv \tan\psi_{ik} e^{j\Delta_{ik}}$ are the generalized ellipsometry parameters. Characterizing an anisotropic sample, in the general case, requires measuring the following six parameters: ψ_{pp}, Δ_{pp}, ψ_{ps}, Δ_{ps}, ψ_{sp}, Δ_{sp}. In THz-TDS, this challenge is easily achieved by illuminating the sample with a TM beam and measuring both TM and TE reflected beams: this gives the respectively the $\tilde{r}_{pp}$ and $\tilde{r}_{ps}$ coefficients. Similarly, the $\tilde{r}_{sp}$ and $\tilde{r}_{ss}$ coefficients are obtained with a TE incident beam. The question is now to derive the material properties of the sample from $\tilde{r}_{ik}$ coefficients, i.e., the dielectric tensor $\overleftrightarrow{\varepsilon}$ (see Annex A.41). In the referential of the principal axes of the crystal, the tensor is diagonal. Unfortunately, in the most general case, the principal axes of the crystal are not oriented in the eigendirections related to the reflection at the sample surface (normal to the surface, and parallel to the surface in the plane of incidence and perpendicular to it). So there exists no simple expression relating $\overleftrightarrow{\varepsilon}$ and $\overleftrightarrow{r}$ [33]. Therefore published work has only dealt with well-oriented samples [34], or with samples of higher symmetry, like uniaxial crystals. For example, the well-known crystal $BaTiO_3$ is a study-case ferroelectric material, which, at room temperature, belongs to the tetragonal crystal system (point group P4 mm-C_{4v}) and thus is biaxial. $BaTiO_3$ was characterized in the far-infrared by FTIR ellipsometry by Hoshina et al. [35]. When the c axis of $BaTiO_3$ is parallel to the surface and incident plane, the reflection coefficients are given by

$$\tilde{r}_{ss} = \frac{\cos\theta - \sqrt{\tilde{n}_e^2 - \sin^2\theta}}{\cos\theta + \sqrt{\tilde{n}_e^2 - \sin^2\theta}},$$

$$\tilde{r}_{pp} = \frac{\sqrt{\tilde{n}_o^2 - \sin^2\theta} - \tilde{n}_o\tilde{n}_e\cos\theta}{\sqrt{\tilde{n}_o^2 - \sin^2\theta} + \tilde{n}_o\tilde{n}_e\cos\theta}, \tag{6.35}$$

while the $\tilde{r}_{sp}$ and $\tilde{r}_{sp}$ coefficients are null. $\tilde{n}_o$ and $\tilde{n}_e$ are respectively the ordinary and extraordinary complex refractive indices.

To our knowledge, THz-TDS generalized ellipsometry has only be applied to materials with an anisotropy induced by a magnetic field [36, 37], namely magneto-optic Kerr effect (MOKE). In [36], Ino et al. characterized an n-doped InAs wafer, under a magnetic field applied perpendicularly to the sample face (z direction). The magnetic field $\vec{H}$ induces a helical movement of the free carriers (cyclotron resonance). Therefore, the displacement of the carriers is not collinear with the EM field, which leads to an anisotropy of the electromagnetic response of the material. In the small signal hypothesis, the dielectric tensor writes:

$$\tilde{\varepsilon} = \begin{pmatrix} \tilde{\varepsilon}_{xx} & \tilde{\varepsilon}_{xy} & 0 \\ -\tilde{\varepsilon}_{xy} & \tilde{\varepsilon}_{xx} & 0 \\ 0 & 0 & \tilde{\varepsilon}_{zz} \end{pmatrix}. \tag{6.36}$$

The x direction is along the surface of the sample, perpendicular to the plane of incidence. The components of $\tilde{\varepsilon}$ may be calculated with the Drude model, into which $\vec{H}$ contributes through the Lorentz force:

$$\begin{cases} \tilde{\varepsilon}_{xx} = \tilde{\varepsilon}_{\infty} - \frac{\omega_p^2\left(\omega^2 + j\omega\Gamma\right)}{\left(\omega^2 + j\omega\Gamma\right)^2 - \omega^2\omega_c^2}, & \tilde{\varepsilon}_{xy} = \frac{j\omega_p^2\omega\omega_c}{\left(\omega^2 + j\omega\Gamma\right)^2 - \omega^2\omega_c^2}, \\ \tilde{\varepsilon}_{zz} = \tilde{\varepsilon}_{\infty} - \frac{\omega_p^2}{\omega^2 + j\omega\Gamma}, & \text{with } \omega_c = \frac{e\left|\vec{H}\right|}{m\,c}. \end{cases} \tag{6.37}$$

Here ω_c and ω_p are respectively the cyclotron and plasma frequencies, and $\tilde{\varepsilon}_{\infty}$ is the dielectric constant at higher frequencies. Note that the influence of the magnetic field appears only through ω_c. Thus the magnetic field induces the non-diagonal components $\tilde{\varepsilon}_{xy} = -\tilde{\varepsilon}_{yx}$, and it perturbs $\tilde{\varepsilon}_{xx}$, but it has no influence on $\tilde{\varepsilon}_{zz}$. It can be shown [37] that

$$\begin{cases} \tilde{\varepsilon}_{xx} = \frac{\left(1+\tilde{r}_{pp}\right)}{2\left(1-\tilde{r}_{pp}\right)^2\cos^2\theta}\left(\frac{1+\tilde{r}_{pp}^2}{1+\tilde{r}_{pp}} + \sqrt{1+\tilde{r}_{pp}^2 - \left(1-\tilde{r}_{pp}\right)^2\sin^2 2\theta}\right), \\ \tilde{\varepsilon}_{xy} = -\frac{\tilde{r}_{sp}}{\cos\theta}\left(\cos\theta + \sqrt{\tilde{\varepsilon}_{xx} - \sin^2\theta}\right)\left(\tilde{\varepsilon}_{xx}\cos\theta + \sqrt{\tilde{\varepsilon}_{xx} - \sin^2\theta}\right). \end{cases} \tag{6.38}$$

Figure 6.20 gives an example of THz-TDS ellipsometric determination of the imaginary part of $\tilde{\varepsilon}_{xx}$ and $\tilde{\varepsilon}_{xy}$ part of n-doped InAs under a 0.48 T magnetic field [36].

6.7 Identification of Absorption Peaks

For several applications, especially in security and environment, a whole characterization of the involved materials is not necessary. One desires mostly to clearly isolate spectral fingerprints (ideally

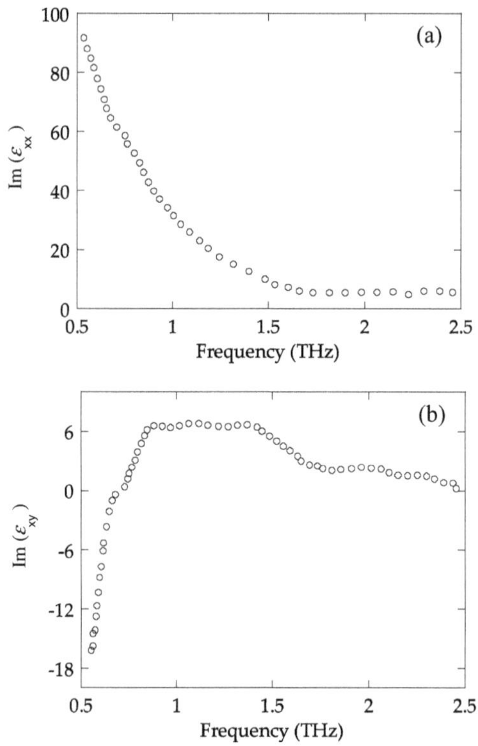

Figure 6.20 Spectra of the imaginary part of ε_{xx} and ε_{xy} of n-type InAs under a magnetic field, with $\omega_c = 2.89$ THz. Adapted from Ino et al. [36].

peaks) in the reflection spectrum and to record their frequencies, from which the chemical composition of the material could be identified. It is known, in visible and infrared spectroscopy, that computing the first and second derivatives of the reflection or transmission spectra may help in analyzing these spectra. When absorption peaks are sharp, the derivation reduces the contributions from broad spectral features [38]. In the THz domain, a smart procedure has been proposed by Zhong et al. [39] based on the following observation: the frequency derivative of the reflection spectrum (modulus) resembles nicely the absorption spectrum of the material. Thus the absorption peaks are clearly observed when computing the derivative of the recorded reflection spectrum, and

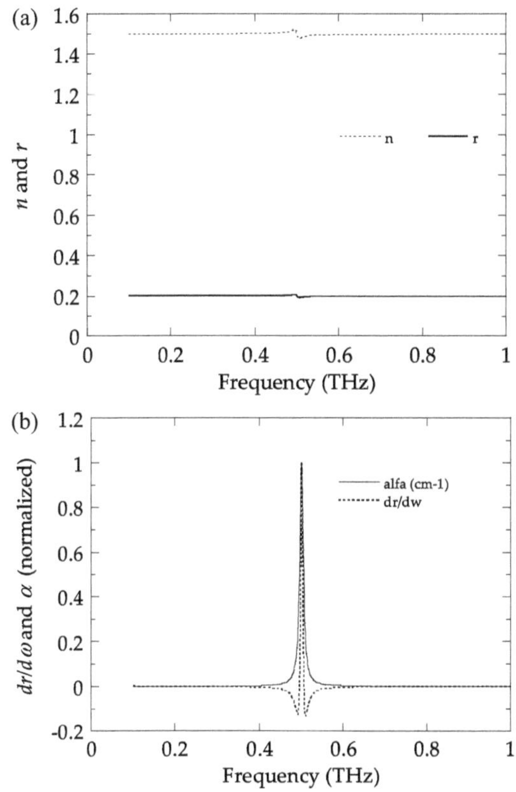

Figure 6.21 (a) Calculated refractive index n and coefficient of reflection r versus frequency for a dielectric material showing a Lorentzian absorption at 0.5 THz. (b) Calculated coefficient of absorption α and derivative $-dr/d\omega$ (both normalized) versus frequency.

the peak identification is easier than from the reflection spectrum. Let us illustrate the method by a calculated example. We consider a material exhibiting a Lorentzian absorption peak at 0.5 THz ($\alpha_{\text{peak}} = 10\ \text{cm}^{-1}$), whose spectral width is 10 GHz. Far from the resonance, the refractive index is $n = 1.5$ (Fig. 6.21).

The coefficient of reflection lies around 0.2, while the resonance induces a small variation (3.5%) of r nearby the resonance (Fig. 6.21a). The derivative ($-dr/d\omega$) of r shows a peak quite similar

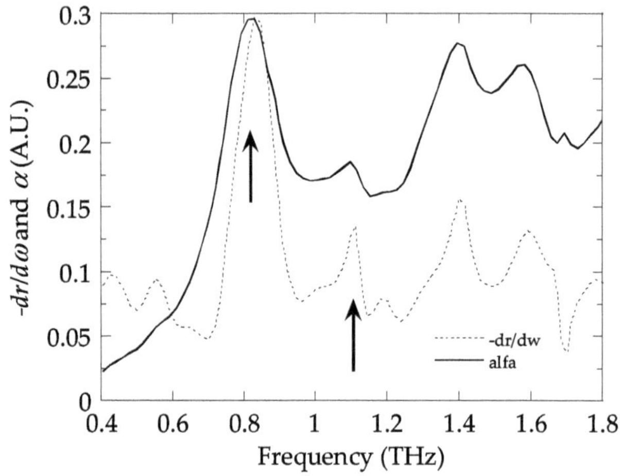

Figure 6.22 The coefficient of absorption α and $-dr/d\omega$ versus frequency of the explosive material RDX measured THZ-TDS in reflection. Adapted from Zhong et al. [39].

to the coefficient of absorption (Fig. 6.21b), at exactly the same frequency. As compared to the r spectrum, the $-dr/d\omega$ peak is noticeably contrasted, which helps in identifying the absorption peaks.

Figure 6.22 presents the experimental validation of the method performed by Zhong et al. [39], using a pellet of RDX. The coefficient of absorption α and the derivative $-dr/d\omega$ are respectively determined from transmission and reflection THz-TDS. Peaks at 0.88, 1.05, 1.33, and 1.55 THz are more contrasted in the $-dr/d\omega$ curve (see arrows in Fig. 6.22).

In the case of weakly absorbing materials ($\kappa \ll n$), the phase of the reflected signal is given by an approached expression of Eq. 6.9:

$$\tan\varphi \approx \frac{-2\kappa}{n^2 - 1}. \tag{6.39}$$

Because $\kappa \ll n$, the refractive index n can be considered constant, i.e., $n \approx n_\infty$. The second derivative of φ is obtained from the

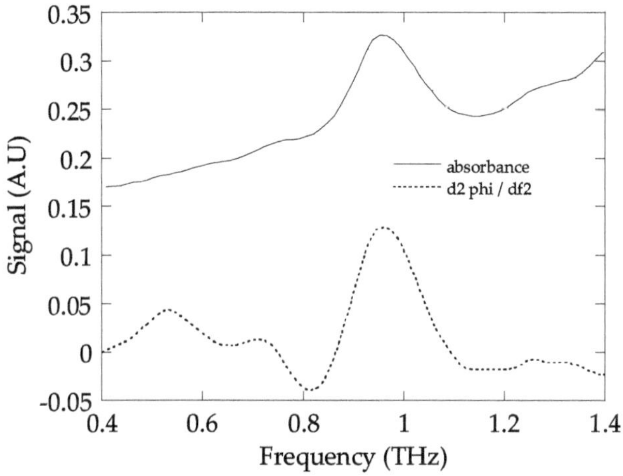

Figure 6.23 Absorbance spectrum (upper curve), measured by transmission THz-TDS and the second order derivative spectrum (lower curve) of the phase measured in reflection. Here the sample is a 1.5 mm thick sample of theophylline powder. Adapted from Zhong et al. [40].

differentiation of Eq. 6.39:

$$\frac{d^2\varphi}{d\omega^2} \approx \frac{-2}{n_\infty^2 - 1}\frac{d^2\kappa}{d\omega^2}. \tag{6.40}$$

From the Kramers–Kronig relations, it can be shown that the second order derivative $d^2\kappa/d\omega^2$ bears a similar shape with κ. Therefore, the second order derivative spectrum of the recorded phase contains and clearly exhibits the absorption features of the sample. This has been demonstrated by Zhong et al. [40], who characterized different explosive and drug samples by THz-TDS in reflection, in view of security applications. Figure 6.23, adapted from this reference, depicts the absorbance spectrum (upper curve) of theophylline (a drug used to treat asthma and other lung diseases), measured by transmission THz-TDS and the second order derivative of the phase measured in reflection. The good contrast of the absorption peak in the phase derivative curve is clearly seen.

References

1. Zaytsev, K. I., Gavdush, A. A., Chernomyrdin, N. V., and Yurchenko, S. O. (2015). Highly accurate in vivo terahertz spectroscopy of healthy skin: variation of refractive index and absorption coefficient along the human body, *IEEE Trans. THz Sci. Technol.*, **5**, pp. 817–827.
2. Kniffin, G. P., and Zurk, L. M. (2012). Model-based material parameter estimation for terahertz reflection spectroscopy, *IEEE Trans. THz Sci. Technol.*, **2**, pp. 231–241.
3. Kniffin, G. P. (2010). Model-based material parameter estimation for terahertz reflection spectroscopy, Master of Science thesis, Portland State University, Paper 241, p. 58.
4. Jepsen, P. U., Jensen, J. K., and Møller, U. (2008). Characterization of aqueous alcohol solutions in bottles with THz reflection spectroscopy, *Opt. Express*, **16**, pp. 9318–9331.
5. Parrott, E. P. J., Sy, S. M. Y., Blu, T., Wallace, V. P., and Pickwell-MacPherson, E. (2011). Terahertz pulsed imaging in vivo: measurements and processing methods, *J. Biomed. Opt.*, **16**, p. 106010.
6. Huang, S., Ashworth, P. C., Kan, K. W., Chen, Y., Wallace, V. P., Zhang, Y.-T., and Pickwell-MacPherson, E. (2009). Improved sample characterization in terahertz reflection imaging and spectroscopy, *Opt. Express*, **17**, pp. 3848–3854.
7. Hirori, H., Yamashita, K., Nagai, M., and Tanaka, K. (2004). Attenuated total reflection spectroscopy in time domain using terahertz coherent pulses, *Jpn. J. Appl. Phys.*, **43**, p. L1287.
8. Nagai, M., Yada, H., Arikawa, T., and Tanaka, K. (2006). Terahertz time-domain attenuated total reflection spectroscopy in water and biological solution, *Int. J. Infrared Millimeter Waves*, **27**(4), pp. 505–515.
9. Newnham, D. A., and Taday, P. (2008). Pulsed terahertz attenuated total reflection spectroscopy, *Appl. Spectrosc.*, **62**, p. 394.
10. Takeya, K., Muto, K., Ishihara, Y., and Kawase, K. (2016). Monitoring theophylline concentrations in saline using terahertz ATR spectroscopy, *Appl. Sci.*, **6**, p. 72.
11. Soltani, A., Busch, S. F., Plew, P., Balzer, J. C., and Koch, M. (2016). THz ATR spectroscopy for inline monitoring of highly absorbing liquids, *Int. J. Infrared Millimeter Waves*, **37**, pp. 1001–1006.
12. Nakanishi, A., Kawada, Y., Yasuda, T., Akiyama, K., and Takahashi, H. (2012). Terahertz time domain attenuated total reflection spectroscopy with an integrated prism system, *Rev. Sci. Instrum.*, **83**, p. 033103.

13. Jones, C. E., and Hilton, A. R. (1968). Optical properties by far infrared ellipsometry, *J. Electrochem. Soc.*, **115**, pp. 106–107.

14. Barth, K.-L., Bhme, D., Kamars, K., Keilmann, F., and Cardona, M. (1993). Far-IR spectroscopic ellipsometer, *Thin Solid Films*, **234**, pp. 314–317.

15. Röseler, A. (2005). Spectroscopic infrared ellipsometry, in *Handbook of Ellipsometry*, Tompkins, H. G., and Irene, E. A., eds. (William Andrew Publishing, Norwich), pp. 763–798.

16. DeNicola, R. O., Saifi, M. A., and Frazee, R. E. (1972). Epitaxial layer thickness measurement by far infrared ellipsometry, *Appl. Opt.*, **11**, pp. 2534–2539.

17. Kircher, J., Henn, R., Cardona, M., Richards, P. L., and Williams, G. P. (1997). Far-infrared ellipsometry using synchrotron radiation, *J. Opt. Soc. Am. B*, **104**, p. 705.

18. Henn, R., Kircher, J., Cardona, M., Wittlin, A., Duijn, V. H. M., and Menovsky, A. A. (1996). Far-infrared c-axis response of $(La,Sr)_2CuO_4$-delta determined by ellipsometry, *Phys. Rev. B*, **53**, pp. 9353–9358.

19. Leibiger, G., Gottschalch, V., Rheinländer, B., Sik, J., and Schubert, M. (2001). Model dielectric function spectra of GaAsN for far-infrared and near-infrared to ultra violet wavelengths, *J. Appl. Phys.*, **89**, pp. 4927–4938.

20. Dobriu, A., and Otani, C. (2009). Ellipsometry in the terahertz range for liquid identification, in *Proceed. of the 34th International Conference on Infrared, Millimeter, and Terahertz Waves*, Busan, Korea, 21–25 Sept. 2009.

21. Sato, M., Higuchi, T., Kanda, N., Konishi, K., Yoshioka, K., Suzuki, T., Misawa, K., and Kuwata-Gonokami M. (2013). Terahertz polarization pulse shaping with arbitrary field control, *Nat. Photonics*, **7**, pp. 724–731.

22. van der Valk, N. C., Wenckebach, T., and Planken, P. C. (2004). Full mathematical description of electro-optic detection in optically isotropic crystals, *J. Opt. Soc. Am. B*, **21**, pp. 622–631.

23. Gaborit, G., Biciunas, A., Bernier, M., and Coutaz, J.-L. (2015). Emitting and receiving terahertz vectorial antennas based on cubic electro-optic crystal, *IEEE Trans. THz Sci. Technol.*, **5**, pp. 828–835.

24. Nagashima, T., and Hangyo, M. (2001). Measurement of complex optical constants of a highly doped Si wafer using terahertz ellipsometry, *Appl. Phys. Lett.*, **79**, pp. 3917–3919.

25. Matsumoto, N., Fujii, T., Kageyama, K., Takagi, H., Nagashima, T., and Hangyo, M. (2009). Measurement of the soft mode dispersion in $SrTiO_3$

by terahertz time-domain spectroscopic ellipsometry, *Jpn. J. Appl. Phys.*, **48**, p. 09KC11.

26. Matsumoto, N., Hosokura, T., Nagashima, T., and Hangyo, M. (2011). Measurement of the dielectric constant of thin films by terahertz time-domain spectroscopic ellipsometry, *Opt. Lett.*, **36**, pp. 265–267.
27. Yatsugi, K., Matsumoto, N., Nagashima, T., and Hangyo, M. (2011). Transport properties of free carriers in semiconductors studied by terahertz time-domain magneto-optical ellipsometry, *Appl. Phys. Lett.*, **98**, p. 212108.
28. Neshat, M., and Armitage, N. P. (2012). Terahertz time-domain spectroscopic ellipsometry: instrumentation and calibration, *Opt. Express*, **20**, pp. 29063–29075.
29. Neshat, M., and Armitage, N. P. (2012). Improved measurement of polarization state in terahertz polarization spectroscopy, *Opt. Lett.*, **37**, pp. 1811–1813.
30. Azzam, R. M. A., and Bashara, N. M. (1972). Generalized ellipsometry for surfaces with directional preference: application to diffraction gratings, *J. Opt. Soc. Am.*, **62**, pp. 1521–1523.
31. Saleh, B. E. A., and Teich, M. C. (1991). *Fundamentals of Photonics* (John Wiley & Sons, Inc., New York), Chapter 6.
32. Schubert, M. (2005). Theory and application of generalized ellipsometry, in *Handbook of Ellipsometry*, Tompkins, H. G., and Irene, E. A., eds. (William Andrew Publishing, Norwich), pp. 637–717.
33. Landry, G. D., and Maldonado, T. A. (1995). Complete method to determine transmission and reflection characteristics at a planar interface between arbitrarily oriented biaxial media, *J. Opt. Soc. Am. A*, **12**, pp. 2048–2063.
34. Schubert, M. (2004). Anisotropic substrates, in *Infrared Ellipsometry on Semiconductor Layer Structures* (Springer-Verlag, Berlin), Chapter 4, pp. 67–79.
35. Hoshina, T., Kanehara, K., Takeda, H., and Tsurumi, T. (2014). Terahertz dielectric response of single-domain $BaTiO_3$ measured by far-infrared spectroscopic ellipsometry, *Jpn. J. Appl. Phys.*, **53**, p. 09PD03.
36. Ino, Y., Shimano, R., Svirko, Y., and Kuwata-Gonokami, M. (2004). Terahertz time domain magneto-optical ellipsometry in reflection geometry, *Phys. Rev. B*, **70**, p. 155101.
37. Yatsugi, K., Matsumoto, N., Nagashima, T., and Hangyo, M. (2011). Transport properties of free carriers in semiconductors studied by

terahertz time-domain magneto-optical ellipsometry, *Appl. Phys. Lett.*, **98**, p. 212108.

38. Whitbeck, M. R. (1981). Second derivative infrared spectroscopy, *Appl. Spectrosc.*, **35**, pp. 93–95.
39. Zhong, H., Redo-Sanchez, A., and Zhang, X.-C. (2006). Identification and classification of chemicals using terahertz reflective spectroscopic focal plane imaging system, *Opt. Express*, **14**, pp. 9130–9141.
40. Zhong, H., Zhang, C., Zhang, L., Zhao, Y., and Zhang, X.-C. (2008). A phase feature extraction technique for terahertz reflection spectroscopy, *Appl. Phys. Lett.*, **92**, p. 221106.

Chapter 7

Pump–Probe THz-TDS

7.1 Introduction

Pump–probe spectroscopy is a method used to study fast processes in dynamic systems based on repetitive laser pulses and time-equivalent techniques that allow to record the fast processes with slow detectors. A pump beam induces a temporary change or excites a material or sample under investigation. A probe beam is then used to interrogate the change in the sample a short time afterwards. Using this technique, induced changes with very short lifetimes can be studied. Because THz-TDS is based on short temporal pulses, it naturally allows one to perform pump-probe experiments to study the dynamic response of matter and devices at THz frequencies. This constitutes a significant advantage over continuous-wave characterization methods, like FTIR. Most published works are based on the optical pump of a sample located in a classical THz-TDS system. The THz pulse probes the sample with a time delay τ compared to the optical pulse. This type of dynamic far-infrared characterization of a sample is either called time-resolved THz spectroscopy (TRTS) [1], or optical pump-THz probe (OPTP) [2]. It was first introduced by Nuss et al. [3] in view of measuring the lifetime of photo-excited carriers in GaAs. Their

Principles of Terahertz Time-Domain Spectroscopy
Jean-Louis Coutaz, Frédéric Garet, and Vincent Wallace

ISBN 978-981-4774-56-7 (Hardcover), 978-0-203-70969-6 (eBook)
www.panstanford.com

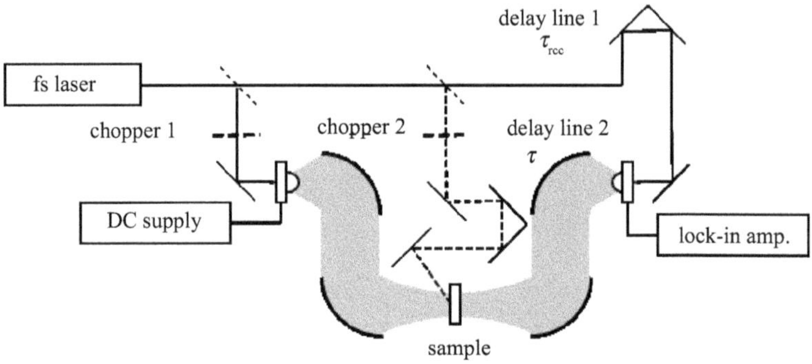

Figure 7.1 Typical optical pump–THz probe spectroscopy system, using THz PC antennas.

experimental scheme was quite different from more modern setups: The THz probe beam was generated as a Cerenkov shock wave in a $LiTaO_3$ crystal onto which the GaAs sample was located. The first optical pump-THz probe experiment, based on a classical THz-TDS system (Fig. 7.1), was published in 1991 by Greene et al. [4], who investigated the photocarriers' dynamics in semi-insulating GaAs. Since these pioneering reports, time-resolved THz-TDS has been the subject of intense research, especially the methodology of the measurements, which has been rigorously studied by Kuzel and his team [2, 5, 6].

7.2 Pump–Probe THz-TDS: Basic Principles

A typical OPTP setup is presented in Fig. 7.1. It is built from a classical THz-TDS system, in which a part of the laser beam illuminates the sample through a second delay line. The main difficulty is to illuminate the sample at the same location by both the THz probe beam and the laser pump beam. Since the laser beam diameter is typically $1 \sim 10$ μm, it is usually more tightly focused than the THz beam and only a limited part of the sample contributes to the photo-variation of the transmitted THz beam, making the signal of interest rather weak. This could be improved by using a double modulation technique. Both the laser beams that

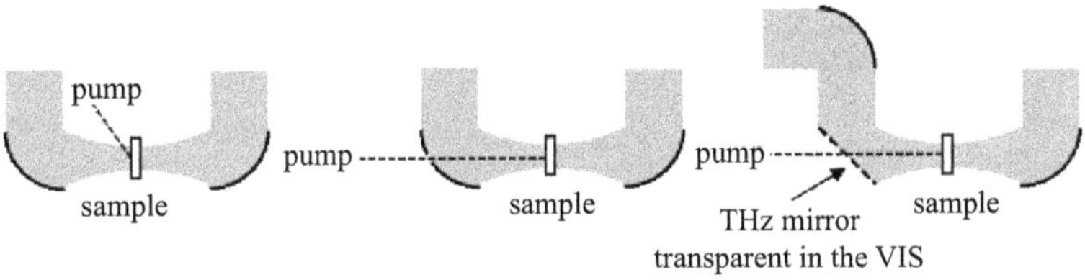

Figure 7.2 Scheme for TRTS where the sample is pumped at oblique incidence (left) and normal incidence (center and right).

excite the emitter and that pump the sample are modulated at two different frequencies, and the recording of signals is performed at the difference of these two frequencies.

Pumping the sample can be done at oblique or normal incidence (Fig. 7.2). In this latter case, THz and optical pulses propagate along the same direction and are superimposed, which simplifies the analysis of the recorded data and also makes the overlapping time shorter for both THz and laser pulses in the sample, leading to a better time resolution. Practically, this could be achieved by propagating the laser beam through a small hole drilled at the center of the THz parabolic mirror (Fig. 7.2, center) or by employing a dichroic mirror, like a glass or quartz sample covered by an indium tin oxide (ITO) film. The ITO film is a transparent conductor; thus, the laser light goes through it, while the THz beam is reflected.

With a view to analyzing the recorded signals, let us again use Eq. 4.46 of the received signal:

$$S_{\text{det}}\left(\tau_{\text{rec}}\right) = \int_{-\infty}^{+\infty}\int_{-\infty}^{+\infty} S_{\text{laser}}\left(t - t' - \tau_{\text{rec}}\right) G\left(t'\right) S_{\text{THz}}(t)\, dt' dt. \quad (7.1)$$

$S_{\text{THz}}(t)$ is the THz waveform that impinges onto the receiver, $G(t)$ is the time response of the receiver, and $S_{\text{laser}}(t)$ is the laser pulse that triggers the receiver. τ_{rec} is the time delay between the emission and detection of the THz pulse. The received THz waveform is equal to the emitted one that is transmitted through the sample:

$$S_{\text{THz}}(t) = \int_{-\infty}^{+\infty} S_{\text{THz,emit}}\left(t - t''\right) T\left(t''\right) dt''. \quad (7.2)$$

$T(t)$ describes the temporal behavior of the sample transmission, i.e., the inverse Fourier transform of its transmission spectrum.

It also includes the effect of propagation from the emitter to the receiver if this propagation suffers from attenuation and dispersion. In the case of a sample that is photo-excited with a time delay τ compared to the THz emission, one gets

$$T(t) \equiv T(t, \tau) = T_o(t) + \Delta T(t - \tau). \quad (7.3)$$

$T_o(t)$ and $\Delta T(t - \tau)$ are the time response of the non-illuminated sample and the photo-induced variation of transmission, respectively. Therefore, the detected THz signal is

$$S_{\text{det}}(\tau_{\text{rec}}, \tau) = \int_{-\infty}^{+\infty}\int_{-\infty}^{+\infty}\int_{-\infty}^{+\infty} S_{\text{laser}}(t - t' - \tau_{\text{rec}})\, S_{\text{THz,emit}}(t - t'') \times T(t'', \tau)\, G(t')\, dt\, dt' dt''. \quad (7.4)$$

The photo-induced time response χ of the sample is not instantaneous, which means that the transmission at time t depends on the photo-induced phenomenon that happens some time before:

$$T(t, \tau) = T_o(t) + \int_{-\infty}^{t} \chi(t - t')\, I_{\text{laser}}(t', \tau)\, dt'. \quad (7.5)$$

$I_{\text{laser}}(t, \tau)$ describes the temporal shape of the pump laser pulse, whose peak reaches the sample with delay time τ compared to the THz pulse. $T_o(t)$ is the time response of the non-illuminated sample, which is simply the inverse Fourier transform of its transmission spectrum:

$$T_o(t) = \frac{1}{\sqrt{2\pi}} \int_{-\infty}^{+\infty} T_o(\omega)\, e^{j\omega t} dt. \quad (7.6)$$

Let us notice that Eq. 7.5 is already simplified, in the sense that it includes the integration of the laser intensity, while a rigorous calculation will imply a third-order nonlinear tensorial interaction between the laser optical field, the THz field, and the material response [2]. Because of causality, $\chi(t)$ is 0 for negative times ($\chi(t < 0) = 0$). Thus, one can enlarge the integration domain in Eq. 7.5 up to $t' \to +\infty$ giving:

$$\Delta T(t - \tau) = \int_{-\infty}^{+\infty} \chi(t - t')\, I_{\text{laser}}(t', \tau)\, dt'. \quad (7.7)$$

The detected signal when the sample is not excited is

$$S_o(\tau_{rec}) = \int_{-\infty}^{+\infty}\int_{-\infty}^{+\infty}\int_{-\infty}^{+\infty} S_{laser}(t - t' - \tau_{rec})\, S_{THz,emit}(t - t'') \times T_o(t'')\, G(t')\, dt\, dt' dt''. \quad (7.8)$$

This expression is greatly shortened by substituting the integrals of the time-shifted terms by the related convolution products:

$$S_o(\tau_{rec}) = \int_{-\infty}^{+\infty} [S_{THz,emit} \otimes T_o](t) \times [S_{laser} \otimes G](t - \tau_{rec})\, dt. \quad (7.9)$$

Similarly, the induced change in the recorded waveform is

$$\Delta S(\tau_{rec}, \tau) = \int_{-\infty}^{+\infty} [S_{THz,emit} \otimes \Delta T](t - \tau) \times [S_{laser} \otimes G](t - \tau_{rec})\, dt. \quad (7.10)$$

The recorded differential signal $\Delta S(\tau_{rec}, \tau)$ does not depend directly on $\Delta T(t)$, which is the required information, but on the convolution product $S_{THz,emit} \otimes \Delta T$.

Therefore, the time resolution of the experiment is limited by the duration of the THz pulses. Typically, in common THz-TDS setups, the pulse duration is of the order of $0.3 \sim 1$ ps. Sub-ps resolution may be achieved by a proper numerical deconvolution of the recorded differential waveform (Eq. 7.10), which requires knowledge of the shape of the THz waveform $S_{THz,emit}(t)$ that impinges on the sample [7–10].

The difference THz waveform is simply measured by modulating the laser pump beam, for example with a chopper, and reading the signal delivered by the receiver with a lock-in amplifier, assuming the decay of the excitation of the sample is much shorter than the chopper period. Figure 7.3 shows the recorded waveforms that are transmitted by a 500 μm thick intrinsic GaAs wafer. The "excited" waveform is measured at $\tau = 2.67$ ps after the excitation of the wafer by an 80 fs laser pulse at 800 nm. The difference between the excited and non-excited waveforms corresponds to the experimental curve $\Delta S(\tau_{rec}, \tau)$, whose expression is given by Eq. 7.10.

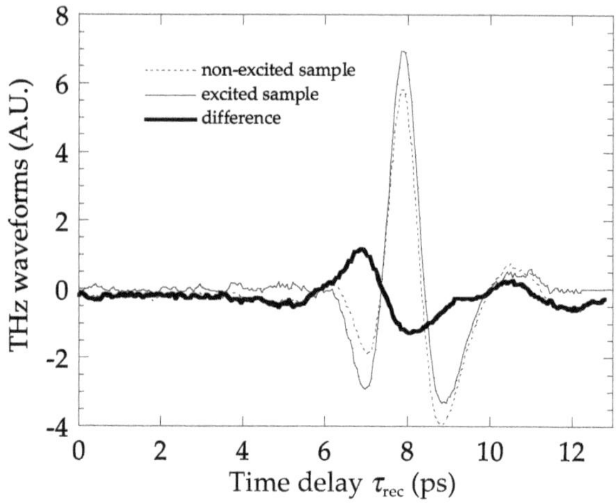

Figure 7.3 THz waveforms transmitted by a 500 μm thick GaAs wafer, which is pumped by a laser beam at $\lambda = 800$ nm (continuous line). The time delay between the optical excitation and the optical probe is $\tau = 2.67$ ps. The dotted line represents the transmitted waveform when the sample is not photo-excited, and the bold line is the difference between the laser being on and off.

This procedure is repeated for different delays τ. Figure 7.4 shows the curves $\Delta S(\tau_{rec}, \tau)$ versus τ_{rec} for different delays τ for the same intrinsic GaAs wafer. At $\tau = 0$ ps (note that the delay time origin is arbitrarily chosen), the optical pulse impinges the sample after the THz probe pulse, so this probe pulse does not encounter any effect of optical excitation of the sample. Therefore, the $\Delta S(\tau_{rec}, \tau)$ curve is null. When τ increases, the temporal overlapping of the pump and probe pulses is bigger as well, and thus the $\Delta S(\tau_{rec}, \tau)$ shows a more pronounced pulse feature.

7.3 Samples with a Slow Photo-Induced Response

Getting the response function $\Delta T(t, \tau)$ of the excited sample from Eq. 7.10 requires moving to the frequency domain, as will be explained below. However, when the photo-induced variation of the

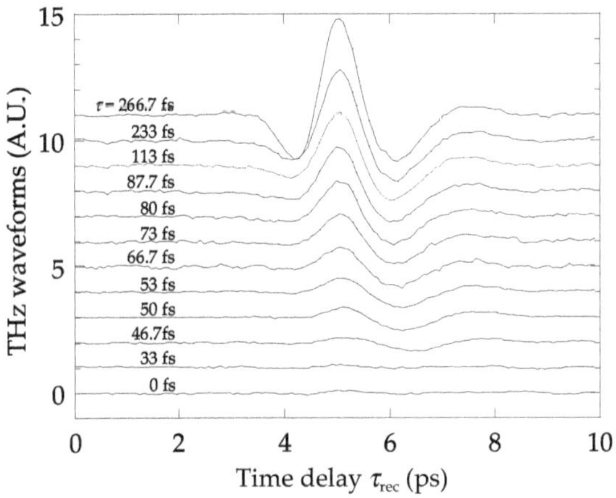

Figure 7.4 Differential THz waveforms transmitted by a 500 μm thick GaAs wafer, which is pumped by a laser beam at $\lambda = 800$ nm. Each curve corresponds to a different value of the time delay τ between the optical excitation and the optical probe. For the sake of legibility, each curve is vertically shifted by 1.

sample transmission is slow compared to the duration of the THz pulses, the first convolution product in Eq. 7.10 becomes a simple product:

$$[S_{\text{THz,emit}} \otimes \Delta T](t-\tau) \approx S_{\text{THz,emit}}(t-\tau) \times \Delta T(t-\tau)$$

Because $S_{\text{THz,emit}}(t-\tau)$ is not normally 0 (only when $t \approx \tau$) and $\Delta T(t)$ is a slowly varying function, one can substitute $\Delta T(t-\tau)$ by $\Delta T(\tau)$ in the latter expression. Finally, we get

$$\Delta S(\tau_{\text{rec}}, \tau) \approx \Delta T(\tau) \int_{-\infty}^{+\infty} S_{\text{THz,emit}}(t)\, [S_{\text{laser}}(t-\tau_{\text{rec}}) \otimes G(t-\tau_{\text{rec}})]\, dt. \tag{7.11}$$

It follows that the time response of the sample is simply obtained by measuring the THz signal variation at any delay τ_{rec} for which there exists a signal, i.e., the integral in Eq. 7.11 is not null. This is demonstrated in Fig. 7.5a. The optically induced variation of THz transmission of a semi-intrinsic GaAs wafer is plotted versus the time delay τ between the THz and optical pulses. The experimental points have been measured for different frequencies, and the general

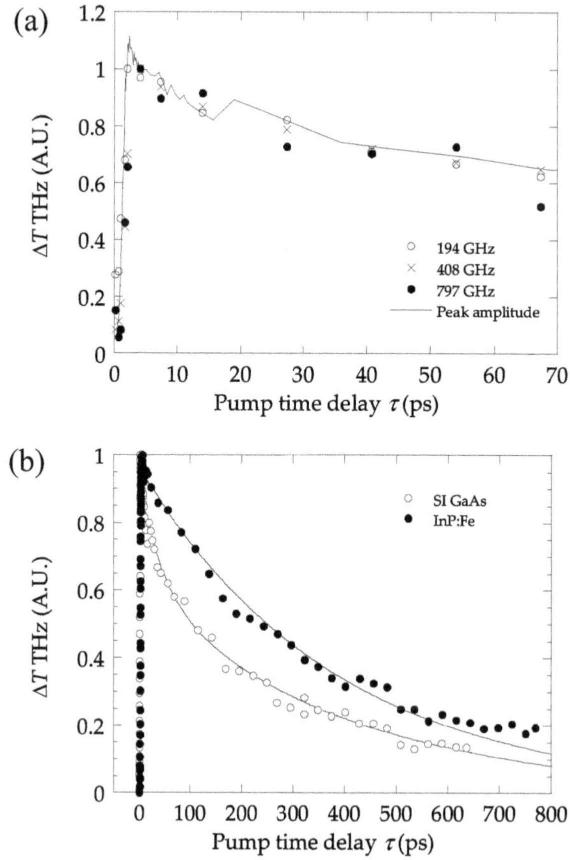

Figure 7.5 (a) Photo-induced variation of the transmission of a SI-GaAs wafer. Points are extracted at different frequencies from the transmission spectra, while the continuous line corresponds to the time-evolution of the THz pulse peak amplitude. (b) Photo-induced variation of the transmission of SI-GaAs and InP:Fe wafers. The continuous lines are exponential fits.

tendency of the plots is the same. Therefore, one can simply record the peak amplitude of the THz waveform [7], without the need of Fourier-transforming the waveforms (choosing the peak amplitude of the THz waveform optimizes the experimental *SNR*). The corresponding curve is plotted as a continuous line in Fig. 7.5a, which exhibits the same behavior as the spectral points. At very early times, i.e., for τ of the order or smaller than the THz waveform duration, the simplification leading to relation Eq. 7.11 is no more valid: the continuous line differs from the spectral points.

Figure 7.5b shows the peak amplitude variation of the THz waveform transmitted by a wafer of GaAs (350 μm thick, semi-intrinsic) and a wafer of InP (395 μm thick, doped with iron [3.5 × 10^{16} cm^{-3}] to compensate for residual conductivity). Both wafers are pumped by a laser pulse at 800 nm. The InP curve is fitted by a single exponential decay curve (decay time 380 ps), which could be related to the free carrier lifetime in this material (400 ps). The GaAs data are fitted by a double exponential curve, whose decay times are 47 ps and 400 ps. The longest decay time could also be linked to the free carrier lifetime in GaAs, while the shortest one could be due to a faster trapping time of the free carriers by surface or bulk defects, and also to a diffusion process: the carriers excited at the surface move into the bulk material [11].

This technique, based on a simple record of the temporal variation of the THz waveform, has been widely used to characterize transient photoconductivity in semiconductors [10, 12, 13], superconductors [7], molecular crystals [9], graphene layers [14], solvation in liquids [15], and so forth.

7.4 Pump-and-Probe Analysis

The Fourier transform of the waveforms 7.9 and 7.10 is easy to perform if one remembers that the Fourier transform of a convolution product is simply the product of the Fourier transform of the involved functions. This Fourier transform can be calculated with either τ_{rec} or τ as the time variable [2]. In the first case, it corresponds to what is done in classical THz-TDS and one gets

$$S_{\mathrm{o}}(\omega) = S_{\mathrm{THz,emit}}(\omega)\, T_{\mathrm{o}}(\omega)\, S^{*}_{\mathrm{laser}}(\omega)\, G^{*}(\omega), \tag{7.12}$$

$$\Delta S(\omega, \tau) = -S_{\mathrm{THz,emit}}(\omega)\, \Delta T(\omega)\, S^{*}_{\mathrm{laser}}(\omega)\, G^{*}(\omega)\, e^{-j\omega\tau}. \tag{7.13}$$

It follows that the spectrum of the photo-induced variation of the transmission is

$$\frac{\Delta T(\omega)}{T_{\mathrm{o}}(\omega)} = -\frac{\Delta S(\omega, \tau)}{S_{\mathrm{o}}(\omega)} e^{j\omega\tau}. \tag{7.14}$$

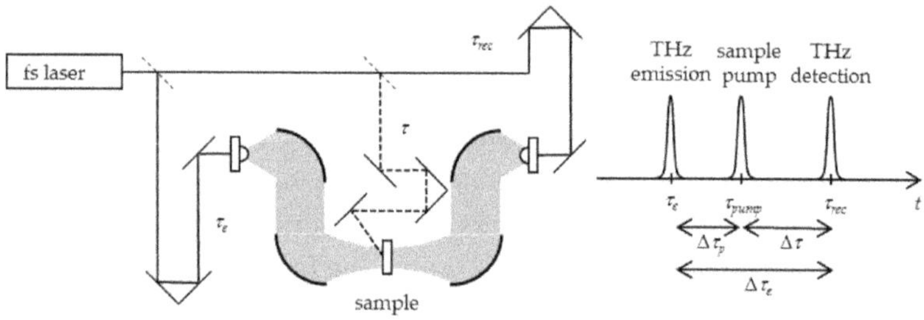

Figure 7.6 Scheme of the principle of a TRTS setup, in which the times of excitation of the emitter, receiver, and pump can be independently varied with two delay lines. The right-side figure shows the chronology of the different laser and THz pulses in the TRTS setup.

Notice that, when taking the modulus of expression 7.14, the phase term $e^{j\omega\tau}$ disappears.

In the second case (τ is the time variable and we use Ω as transform variable instead of ω to avoid any misunderstanding), only $\Delta S(\tau_{\text{rec}}, \tau)$ can be Fourier-transformed because $S_0(\tau_{\text{rec}})$ does not depend on τ:

$$\Delta S(\tau_{\text{rec}}, \Omega) = -S^*_{\text{THz,emit}}(\Omega)\Delta T^*(\Omega)S_{\text{laser}}(\Omega)G(\Omega)e^{-j\Omega\tau_{\text{rec}}} \quad (7.15)$$

It could be emphasized that Eqs. 7.13 and 7.15 bear the same information about the photo-induced phenomenon and they only slightly differ from a mathematical point of view. In both cases, retrieving $\Delta T(\omega)$ requires the knowledge of the incoming THz spectrum $S_{\text{THz,emit}}(\omega)$ as well as the response of the receiver $S_{\text{laser}}(\omega)G(\omega)$.

The experimental procedure and the data analysis have been rigorously studied by Nemec and coworkers [2, 5, 6]. These authors consider a general setup in which the laser pulses that excite the THz emitter and receiver and pump the sample can be time-delayed (see Fig. 7.6). Thus three experimental schemes are possible: varying $\Delta\tau$ and $\Delta\tau_e$ (see definitions of these time intervals on Fig. 7.6), or $\Delta\tau$ and $\Delta\tau_p$, or $\Delta\tau_e$ and $\Delta\tau_p$. Because these time intervals are dependent, only the two first cases are of interest. Figure 7.7 (top) shows a 2D temporal plot of the reachable phenomena for the first and second cases (left and right columns, respectively): for example, when varying $\Delta\tau$ and $\Delta\tau_e$ (left column), the sample is excited for positive $\Delta\tau_e$ and the THz signal is measurable for positive $\Delta\tau$. Of

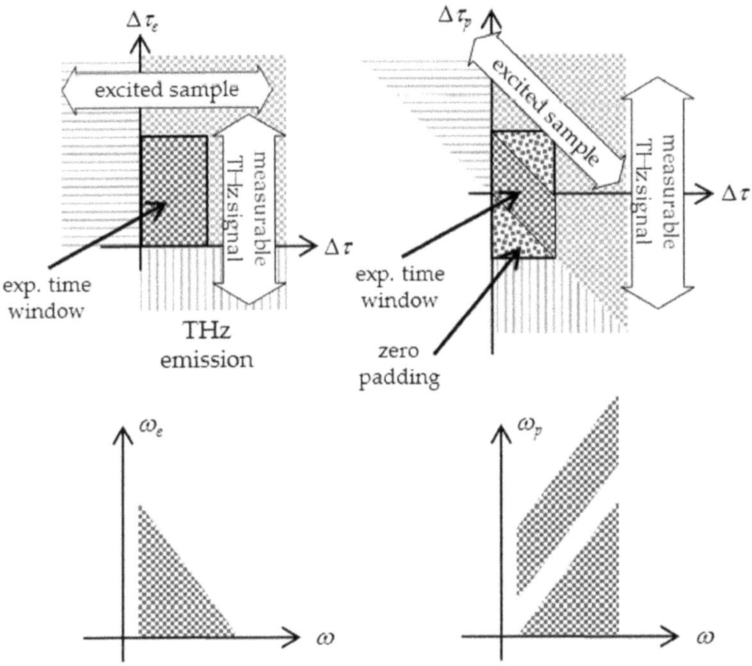

Figure 7.7 Top: 2D temporal plots of the reachable phenomena in TRTS when both $\Delta\tau$ and $\Delta\tau_e$ (left plot), and both $\Delta\tau$ and $\Delta\tau_p$ (right plot) are swept (left plot). Bottom: Corresponding plots in the frequency space. Frequencies ω, ω_e, and ω_p are the conjugate variables of respectively $\Delta\tau$, $\Delta\tau_e$, and $\Delta\tau_p$.

course, experimentally, both $\Delta\tau$ and $\Delta\tau_e$ are not measured up to infinity, but over a range in a given time window. The bottom 2D maps (Fig. 7.7) are plotted in the spectral space. The gray parts correspond to the spectral regions in which information on $\Delta T(\omega)$ can be obtained. For a more detailed explanation, see Nemec et al. [2, 5].

Inversing Eqs. 7.13 or 7.15 helps determine the experimental $\Delta T(\omega)$ spectrum. Extracting the useful information on the dynamic of the photo-induced phenomena in the sample material is not an easy task. Therefore, it is preferable to model $\Delta T(\omega)$ starting from with well-known theories of light–matter interaction, and then to fit the experimental data with the model. Rigorous modeling [2]

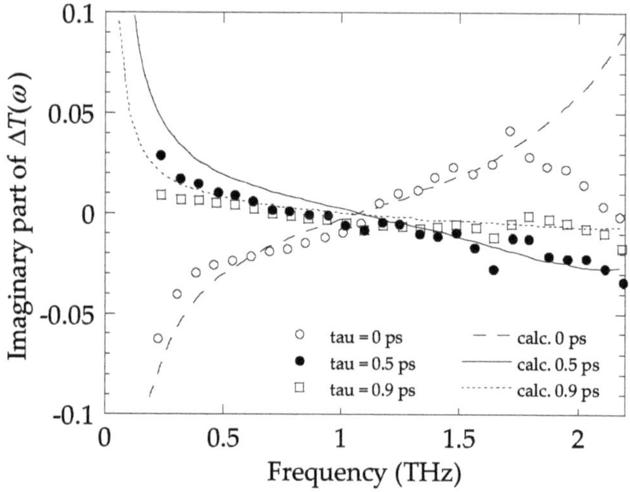

Figure 7.8 The imaginary part of $\Delta T(\omega)$ recorded and calculated for a sample of radiation-damaged silicon deposited on sapphire (RD-SOS). The symbols are experimental data, and the lines are calculated with the Drude model (adapted from Nemec [6]). Note that $\Delta\tau_p$ is denoted as "tau" in the figure.

requires us to consider the photo-induced change of the material parameters, and also the propagation effect through the sample of both the THz probe pulse and the laser pump pulse. When dealing with photo-excited free carriers in the samples, the Drude model, even if simple and rather crude, is widely used because it permits a reliable description of the observed photoconductivity effects [5, 16, 17]. For optical excitation of resonance modes in dielectrics (crystals, liquids, etc.), the Lorentz/Debye model (damped harmonic oscillator) fits the data well [1, 17]. In superconductors, one may use a two-fluids conductivity model [7]. Figure 7.8 presents the experimental and calculated imaginary parts of $\Delta T(\omega)$ for a sample of radiation-damaged silicon deposited on sapphire (RD-SOS) [6]. The irradiation dose ($<10^{14}$ cm^{-2} using 100 keV O^+ ions) creates deep-level defects in silicon which make the material ultrafast (free carriers are efficiently trapped by the defects), while keeping a rather high carrier mobility. The experimental data (dots) (shown in Fig. 7.8) are recorded for different $\Delta\tau_p$ (0, 0.5, and 0.9 ps), i.e., first when THz and pump pulses overlap in the sample, and then when the THz pulse probes the sample 0.5 and 0.9 ps after excitation. The corresponding lines are the best fits of the experimental data calculated with the Drude model and by considering the mixing of optical and THz frequencies in the sample. Here it leads to a free carrier decay time of about 400 fs and $\mu \approx 340 \ \ cm^2/V/s$ mobility.

References

1. Kindt, J. T., and Schmuttenmaer, C. A. (1999). Theory for determination of the low-frequency time-dependent response function in liquids using time-resolved terahertz pulse spectroscopy, *J. Chem. Phys.*, **110**, pp. 8589–8596.
2. Nemec, H., Kadlec, F., and Kuzel, P. (2002). Methodology of an optical pump-terahertz probe experiment: an analytical frequency-domain approach, *J. Chem. Phys.*, **117**, pp. 8454–8467.
3. Nuss, M. C., Auston, D. H., and Capasso, F. (1987). Direct subpicosecond measurement of carrier mobility of photoexcited electrons in gallium arsenide, *Phys. Rev. Lett.*, **58**, p. 2355.
4. Greene, B. I., Federici, J. F., Dykaar, D. R., Levi, A. F. J., and Pfeiffer, L. (1991). Picosecond pump and probe spectroscopy utilizing freely propagating terahertz radiation, *Opt. Lett.*, **16**, pp. 48–50.
5. Nemec, H., Kadlec, F., Surendran, S., Kuzel, P., and Jungwirth, P. (2005). Ultrafast far-infrared dynamics probed by terahertz pulses: A frequency domain approach. I. Model systems, J. Chem. Phys., **122**, p. 104503.
6. Nemec, H., Kadlec, F., Kadlec, C., Kuzel, P., and Jungwirth, P. (2005). Ultrafast far-infrared dynamics probed by terahertz pulses: a frequency-domain approach. II. Applications, *J. Chem. Phys.*, **122**, p. 104503.
7. Averitt, R., Rodriguez, G., Siders, J. L. W., Trugman, S. A., and Taylor, A. J. (2000). Conductivity artifacts in optical-pump THz-probe measurements of $YBa_2Cu_3O_7$, *J. Opt. Soc. Am. B*, **17**, pp. 327–321.
8. Lui, K. P. H., and Hegmann, F. A. (2001). Ultrafast carrier relaxation in radiation-damaged silicon on sapphire studied by optical-pump-terahertz-probe experiments, *Appl. Phys. Lett.*, **78**, p. 3478.
9. Hegmann, F. A., Tykwinski, R. R., Lui, K. P. H., Bullock, J. E., and Anthony, J. E. (2002). Picosecond transient photoconductivity in functionalized pentacene molecular crystals probed by terahertz pulse spectroscopy, *Phys. Rev. Lett.*, **89**, p. 227403.

10. Beard, M. C., Turner, G. M., and Schmuttenmaer, C. A. (2001). Subpicosecond carrier dynamics in low-temperature grown GaAs as measured by time-resolved terahertz spectroscopy, *J. Appl. Phys.*, **90**, p. 5915.
11. Miranda, R. S., Tom, H. W. K., Johnson, A. M., Bridges, T. J., and Aumiller, G.D. (1992). Study of carrier dynamics in InP:Fe using time-resolved infrared reflexion and transmission, *Appl. Phys. Lett.*, **60**, p. 1105.
12. Beard, M. C., Turner, G. M., and Schmuttenmaer, C. A. (2000). Transient photoconductivity in GaAs as measured by time-resolved terahertz spectroscopy, *Phys. Rev. B*, **62**, p. 15764.
13. Prabhu, S. S., Ralph, S. E., Melloch, M. R., and Harmon, E. S. (1997). Carrier dynamics of low-temperature-grown GaAs observed via THz spectroscopy, *Appl. Phys. Lett.*, **70**, p. 2419.
14. George, P. A., Strait, J., Dawlaty, J., Shivaraman, S., Chandrashekhar, M., Rana, F., and Spencer, M. G. (2008). Ultrafast optical-pump terahertz-probe spectroscopy of the carrier relaxation and recombination dynamics in epitaxial graphene, *Nano Lett.*, **8**, pp. 4248–4251.
15. Haran, G., Sun, W.-D., Wynne, K., and Hochstrasser, R. M. (1997). Femtosecond far-infrared pump-probe spectroscopy: a new tool for studying low-frequency vibrational dynamics in molecular condensed phases, *Chem. Phys. Lett.*, **274**, p. 365.
16. Saeta, P. N., Federici, J. F., Greene, B. I., and Dykaar, D. R. (1992). Intervalley scattering in GaAs and InP probed by pulsed far-infrared transmission spectroscopy, *Appl. Phys. Lett.*, **60**, p. 1477.
17. Fekete, L., Kužel, P., Němec, H., Kadlec, F., Dejneka, A., Stuchlík, J., and Fejfar, A. (2009). Ultrafast carrier dynamics in microcrystalline silicon probed by time-resolved terahertz spectroscopy, *Phys. Rev. B*, **79**, p. 115306.

Chapter 8

Waveguides, Metamaterials, and Plasmonic Devices

8.1 Introduction

In spectroscopy, waveguides, metamaterials, and plasmonic structures are employed to increase the length of propagation of light in matter, or/and to enhance electromagnetic fields in matter. In any case, it improves the interaction between matter and the incoming light and thus makes spectroscopy more efficient. This allows us to detect and characterize very small amounts of matter, like thin films, or materials that have very low absorption. In this chapter, we will present THz-TDS using dielectric or metallic waveguides, in which or onto which the material or film to be characterized is located. We will also address metallic/conducting structures that support localized or propagating surface plasmons (SP). The confinement of the THz field in the vicinity of the metal surfaces, reliefs, or sub-wavelength features makes transmission or reflection by the samples very sensitive to any matter deposited on them.

Principles of Terahertz Time-Domain Spectroscopy
Jean-Louis Coutaz, Frédéric Garet, and Vincent Wallace

ISBN 978-981-4774-56-7 (Hardcover), 978-0-203-70969-6 (eBook)
www.panstanford.com

8.2 THz-TDS with Waveguides

8.2.1 The Advantage of THz-TDS with Waveguiding Structures

Spectroscopy of small amounts of matter is difficult to perform because the signals are weak. If one deals with a thin film, either in transmission or in reflection, the interaction between the THz beam and the thin film is proportional to $\exp\left(-j\tilde{k}_z d\right)$, where d is the film thickness and $\tilde{k}_z$ is the component of the wavevector along the propagation direction (here z). As d is small, this expression can be approximated by

$$\exp\left(-j\tilde{k}_z d\right) \approx 1 - j\tilde{k}_z d. \tag{8.1}$$

Thus, if we consider normal incidence transmission THz-TDS and if we neglect the contribution of the Fresnel coefficients as well as the Fabry–Pérot internal reflections in the film, the influence of the film on the modulus and phase of the detected signal is

$$\Delta t_{\text{THz-TDS}} \approx -\frac{\omega}{c}\kappa d, \quad \Delta\varphi_{\text{THz-TDS}} = -\frac{\omega}{c} n d. \tag{8.2}$$

At 1 THz, if one is able to measure a variation of transmission $\Delta t = 10^{-3}$, the minimum thickness of the film that can be studied is d_{min} (μm) $\approx 10/\alpha(\text{cm}^{-1})$. Thus, for transparent materials, the films must be quite thick (e.g., $\alpha \approx 0.1$ cm$^{-1} \Rightarrow d_{\text{min}} \approx 100$ μm). Therefore, classical transmission THz-TDS does not permit to precisely characterize micrometer-thick or even thinner films.

As we have seen previously, more sophisticated techniques like ellipsometry or ATR lead more or less to the same sensitivity as classical THz-TDS. Thus the only way to study and characterize very thin films is to excite electromagnetic resonances within the film, which makes the effective thickness of the film much bigger than d_{min}. Among these resonances, waveguide modes may greatly enhance the sensitivity of THz-TDS measurements by increasing the interaction length between the THz beam and the matter under study. The basic structure, if the film thickness is large enough to make it robust, is based on using the stand-alone film as a waveguide (Fig. 8.1) or, if the film is very thin, to deposit it over or in the waveguide (Fig. 8.2). Therefore, the guided THz beam interacts

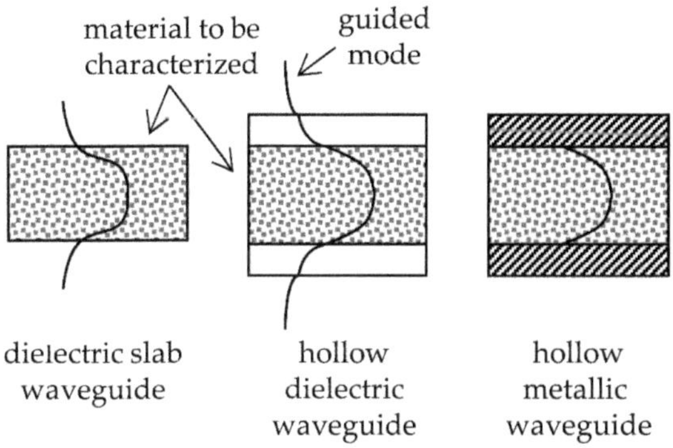

Figure 8.1 Different types of waveguides for THz waves, in which the core is made of the material to be characterized. The curves represent the guided THz field.

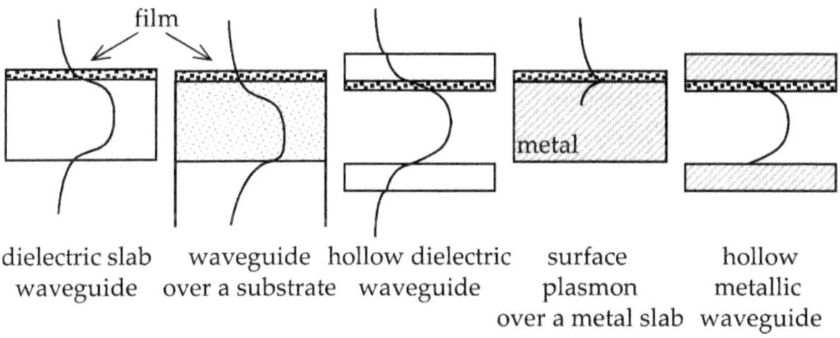

Figure 8.2 Different types of waveguides for THz waves, which comprise a thin film of matter to be characterized. The curves represent the guided THz field.

with the film over a very long distance. In such experiments, one measures the modulus and phase of the THz beam transmitted by the waveguide. The reference is measured with the waveguide that does not include the film, or with a shorter waveguide.

To guide the THz light, the waveguide must exhibit a larger refractive index than the surrounding media (low-index slabs can propagate guided light but with large radiation losses—the so-called leaky modes). Moreover, the waveguide material must be transparent to avoid absorption losses. In the THz domain, this last constraint is more difficult to fulfill than in optics, and the best material to guide THz is dry air. It is why hollow or photonic crystal THz waveguides are quite popular. Figures 8.1 and 8.2 show the cross section (from the side) of the main types of waveguides that could be used for THz-TDS of films or gases (here the light is guided from left to right). In Fig. 8.1, the core medium of the waveguides (semi-infinite planar structures, rectangular or cylindrical waveguides) is made of the material to be characterized. In Fig. 8.2, the thin film to be characterized is deposited against the waveguide: from the left to the right, a simple slab or cylinder of high-index dielectric material (here the surrounding medium is air or vacuum), a high-index dielectric waveguide deposited over a low-index dielectric substrate (the upper medium is air), a hollow dielectric waveguide (the inner material is air, while the walls of the waveguide are made of a dielectric transparent material), an air–metal interface (either flat or cylindrical wire) that supports the propagation of a surface plasmon wave, and finally a hollow metallic waveguide. Other guided structures for THz spectroscopy are mostly based on 2D devices, i.e., microwave-like integrated circuits. The THz wave is guided along coplanar, slot, or microstrip lines. Such devices are very helpful to design compact THz sensors, including active components (sources, detectors, etc.), but they better run at a few hundreds of GHz than over 1 THz.

The advantage of the waveguide device can be crudely addressed as follows. If neglecting the Fresnel coefficients (transmission and reflection at an interface), transmission in a THz-TDS experiment is proportional to

$$t_{\text{THz-TDS}} \propto 1 - e^{-\frac{\omega}{c}\kappa d},$$

in which d is the film thickness. When the film is located in a waveguide (Fig. 8.10), the transmission is given by

$$\begin{aligned} t_{\text{waveguide(covered)}} - t_{\text{waveguide(bare)}} &\propto \left(e^{-\frac{\omega}{c}\kappa L} - e^{-\frac{\omega}{c}\kappa_{\text{w}} L}\right) \\ &\quad \times \frac{\int_{S_{\text{film}}} E_{\text{guided,THz}}(x, y)\, dx\, dy}{\int_{S_{\text{waveguide}}} E_{\text{guided,THz}}(x, y)\, dx\, dy} \\ &\approx \left(e^{-\frac{\omega}{c}\kappa L} - e^{-\frac{\omega}{c}\kappa_{\text{w}} L}\right) \frac{S_{\text{film}}}{S_{\text{waveguide}}}, \end{aligned}$$

which leads to, when $\kappa L \ll 1$,

$$\Delta t_{\text{waveguide}} = t_{\text{waveguide(covered)}} - t_{\text{waveguide(bare)}} \approx \frac{\omega}{c}(\kappa_{\text{w}} - \kappa)\frac{S_{\text{film}}}{S_{\text{waveguide}}}L. \quad (8.3)$$

Here L is the length of the waveguide, κ_{w} is the imaginary part of the waveguide effective index, and S_{film} and $S_{\text{waveguide}}$ are the sections of the film and of the waveguide. Expression 8.3 is obtained by crudely assuming that the THz field amplitude is constant over the waveguide cross-section and does not vary when inserting the film in the waveguide. We make the hypothesis that the minimum detectable variation Δt of transmission, induced by the absorption κ of the material, is the same for both techniques (waveguide and classical transmission THz-TDS):

$$\Delta t_{\text{waveguide}} \approx -\frac{\omega}{c}L(\kappa - \kappa_{\text{w}})\frac{S_{\text{film}}}{S_{\text{waveguide}}} = \Delta t_{\text{THz-TDS}} \approx -\frac{\omega}{c}\kappa d. \quad (8.4)$$

Therefore, the ratio of the smallest κ measurable with the classical scheme ($\kappa_{\text{THz-TDS}}$) and with the waveguide structure ($\kappa_{\text{waveguide}}$) is

$$G = \frac{\kappa_{\text{THz-TDS}}}{\kappa_{\text{waveguide}}} \approx \frac{S_{\text{film}}}{S_{\text{waveguide}}}\frac{L}{d}. \quad (8.5)$$

This is the gain, G, of sensitivity delivered by a waveguide technique compared to the classical THz-TDS setup. If we consider a hollow cylindrical waveguide with an inner radius $R \approx \lambda$ and that the film is stuck against the waveguide's inner surface ($S_{\text{film}} = 2\pi Rd$), the gain in using the waveguide technique is

$$G \approx \frac{2L}{\lambda} = \frac{2L}{c}f. \quad (8.6)$$

Consequently, at 1 THz ($\lambda = 300$ μm), a 3 cm waveguide improves the sensitivity by 200. Of course, expression 8.5 has been determined with too crude simplifications. Especially, the overlapping integral leading to Eq. 8.3 has been strongly overestimated. Thus, the gain is of the order of $20 \sim 100$, or, in other words, the waveguide technique can characterize films $20 \sim 100$ times less absorbing or thinner than the classical transmission THz-TDS technique. The results from a rigorous calculation for hollow waveguides, established by Gallot and coworkers [1], are given below.

THz-TDS in wave-guiding conditions has been initiated [1] and largely studied by D. Grischkowsky and his colleagues [2–8]. Especially, the absorption of nanometer-thick water layers has been remarkably well determined [2].

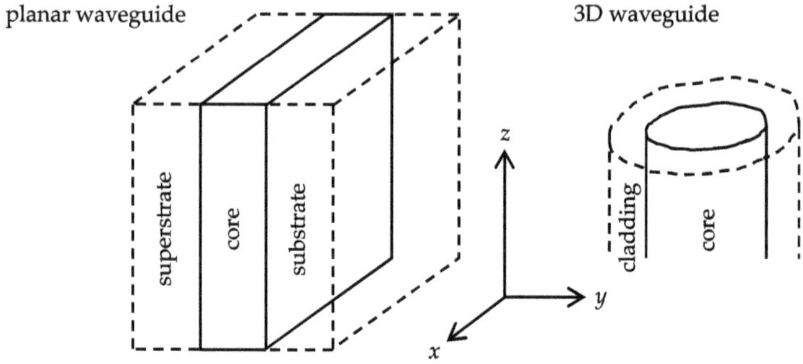

Figure 8.3 Planar (left) and 3D (right) waveguides. Superstrate, substrate, or cladding could be semi-infinite media. For planar structures, superstrate and substrate could be identical, especially in the case of stand-alone waveguides.

8.2.2 Basic Notions of Propagation in Waveguides

Figure 8.3 presents a 2D (planar) waveguide and a 3D waveguide. The beam is guided along the z direction. We assume the waveguide is not bent, meaning that the z direction forms a straight line. It follows that the propagation of the wave along the waveguide, i.e., along z, is described by $\exp\left(-j\, k_{gz} z\right) = \exp\left(-j\, n_{\text{eff}} \omega z / c\right)$. n_{eff} is the effective index of the guided light. Here, in this introduction, we consider only lossless waveguides (attenuation in waveguides will be presented later on), and thus n_{eff} and k_{gz} are real. In both structures, either the superstrate and substrate, or the cladding, could be either semi-infinite media or layers, but these latter ones must be thicker than the decay length of the guided field in the layer material.

In the planar waveguide, the geometrical symmetry leads to two eigen polarizations of the guided field, either transverse electric (TE or s), $E_z = 0$, or transverse magnetic (TM or p), $H_z = 0$. In the 3D waveguide, in addition to the TE and TM polarizations, there exists a transverse electromagnetic (TEM) eigenmode, for which $E_z = H_z = 0$. Any guided light is a linear combination of TE, TM, and TEM fields, namely a hybrid mode. In metallic hollow waveguides, both TE and TM modes exhibit a cutoff frequency, below which light cannot be guided, while the TEM mode has no cutoff.

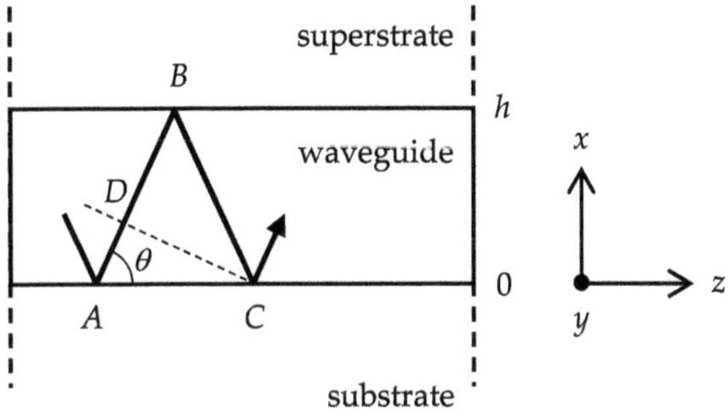

Figure 8.4 Zigzag model in a planar waveguide. The propagating ray is plotted as a bold line.

The properties of propagation of the guided light must be studied in the framework of electromagnetism. However, a simple ray model helps in comprehending these properties. Let us consider the planar structure depicted in Fig. 8.4, where a guided ray is plotted. The refractive index of the guiding layer is n and the angle between the ray and the waveguide interfaces is θ. If there is no phase variation when the ray is reflected at the interfaces, the light can be guided if the phase in D is the same as in C (modulo $2m\pi$, m is an integer), i.e., if the DC plane is a wavefront.

A simple geometrical analysis gives

$$DB + BC = 2\,h \sin\theta. \tag{8.7}$$

Therefore, the phase condition is

$$\frac{2\pi}{\lambda} n\,(DB + BC) = 2m\pi \;\;\Rightarrow\;\; \sin\theta = m\frac{\lambda}{2nh} = \frac{m}{2nh}\frac{c}{f}. \tag{8.8}$$

The sine function in Eq. 8.8 is always smaller than 1; thus, the inequality follows:

$$m\frac{\lambda}{2nh} = \frac{m}{2nh}\frac{c}{f} \leq 1 \;\;\Rightarrow\;\; f \geq f_c = m\frac{c}{2nh}. \tag{8.9}$$

This expression defined the cutoff frequency f_c, below which no TE/TM modes are guided in the hollow planar metallic waveguide. The TEM mode cannot be described by the simple ray model: it does not exhibit a spectral cutoff.

The component along z of the wavevector of the guided ray is

$$k_z = \frac{2\pi f}{c} n \cos\theta. \tag{8.10}$$

Here, care must be taken: this is not the guided wavevector along z. Let us calculate the velocity, along the waveguide, of the guided light. From Eq. 8.10, we get

$$v_z = \frac{2\pi f}{k_z} = \frac{c}{n\cos\theta} \quad \Rightarrow \quad n_{\text{eff}} = \frac{c}{v_z} = n\cos\theta. \tag{8.11}$$

However, we can also calculate this velocity by dealing with the time for the wave to propagate from A to C (Fig. 8.4):

$$v_g = \frac{AC}{t_{AB} + t_{BC}} = \frac{AC}{AB\frac{n}{c} + BC\frac{n}{c}} = \frac{c}{n}\cos\theta \quad \Rightarrow \quad n_{\text{eff}} = \frac{n}{\cos\theta}. \tag{8.12}$$

Equation 8.12 gives the correct expression of the velocity, which in fact corresponds to the velocity of the electromagnetic energy along the waveguide. Consequently, it can be derived from Eq. 8.10 by considering the group velocity v_G:

$$\frac{1}{v_G} = \frac{\partial k_z}{\partial \omega} = \frac{\partial}{\partial \omega}\left(\frac{\omega}{c} n\cos\theta\right) = \frac{n}{c\cos\theta}. \tag{8.13}$$

Therefore, the guided velocity (Eq. 8.12) is equal to the group velocity (Eq. 8.13). The guided wavevector along z is

$$k_{gz} = \frac{\omega}{v_g} = \frac{2\pi f}{c}\frac{n}{\cos\theta}. \tag{8.14}$$

The attenuation of the waveguide is defined with respect to the guided light energy lost when propagating over a unity distance. Consider here propagation from A to C, without energy loss at the interface, but with absorption α in the core material. We obtain

$$P(z = C) = P(z = A)\, e^{-\alpha(AB+BC)} = P(z = A)\, e^{-\frac{\alpha}{\cos\theta}AC}. \tag{8.15}$$

Therefore, the effective attenuation of the waveguide is

$$\alpha_{\text{eff}} = \frac{\alpha}{\cos\theta}. \tag{8.16}$$

8.2.3 Planar Dielectric THz Waveguides

We consider a planar waveguide made of three media separated by two flat parallel interfaces (Fig. 8.5). We will give here an electromagnetic treatment of this waveguide, which applies to both dielectric and metallic waveguides.

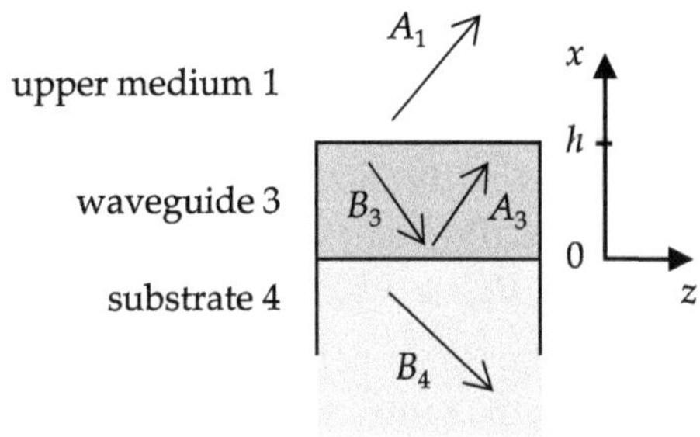

Figure 8.5 Planar waveguide structure.

The THz beam is guided along the direction z in the waveguide, which means that the maximum of the field in the plane (x, y) perpendicular to z is located in the waveguide. Each of the three media is homogeneous and isotropic; therefore, the solution in each medium of the Helmholtz propagation equation is a plane wave composed of a wave propagating upwards and a wave propagating downwards. Moreover, as described in the Annex, the electrical field propagating in this three-media structure can be projected along the TE (here y) and TM eigen-polarizations. Both TE and TM cases may be solved simultaneously by setting (see A.18):

$$F^o = E_y^o\,(\mathrm{TE}), \quad F^o = H_y^o\,(\mathrm{TM}). \tag{8.17}$$

The superscript o is used to label the waveguide without the layer. In medium i $(i = 1, 3, 4)$, the field is

$$F_i^o = \left(A_i^o\, e^{-jk_{ix}^o x} + B_i^o\, e^{+jk_{ix}^o x}\right) e^{-jk_{iz}^o z}, \quad k_{ix}^{o2} + k_{iz}^{o2} = \varepsilon_i \frac{\omega^2}{c^2}. \tag{8.18}$$

In media 1 and 4 (upper medium and substrate), there is no incoming field from $x \to \pm\infty$, and thus $A_4^o = B_1^o = 0$. In addition, for the fields not to diverge at $x \to \pm\infty$, i.e., the field is evanescent in the surrounding media, the imaginary part of k_{ix}^o must be negative:

$$\mathrm{Im}\left(k_{ix}^o\right) = \mathrm{Im}\left(\sqrt{\varepsilon_i \frac{\omega^2}{c^2} - k_{iz}^{o2}}\right) > 0. \tag{8.19}$$

This enforces the refractive index of the guiding layer to be higher than the ones of the surrounding media. At the two interfaces, the tangential components of the $\vec{E}$ and $\vec{H}$ fields are conserved (see A.19):

$$\begin{aligned} x = h: &\quad \begin{cases} A_1^o e^{-jk_{1x}^o h} = A_3^o e^{-jk_{3x}^o h} + B_3^o e^{jk_{3x}^o h}, \\ k_{1x}^o A_1^o e^{-jk_{1x}^o h} = k_{3x}^o \alpha_{13} \left(A_3^o e^{-jk_{3x}^o h} - B_3^o e^{jk_{3x}^o h}\right), \end{cases} \\ x = 0: &\quad \begin{cases} A_3^o + B_3^o = B_4^o, \\ k_{3x}^o \left(A_3^o - B_3^o\right) = -k_{4x}^o \alpha_{34} B_4^o. \end{cases} \end{aligned} \tag{8.20}$$

In addition to Eq. 8.20, the component k_z of the wavevector along the propagation direction is the same in each medium. We write it as

$$k_z^o = \frac{\omega}{c} n_{\text{eff}}^o. \tag{8.21}$$

n_{eff}^o is the effective index of the guided light, which determines its phase velocity as well as its absorption. From Eq. 8.20, one easily gets two equations between the fields in the upper medium and in the substrate:

$$\begin{cases} B_4^o = \frac{e^{-jk_{1x}^o h}}{2} \left(e^{jk_{3x}^o h}\left(1 + \frac{k_{1x}^o}{k_{3x}^o \alpha_{13}}\right) + e^{-jk_{3x}^o h}\left(1 - \frac{k_{1x}^o}{k_{3x}^o \alpha_{13}}\right)\right) A_1^o, \\ B_4^o = -\frac{e^{-jk_{1x}^o h}}{2} \frac{k_{3x}^o}{k_{4x}^o \alpha_{34}} \left(e^{jk_{3x}^o h}\left(1 + \frac{k_{1x}^o}{k_{3x}^o \alpha_{13}}\right) - e^{-jk_{3x}^o h}\left(1 - \frac{k_{1x}^o}{k_{3x}^o \alpha_{13}}\right)\right) A_1^o. \end{cases} \tag{8.22}$$

As these two expressions must be equivalent, one obtains

$$\tan\left(k_{3x}^o h\right) = j \frac{k_{3x}^o \left(\alpha_{13}\alpha_{34} k_{4x}^o + k_{1x}^o\right)}{\alpha_{13} {k_{3x}^o}^2 + \alpha_{34} k_{1x}^o k_{4x}^o}. \tag{8.23}$$

In Eq. 8.23, k_{1x}^o and k_{4x}^o are mostly imaginary because $\varepsilon_{\text{i}}\omega^2/c^2 < k_{\text{iz}}^{o2}$. Thus, it is preferable to write

$$k_{\text{ix}}^o = \sqrt{\varepsilon_{\text{i}}\frac{\omega^2}{c^2} - k_{\text{iz}}^{o2}} = j\sqrt{k_{\text{iz}}^{o2} - \varepsilon_{\text{i}}\frac{\omega^2}{c^2}} = jk_{\text{ix}}^{o*},$$

$$\Rightarrow \tan\left(k_{3x}^o h\right) = \frac{k_{3x}^o \left(\alpha_{13}\alpha_{34} k_{4x}^{o*} + k_{1x}^{o*}\right)}{\alpha_{13} {k_{3x}^o}^2 - \alpha_{34} k_{1x}^{o*} k_{4x}^{o*}}. \tag{8.24}$$

This is the dispersion equation of the waveguide, which permits to determine the effective index n_{eff}. This equation has no analytical solution and must be solved numerically. Because the tangent function is periodic, Eq. 8.24 exhibits a large number of discrete solutions, namely the modes.

Figure 8.6 shows the effective index n_{eff}^o of a waveguide made of a 100 μm thick HR-Si wafer ($n = 3.42$, no loss) in air. The THz

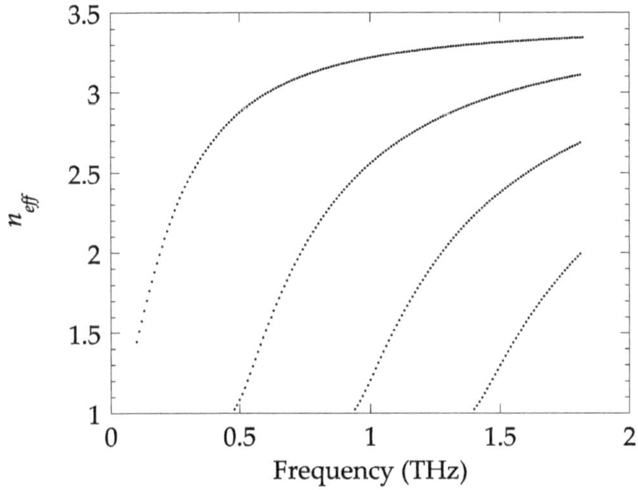

Figure 8.6 Dispersion of the TE guided modes in a 100 μm thick HR Si wafer.

light is guided into the wafer. Here, its polarization is TE. The four first modes are plotted from 0.1 to 1.8 THz. The first mode (TE_0) dispersion curve starts at $f = 0$, i.e., there is no cut-off frequency for this mode. Even with a very small wafer thickness, there still exists a guided mode, because the guided field could spread far in the surrounding media. For the higher modes, there exists a cut-off frequency below which the mode does not propagate. All the curves saturate to the refractive index of the guiding layer; here $n = 3.42$, which means that most of the THz field is concentrated in the guiding layer at higher frequencies. When the material of the guiding layer exhibits absorption, the effective index is slightly modified, and the waveguide presents propagation losses. Figure 8.7 shows the propagation losses (α_w in dB/cm) of the waveguide and the ratio α_w / α_{mat}, where α_{mat} is the absorption coefficient of the waveguide material. The structure is the same as the previous one (Fig. 8.6), but the refractive index of the Si material is $n = 3.42 - j\,0.146 \times 10^{-3}$. For this rather weak absorption ($\alpha_{mat} \approx 0.03 \rightarrow 0.5\ \text{cm}^{-1}$ in the THz range), there is almost no effect on n_{eff} ($\Delta n_{eff} < 10^{-7}$). α_w tends to zero at the cut-off frequency, because the field is less and less concentrated in the waveguide, and then α_w

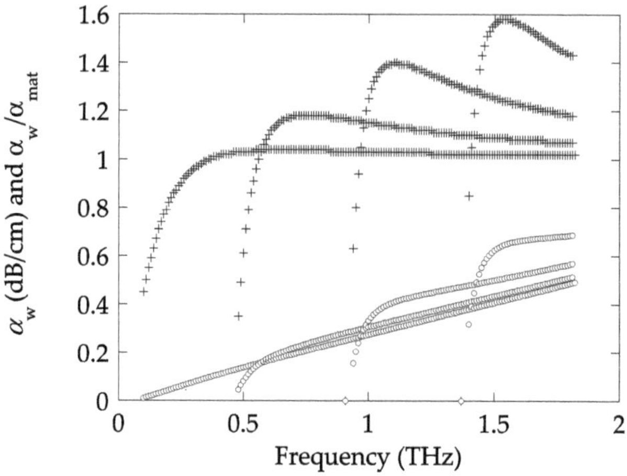

Figure 8.7 Dispersion (circles) of the attenuation (in dB/cm) and dispersion (crosses) of the ratio α_w / α_{mat} for the TE guided modes in a 100 μm thick HR-Si wafer with $n = 3.42 - j\,0.146 \times 10^{-3}$.

increases with the frequency. It is slightly larger for higher modes: in a ray-tracing model, this is simply explained by the longer path in the waveguide layer of the higher modes when propagating over the same distance along the waveguide.

More interesting are the α_w/α_{mat} curves. They correspond to the increase in sensitivity of the waveguide technique compared to the classical THz-TDS, normalized to the same propagation length, i.e., for a waveguide length which is equal to the sample thickness in a classical THz-TDS scheme. Close to the cut-off frequency, most of the light spreads outside the guiding layer, and thus the interaction between the THz wave and the guiding material is weak: the ratio tends to zero. At higher frequencies, the ratio saturates to a value slightly larger than 1, because both of the concentration of the guided light in the waveguide and of the zigzag ray propagation effect. At these higher frequencies, the advantage of the waveguide technique is obvious. The gain of sensitivity of the waveguide technique, assuming that the in- and out-coupling coefficients between the guided wave and the freely propagating wave are 100%, is roughly speaking L/d, where L is the length of the waveguide and d is the sample thickness. This gain could be extremely high, e.g., 10^3 for a 10 cm waveguide and a 100 μm sample. Even if the coupling coefficients are of the order of a few tens of percent, the gain of the waveguide technique is substantial.

Let us notice the important restriction when performing waveguided THz-TDS. It is preferable to excite only one guided mode, and thus to work below the cutoff frequency of the second propagating mode. If not, the guided field is written as a linear combination of the modes:

$$F = \sum_{i=0}^{\infty} a_i F_i \tag{8.25}$$

Each mode exhibits specific effective index and attenuation, which are related to the guiding material properties through Eq. 8.24. Inverting Eq. 8.24 with a multimode excitation (Eq. 8.25) is more difficult than in the single mode case, but it can be done numerically. However, the practical difficulty is to know the experimental a_i coefficients, i.e., the weight of each mode. Generally, this makes the problem impossible to solve with any great accuracy.

8.2.4 Parallel Plate Metallic THz Waveguides

Planar metallic waveguides are made of two thick metal slabs with flat surfaces, which are arranged to form a parallel structure with a vacuum or dielectric gap in between. The light is guided in the gap. The calculations given previously are still valid, but the results are different because of the high susceptibility of metals in the THz domain. A simplified description of the propagation in such waveguides may help in understanding the waveguides properties. Let us assume the metal is a perfect conductor, and therefore the field at the surface of the metal walls is null. So we set in Eq. 8.18 the amplitudes A_1 and B_4 to zero, and we get the field guided in the hollow part of the waveguide:

$$F_m^o = 2jA\sin\left(k_x^o x\right) e^{-jk_z^o z}, \quad k_x^o d = m\pi. \tag{8.26}$$

m is an integer that labels the guided modes. The effective index is easily obtained:

$$k_z^o = \frac{\omega}{c}\sqrt{1-\left(\frac{mc}{2df}\right)^2} \quad \Rightarrow \quad n_{\text{eff}}^o = \sqrt{1-\left(\frac{mc}{2df}\right)^2}. \tag{8.27}$$

The effective index is thus smaller than 1, i.e., the phase velocity exceeds the speed of light in vacuum. This is not violating relativity because the phase velocity is not the velocity of the electromagnetic energy. This later one is equal to the group velocity which, for the guided wave, is

$$v_{\mathrm{g}}^{o} = \left(\frac{\partial\, k_{\mathrm{z}}^{o}}{\partial\, \omega}\right)^{-1} = c\sqrt{1-\left(\frac{mc}{2df}\right)^{2}}. \tag{8.28}$$

The effective index at which the guided energy propagates along the waveguide is

$$n_{\mathrm{g,eff}} = \frac{c}{v_{\mathrm{g}}^{o}} = \left(1-\left(\frac{mc}{2df}\right)^{2}\right)^{-1/2}. \tag{8.29}$$

There exists a cutoff frequency, namely $f_{\mathrm{c}} = mc/2d$, below which the THz mode is not guided. At higher frequencies, $n_{\mathrm{g,eff}}$ tends to 1, while it becomes very large when f is close to f_{c}. Figure 8.8 shows the dispersion curves of the effective group index $n_{\mathrm{g,eff}}$ (a) as well as (b) the corresponding attenuation for a planar waveguide made of two aluminum slabs ($\varepsilon_{\mathrm{Al}} = -(3 \times 10^{4} - j\,1 \times 10^{6})$ [9]) separated by a 1 mm thick air gap, in which the THz beam is guided. The polarization is TE and the four first modes are seen. Close to the cutoff frequency, both effective index and attenuation of each mode increase abruptly. At higher frequencies, $n_{\mathrm{g,eff}}$ tends to 1 and the attenuation vanishes. At any frequency, except at f_{c}, the attenuation due to the ohmic losses in the metal is small. For example, at 0.5 THz, $n_{\mathrm{g,eff}}^{o} \approx 1.25$ and $\alpha_{\mathrm{g,eff}}^{o} \approx 0.15$ dB/cm. The great difference with the dielectric waveguide is that the first mode exhibits a cutoff frequency. Below this frequency, there is no guided light in the structure.

In the TM polarization case, the dispersion curves (Fig. 8.9a) are rather similar to the TE case but are slightly shifted in frequency. Also, there exists a transverse electromagnetic (TEM) mode (TM_0) whose propagates in the core material at the velocity of a plane wave in the same material, and without any cutoff frequency. In addition, its dispersion and attenuation are very low, making it a nice way of carrying and guiding information at THz frequencies [10]. Figure 8.9b shows the measured attenuation of such a TEM mode for a copper–air–copper waveguide with $d = 108$ μm.

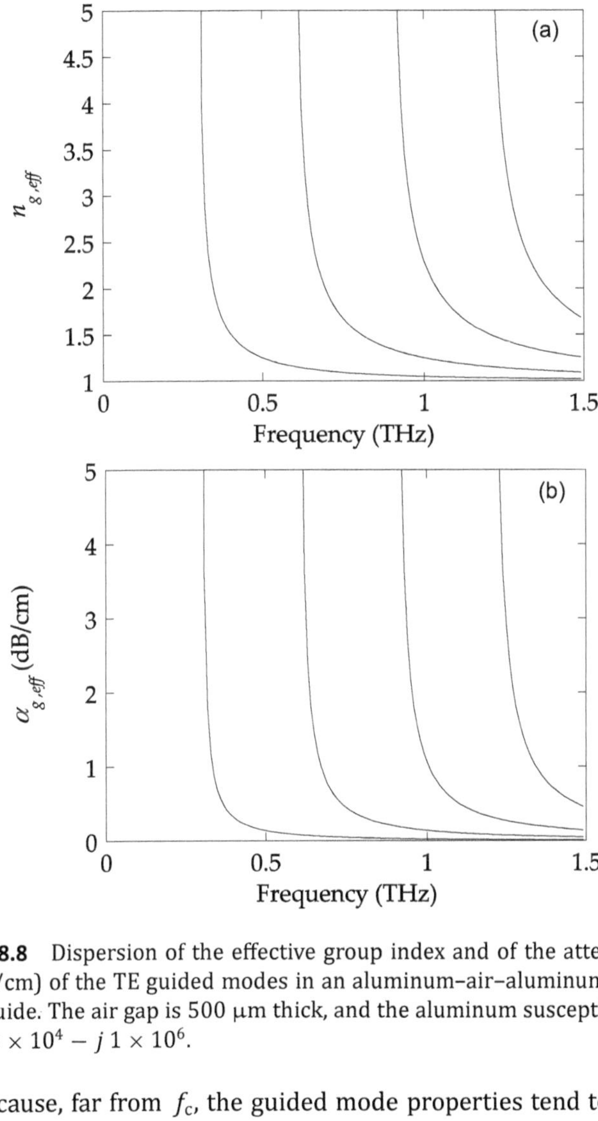

Figure 8.8 Dispersion of the effective group index and of the attenuation (in dB/cm) of the TE guided modes in an aluminum–air–aluminum planar waveguide. The air gap is 500 μm thick, and the aluminum susceptibility is $\varepsilon_{Al} = 3 \times 10^4 - j\,1 \times 10^6$.

Because, far from f_c, the guided mode properties tend towards the ones of a plane wave in the core material, inserting a transparent material inside the metallic waveguide will make the spectroscopy technique very sensitive to the material properties. As an example, we calculate here the waveguide properties of the previous structure (Fig. 8.8), but the central part is filled with a material

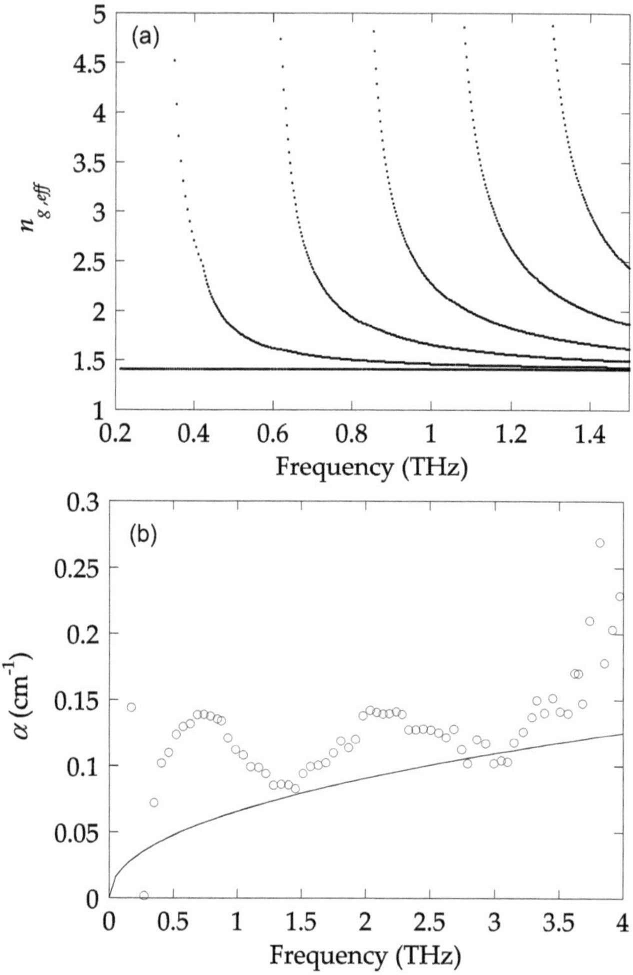

Figure 8.9 (a) Dispersion of the effective group index of the TM guided modes in an aluminum–dielectric–aluminum planar waveguide. The dielectric layer is 500 μm thick ($n \approx 1.414$) and the aluminum susceptibility is $\varepsilon_{\text{Al}} = 3 \times 10^4 - j \times 10^6$. The mode with $n_{\text{g,eff}} = 1.414$ is the TEM mode. (b) Measured attenuation (circles) together with the model (continuous lines) of the TEM mode of a copper–air–copper waveguide (air thickness 108 μm). Adapted from Mendis and Grischkowsky [10].

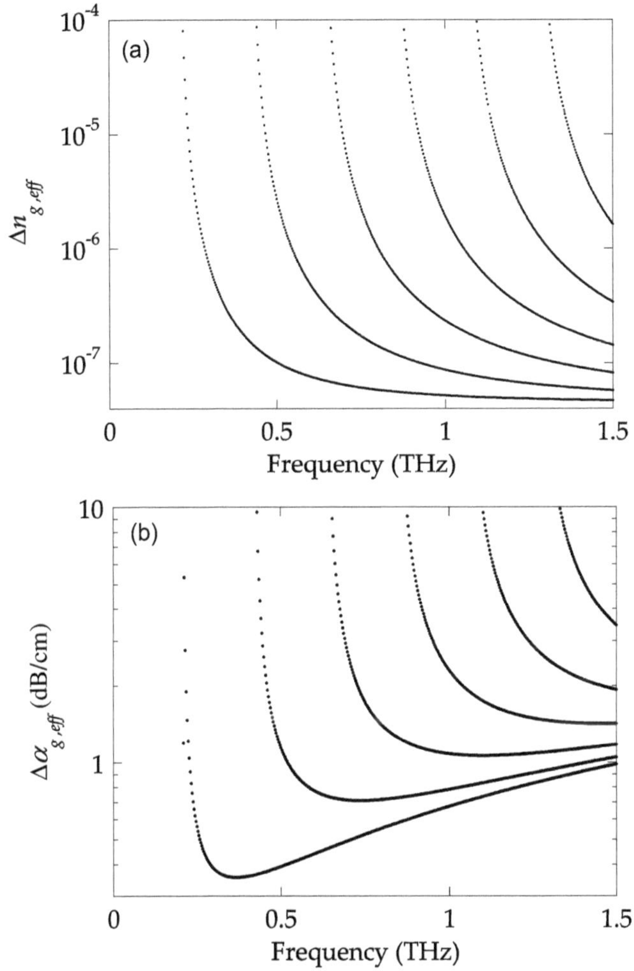

Figure 8.10 Dispersion of the difference of the effective group index (a) and attenuation (b) for the filled hollow metallic hollow waveguide, when the inner material becomes absorbing (see text). The field polarization is TE.

of susceptibility $\varepsilon_{\mathrm{mat}} = 2$. Then we simulate the effect of a weak absorption in this material by setting $\varepsilon_{\mathrm{mat}} = 2 - j\,1 \times 10^{-3}$. We plot in Fig. 8.10 the difference between the transparent and absorbing core cases of the index (top) and attenuation (bottom) of the waveguide. The difference $\Delta n^{o}_{\mathrm{g,eff}}$ remains small far from the cutoff frequencies.

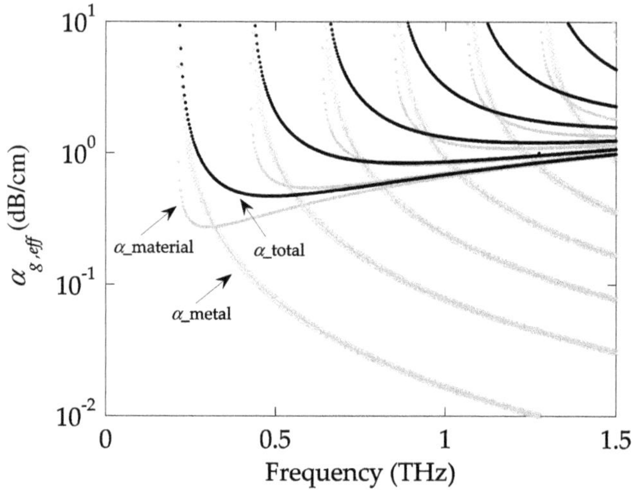

Figure 8.11 Dispersion of the attenuation of the filled waveguide (Fig. 8.10): total attenuation (black dots), attenuation due to absorption in the core material (grey dots) and attenuation through ohmic losses in the metal (grey crosses).

The effect on the attenuation is more pronounced: it is always larger than 0.3 dB/cm and it increases both close to f_c and at higher frequencies.

This behavior is understood when looking at Fig. 8.11, where we plot the attenuation of the waveguide filled with the absorbing material, together with the attenuations only due respectively to ohmic losses in the metal and to absorption in the core material. Ohmic losses are extremely large at f_c and then strongly decrease with frequency. Absorption losses are also very large at f_c, then, when increasing the frequency, they first decrease abruptly; secondly, they reach a minimum and finally they grow again.

Therefore, the total attenuation of each waveguide mode, which is the sum of the ohmic and dielectric losses, is dominated by the ohmic losses nearby f_c and by the absorption losses in the core material far from f_c.

Instead of filling the whole space in between the metallic plate with the material to be studied, a very practical experimental way is to deposit a film of this material against one of the metal

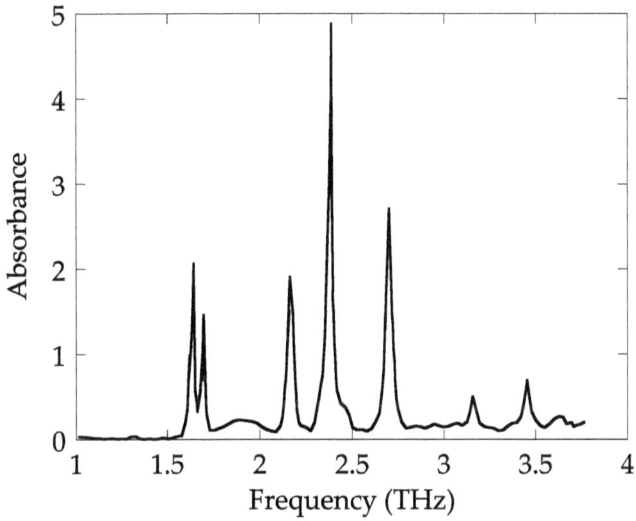

Figure 8.12 Spectral absorbance at $T = 13.8$ K of a sublimated Tris thin film in a polished aluminum parallel plate waveguide. Adapted from Sree Harsha and Grischkowsky [11].

plates. This scheme has been nicely addressed by Gallot et al. [1], who studied a rectangular hollow waveguide (the parallel plate waveguide could be seen as the limit of the rectangular waveguide when its lateral width tends toward infinity). In addition to the electromagnetic calculation of the propagation in such a complex waveguide structure, they also treated the question of coupling the guided mode to the free space beam in a classical THz-TDS system. They showed that the enhancement G of sensitivity compared to classical THz-TDS when determining the absorption coefficient of the material, is given by

$$G = \frac{1}{n_{\mathrm{mat}}^3}\frac{L}{h} \tag{8.30}$$

L is the length of the waveguide, and h is the width of the waveguide gap.

The film thickness is supposed to be much smaller than h, and only the first mode is excited. If the classical THz-TDS characterization is performed in reflection, with the film deposited over a metallic mirror, G is divided by 2 because the THz beam crosses twice the film. Typically, with $n_{\text{mat}} = 2$, $L = 2.5$ cm, and $h = 125$ μm, $G = 102$ [4]. Figure 8.12 shows the spectral absorbance of a film of Tris (an amino alcohol extensively used in buffer solutions in biochemistry and molecular biology), recorded by Grischkowsky's team [11] using a parallel plate waveguide made of polished aluminum ($h = 50$ μm).

The same team has published astonishing results regarding the characterization of nanometer-thick water layers adsorbed onto the metal surface [1]. The waveguide was made of planar copper plates separated by a gap of 52 μm. Both refractive index and absorption coefficient of a ~20 nm thick water layer have been estimated up to 2.5 THz.

8.2.5 Cylindrical THz Waveguides

In a way, propagation in cylindrical waveguides is comparable to that in planar waveguides, but the cylindrical geometry induces an additional boundary condition to the electromagnetic problem: when the azimuthal angle ϕ reaches 2π, the situation is the same as for $\phi = 0$. Thus, the field is quantified not only along the radius r but also versus the azimuth angle ϕ. Usually, the guided field is decomposed into the radial, azimuthal, and longitudinal components, namely F_{r}, F_{ϕ}, and F_{z}.

In the simplest case of a hollow metallic waveguide, of internal radius a, and whose metal is considered as a perfect conductor, the longitudinal component of the guided field is

$$F_{\nu,\text{m,z}} = A\cos(\nu\phi)\, J_{\nu}(k_{\text{c}} r)\, e^{-jk_{\text{z}}z}, \tag{8.31}$$

with

$$k_{\text{z}}^2 = k^2 - k_{\text{c}}^2, \quad k_{\text{c,TM}} = \frac{p_{\nu\,\text{m}}}{a},\; k_{\text{c,TE}} = \frac{p_{\nu m}'}{a}. \tag{8.32}$$

$J_{\nu}(x)$ is a first kind Bessel function of order ν; $p_{\nu m}$, and $p_{\nu m}'$ are the m-th zeros of $J_{\nu}(x)$ and of its derivative (see Table 8.1).

In the TM polarization case, $F_{\nu,\text{m,z}} = E_{\nu,\text{m,z}}$, while $F_{\nu,\text{m,z}} = H_{\nu,\text{m,z}}$ in the TE case. There exists no TEM guided wave. The cutoff frequency is deduced from

$$k_{\text{z}} = 0 \quad \Rightarrow \quad f_{\text{cutoff}}^{\text{TE}} = \frac{c}{2\pi n_{\text{core}}}\frac{p_{\nu m}'}{a}, \quad f_{\text{cutoff}}^{\text{TM}} = \frac{c}{2\pi n_{\text{core}}}\frac{p_{\nu m}}{a} \tag{8.33}$$

Table 8.1 Zeroes of the first-kind Bessel function and of its derivative

		$m = 1$	$m = 2$	$m = 3$
$\nu = 0$	$p_{\nu,m}$	0	5.5201	8.6537
	$p'_{\nu,m}$	0	3.8317	7.0156
$\nu = 1$	$p_{\nu,m}$	3.8317	7.0156	10.1735
	$p'_{\nu,m}$	1.8412	5.3314	8.5363
$\nu = 2$	$p_{\nu,m}$	5.1356	8.4172	11.6198
	$p'_{\nu,m}$	3.0542	6.7061	9.9695

n_{core} is the refractive index of the material inside the metallic waveguide.

The propagation losses in the waveguide can be treated in an elegant manner. Because the absorption coefficient appears in the propagation exponent, the contributions of the ohmic losses in the metal (α_{metal}) and of the dielectric losses in the core material (α_{diel}) are added.

Let us first estimate α_{metal}. We define the time-averaged guided power at location z in the waveguide:

$$\overline{P_z}(z) = P_o\, e^{-\alpha z} \Rightarrow \frac{d}{dz}\overline{P_z}(z) = -\alpha \overline{P_z}(z) \tag{8.34}$$

$$\Rightarrow \alpha = -\frac{\frac{d}{dz}\overline{P_z}(z)}{\overline{P_z}(z)} \approx \frac{\overline{P_z}(z) - \overline{P_z}(z+\Delta z)}{\overline{P_z}(z)\,\Delta z} = \frac{\overline{P_{\text{loss}}}(z, \Delta z)}{\overline{P_z}(z)\,\Delta z} \tag{8.35}$$

$\overline{P_{\text{loss}}}(z, \Delta z) = \overline{P_z}(z) - \overline{P_z}(z+\Delta z)$ is the absorbed power when the guided wave propagates over the distance Δz. The absorption in the metal is due to the Joule effect (i.e., the heating of metal due to the flow of current):

$$\overline{P_{\text{loss}}}(z, \Delta z) = R\, I^2 = \frac{\Delta z}{2\pi a\, \delta\, \sigma} I^2 = \frac{\Delta z}{2\pi a} R_S I^2, \tag{8.36}$$

where I is the current flowing at the metal surface, R is the resistance of the waveguide over the distance Δz, $\delta = (\pi f \sigma \mu_o)^{1/2}$ is the penetration depth of the field in the metal, σ is the metal conductivity, and R_S is the metal sheet resistance. I, or more exactly the current density J, is determined by applying the boundary conditions at the metal surface, i.e.

$$\vec{J} = \vec{u}_{\text{diel}\rightarrow\text{met}} \times \left(\vec{H}_{\text{met}} - \vec{H}_{\text{diel}}\right) \quad \Rightarrow \quad I = \int_{\text{metal surface}} \vec{J} \cdot d\vec{l}. \quad (8.37)$$

This last expression depends on the shape of the waveguide. For the cylindrical waveguide treated here, the calculation is quite easy and gives

$$\alpha_{\nu m}^{\text{TE}} = \frac{R_S}{\eta a} \frac{\left(\frac{f_{\text{cutoff}}^{\text{TE}}}{f}\right)^2 + \frac{\nu^2}{p_{\nu m}^{\prime 2} - \nu^2}}{\sqrt{1 - \left(\frac{f_{\text{cutoff}}^{\text{TE}}}{f}\right)^2}}, \; \alpha_{\nu m}^{\text{TM}} = \frac{R_S}{\eta a \sqrt{1 - \left(\frac{f_{\text{cutoff}}^{\text{TM}}}{f}\right)^2}}. \quad (8.38)$$

η is the waveguide core material impedance. The dielectric losses are derived from Eq. 8.32 and are simply written as

$$\alpha_{\text{diel}}^{\text{TE,TM}} = \alpha \Bigg/ \sqrt{1 - \left(\frac{f_{\text{cutoff}}^{\text{TE,TM}}}{f}\right)^2}, \quad (8.39)$$

where α is the coefficient of absorption of the core material.

When the hollow circular waveguide is filled with a lossy dielectric material, the total propagation losses are the sum of the ohmic and dielectric losses. The relative contribution of the dielectric losses is

$$\frac{\alpha_{\text{diel}}^{\text{TM}}}{\alpha_{\text{total}}^{\text{TM}}} = \frac{\alpha_{\text{diel}}^{\text{TM}}}{\alpha_{\text{ohm}}^{\text{TM}} + \alpha_{\text{diel}}^{\text{TM}}} \approx \frac{1}{1 + n \frac{R_S}{Z_o a \alpha}}, \quad (8.40)$$

where n is the refractive index of the dielectric material, and Z_o is the vacuum impedance. In the THz range, the sheet resistance of noble metals is of the order of $0.1 \sim 1\ \Omega/\text{sq}$, and thus ohmic losses become preponderant when the dielectric absorption is smaller than a few hundredths of cm^{-1}.

Significant progress on propagating guided THz signals was done by D. Mittleman and his team in 2004 [12]. They made use of a simple metal wire, which is well known, since the first calculation by Sommerfeld [13], to support and propagate surface plasmon-like wave, up to microwave frequencies. The metal wire geometry is a simple cylinder (diameter $\approx$1 mm) of metal located in a dielectric medium (generally air). The guided field is associated with collective oscillations of free charges in the radial direction at the metal surface. Such Sommerfeld waves propagate over long distances ($\alpha \approx 0.01\ \text{cm}^{-1}$) with an almost nondispersive group velocity [12]. Because the THz field does not enter deep in the metal, it spreads quite far (a few millimeters) in the ambient medium, and it is rather sensitive to any material deposited on the wire surface. This property has been employed by Walther et al. [14] to retrieve the absorption spectrum of polycrystalline lactose powder dispersed on top of the wire. Even if the signals are rather noisy, the strongest lactose line at 0.53 THz is well observed. This experimental scheme is quite easy to setup and can be used to rapidly characterize powders and liquids.

8.2.6 THz-TDS with Propagating Surface Plasmons

The electrical coupling between light (photons) and elementary excitation in solid state matter is called a polariton. These elementary excitations comprise free carriers (plasmon), phonons, or excitons (bound electron–hole pair in a semiconductor). Depending on the excitation involved, the coupled state is called phonon polariton or plasmon polariton. When dealing with metallic surfaces, plasmon polaritons (or simply speaking, surface plasmons [SPs]) can be excited. They correspond to a collective oscillating motion of the free carriers in the metal. For frequencies below the plasma frequency, light does not propagate inside the metal but decays exponentially from the surface over a penetration depth smaller than the wavelength. Consequently, the excitation of a plasmon in metals or other conducting materials takes place at the surface and the main movement of the free charges occurs perpendicularly to the surface. Because the SP field decays exponentially in both metal and upper medium, the surface acts as a waveguide along which SP propagates. In the THz range, such SPs are expected to propagate over long distances (tens of centimeters), making it possible to design 2D THz circuits and sensitive sensing devices.

The properties of propagating SP can be easily deduced from the electromagnetic study of transmission/reflection of a plane wave at the interface between two media, whose dielectric constants are $\tilde{\varepsilon}_1$ and $\tilde{\varepsilon}_2$ respectively. The transmitted and reflected fields obey the following matrix equation (see Annex, Eq. A.22):

$$\begin{pmatrix} -1 & 1 \\ 1 & \frac{\tilde{k}_{2z}}{\tilde{k}_{1z}}\tilde{\alpha}_{12} \end{pmatrix} \begin{pmatrix} F_r \\ F_t \end{pmatrix} = \begin{pmatrix} 1 \\ 1 \end{pmatrix} F_i. \tag{8.41}$$

Here $F = E_y$ and $\tilde{\alpha}_{12} = \tilde{\mu}_1/\tilde{\mu}_2$ in the TE polarization case, while $F = H_y$ and $\tilde{\alpha}_{12} = \tilde{\varepsilon}_1/\tilde{\varepsilon}_2$ in the TM one. z is the direction normal to the interface, and the plane of incidence is (x, O, z). If there is no exciting field ($F_i = 0$), the solution of Eq. 8.41 determines the proper modes of the structure, which are obtained by setting the following determinant to zero:

$$\det\begin{pmatrix} -1 & 1 \\ 1 & \dfrac{\tilde{k}_{2z}}{\tilde{k}_{1z}}\tilde{\alpha}_{12} \end{pmatrix} = 0 \quad \Rightarrow \quad \frac{\tilde{k}_{2z}}{\tilde{k}_{1z}}\tilde{\alpha}_{12} + 1 = 0. \tag{8.42}$$

This equation has no solution in the TE case. In the TM case, we get

$$\tilde{k}_x = \tilde{k}_{SP} = \frac{\omega}{c}\sqrt{\frac{\tilde{\varepsilon}_1\tilde{\varepsilon}_2}{\tilde{\varepsilon}_1 + \tilde{\varepsilon}_2}} = \frac{\omega}{c}\tilde{n}_{eff} \equiv \frac{\omega}{c}\tilde{n}_{SP}. \tag{8.43}$$

Let us examine this expression in the simple case of no loss ($\tilde{\varepsilon}_2 = \varepsilon$) and with air/vacuum as above medium ($\varepsilon_1 = 1$). Figure 8.13a shows k_x normalized to ω/c (i.e., the effective index n_{eff} of the mode) versus ε.

For $\varepsilon > 0$, n_{eff} is smaller than 1, which is physically impossible for a propagating wave. For $\varepsilon < 0$, the mode exists for values of ε smaller than -1. In this domain of existence ($\varepsilon < -1$), the wavevector component perpendicular to the interface is

$$\tilde{k}_{1z} = \frac{\omega}{c}\frac{1}{\sqrt{1+\varepsilon}}, \quad \tilde{k}_{2z} = \frac{\omega}{c}\frac{\varepsilon}{\sqrt{1+\varepsilon}}. \tag{8.44}$$

Because $\varepsilon < -1$, the square-roots of the denominators, and thus the wavevector components, are purely imaginary, i.e., $k_{1z} = j\left|\tilde{k}_{1z}\right|$, $k_{2z} = j\left|\tilde{k}_{2z}\right|$. Therefore, the mode is guided along the interface and is evanescent in the two surrounding media. Such a mode exists only if $\varepsilon < -1$. This is the case of metals and doped semiconductors up to their plasma frequencies. Usually, the metal or semiconductor conductivity is rather well described by a classical Drude model, and

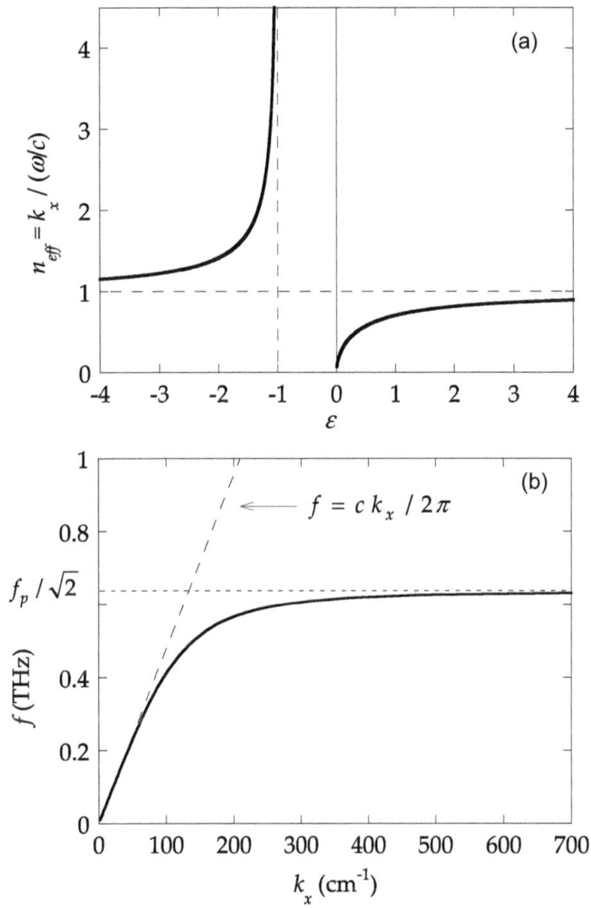

Figure 8.13 (a) $n_{\rm eff}$ versus ε for a SP propagating at the interface between a medium of permittivity ε (real) and air ($\varepsilon = 1$). (b) Dispersion curve of an SP propagation on a Drude-like medium with $\varepsilon_\infty = 1$ and $f_{\rm p} = 0.903$ THz.

thus the SP wavevector is

$$\tilde{\varepsilon} = \varepsilon_\infty - \frac{\omega_{\rm p}^2}{\omega\,(\omega + j\Gamma)} \quad \Rightarrow \quad \tilde{k}_{\rm SP} = \frac{\omega}{c}\sqrt{\frac{\omega_{\rm p}^2 - \varepsilon_\infty\omega\,(\omega + j\Gamma)}{\omega_{\rm p}^2 - (1+\varepsilon_\infty)\,\omega\,(\omega + j\Gamma)}}. \tag{8.45}$$

$f_{\rm p} = \omega_{\rm p}/2\pi$ is the plasma frequency, Γ is the damping frequency, and ε_∞ is the contribution to ε of bound electrons. The plasma frequency is related to the density N of free carriers in the conducting material, $\omega_p^2 = Ne^2/m^*\varepsilon_o$, where $-e$ is the charge of the electron, m^* is its effective mass, and ε_o is the permittivity of vacuum. Figure 8.13 (right) presents the curve f versus k_x with $\varepsilon_\infty = 1$, $\Gamma = 0$, and $f_p = 0.903$ THz, i.e., $N = 10^6$ cm^{-3} and $m^* = m_e \approx 9 \times 10^{-31}$ kg, which corresponds to a doped semiconductor. For small values of f, the SP dispersion curve coincides almost with the light line: In fact, most of the electromagnetic SP energy is located in air. Its field in air looks like a plane wave and the SP is said to be photon-like. On the opposite, at higher k_x values, f tends to the asymptotic value $f_p/\sqrt{2}$. The SP propagation properties are driven by the parameters of the conducting material: this is the plasmon-like regime. There also exists an upper branch of the dispersion curve, above f_p (not drawn in Fig. 8.13), which describes the dispersion of bulk plasmons. Note that the dispersion curve keeps almost the same shape when changing N, but with the f scale varying as $\sqrt{N}$. Thus, for doped semiconductor, the frequency at which there is a transition from the photon-like regime to the plasmon-like belongs to the THz range, while it is in the visible for metals ($N \sim 10^{22}$ cm^{-3}).

For pure and especially noble metals, propagating SP at THz frequencies are always excited in the photon-like regime. The plasma frequency belongs to the UV range, and thus it is much larger than the THz frequency. It follows that

$$\tilde{n}_{SP} \approx 1 + \frac{\omega^2}{2\omega_p^2} + j\frac{\omega\Gamma}{2\omega_p^2}, \quad \tilde{k}_{1z} \approx j\frac{\omega}{c}\frac{\omega}{\omega_p}, \quad \tilde{k}_{2z} = j\frac{\omega}{c}\frac{\omega_p}{\omega}. \tag{8.46}$$

Therefore, because of the high THz permittivity of metals, the SP effective index value is close to 1, and consequently, the SP field is mostly located in air. The penetration depth d of the SP field in the surrounding media is the inverse of $|\tilde{k}_z|$ and is

$$d_{1z} = \frac{c\,\omega_p}{\omega^2}, \quad d_{2z} = \frac{c}{\omega_p} \quad \Rightarrow \quad \frac{d_{1z}}{d_{2z}} = \frac{\omega_p^2}{\omega^2} \approx |\varepsilon| \tag{8.47}$$

For example, according to Ordal et al. [9], the permittivity of bulk aluminum at 1 THz is $\varepsilon_{Al} = -\left(3 \times 10^4 - j\,1 \times 10^6\right)$. The effective index and propagation loss coefficient are respectively $n_{SP} \approx 1 + 1.5 \times 10^{-8}$ and $\alpha \approx 2 \times 10^{-4}$ cm^{-1}. Therefore, the SP is expected

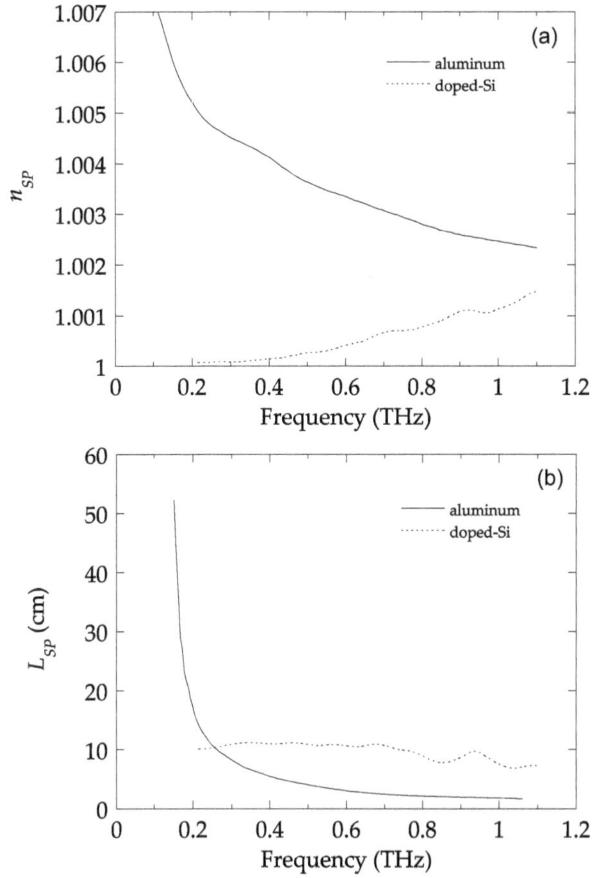

Figure 8.14 n_{SP} (a) and L_{SP} (b) versus frequency for a SP propagating over aluminum (continuous line) and n-doped Si (dashed line). These curves are deduced from measured THz-TDS data. Adapted from Nazarov and Coutaz [22].

to propagate over a distance $L_{SP} = 1/\alpha \approx 50$ m! It spreads over 7 cm in air, but remains confined to 70 nm in the metal; however, the corresponding experimental values are very different. Typically, at THz frequencies, SP propagates over a few centimeters instead of meters, and the decay length in air is of the order of millimeters. This has been reported for metals [15–18] and for doped semiconductors [19–21]. This is observed in Fig. 8.14, where the spectra of effective

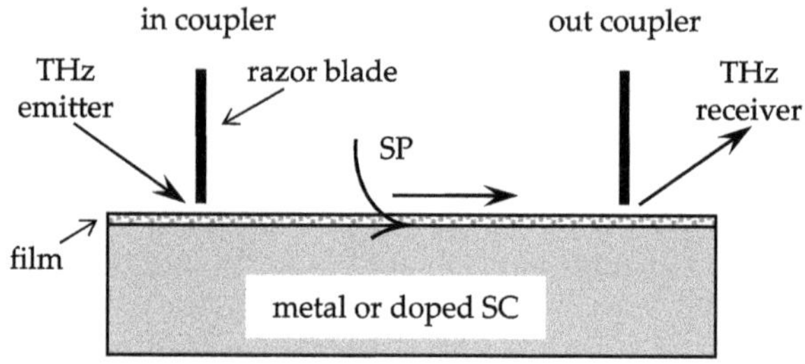

Figure 8.15 Scheme of a THz-TDS experiment involving the excitation of SP. Here the SP couplers are razor blades, but they could be substituted by grating or prism couplers.

index n_{SP} and decay length L_{SP} of SP propagating on aluminum and n-doped Si have been evaluated experimentally (adapted from Nazarov and Coutaz [22]). Here L_{SP} is of the order of centimeters instead of meters. The physical origin of such discrepancy is still controversial [22–24].

SP sensors are based on the change of SP propagation parameters induced by a film deposited over the conducting medium. Two experimental schemes may be employed: (i) an SP coupler (prism or grating), which will be described in the next paragraph; (ii) a device (see Fig. 8.15) with an input SP coupler and an output one in between which the SP propagates. In this region between the two couplers, the metal (or conduction semiconductor) substrate will be covered by the film to be characterized. This device is inserted in a common THz-TDS setup. Razor blades (Fig. 8.15) or metal tips, with which coupling occurs thanks to near-field diffraction at the metal edge, are easy to implement. They have been employed to efficiently excite surface plasmons at metal [25] or doped semiconductor [26] surfaces, or guided mode along a Goubau line or a wire [27].

The SP propagation parameters are strongly perturbed by the thin film [28]. The determination of the film parameters could be done by using the bare device as reference. This technique has been nicely demonstrated by Isaac et al. [29]. A thin layer of polystyrene is spin coated over either a gold film or a conducting InSb substrate, as depicted in Fig. 8.15. The distance between the blades is 1 cm. The THz waveforms are measured with and without the polystyrene

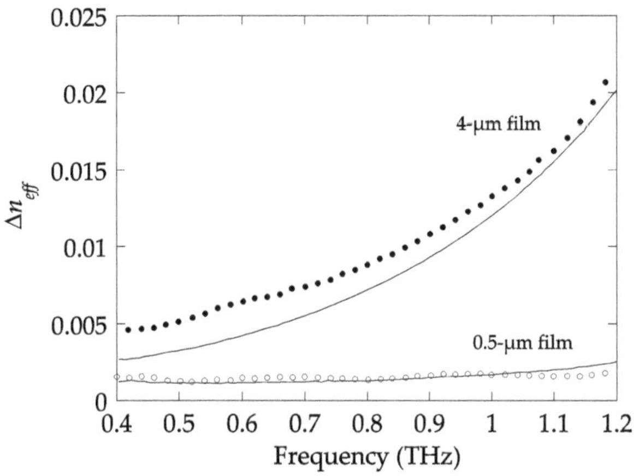

Figure 8.16 The difference of SP effective index between the covered and bare InP cases. The circles are experimental data, and the curves are the best-calculated fits which lead to determine the refractive index of polystyrene. The upper results are obtained with a 4 μm thick polystyrene film, and the lower ones with a 0.5 μm film. Adapted from Isaac et al. [29].

layer. Because loss due to absorption in the polystyrene layer is weak, the authors extracted only the refractive index of the layer, which is accurately determined thanks to the long SP propagation length between the two coupling blades. Figure 8.16 shows the difference of SP effective index induced by the thin film, compared to the bare InP substrate. The best fit with an electromagnetic modeling allows one to determine the refractive index of the film, which is the single unknown parameter. Here, one gets $n_{\text{polystyrene}} = 1.7 \pm 0.6$, which is fairly constant over the studied frequency range because of extremely weak absorption.

Another nice example of SP-mediated THz-TDS of thin films is given by Ng et al. [30]. Here the scheme is the same as previously, but the SP propagates over a gold-coated surface that exhibits a grating shape. Each period ($d = 60$ μm) of the grating shape includes a slightly trapezoidal relief of height 31 μm and width ~30 μm (see inset in Fig. 8.17). The very deep relief (the groove depth is about half the period) causes the SP to propagate at a reduced speed (spoof SP). As a result, the transition between the light-line regime and the plasmon regime (see Fig. 8.13) happens around

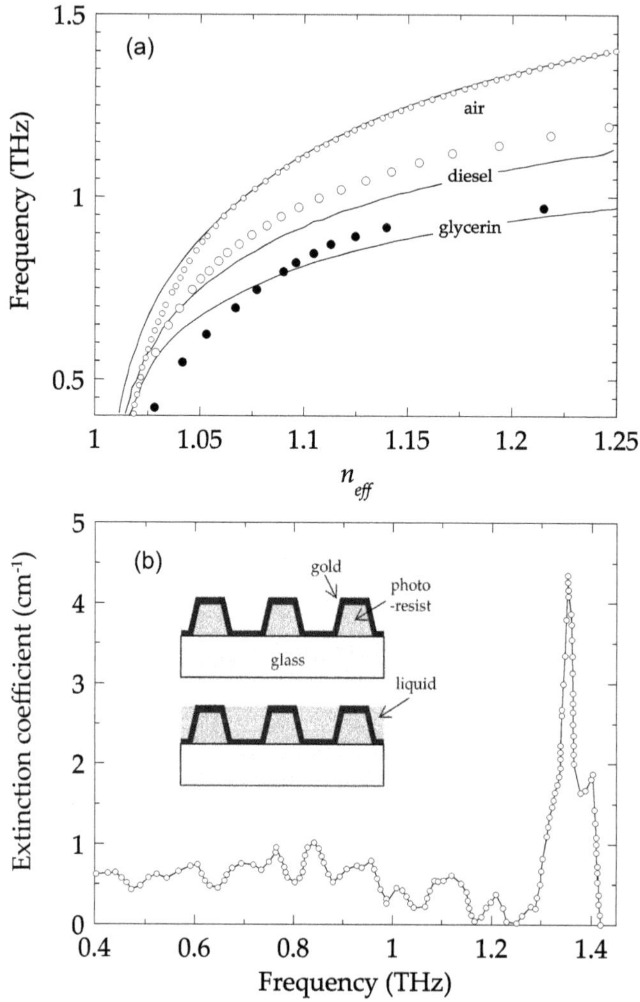

Figure 8.17 (a) Dispersion curve of spoof SP propagating over the grating depicted on the right when the grating grooves are empty or filled with diesel or glycerin. Circles are experimental data and curves are the best-calculated fits which permit the determination of the refractive index of the liquid. (b) Extinction coefficient spectrum of the spoof SP when the grating is covered with a thin layer of α-lactose powder. Adapted from Ng et al. [30].

1.5 THz, without the need of decreasing the free carrier density in the metal. When the groove depth is filled with a liquid, the spoof SP propagation is strongly perturbed. Figure 8.17a presents the dispersion curves of this spoof SP, recorded (circles) and simulated (curves). The data labeled "air" are recorded without liquid in the grooves, while other data have been obtained with diesel or glycerin filling the grooves. The refractive index of the liquids is retrieved by fitting the experimental data with the numerical curves. Figure 8.17b shows an interesting result: α-lactose monohydrate powder is sprinkled onto the spoof SP device, with a surface mass density of the order of 300 ng/mm^2. The spectrum of the device extinction coefficient ($= -\ln(T)/L$, where T is the transmission of the device from one blade coupler to the other one, and L is the distance between the two blades) exhibits a strong absorption line around 1.4 THz, which is a known fingerprint of α-lactose. When the α-lactose powder is deposited over a flat gold film, no fingerprint is observed, demonstrating the sensitivity enhancement by excitation of spoof SP.

8.2.7 THz-TDS with Waveguide Couplers

Waves that are guided, either in a dielectric waveguide or surface plasmon, exhibit a wavevector larger than the one of a plane wave in the surrounding media. Because of the conservation of the wavevector component parallel to the waveguide surface, exciting a guided wave with a plane wave incoming from the side requires a coupler that matches the wavevectors. In the THz regime, three different couplers have been used (Fig. 8.18): prism [31], grating [32], and blade [33]. THz spectroscopy can be performed by analyzing the coupling conditions versus the properties of the studied material, which can be either the waveguide material or an additional layer at the coupler location.

8.2.7.1 Prism couplers

The prism is made of a material whose refractive index, n_P, is larger than that, n_G, of the guiding layer. The prism is located at a distance δ

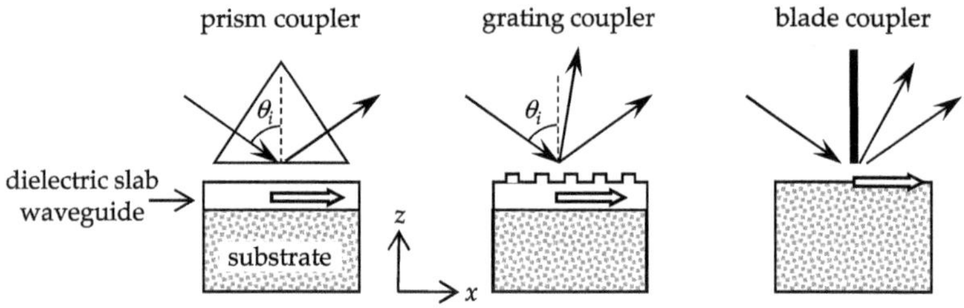

Figure 8.18 Main couplers used to excite a waveguide mode with a plane wave incident from the side.

from the top side of the waveguide. We suppose that the surrounding medium is air. Inside the prism, the incident plane wave impinges the bottom side under an angle θ_i. Because n_P is also larger than the refractive index of air that occupies the gap between the prism and the waveguide, total reflection occurs at the bottom side of the prism when $n_P \sin\theta_i > 1$.

In other words, the parallel component of the incoming wavevector, i.e., $k_{ix} = k_x = n_P \sin\theta_i\, \omega/c$, is the same in the four different media (prism, air gap, waveguide, and substrate). In the air gap, the wavevector component normal to the waveguide is $k_z = \omega/c\sqrt{1 - n_P^2 \sin^2\theta_i}$. The value under the square root is negative; thus k_z is imaginary: the wave is evanescent in the air gap, i.e., its longitudinal wavevector is larger than those of a plane wave in air. Therefore, excitation of a guided mode with effective index n_{eff} is possible, and it occurs when the following relationship is fulfilled:

$$k_{ix} = k_G \quad \Leftrightarrow \quad n_P \sin\theta_i = n_{eff}. \tag{8.48}$$

Note that this expression is approached but not rigorous because the presence of the prism perturbs the guided propagation in the layer. However, it gives a coupling condition close to the actual one. When coupling occurs, sharp dips are seen in the reflection spectrum which are called black lines; these appear if either the waveguide exhibits losses, or if the exciting field is not a plane wave but focused. Measuring, for a given frequency (angle), the angle (frequency) at which the black line is observed leads to determine n_{eff} through relation 8.48. It should be emphasized that only guided modes whose effective index n_{eff} is smaller than n_P can be excited.

This is a practical limitation of this technique, which requests a very high index prism made of a transparent material. The prism coupler has also been used to excite surface plasmons propagating over a metal surface [34]. Here the constraint on the refractive index of the prism is not difficult to fulfill, because the effective index of the surface plasmon is close to 1.

These black lines are clearly observed in Fig. 8.19a. Here a 495 μm thick GaAs wafer serves as a waveguide. The prism is made of HR silicon and is located 40 μm above the waveguide. In the same original paper [31], Rau and coauthors reported on the effect of a thin oil film (thickness less than 1 μm) deposited onto the GaAs waveguide.

The oil film slightly changes the properties of the waveguide and thus shifts the black lines. This shift is clearly seen when the ratio of the spectra of the covered and bare waveguides is plotted (Fig. 8.19b). However, to our knowledge, this method as well as the blade coupling technique have not been employed to determine the parameters of a film overcoat. In the blade technique, coupling is achieved through the excitation of evanescent waves at the gap between the blade tip and the waveguide.

8.2.7.2 Diffraction grating couplers

Diffraction gratings are most of the time a 1D periodic arrangement of similar patterns, like strips. When illuminated by an incoming beam, the additive or subtractive sum of all the waves diffracted by each periodic pattern leads to the well-known grating diffractive effect: rays of light propagate in several discrete directions, which depend on the light wavelength (Fig. 8.20).

This is easily explained using an electromagnetic theory developed by Rayleigh [35]. Let a plane wave be incident on a grating made of strips of period d located in air, under oblique incidence (angle of incidence θ_i; see Fig. 8.20). The incoming E-field is

$$E_i\,(x, y, z) = E_i\, e^{-j(k_{ix}x - k_{iz}z)}, \tag{8.49}$$

where the components of the wavevector are $k_{ix} = \omega \sin\theta_i / c$ and $k_{iz} = \omega \cos\theta_i / c$. In the upper medium, the total field is the sum of the incident one and the reflected one (this reflected wave is also a

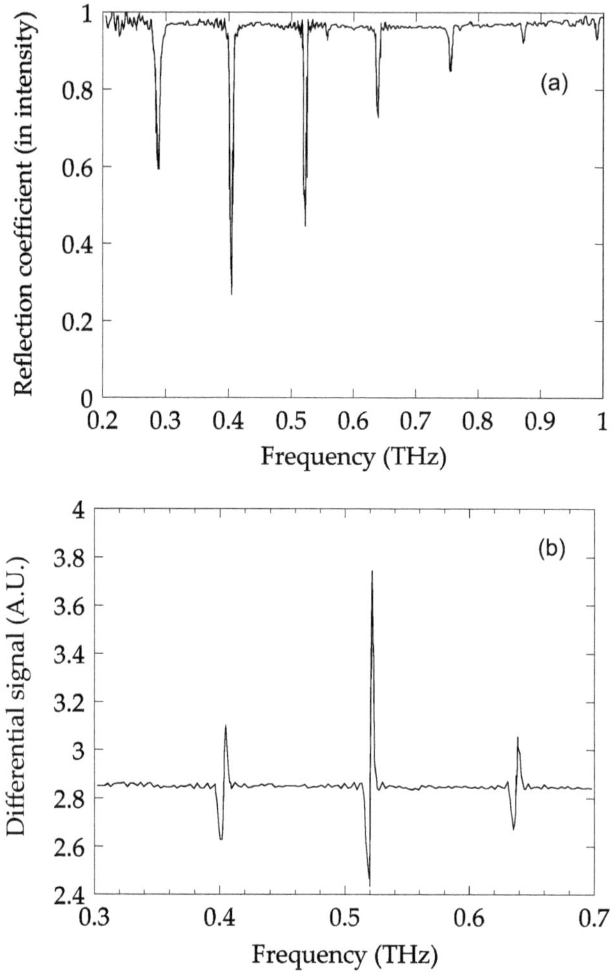

Figure 8.19 (a) Reflection spectrum (in intensity, TE polarization) of a prism coupler. The waveguide is a GaAs wafer. (b) Ratio of the spectra of the oil-covered waveguide and bare waveguide. Adapted from Rau et al. [31].

plane wave as the upper medium is homogeneous):

$$E_{\mathrm{i}}\, e^{-j(k_{\mathrm{ix}}x - k_{\mathrm{iz}}z)} + E_{\mathrm{r}}\,(x, y, z) = E_{\mathrm{i}}\, e^{-j(k_{\mathrm{ix}}x - k_{\mathrm{iz}}z)} + E_{\mathrm{r}} e^{-j(k_{\mathrm{rx}}x + k_{rz}z)}. \tag{8.50}$$

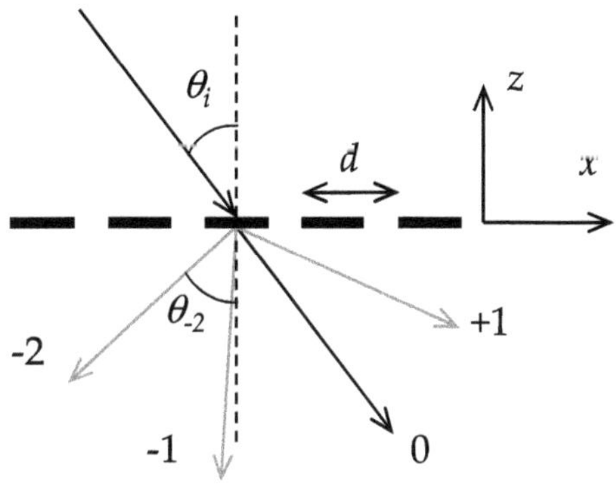

Figure 8.20 Diffraction (here in transmission) by a grating.

Due to the periodicity of the grating structure, this total field is the same at x and $x + d$. Thus,

$$E_{\mathrm{i}}\, e^{-j(k_{\mathrm{ix}}x-k_{\mathrm{iz}}z)} + E_{\mathrm{r}}e^{-j(k_{\mathrm{rx}}x+k_{rz}z)} = E_{\mathrm{i}}\, e^{-j(k_{\mathrm{ix}}(x+d)-k_{\mathrm{iz}}z)} + E_{\mathrm{r}}e^{-j(k_{\mathrm{rx}}(x+d)+k_{rz}z)},$$

$$\Rightarrow E_{\mathrm{i}} + E_{\mathrm{r}}e^{-j\Delta k_x x} = e^{-jk_{\mathrm{ix}}d}\left(E_{\mathrm{i}} + E_{\mathrm{r}}e^{-j\Delta k_x(x+d)}\right),$$
$$\Delta k_x = k_{\mathrm{rx}} - k_{\mathrm{ix}}. \tag{8.51}$$

This equation should be valid whatever the value of x. Therefore,

$$e^{-j\Delta k_x x} = e^{-j\Delta k_x(x+d)} \quad \Rightarrow \quad \Delta k_x d = (k_{\mathrm{rx}} - k_{\mathrm{ix}})\, d = 2m\pi. \tag{8.52}$$

It follows that the parallel component of the reflected wavevector can take several values depending on the integer m. We write

$$\begin{cases} k_{\mathrm{rx}} \equiv k_{\mathrm{mx}} = k_{\mathrm{ix}} + m\,\frac{2\pi}{d} \equiv k_{\mathrm{ix}} + m\,k_{\mathrm{grating}} \Leftrightarrow \sin\theta_m = \sin\theta_{\mathrm{i}} + m\frac{\lambda}{d}, \\ k_{\mathrm{mz}}^2 = \frac{\omega^2}{c^2} - k_{\mathrm{mx}}^2. \end{cases} \tag{8.53}$$

The first equation in 8.53 is the grating diffraction equation. θ_m is the angle between the direction of propagation of order m and the direction normal to the plane of the grating. Note that if k_{mx}^2 is larger than ω^2/c^2, then k_{mz}^2 is negative. It means that k_{mz} is imaginary: the order does not propagate in the ambient medium, but remains bound to the grating. The corresponding field is evanescent.

When the grating is etched on the surface of a bulk conducting medium, or of a dielectric waveguide (see Fig. 8.21), it serves as

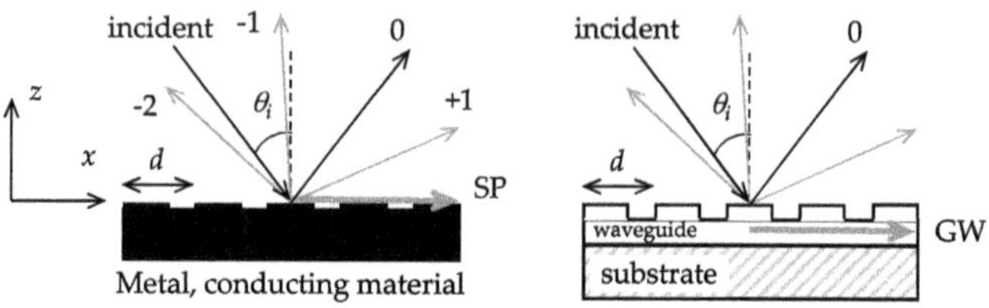

Figure 8.21 Gratings used as couplers to excite an SP on a metallic slab (left) or a guided mode (GW) in a dielectric waveguide (right). Here the coupling is achieved through the second diffraction order.

a coupler to excite guided modes, either SP or waveguide modes. The wavevector k_g of the latter is larger than the wavevector of any plane wave propagating in the ambient media or in the substrate. Thus, to excite these guided waves with an incident plane wave, the necessary additional wavevector to realize the phase matching is brought by the grating:

$$k_{\mathrm{g}} = k_{\mathrm{ix}} + m\, k_{\mathrm{grating}} \quad \Leftrightarrow \quad n_{\mathrm{eff}} = \sin\theta_{\mathrm{i}} + m\frac{\lambda}{d}. \tag{8.54}$$

Equation 8.54 is an approached relation in the sense we did not take into account absorption losses in the materials, as well as the influence of the grating corrugation on the guided wave properties as compared to the noncorrugated geometry. When a guided mode is excited, one clearly observes black lines in the transmitted or reflected spectra for a given angle of incidence (or in the angular plot at a given frequency).

They result from the absorption of the excited guided waves and the related missing energy in transmission or reflection. Without any loss, black lines do not exist because the guided waves are also diffracted by the grating and thus give back the guided energy to directly transmitted or reflected beams. Figure 8.22 (adapted from Roux et al. [36]) shows the transmission (TE polarization) of a 210 μm thick HR silicon wafer, which serves as a THz dielectric waveguide. A rectangular grating (period 322 μm, groove depth 32 μm) has been etched on the wafer surface and acts as a coupler to excite the modes of the waveguide. Corresponding black lines are clearly seen as narrow dips in the overall curve, while the broad oscillations are due the Fabry–Pérot effect inside the wafer. Data

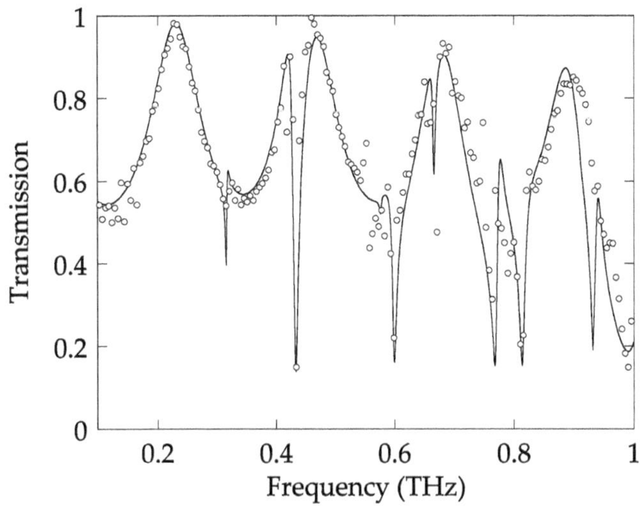

Figure 8.22 TE transmission spectrum of an HR silicon wafer that serves as a waveguide. A grating relief is etched on one side of the wafer. A guided wave is excited at each narrow dip observed in the spectrum, while the broader oscillations correspond to a Fabry–Pérot effect inside the wafer. Open circles are experimental data; continuous line is calculated. Adapted from Roux et al. [36].

have been measured by THz-TDS, while the continuous curve is calculated with a differential electromagnetic method.

It is possible to characterize a tiny amount of matter, by depositing it over the grating coupler. This thin film will change the waveguide properties, and thus it will shift the black line frequency and modify their shape. The refractive index of the film and its absorption coefficient may be retrieved by performing an inverse electromagnetic modeling of the grating coupler. This has been first demonstrated by Laamiri and coworkers [37], who used a 285 μm thick HR silicon wafer with a rectangular grating (d = 250 μm) etched on its top face (this corresponds to the right-side drawing of Fig. 8.23, without any substrate, because the wafer is solid enough to stand alone). On its bottom face, a 7.1 μm thick layer of photoresist has been spin-coated. A black line recorded by THz-TDS in transmission is plotted in Fig. 8.23a. The full circles are data recorded with the bare device, while the open circles are related to

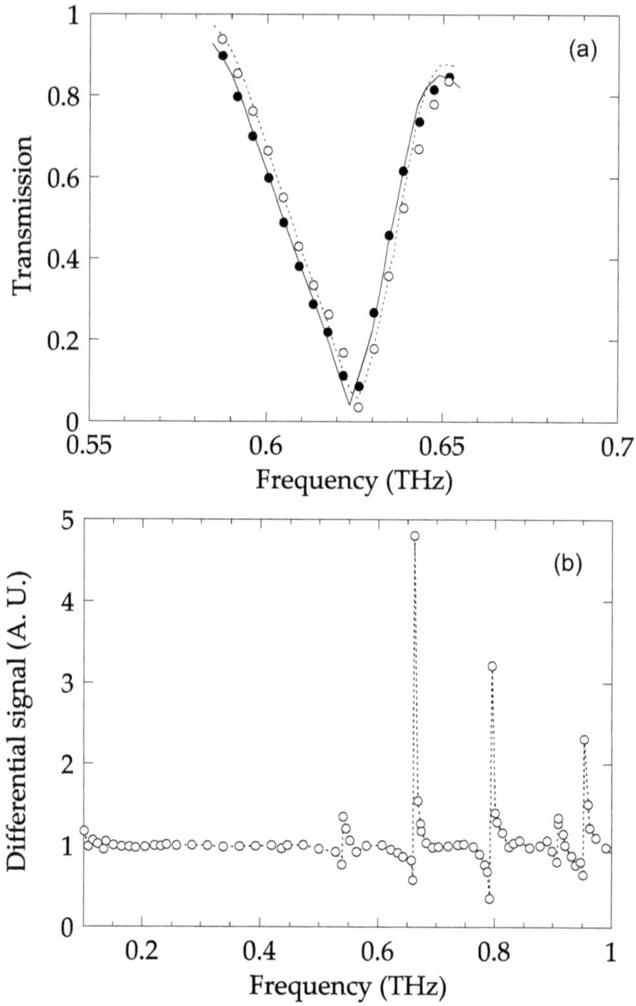

Figure 8.23 (a) TE transmission spectrum of the bare and covered grating device: full circles and continuous line are respectively experimental data and calculated curve for the bare device. Open circles and dashed curve are obtained for the covered device. (b) Ratio of the transmission spectra of the covered and bare devices. Adapted from Laamiri et al. [37].

the covered device. Both data are fitted (respectively continuous and dashed lines) with a differential electromagnetic method, in which only the refractive index and the absorption coefficient of the film are adjustable parameters. The best fit gives $n_{\mathrm{film}} = 1.62 \pm 0.03$ and $\alpha = 34 \pm 9\ \mathrm{cm}^{-1}$ at $f = 0.6$ THz. The advantage of exciting guide modes is illustrated in Fig. 8.23b. The ratio of the experimental transmission curves of the covered and bare samples are plotted. Out of resonance, the film does not noticeably change the transmission, and the ratio is around 1. At the resonance, when a guided mode is excited, the ratio increases up to 5. In other words, due to the coupling phenomenon, the incoming wave is transferred to the waveguide mode, which propagates over a distance equal to about five times the thin film thickness before being coupled out by the grating. Thus, this wave interacts with about five times more film matter than without the wave-guiding effects.

8.2.8 On-Chip Waveguide THz Spectroscopy

Another way of guiding THz waves is extrapolated from microwave technology. 2D microwaves lines are metallic strips deposited over a substrate. They can be sorted by four categories, namely strip (CPS), slot (CPSlot), coplanar (CPW), and Goubau and microstrip (MSL) lines, depicted in Fig. 8.24. Such waveguides have been widely studied at THz frequencies not only for propagation applications but also in view of on-chip THz spectroscopy.

Among coplanar structures, MSL and CPW are the most widespread waveguides at sub-THz and THz frequencies. The guiding parameters of microstrip lines are [38].

$$\varepsilon_{\mathrm{eff,MSL}} = \frac{\varepsilon+1}{2} + \frac{\varepsilon-1}{2}\frac{1}{\sqrt{1+12\frac{d}{w}}}, \tag{8.55}$$

$$\alpha_{\mathrm{eff,MSL}} = \frac{\alpha}{2}\frac{\varepsilon_{\mathrm{eff}}-1}{\varepsilon-1}\sqrt{\frac{\varepsilon}{\varepsilon_{\mathrm{eff}}}} + \frac{1}{Z\,w}\sqrt{\frac{\pi f \mu_{\mathrm{o}}}{\sigma}}. \tag{8.56}$$

Here ε and α are the real part of the relative permittivity and the absorption coefficient of the dielectric medium in between the strip and the ground. σ is the conductivity of the metal, and Z is the

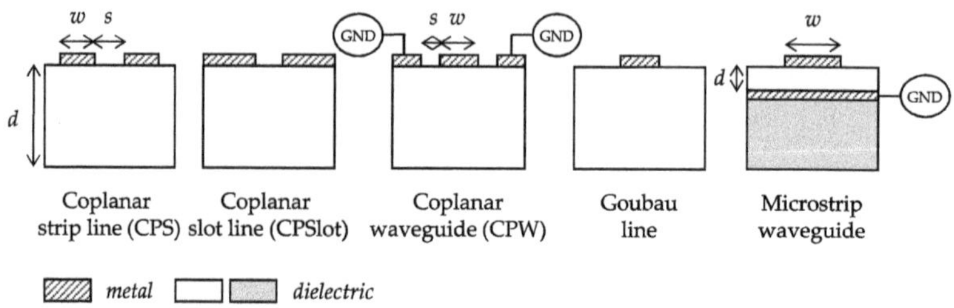

Figure 8.24 Some typical 2D microwave lines that are also employed in the THz range (cross sections).

impedance of the line defined by

$$\begin{cases} w \leq d: & Z = \dfrac{60}{\sqrt{\varepsilon_{\text{eff}}}} \ln\left(\dfrac{8d}{w} + \dfrac{w}{4d}\right), \\ w \geq d: & Z = \dfrac{120\,\pi}{\sqrt{\varepsilon_{\text{eff}}}\sqrt{\dfrac{w}{d} + 1.393 + \dfrac{2}{3}\ln\left(1.444 + \dfrac{w}{d}\right)}}. \end{cases} \tag{8.57}$$

Expression 8.56 is valid for the quasi TEM mode. In Eq. 8.56, the first term corresponds to losses in the dielectric medium, while the second term depicts the ohmic losses in the metal. To reduce the dielectric losses, the dielectric material must be as transparent as possible, as, in the THz range, Topas® [39] or polypropylene [40].

Figure 8.25 shows the experimental spectra of the attenuation and ε_{eff} of a microstrip waveguide with polypropylene as a dielectric, attached to the wafer with a copolymer acrylate adhesive layer. At higher frequencies, attenuation increases because of the radiative losses that vary as f^3. At lower frequencies, attenuation is dominated by the dielectric losses, and the line exhibits a very weak dispersion.

MSLs have been used to perform THz-TDS of polycrystalline lactose monohydrate [41]. A pellet of the sample is located over the waveguide and perturbs the guided mode. The absorption line at 534 GHz was clearly observed in the spectrum transmitted by the covered line. However, the microstrip structure is more useful to characterize absorbing liquids, and especially small amounts of water in which some matter is dissolved, for example biological materials [42]. When the microstrip line is covered by the sample

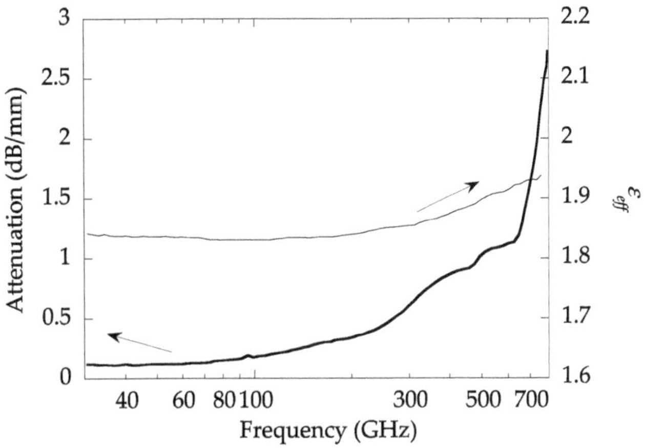

Figure 8.25 Spectra of the attenuation and the effective real permittivity of an MSL deduced from electro-optic sampling measurement. The dielectric layer is made of polypropylene acrylate. Adapted from Nagel et al. [41].

material, including a dielectric buffer layer, the expressions of the induced modifications of the effective index and of the attenuation are obtained analytically, but take complicated mathematical expressions [43, 44]. Nevertheless, they allow the retrieval of both the refractive index and the absorption coefficient of liquids located in a cuvette over the microstrip waveguide. Figure 8.26 presents the variation of the effective index of a microstrip line induced by water located 7 $\sim$ 10 μm over the line (a dielectric layer of polyimide separates the microstrip line and water) [45]. Microstrip lines could also be designed to form a waveguide resonator, which enhances the sensitivity to the parameters of the studied material [46, 47].

CPS and CPW coplanar waveguides have also been extensively studied at THz frequencies in view designing transmission lines [48, 49].

The guiding parameters are

$$n_{\text{eff}} = \sqrt{\varepsilon_{\text{eff}}} = \sqrt{\frac{\varepsilon + 1}{2}} + \frac{\sqrt{\varepsilon} - \sqrt{\frac{\varepsilon+1}{2}}}{1 + a\left(\frac{4d}{cf\sqrt{\varepsilon-1}}\right)^{b}}, \tag{8.58}$$

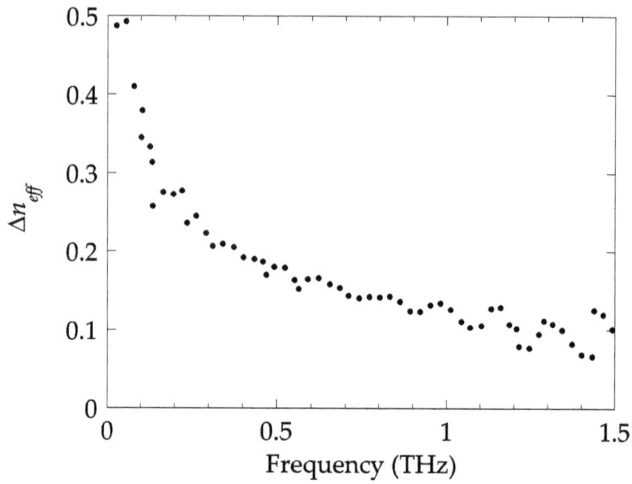

Figure 8.26 The spectrum of the variation Δn_{eff} of the effective index of a microstrip line induced by water placed over a 7 ~ 10 μm thick layer of polyimide that covers the line. Adapted from Ohkubo et al. [45].

$$\begin{cases} \alpha_{\text{CPW}} = \dfrac{\pi^5}{16} \dfrac{(\varepsilon - \varepsilon_{\text{eff}})^2}{\sqrt{\varepsilon_{\text{eff}}}} \dfrac{(s+2w)^2}{K'(k)\,K(k)} \dfrac{f^3}{c^3}, \\ \alpha_{\text{CPS}} = 8\left(3 - \sqrt{8}\right) \dfrac{\varepsilon_{\text{eff}}}{\varepsilon} \alpha_{\text{CPW}}. \end{cases} \tag{8.59}$$

$K(k)$ is the complete elliptic integral of the first kind, $K'(k) = K(\sqrt{1-k^2})$ and $k = s/(s+2w)$. a and b are geometrical coefficients that are almost independent of frequency [50]. Here the attenuation is supposed to be dominated over 200 GHz by the radiative losses, and both ohmic losses in the metallic strips and dielectric losses in the substrate are neglected.

CPS lines have been utilized only for spectroscopy of solid materials [50, 51]. In fact, they cannot sustain an over coating liquid that would electrically shorten the metal strips of the CPS. The spectroscopy of iron erbium garnet [50] and polycrystalline biotin [51] powders has been published. In the latter case, the peaks of absorption of biotin are clearly identified (see Fig. 8.27b).

Goubau lines also exhibit interesting propagation properties at sub-THz and THz frequencies. A planar Goubau line is a single conductor strip deposited on or embedded in a dielectric substrate.

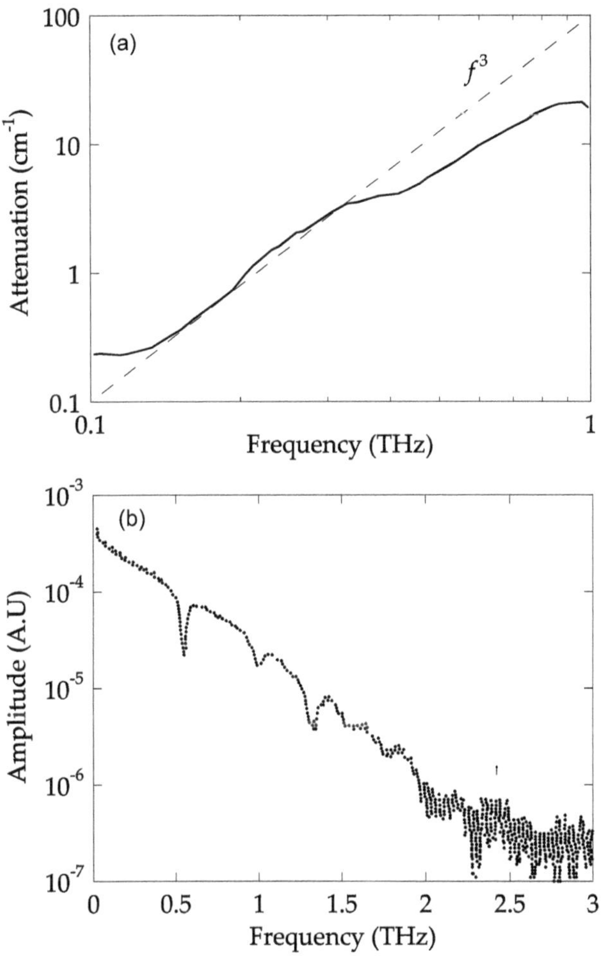

Figure 8.27 (a) Measured attenuation versus frequency of a CPS line on sapphire. Adapted from Frankel et al. [48]. (b) Amplitude of the signal transmitted by a CPS line covered with a layer of biotin powder. Adapted from Kadoya et al. [51].

Such a line can propagate waves up to 1 ~ 2 THz [52, 53]. It is used to make a direct spectroscopy of materials deposited over the line, for example lactose monohydrate [53]. Here again, as in Fig. 8.27b, the absorption peaks of the material are clearly identified, but the material parameters (refractive index and absorption coefficient) are not extracted. The main advantage of Goubau lines is the facility of designing and manufacturing complex 2D circuits [54], like 2D interferometers or circuits including some microfluidic channels to carry low amounts of liquids close to the Goubau line.

8.3 THz-TDs with Metamaterials

8.3.1 Introduction

Metamaterials are by definition artificial materials made with repetitive sub-wavelength structures, whose period is also sub-wavelength. An incident wave interacts with the metamaterial as if this material would have been a continuous medium. The electromagnetic parameters (refractive index, absorption) of the metamaterial are those of the related continuous medium, but they depend on the response of each sub-wavelength structure. The "macro" properties of the material are thus driven by its "micro" properties. This is especially true when each sub-wavelength structure is resonantly excited by the incoming wave. Thus, it is easy to understand that the whole response of the metamaterials will be strongly perturbed by any layer or amount of matter located at the sub-wavelength structure, especially in the resonance regime. This occurs especially with metallic metamaterials because of the high dielectric contrast between metal and air. Practically, it is easier to employ 2D metasurfaces instead of 3D bulky metamaterials. Indeed, matter to be characterized can be deposited without difficulty as a whole (films, layers) over the sub-wavelength structures. This will dramatically change the transmission or reflection of the metasurface, more so than with a bulk substrate, for which the change would be proportional to the relative amount of additional matter. This enhancement effect can be used to detect very small amounts of matter localized at or nearby the metasurface. When dealing with metallic metasurfaces, the sensitivity enhancement originates in electromagnetic resonances occurring in each sub-wavelength structure, namely localized surface plasmons (SPs).

These devices are also used with THz wavelengths of the order of the periodicity of the sub-wavelength structures. This leads to strong diffractive effects, as with diffraction gratings. They allow one to excite propagating (or delocalized) SPs that also participate in the enhancement phenomenon. Both delocalized and localized SPs are involved in metasurfaces. Metallic diffraction gratings are perhaps the simplest metasurface, and they permit an efficient excitation of propagating SPs.

In fact, only a very few papers report actual spectroscopy using metamaterials. Most of the published work is dedicated to sensing the presence of a tiny amount of material over the metamaterial, and sometimes to quantifying this involved amount by observing the THz transmission at a given frequency. Identifying some absorption lines could also help in determining the chemical composition of the matter.

Localized SPs (LSPs) are oscillations of the charge density similar to the ones of propagating SPs, but that occur at metallic particles, or sub-wavelength geometries and devices. The related field enhancement is thus observed in limited areas (sub-wavelength size) of the samples. A rigorous electromagnetic modeling of the response of a metallic surface that supports LSP resonances is rather difficult and should be performed using numerical methods. However, some approximations may help in getting information on this response even from analytical expressions. When dealing with nanoparticles of simple shapes (spheres, cylinders, etc.), as the wavelength is much larger than the region, a quasi-static approximation gives good results [55]. The field in the vicinity of the particle is the sum of the incident field and the field scattered by the particle. This last case is calculated by solving the Laplace equation, and it includes dipolar and even multipolar contributions. For more complicated geometries, there exists several numerical methods [56]. For the case of composite materials, where the number of randomly arranged particles is extremely large, a solution can be found using effective medium theory, which will be described in Chapter 10.

8.3.2 Hole Arrays

Metal sheets perforated with an array of holes are among the most popular metasurfaces, because they are easily manufactured, at least for the THz range, and their transmission can be measured with a classical THz-TDS setup. Therefore, they have been extensively studied even in the far infrared. However, their electromagnetic response is not obvious to understand and explain. Transmission through a single hole involves firstly excitation of LSPs near to the hole at the entrance face of the metal sheet, then propagation of the wave in the hole, excitation of LSPs at the output face of the sheet, and finally diffraction toward the far-field region. If the hole diameter D is larger than the wavelength, the hole can be treated as a metallic waveguide, for which there exists a cut-off frequency f_c (see relation 8.33). For cylindrical holes of diameter D, f_c (THz)$\approx$ $0.1758/D$ (mm). Below this cut-off frequency, the THz field is transmitted by the hole through evanescent waves. The reflected and transmitted fields include both propagating waves, which contribute to the far-field signals, and evanescent waves bound to the metal. It should be emphasized that the amplitudes of both propagating and evanescent fields are strongly interdependent. These evanescent waves are at the origin of what is called supertransmission, in the sub-cut-off spectral range, corresponding to an observed far field transmission bigger than the one expected from a simple diffraction calculation. Let us notice that the transmission by such a hole, even in the super-transmission regime, remains quite weak because of the small diameter of the hole as compared to the wavelength. For spectroscopic studies of some matter bound to the hole array, it is of prime importance to know the distance, z, from the hole array the evanescent waves can be neglected. Masson and Gallot [57, 58] have shown, for a cylindrical hole of diameter D drilled in a metal sheet in the super-transmission regime, that there exist three regions. The far-field region is defined by $z > 1.1 \times D$ (the evanescent wave contribution is null). The true near-field region is located for $z < 0.6 \times D$: in this region, the field amplitude variation with the distance z is rather cluttered. In between, $0.6 \times D < z < 1.1\times D$, the field decreases exponentially with z: this the contrast near-field region (see Fig. 8.28).

When dealing with periodic arrays instead of a single hole, coupling effects between holes may occur through SP or LSP. Also, far-field interferences are observed, like the ones produced by diffraction gratings. Thus, the hole array may serve as a grating

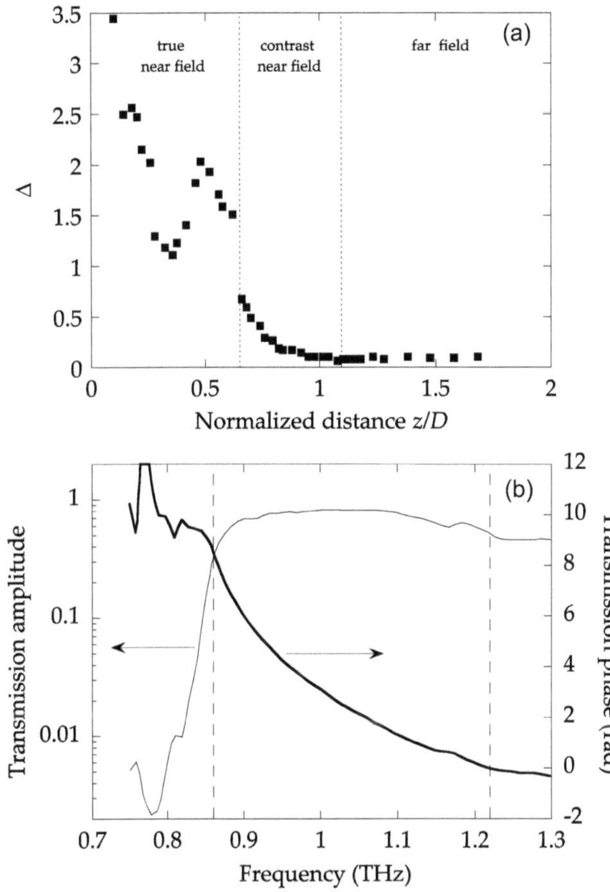

Figure 8.28 (a) Δ (a value proportional to the energy of the THz field) versus the normalized distance z/D from the hole array (adapted from Masson and Gallot [58]). The dashed vertical lines separate the three near-field and far-field regions. (b) Amplitude and phase spectra of the transmission through a hole array (adapted from Ganz et al. [59]). The first vertical dashed line corresponds to f_c (0.86 THz), and the second one (1.22 THz) to the first Wood's anomaly.

coupler to excite SP propagating over the metal sheet. Typically, the distance, d, between holes should be of the order of the THz wavelength, and the propagation length of the SP (limited by losses, due to either absorption in the metal or scattering by roughness) should be longer than this distance. The SP excitation obeys the grating diffraction equation (8.53), i.e., for a square array of period d:

$$k_{ix} + \sqrt{m^2 + p^2}\, k_{grating} = k_{SP} \Leftrightarrow \sin\theta + \sqrt{m^2 + p^2}\frac{c}{d\, f} = n_{eff} \approx 1. \tag{8.60}$$

Here $k_{grating} = 2\pi/d$, θ is the angle of incidence, and m and p are integers that account for the periodicity of the array in the direction (x, y), with $x = m \times d$, $y = p \times d$. When such excitation occurs, a part of the incoming energy is coupled to the SP and is lost by Joule effect in the metal. This missing energy leads to absorption lines in the reflected or transmitted spectra, known as Wood's anomalies. At normal incidence, the first Wood's anomaly ($m = 1$, $p = 0$) is observed at $f \approx c/d$. This is illustrated in Fig. 8.28b, adapted from Ganz et al. [59], where the amplitude and phase of the transmission of a metallic hole array are plotted versus frequency. Data have been obtained from THz-TDS measurements. The sample is a 1 mm thick brass plate perforated with cylindrical holes ($D = 204$ μm; thus $f_c = 0.86$ THz). The holes form a square lattice (the square length, i.e., the period, is $d = 246$ μm). Below f_c, the transmission decreases abruptly while the phase is large, which can be explained by a superluminal transmission by the holes. In between f_c and the first Wood's anomaly (1.22 THz), no grating-like diffraction happens, and transmission is high (82%, i.e., about 20% more than expected from a geometrical calculation). Above 1.22 THz, transmission decreases due to the existence of propagating diffracted orders.

When the filling factor (i.e., the percentage of the surface that is covered by metal) of the metasurface decreases, the effective index n_{eff} of the excited SP (Eq. 8.60) differs more and more from that of a SP propagating on a bulky flat metallic sample. When dealing with sub-wavelength structures, one can simply derive the actual n_{eff} value from an effective medium description (see Chapter 10) of the heterogeneous metallic surface. Typically, the effective permittivity of the metasurface decreases when the filling factor diminishes. Thus, SP propagates at a reduced speed compared to the flat continuous metal surface case: one speaks of spoof SP [60].

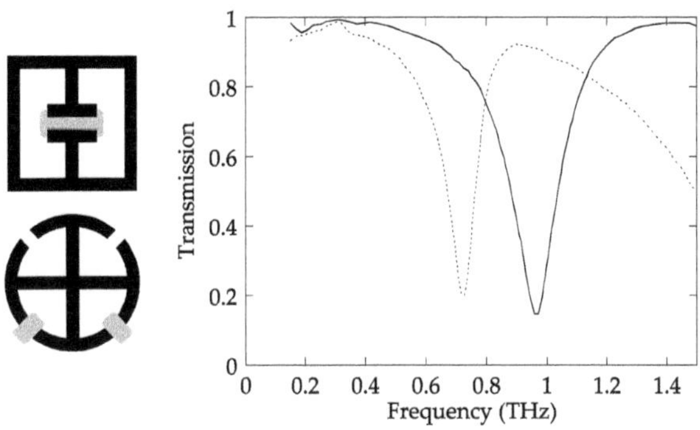

Figure 8.29 The transmission spectrum of an array of metallic patterns shown on the left, measured by THz-TDS (adapted from Padilla et al. [62]). The dotted line corresponds to the upper pattern, the continuous one to the lower one. The gray areas in the pattern drawing are the regions where the THz E-field is strongly enhanced at the resonance.

8.3.3 Metasurfaces

Metasurfaces are commonly referred to [61], in the THz range, as more complicated metallic structures deposited over a substrate. Generally, each pattern includes (i) a loop in which light-induced currents generate a magnetic field and (ii) a dipole-like pattern that interacts strongly with the incident E field. Both elements can be resonantly excited. The main goal of such design is to lead to simultaneous electric and magnetic resonances occurring at the same frequency, in the vicinity of which both the permittivity and the susceptibility could be negative, leading to a negative refractive index.

Figure 8.29 presents the transmission spectra of metasurfaces made of an array of the metallic pattern shown on the left side of the figure. These 200 nm thick gold patterns are deposited on SI GaAs substrate. Their outer dimensions are 36 μm, leading to a periodicity of 50 μm. The transmission spectra show a pronounced resonance. At the resonance, the field is strongly enhanced in the region, indicated by a gray color in the pattern drawing. Here the resonance occurs for both electric and magnetic fields, leading to a negative value of the effective index in the spectral range corresponding to the positive slope of the resonance dip [62].

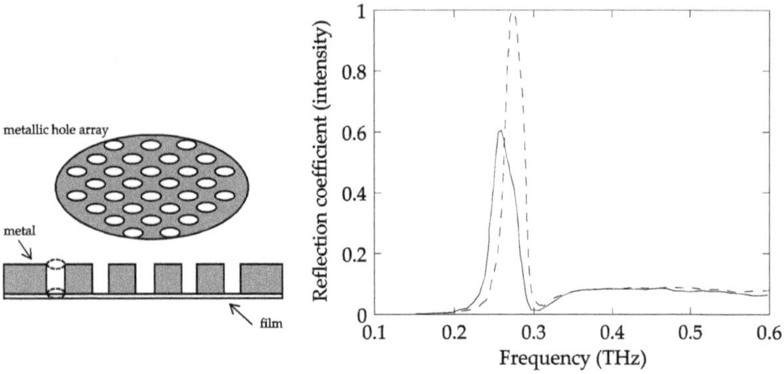

Figure 8.30 Metallic hole array (left) and its measured transmission (right) with (continuous line) and without (dashed line) a 50 μm thick polypropylene film pressed against the metal slab. Adapted from Tanaka et al. [63].

8.3.4 THz-TDS Involving Localized SP

As mentioned above, localized SPs are involved in the sub-wavelength transmission properties of hole arrays perforated in metallic slabs. Any small amount of material located nearby the holes will dramatically change the transmitted THz field.

This has been nicely exploited by Miyamaru et al. [63, 64], who studied a film of dielectric material pressed against a metallic hole array (Fig. 8.30). The peak observed in the transmission spectrum is due to the first mode of propagation in each cylindrical hole, whose diameter is 0.68 μm. This peak is about three times higher than expected from a geometrical calculation. When a 50 μm thick polypropylene film is deposited on the metal, the super-transmission peak is attenuated and slightly shifted toward lower frequencies. When the film thickness increases (not drawn here), the frequency shift saturates. This is in agreement with a model proposed by Pockrand (note that for finite thickness films, there exists no analytical expression of the transmission difference induced by the film). Miyamaru et al. [63, 64] have also studied other films, like clean and printed papers, nylon membrane impregnated with glycerin, etc. Thanks to the super-transmission phenomenon and thus to the role of localized SP, this technique is very sensitive

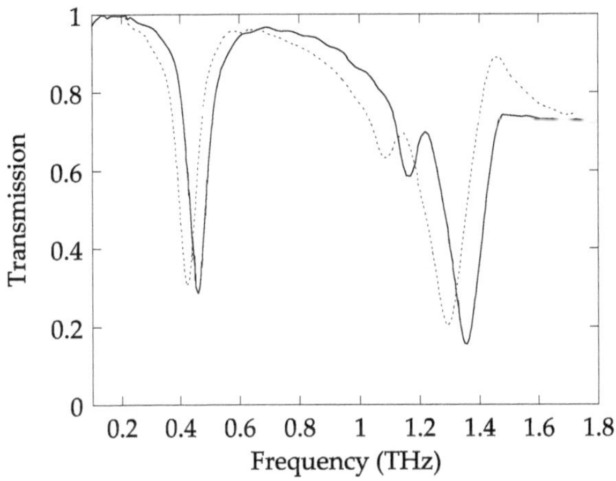

Figure 8.31 The transmission spectrum of an array of double split ring metallic resonators (SRRs) deposited over an HR silicon wafer. Dashed line: bare SRR. Continuous line: SRR covered by a 16 μm thick layer of photoresist. Adapted from O'Hara et al. [68].

to sensing any minute amount of matter. Therefore, since the pioneering work of Miyamaru et al., THz sensors based on metallic hole arrays have been widely employed (see references in the review paper by Reinhard et al. [65]). The technique has also been enlarged to dielectric periodic hole structures [66].

The influence of a thin film deposited over a more complicated metasurface has also been widely studied [67–72]. The basic principle is the same as for the hole array devices: localized SPs are strongly perturbed by some matter deposited on the metasurface. A nice example is shown in Fig. 8.31. The metasurface is formed with an array of double split ring resonators (SRRs) made by photolithography of aluminum film deposited over an HR silicon wafer. The transmission spectrum exhibits three resonances, at 0.46 THz (LC resonance), 1.356 THz (electrical dipole), and 1.16 THz (inner ring). When the metasurface is covered by a 16 μm thick layer of photoresist, the resonances are clearly frequency-shifted by 36, 60, and 78 GHz, respectively. O'Hara and his colleagues [68] were even able to detect the influence of a 100 nm thick film of BeO_2; this is probably the ultimate limit of sensitivity of such sensors.

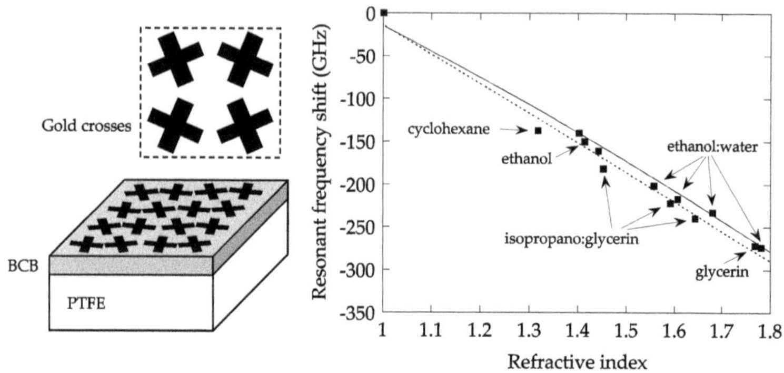

Figure 8.32 The shift of the resonance frequency of the metasurface shown on the left versus the refractive index of a liquid put against the metasurface. The shift is defined as compared to the bare metasurface. The continuous line is a full-wave calculation, while the dashed line is obtained with an analytical modeling. Adapted from Reinhard et al. [72].

As clearly explained in [68], many factors should be optimized to make the sensor very sensitive, such as resonant pattern geometry, substrate composition, analyte composition, and deposition. On the other hand, the technique is limited in sensitivity by the precision in the metasurface fabrication, the control of the film deposition process which monitors the quality of the deposited film, as well as by the resolution of the THz-TDS setup. Because this technique is based on the resonant response of the metasurfaces, it is extremely difficult to apply it to broadband spectroscopy. The metasurface should be designed to the material to be sensed and to the relevant frequency. They are limited to detect a given material, and at best, to quantify the amount of deposited material. When the material permittivity does not show any sharp feature in the THz spectral range, but an almost constant refractive index, measuring the frequency shift induced by a film of this material allows for the determination of its refractive index. This is illustrated by Fig. 8.32. The resonance of a metasurface, made of gold crosses deposited over a BCB layer on top of a PTFE substrate, is measured at 1.6 THz by reflection THz-TDS (the incoming THz beam impinges from the substrate side). Then a liquid is put in contact with the metasurfaces, and the shift of the resonance is observed. Figure 8.32 shows the frequency shift induced by different liquids, versus the refractive index of these liquids. The shift varies almost linearly with the refractive index.

References

1. Zhang, J., and Grischkowsky, D. (2004). Waveguide terahertz time-domain spectroscopy of nanometer water layers, *Opt. Lett.*, **29**, pp. 1617–1619.
2. Zhang, J., and Grischkowsky, D. (2005). Adiabatic Compression of parallel-plate metal waveguides for sensitivity enhancement of waveguide THz time-domain spectroscopy (THz-TDS), *Appl. Phys. Lett.*, **86**, p. 061109.
3. Melinger, J. S., Laman, N., Sree Harsha, S., and Grischkowsky, D. (2006). Line narrowing of terahertz vibrational modes for organic thin polycrystalline films within a parallel plate waveguide, *Appl. Phys. Lett.*, **89**, p. 251110.
4. Laman, N., Sree Harsha, S., Grischkowsky, D., and Melinger J. S. (2008). 7 GHz resolution waveguide THz spectroscopy of explosives related solids showing new features, *Opt. Express*, **16**, pp. 4049–4105.
5. Laman, N., Sree Harsha, S., and Grischkowsky, D. (2008). Narrow-line waveguide terahertz time-domain spectroscopy of Aspirin and Aspirin precursors, *Appl. Spectrosc.*, **62**, pp. 319–326.
6. Sree Harsha, S., and Grischkowsky, D. (2010). THz (far infrared) characterization of Tris (hydroxymethyl) aminomethane (TRIS) using high resolution waveguide THz-TDS, *J. Phys. Chem. A*, **114**, p. 3489.
7. Theuer, M., Sree Harsha, S., and Grischkowsky, D. (2010). Flare coupled metal parallel-plate waveguides for high resolution THz time-domain spectroscopy, *J. Appl. Phys.*, **108**, p. 113105.
8. Sree Harsha, S., Melinger, J. S., Qadri, S. B., and Grischkowsky, D. (2012). Substrate independence of THz vibrational modes of polycrystalline thin films of molecular solids in waveguide THz-TDS, *J. Appl. Phys.*, **111**, p. 023105.
9. Ordal, M. A., Bell, R. J., Alexander, R. W., Long, L. L., and Querry, M. R. (1985). Optical properties of fourteen metals in the infrared and far infrared: Al, Co, Cu, Au, Fe, Pb, Mo, Ni, Pd, Pt, Ag, Ti, V, and W, *Appl. Opt.*, **24**, pp. 4493–4499.

10. Mendis, R., and Grischkowsky, D. (2001). Undistorted guided-wave propagation of subpicosecond terahertz pulses, *Opt. Lett.*, **26**, pp. 846–848.
11. Sree Harsha, S., and Grischkowsky, D. (2011). Terahertz (far-infrared) characterization of Tris(hydroxymethyl)aminomethane using high-resolution waveguide THz-TDS, *J. Phys. Chem. A*, **114**, pp. 3489–3494.
12. Wang, K. L., and Mittleman, D. M. (2004). Metal wires for terahertz wave guiding, *Nature (London)*, **432**, p. 376.
13. Sommerfeld, A. (1899). Ueber die Fortpflanzung elektrodynamischer Wellen längs eines Drahtes, *Ann. Phys.*, **303**, pp. 233–290.
14. Walther, M., Freeman, M. R., and Hegmann, F. A. (2005). Metal-wire terahertz time-domain spectroscopy, *Appl. Phys. Lett.*, **87**, p. 261107.
15. Begley, D. L., Alexander, R. W., Ward, C. A., Miller, R., and Bell, R. J. (1979). Propagation distances of surface electromagnetic waves in the far infrared, *Surf. Sci.*, **81**, pp. 245–251.
16. O'Hara, J. F., Averitt, R., and Taylor, A. (2005). Prism coupling to terahertz surface plasmon polaritons, *Opt. Express*, **13**, pp. 6117–6126.
17. Mukina, L. S., Nazarov, M. M., and Shkurinov, A. P. (2006). Propagation of THz plasmon pulse on corrugated and flat metal surface, *Surf. Sci.*, **600**, pp. 4771–4776.
18. Nazarov, M., Coutaz, J.-L., Shkurinov, A., and Garet, F. (2007). THz surface plasmon jump between two metal edges, *Opt. Commun.*, **277**, p. 33.
19. Gomez Rivas, J., Kuttge, M., Haring Bolivar, P., Kurz, H., and Sanchez-Gil, J. A. (2004). Propagation of surface plasmon polaritons on semiconductor gratings, *Phys. Rev. Lett.*, **93**, p. 256804.
20. Hirori, H., Nagai, M., and Tanaka, K. (2005). Destructive interference effect on surface plasmon resonance in terahertz attenuated total reflection, *Opt. Express*, **13**, p. 10801.
21. Lo, S.-Z. A., and Murphy, T. E. (2010). Terahertz surface plasmon propagation in nanoporous silicon layers, *Appl. Phys. Lett.*, **96**, p. 201104.
22. Nazarov, M., and Coutaz, J.-L. (2011). Terahertz surface waves propagating on metals with sub-wavelength structure and grating reliefs, *J. Infrared Millimeter THz Waves*, **32**, pp. 1054–1073.
23. Laman, N., and Grischkowsky, D. (2007). Reduced conductivity in the terahertz skin-depth layer of metals, *Appl. Phys. Lett.*, **90**, p. 122115.

24. Gong, M., Jeon, T.-I., and Grischkowsky, D. (2009). THz surface wave collapse on coated metal surfaces, *Opt. Express*, **17**, pp. 17088–17101.

25. Saxler, J., Gomez Rivas, J., Janke, C., Pellemans, H. P. M.;, Haring Bolivar, P., and Kurz, H. (2004). Time-domain measurements of surface plasmon polaritons in the terahertz frequency range, *Phys. Rev. B*, **69**, p. 155427.

26. Gómez Rivas, J., Kuttge, M., Kurz, H., Haring Bolivar, P., and Sánchez-Gil, J. A. (2006). Low-frequency active surface plasmon optics on semiconductors, *Appl. Phys. Lett.*, **88**, p. 082106.

27. Wang, K., and Mittleman, D. M. (2004). Metal wires for terahertz waveguiding, *Nature*, **432**, pp. 376–379.

28. Saxler, J., Gomez Rivas, J., Janke, C., Pellemans, H. P. M., Haring Bolivar, P., and Kurz, H. (2004). Time-domain measurements of surface plasmon polaritons in the terahertz frequency range, *Phys. Rev. B*, **69**, p. 155427

29. Isaac, T. H., Barnes, W. L., and Hendry, E. (2008). Determining the terahertz optical properties of subwavelength films using semiconductor surface plasmons, *Appl. Phys. Lett.*, **93**, p. 241115.

30. Ng, B., Hanham, S. M., Wu, J., Fernaìndez-Domínguez, A. I., Klein, N., Liew, Y. F., Breese, M. B. H., Hong, M., and Maier, S. A. (2014). Broadband terahertz sensing on spoof plasmon surfaces, *ACS Photonics*, **1**, pp. 1059–1067.

31. Rau, C., Torosyan, G., Beigang, R., and Nerkararyan, Kh. (2005). Prism coupled terahertz waveguide sensor, *Appl. Phys. Lett.*, **86**, p. 211119.

32. Roux, J.-F., Aquistapace, F., Garet, F., Duvillaret, L., and Coutaz, J.-L. (2002). Grating-assisted coupling of terahertz waves into a dielectric waveguide studied by terahertz time-domain spectroscopy, *Appl. Opt.*, **41**, pp. 6507–6513.

33. Alieva, E. V., Beitel, G., Kuzik, L. A., Sigarev, A. A., Yakovlev, V. A., Zhizhin, G. N., van der Meer, A. F. G., and van der Wiel, M. (1997). Linear and nonlinear FEL–SEW spectroscopic characterization of nanometer-thick films, *J. Appl. Spectrosc.*, **51**, p. 584.

34. O'Hara J. F., Averitt R. D., and Taylor A. J. (2005). Prism coupling to terahertz surface plasmon polaritons, *Opt. Express*, **13**, p. 6117.

35. Rayleigh, L. (1907). On the dynamical theory of gratings, *Proc. R. Soc. London, Ser. A*, **79**, pp. 399–416.

36. Roux, J.-F., Aquistapace, F., Garet, F., Duvillaret, L., and Coutaz, J.-L. (2002). Grating-assisted coupling of terahertz temporal waves into a dielectric waveguide studied by terahertz time-domain spectroscopy, *Appl. Opt.*, **41**, pp. 6507–6513.

37. Laamiri, Y., Garet, F., and Coutaz, J.-L. (2009). Thin-film characterization by terahertz time-domain spectroscopy using grating-assisted excitation of guided modes, *Appl. Phys. Lett.*, **94**, p. 071106.

38. Pozar, D. M. (2012). *Microwave Engineering*, 4th edn. (John Wiley & Sons, Inc., New York).

39. Peytavit, E., Donche, C., Lepilliet, S., Ducournau, G., and Lampin, J. F. (2011). Thin-film transmission lines using cyclic olefin copolymer for millimetre-wave and terahertz integrated circuits, *Electron. Lett.*, **47**, pp. 453–454.

40. Nagel, M., Dekorsy, T., Brucherseifer, M., Haring Bolivar, P., and Kurz, H. (2001). Characterization of polypropylene thin-film microstrip lines at millimeter and submillimeter wavelengths, *Microwave Opt. Technol. Lett.*, **29**, pp. 97–100.

41. Byrne, M. B., Cunningham, J., Tych, K., Burnett, A. D., Stringer, M. R., Wood, C. D., Dazhang, L., Lachab, M., Linfield, E. H., and Davies, A. G. (2008). Terahertz vibrational absorption spectroscopy using microstrip-line waveguides, *Appl. Phys. Lett.*, **93**, p. 182904.

42. Kitagawa, J., Ohkubo, T., Onuma, M., and Kadoya Y. (2006). THz spectroscopic characterization of biomolecule/water systems by compact sensor chips, *Appl. Phys. Lett.*, **89**, p. 041114.

43. Verma, A. K., and Sadr, G. H. (1992). Unified dispersion model for multilayer microstrip line, *IEEE Trans. Microwave Theory Tech.*, **40**, p. 1587.

44. Verma, A. K., and Bhupal, A. (1998). Dielectric loss of multilayer microstrip line, *Microwave Opt. Technol. Lett.*, **17**, p. 368.

45. Ohkubo, T., Onuma, M., Kitagawa, J., and Kadoya, Y. (2006). Microstrip-line-based sensing chips for characterization of polar liquids in terahertz, *Appl. Phys. Lett.*, **88**, p. 212511.

46. Cunningham, J., Byrne, M., Upadhya, P., Lachab, M., Linfield, E. H., and Davies, A. G. (2008). Terahertz evanescent field microscopy of dielectric materials using on-chip waveguides, *Appl. Phys. Lett.*, **92**, p. 032903.

47. Nagel, M., Bolivar, P. H., Brucherseifer, M., Kurz, H., Bosserhoff, A., and Buttner, R. (2002). Integrated THz technology for label-free genetic diagnostics, *Appl. Phys. Lett.*, **80**, pp. 154–156.

48. Frankel, M. Y., Gupta, S., Valdmanis, J., and Mourou, G. A. (1991). Terahertz attenuation and dispersion characteristics of coplanar transmission lines, *IEEE Trans. Microwave Theory Tech.*, **39**, pp. 910–916.

49. Cheng, H., Whitaker, J. F., Weller, T. M., and Katehi, L. P. B. (1994). Terahertz-bandwidth characteristics of coplanar transmission lines on low permittivity substrates, *IEEE Trans. Microwave Theory Tech.*, **42**, pp. 2399–2406.

50. Sprik, R., Duling, III, I. N., Chi, C. C., and Grischkowsky, D. (1987). Far infrared spectroscopy with subpicosecond electrical pulses on transmission lines, *Appl. Phys. Lett.*, **51**, pp. 548–550.

51. Kadoya, Y., Onuma, M., Yanagi, S., Ohkubo, T., Sato, N., and Kitagawa, J. (2008). THz wave propagation on strip lines: devices, properties, and applications, *Radioengineering*, **17**, pp. 48–55.

52. Gacemi, D., Mangeney, J., Laurtent, T., Lampin, J.-F., Akalin, T., Blary, K., Degiron, A., Crozat, P., and Meng, F. (2012). THz surface plasmon modes on planar Goubau lines, *Opt. Express*, **20**, pp. 8466–8471.

53. Russell, C., Wood, C. D., Burnett, A. D., Li, L., Linfield, E. H., Davies, A. G., and Cunningham, J. E. (2013). Spectroscopy of polycrystalline materials using thinned-substrate planar Goubau line at cryogenic temperatures, *Lab Chip*, **13**, pp. 4065–4070.

54. Laurette, S., Treizebre, A., Elagli, A., Hatirnaz, B., Froidevaux, R., Affouard, F., Duponchel, L., and Bocquet, B. (2012). Highly sensitive terahertz spectroscopy in microsystem, *RSC Adv.*, **2**, pp. 10064–10071.

55. Kreibig, U., and Vollmer, M. (1995). *Optical Properties of Metal Clusters* (Springer-Verlag, Berlin).

56. Pitarke, J. M., Silkin, V. M., Chulkov, E. V., and Echenique, P. M. (2007). Theory of surface plasmons and surface-plasmon polaritons, *Rep. Prog. Phys.*, **70**, pp. 1–87.

57. Masson, J.-B., and Gallot, G. (2006). True near field versus contrast near field imaging, *Opt. Express*, **14**, p. 11566.

58. Masson, J.-B., and Gallot, G. (2007). True near field versus contrast near field imaging. II. Imaging with a probe, *Opt. Express*, **15**, pp. 3078–3084.

59. Ganz, T., Brehm, M., von Ribbeck, H. G., van der Weide, D. W., and Keilmann, F. (2008). Vector frequency-comb Fourier-transform spectroscopy for characterizing metamaterials, *New J. Phys.*, **10**, p. 123007.

60. Pendry, J. B., Martin-Moreno, L., and Garcia-Vidal, F. J. (2004). Mimicking surface plasmons with structured surfaces, *Science*, **135**, p. 847.

61. Withayachumnankul, W., and Abbott, D. (2009). Metamaterials in the terahertz regime, *IEEE Photonics J.*, **1**, pp. 99–118.

62. Padilla, W. J., Aronsson, M. T., Highstrete, C., Lee, M., Taylor, A. J., and Averitt, R. D. (2007). Electrically resonant terahertz metamaterials: Theoretical and experimental investigations, *Phys. Rev. B*, **75**, p. 041102-R.

63. Tanaka, M., Miyamaru, F., Hangyo, M., Tanaka, T., Akazawa, M., and Sano, E. (2005). Effect of a thin dielectric layer on terahertz transmission characteristics for metal hole arrays, *Opt. Lett.*, **30**, pp. 1210–1212.

64. Miyamaru, F., Hayashi, S., Otani, C., Kawase, Ogawa, K. Y., Yoshida, H., and Kato, E. (2006). Terahertz surface-wave resonant sensor with a metal hole array, *Opt. Lett.*, **31**, pp. 1118–1120.

65. Reinhard, B., Paul, O., and Rahm, M. (2013). Metamaterial-based photonic devices for terahertz technology, *IEEE J. Sel. Top. Quantum Electron.*, **19**, p. 8500912.

66. Fan, F., Zhang, X., Li, S., Deng, D., Wang, N., Zhang, H., and Chang, S. (2005). Terahertz transmission and sensing properties of micro structured PMMA tube waveguide, *Opt. Express*, **23**, p. 27204.

67. Debus, C., and Haring Bolivar, P. (2007). Frequency selective surfaces for high sensitivity terahertz sensing, *Appl. Phys. Lett.*, **91**, p. 184102.

68. O'Hara, J. F., Singh, R., Brener, I., Smirnova, E., Han, J., Taylor, A. J., and Zhang, W. (2008). Thin-film sensing with planar terahertz metamaterials: Sensitivity and limitations, *Opt. Express*, **16**, pp. 1786–1795.

69. Al-Naib, I. A. I., Jansen, C., and Koch, M. (2008). Thin-film sensing with planar asymmetric metamaterial resonators, *Appl. Phys. Lett.*, **93**, p. 083507.

70. Tao, H., Strikwerda, A. C., Liu, M., Mondia, J. P., Ekmekci, E., Fan, K., Kaplan, D. L., Padilla, W. J., Zhang, X., Averitt, R. D., and Omenetto, F. G. (2010). Performance enhancement of terahertz metamaterials on ultrathin substrates for sensing applications, *Appl. Phys. Lett.*, **97**, p. 261909.

71. Tao, H., Chieffo, L. R., Brenckle, M. A., Siebert, S. M., Liu, M., Strikwerda, A. C., Fan, K., Kaplan, D. L., Zhang, X., Averitt, R. D., and Omenetto, F. G. (2011). Metamaterials on paper as a sensing platform, *Adv. Mater.*, **23**, pp. 3197–3201.

72. Reinhard, B., Schmitt, K. M., Wollrab, V., Neu, J., Beigang, R., and Rahm, M. (2012). Metamaterial near-field sensor for deep-subwavelength thickness measurements and sensitive refractometry in the terahertz frequency range, *Appl. Phys. Lett.*, **100**, p. 221101.

Chapter 9

Characterization of Anisotropic Materials

9.1 Introduction

In anisotropic materials, the response of the materials to electromagnetic waves is different depending on the direction of the exciting field. It concerns both electric and magnetic fields, and it can be simply expressed with tensorial relations:

$$\vec{D} = \overleftrightarrow{\varepsilon} : \vec{E}, \quad \vec{B} = \overleftrightarrow{\mu} : \vec{H}. \tag{9.1}$$

$\vec{E}$ and $\vec{B}$ are the exciting electric and magnetic fields, respectively. $\vec{D}$ and $\vec{H}$ are the displacement field and the magnetic field strength, respectively.[a] $\overleftrightarrow{\varepsilon}$ and $\overleftrightarrow{\mu}$ are the corresponding permittivity and permeability tensors (third-rank matrices). For isotropic materials, these tensors are third-rank identity matrices. For lossless materials, the matrix elements are real.

When illuminated by an electromagnetic wave, the basic constituents of matter, i.e., atoms or molecules, become polarized, i.e.,

[a] Note that $\vec{H}$ is very often named magnetic field, but it must not be confused with $\vec{B}$. Note also the opposite excitation/response relations for electric and magnetic cases, due to historical reasons.

Principles of Terahertz Time-Domain Spectroscopy
Jean-Louis Coutaz, Frédéric Garet, and Vincent Wallace

ISBN 978-981-4774-56-7 (Hardcover), 978-0-203-70969-6 (eBook)
www.panstanford.com

they acquire a dipolar moment or their dipolar moment changes (multipolar effects are also observed, but most of the time are weaker than the dipolar ones). These phenomena are only related to bound charges. One can take them into account by using the polarization $\vec{P}$ and the magnetization $\vec{M}$ of matter, which are defined as the volume density of individual dipolar moments. Hence one gets

$$\vec{D} = \varepsilon_o \vec{E} + \vec{P}, \quad \vec{B} = \mu_o \vec{H} + \vec{M}. \tag{9.2}$$

$\vec{P}$ and $\vec{M}$ are functions of the $\vec{E}$ and $\vec{B}$ fields. For intense fields, for example when dealing with pulsed laser beams, $\vec{P}$ and $\vec{M}$ are no more linear functions of respectively $\vec{E}$ and $\vec{B}$, as they saturate when they reach a particular magnitude of the exciting field. However, in most of the cases, the linear variation of $\vec{P}$ and $\vec{M}$ is stronger than the nonlinear one, and thus these vectors can be written as a series expansion of the exciting fields:

$$\begin{cases} \vec{D} = \overleftrightarrow{\varepsilon} : \vec{E} = \varepsilon_o \left(\vec{E} + \overleftrightarrow{\chi}^{(1)} : \vec{E} + \overleftrightarrow{\chi}^{(2)} : \vec{E} : \vec{E} \ldots \right), \\ \vec{B} = \overleftrightarrow{\mu} : \vec{H} = \mu_o \left(\vec{H} + \overleftrightarrow{\chi}_m^{(1)} : \vec{H} + \overleftrightarrow{\chi}_m^{2} : \vec{H} : \vec{H} \ldots \right). \end{cases} \tag{9.3}$$

$\overleftrightarrow{\chi}^{(i)}$ and $\overleftrightarrow{\chi}_m^{(i)}$ are respectively the electric and magnetic susceptibilities of rank i. The two first terms in Eq. 9.3 describe the linear response of the material.

If the matter contains free charges, the impinging wave will induce currents that contribute to the macroscopic response of matter. Because the Laplace force mixes the influence of both $\vec{E}$ and $\vec{B}$, the series expansions (Eq. 9.3) should include additional terms. One can summarize all the linear and nonlinear electromagnetic phenomena by the following chart [1]:

$$\varepsilon\left(\vec{k}, \vec{E}, \vec{B}\right) \propto \varepsilon + \underbrace{O\left(\vec{k}\right)}_{\substack{\text{natural} \\ \text{optical} \\ \text{activity}}} + \underbrace{O\left(\vec{E}\right)}_{\substack{\text{electro-} \\ \text{optical} \\ \text{effects}}} + \underbrace{O\left(\vec{B}\right)}_{\substack{\text{magneto-} \\ \text{optical} \\ \text{effects}}} + \underbrace{O\left(\vec{E} : \vec{E}\right)}_{\substack{\text{nonlinear} \\ \text{optical} \\ \text{effects}}}$$

$$+ \underbrace{O\left(\vec{B} : \vec{B}\right)}_{\substack{\text{magneto-} \\ \text{optical} \\ \text{effects}}} + \underbrace{O\left(\vec{E} : \vec{B}\right)}_{\substack{\text{magneto-} \\ \text{electric} \\ \text{effects}}} + \ldots$$

$\vec{k}$ is the wavevector of the THz beam and $O(x)$ means terms depending linearly on x.

Let us examine a plane wave propagating in an anisotropic medium. We write its electrical $\vec{E}$ or magnetic $\vec{H}$ field under the general propagating form:

$$\vec{F}\,(\vec{r},t) = \vec{F}\; e^{j\left(\omega t - \vec{k}\cdot\vec{r}\right)},\ \vec{F}\,(\vec{r},t) = \vec{E}\,(\vec{r},t) \text{ or } \vec{F}\,(\vec{r},t) = \vec{H}\,(\vec{r},t) \tag{9.4}$$

$$\Rightarrow \vec{\nabla} \times \vec{F}\,(\vec{r},t) = -j\,\vec{k} \times \vec{F}\; e^{j\left(\omega t - \vec{k}\cdot\vec{r}\right)}. \tag{9.5}$$

Then we enter these expressions into Maxwell's equations (A.2):

$$\begin{cases} \vec{k} \times \vec{E} = \omega\,\overleftrightarrow{\mu} : \vec{H}, \\ \vec{k} \times \vec{H} = j\left(j\,\omega\overleftrightarrow{\varepsilon} + \overleftrightarrow{\sigma}\right) : \vec{E}. \end{cases} \tag{9.6}$$

Here $\overleftrightarrow{\sigma}$ is the conductivity tensor. These two coupled equations cannot be solved analytically for the general case, even if one does not address the nonlinear regime. Commonly, in the linear regime, one treats propagation independently either in dielectric or magnetic materials. Because the different vectorial components of a wave do not propagate at the same speed, the state of polarization of the THz beam varies when transmitted or reflected by the anisotropic sample. The general case (any incident angle, incoming polarization state, and orientation of the principal axes) yields complex equations (see, e.g., [2]) that can only be solved analytically for a limited number of simple schemes. In the case of magnetic or magneto-optical materials, the study of magneto-optical effects is of importance in the THz range, as it gives useful information on the origin and dynamics of magnetism. Most of these effects could be addressed by measuring the change of polarization of the probe beam occurring either in transmission or reflection (polarimetry), or the differential absorption of the different polarization components of the probe beam (dichroism). In addition, a spectral study may emphasize resonances of the physical phenomena that participate to the electromagnetism response of matter. As well, time-resolved experiments can give information on the dynamics of these phenomena.

9.2 Some Basic Principles of Propagation in Anisotropic Media

9.2.1 Dielectric Crystals

Let us first consider transparent nonmagnetic dielectric crystals, which exhibit a linear optical response. In Eq. 9.6, $\overleftrightarrow{\mu} = \mu_o \times \overleftrightarrow{1}$ and $\overleftrightarrow{\sigma} = 0$. Moreover, the permittivity tensor $\overleftrightarrow{\varepsilon}$ is real and is diagonal when defined in the principal coordinate reference frame (the principal axes X, Y, Z correspond to the crystallographic axes, except for monoclinic, orthorhombic and triclinic crystals [3]). A plane wave propagating in the crystal obeys Maxwell's equations, which are simplified to

$$\vec{k} \times \vec{E} = \omega \vec{B}, \quad \vec{k} \times \vec{B} = -\mu_o \omega \vec{D}, \quad \vec{k} \cdot \vec{D} = 0, \quad \vec{k} \cdot \vec{B} = 0. \qquad (9.7)$$

It follows that the $\vec{D}$, $\vec{B}$ plan is perpendicular to $\vec{k}$, but $\vec{E}$ is not. Also, the Poynting vector that gives the direction of the energy flow is not parallel to $\vec{k}$. We easily derive from Eq. 9.7:

$$\vec{D} = -\frac{\left(\vec{k} \cdot \vec{E}\right)\vec{k} - k^2 \vec{E}}{\mu_o \omega^2} \Rightarrow E_i = \frac{\vec{k} \cdot \vec{E}}{k^2 - \varepsilon_o \mu_o \omega^2 \varepsilon_i} k_i \quad (i = X, Y, Z). \qquad (9.8)$$

Here ε_i are elements of the relative permittivity tensor (i.e., $\overleftrightarrow{\varepsilon}/\varepsilon_o$). By multiplying Eq. 9.8 by k_i and then adding all the i terms, we get

$$\sum_i \frac{v^2}{v_i^2 - v^2} u_i^2 = 0, \quad \text{with } \vec{u} = \frac{\vec{k}}{k}, \; v = \frac{\omega}{k} \text{ and } v_i = \frac{c}{\sqrt{\varepsilon_i}} \equiv \frac{c}{n_i}. \qquad (9.9)$$

This is the Fresnel equation. v is the phase velocity of the wave that propagates in the crystal with a wavevector $\vec{k}$. Equation 9.9 exhibits only two solutions, namely v' and v'', whose values are lying in between the minimum and maximum values of v_i $(i = X, Y, Z)$. For each of these two solutions, Eq. 9.8 can be solved and it can be shown that they correspond a $\vec{D}$ vector linearly polarized along two orthogonal directions, which are called the neutral directions of the crystal. These two fields propagate like in an isotropic medium, at respective velocities v' and v''. Let us examine Eq. 9.8 again. Because $\vec{D}\perp\vec{k}$, we can calculate and simplify the product $\vec{D}\cdot\vec{D}$:

$$\vec{D}\cdot\vec{D} = D^2 = -\frac{\left(\vec{k}\cdot\vec{E}\right)\left(\vec{D}\cdot\vec{k}\right) - k^2\left(\vec{D}\cdot\vec{E}\right)}{\mu_0\omega^2} = \frac{k^2}{\mu_0\omega^2}\vec{D}\cdot\vec{E},$$

$$\Rightarrow D_X^2 + D_Y^2 + D_Z^2 = \frac{k^2}{\mu_0\varepsilon_0\omega^2}\left(\frac{D_X^2}{\varepsilon_X} + \frac{D_Y^2}{\varepsilon_Y} + \frac{D_Z^2}{\varepsilon_Z}\right).$$

Thus the refractive index for a wave polarized in the i direction is $n_i = \sqrt{\varepsilon_i}$. Then one sets $I = D_I/D$ $(I = X, Y, Z)$ and we define the refractive index $n = k/\sqrt{\mu_0\varepsilon_0}\omega$, which leads to

$$\frac{1}{n^2} = \frac{X^2}{n_X^2} + \frac{Y^2}{n_Y^2} + \frac{Z^2}{n_Z^2}, \tag{9.10}$$

which one will recognize as the equation of an ellipse. The refractive index n seen by the wave is given by the intersection of the $\vec{D}$ vector with the ellipsoid. At this location, the field $\vec{E}$ is normal to the surface of the ellipsoid.

In most of THz characterization of anisotropic crystals, these crystals are uniaxial and are illuminated under normal incidence. If the optical axis is normal to the faces of a flat crystal, i.e., the beam propagates along the optical axis, the crystal behaves like an isotropic medium (excepted in chiral materials treated below). If the optical axis belongs to the input face of a flat sample, the $\vec{D}$ vector must be decomposed into components along (extraordinary component) and normal (ordinary) to the optical axis, which travel respectively at speeds c/n_e and c/n_o. Let us call θ the angle between $\vec{E}$ and the optical axis (here we can use $\vec{E}$ instead of $\vec{D}$ because the beam propagates along a principal axis). The beam travels in the X direction, and the optical axis is oriented along Z. At the entrance face of the crystal, $\vec{E}$ is written as

$$\vec{E}_o = (E_o \sin\theta\, \vec{u}_Y + E_o \cos\theta\, \vec{u}_Z)\cos(\omega t). \tag{9.11}$$

$\vec{u}_I$ is the unit vector along direction I. Using cosine and sinus functions helps in analyzing the phenomena. After propagation over a distance $X = d$, the field becomes

$$\begin{aligned}\vec{E}(d) &= \vec{E}_Y(d) + \vec{E}_Z(d)\\ &= E_o \sin\theta \cos\left(\omega t - \frac{\omega}{c}n_o d\right)\vec{u}_Y\\ &\quad + E_o\cos\theta\cos\left(\omega t - \frac{\omega}{c}n_e d\right)\vec{u}_Z.\end{aligned} \tag{9.12}$$

This expression can be simplified by introducing the phase φ between the two transmitted components:

$$\sin^2\varphi = \left(\frac{E_Y(d)}{E_o\sin\theta}\right)^2 + \left(\frac{E_Z(d)}{E_o\cos\theta}\right)^2 - 2\frac{E_Y(d)}{E_o\sin\theta}\frac{E_Z(d)}{E_o\cos\theta}\cos\varphi, \tag{9.13}$$

with $\varphi = \frac{\omega}{c}(n_o - n_e)\,d$. This is the equation of an ellipse that is drawn by the tip of the $\vec{E}$ field while this field rotates periodically in the $Y - Z$ plane. The shape of the ellipse depends on d through φ. If $\varphi = 0\ [\pi]$, the output field shows the same polarization as the input field. If $\varphi = \pm\pi/2\ [\pi]$, Eq. 9.13 turns into the equation of a circle: the beam is circularly polarized, its tip traces out a circle.

9.2.2 THz Optical Activity

Optical active (or chiral) media exhibit a birefringence when illuminated by circularly polarized light: left- or right-circular waves travel at different speeds. Such a phenomenon occurs in materials like sugar solutions, proteins, nucleic acids, etc., and is named natural optical activity. A similar effect can be stimulated by applying a magnetic field to ordinary isotropic materials (glass, metal, water, etc.): in such gyroelectric materials, the magnetic field breaks their isotropy. As well, the effect is also observed with materials in which a magnetization exists, like ferrites (gyromagnetic materials). At the molecular or crystal system level, the chiral material does not show any mirror symmetry (roto-inversion) [4]: such symmetry rule also applies to metamaterials.

Natural optical activity is an effect due to nonlocality in anisotropic media [5]. Nonlocality means that the electrical response of matter to an electrical excitation at a location $\vec{r}$ depends on the field at nearby locations $\vec{r}'$. Let us recall the usual relation between the response, namely the displacement vector $\vec{D}(\vec{r})$ and the excitation $\vec{E}(\vec{r})$, when this response is local:

$$\vec{D}(\vec{r}) = \vec{\vec{\varepsilon}} : \vec{E}(\vec{r}) \quad \Rightarrow \quad D_i = \sum_m \varepsilon_{im} E_m. \tag{9.14}$$

When the response is nonlocal, expression 9.14 becomes

$$D_i(\vec{r}) = \sum_m \frac{1}{2\pi^3} \int \varepsilon_{im}(\vec{r}')\, E_m(\vec{r} - \vec{r}')\, d\vec{r}'. \tag{9.15}$$

In fact, $\vec{D}(\vec{r})$ is only determined by the response of the material in the very close vicinity of $\vec{r}$, and thus we can expend the field in series:

$$E_m\left(\vec{r}-\vec{r}'\right) \approx E_m(\vec{r})\delta_{\vec{r}-\vec{r}'} + \sum_p \frac{\partial\, E_m(\vec{r})}{\partial\, x_p} d\, x_p. \qquad (9.16)$$

It follows that Eq. 9.15 can be approximated by

$$D_i(\vec{r}) = \sum_m \varepsilon_{im}(\vec{r})E_m(\vec{r}) + \sum_m \sum_p \underbrace{\frac{1}{2\pi^3}\int \varepsilon_{im}\left(\vec{r}'\right) d\vec{r}'d\, x_p}_{\equiv \varepsilon_o \gamma_{imp}} \frac{\partial\, E_m(\vec{r})}{\partial\, x_p},$$

$$\Rightarrow\quad D_i(\vec{r}) = \sum_m \left(\varepsilon_{im}(\vec{r})E_m(\vec{r}) + \varepsilon_o \sum_p \gamma_{imp} \frac{\partial\, E_m(\vec{r})}{\partial\, x_p}\right). \qquad (9.17)$$

For transparent materials, $\overleftrightarrow{\gamma}$ is a real tensor that obeys the following symmetry relation: $\gamma_{imp} = -\gamma_{mip}$, $\gamma_{iip} = 0$.

We now consider the propagation of a plane wave in the nonlocal medium:

$$\begin{aligned}\vec{E}\left(\vec{r}, t\right) &= \vec{E}_o\, e^{j\left(\omega t - \vec{k}\cdot\vec{r}\right)}\\ &\Rightarrow \frac{\partial\, E_m(\vec{r})}{\partial\, x_p} = -j\, k_p E_m(\vec{r}) = -j\, n\frac{\omega}{c} u_p E_m\left(\vec{r}\right),\\ &\Rightarrow D_i(\vec{r}) = \sum_m \left(\varepsilon_{im} - j\,\varepsilon_o n \sum_p \frac{\omega}{c}\gamma_{imp} u_p\right) E_m(\vec{r}). \qquad (9.18)\end{aligned}$$

n is the refractive index for a beam propagating along $\vec{k}_p$ and u_p is the direction cosine of $\vec{k}_p$. We then introduce the pseudo-tensor $\overleftrightarrow{g}$ defined by

$$\gamma_{imp} = -\frac{c}{\omega}\sum_l e_{iml} g_{lp}. \qquad (9.19)$$

e_{iml} is the Levi–Civita symbol.[b] Hence,

$$D_i(\vec{r}) = \sum_m \left(\varepsilon_{im} + j\,\varepsilon_o n \sum_l e_{iml} g_l\right) E_m(\vec{r}), \quad \text{with} \quad g_l = \sum_p g_{lp} u_p. \qquad (9.20)$$

---

[b] $e_{xyz} = e_{zxy} = e_{yzx} = 1$, $e_{zyx} = e_{xzy} = e_{yzx} = -1$, otherwise $e_{iml} = 0$.

In this expression, the term with the sum over l is a 3×3 anti-symmetric matrix:

$$\sum_l e_{iml} g_l = \begin{pmatrix} 0 & e_{xyz} g_z & e_{xzy} g_y \\ e_{yxz} g_z & 0 & e_{yzx} g_x \\ e_{zxy} g_y & e_{zyx} g_x & 0 \end{pmatrix} = -\begin{pmatrix} 0 & -g_z & g_y \\ g_z & 0 & -g_x \\ -g_y & g_x & 0 \end{pmatrix}. \tag{9.21}$$

Let us recall that such a matrix is associated with the cross product:

$$\vec{A} \times \vec{B} = \begin{pmatrix} A_y B_z - A_z B_y \\ A_z B_x - A_x B_z \\ A_x B_y - A_y B_x \end{pmatrix} = \overleftrightarrow{G} : \vec{B}, \text{ with } \overleftrightarrow{G} = \begin{pmatrix} 0 & -A_z & A_y \\ A_z & 0 & -A_x \\ -A_y & A_x & 0 \end{pmatrix}. \tag{9.22}$$

Therefore, $\vec{D}(\vec{r})$ takes the following expression:

$$\vec{D}(\vec{r}) = \overleftrightarrow{\varepsilon} : \vec{E}(\vec{r}) - j\, \vec{G} \times \vec{E}(\vec{r}). \tag{9.23}$$

$\vec{G}$ is the gyration vector whose components are $g_l = \varepsilon_o n \sum_p g_{lp} u_p$ (see Eq. 9.20). Optical activity does not exist in centro-symmetric media. In isotropic materials like liquids, which are made of two stereo-isomeric forms [5], activity exists, and $\vec{g}$ and $\vec{k}$ are parallel. In crystalline α-quartz, $g_{xx} = g_{yy} \neq g_{zz}$, otherwise $g_{lp} = 0$.

Let us now look at the case of materials in which there exists a natural or induced magnetization $\vec{M}$. Both permittivity and permeability are tensors that depend on $\vec{M}$. However, $\overleftrightarrow{\mu}$ is very close the vacuum permeability [4] and, in a first approximation, we write $\overleftrightarrow{\mu} = \mu_o \times \overleftrightarrow{1}$. For the sake of simplicity, we consider an isotropic medium with $\vec{M}$ aligned along the z direction. The tensor $\overleftrightarrow{\varepsilon}(\vec{M})$ must be invariant by $\pi/2$ rotations around z. It follows that

$$\overleftrightarrow{\varepsilon}\left(\vec{M}\right) = C_4^{-1} : \overleftrightarrow{\varepsilon}\left(\vec{M}\right) : C_4 \quad \Rightarrow \quad \overleftrightarrow{\varepsilon}\left(\vec{M}\right) = \varepsilon_o \begin{pmatrix} \varepsilon_{xx} & \varepsilon_{xy} & 0 \\ -\varepsilon_{xy} & \varepsilon_{xx} & 0 \\ 0 & 0 & \varepsilon_{zz} \end{pmatrix}. \tag{9.24}$$

$\overleftrightarrow{\varepsilon}(\vec{M})$ is the sum of a diagonal matrix and an anti-symmetric matrix:

$$\overleftrightarrow{\varepsilon}\left(\vec{M}\right) = \varepsilon_o \begin{pmatrix} \varepsilon_{xx} & 0 & 0 \\ 0 & \varepsilon_{xx} & 0 \\ 0 & 0 & \varepsilon_{zz} \end{pmatrix} + \varepsilon_o \begin{pmatrix} 0 & \varepsilon_{xy} & 0 \\ -\varepsilon_{xy} & 0 & 0 \\ 0 & 0 & 0 \end{pmatrix}. \tag{9.25}$$

The off-diagonal elements result from the physical origin of magneto-optics, e.g., Hall or cyclotron effects that make the movement of electrons not collinear with the E-field of the wave. From a mathematical point of view, the situation is similar to natural activity.

A plane wave propagating ($e^{j(\omega t - \vec{k}\cdot\vec{r})}$) in such an active medium (either natural or magneto-induced activity), obeys the following Maxwell equations:

$$\nabla \times \vec{E} = -j\vec{k} \times \vec{E} = -j\omega\mu_o \vec{H}, \quad \nabla \times \vec{H} = -j\vec{k} \times \vec{H} = j\omega\overleftrightarrow{\varepsilon} : \vec{E},$$
$$\Rightarrow \nabla \times \nabla \times \vec{E} = -j\omega\mu_o \left(\nabla \times \vec{H}\right)$$
$$\Rightarrow k^2\vec{E} - \left(\vec{k}\cdot\vec{E}\right)\vec{k} = \mu_o\omega^2\overleftrightarrow{\varepsilon} : \vec{E}. \qquad (9.26)$$

For a wave propagating along the z direction, i.e., $\vec{k} = k\vec{u}_z$ with $k \equiv n\,\omega/c$, Eq. 9.26 simplifies into the following matrix form:

$$\frac{\omega^2}{c^2}\begin{pmatrix} n^2 - \varepsilon_{xx} & -\varepsilon_{xy} & 0 \\ \varepsilon_{xy} & n^2 - \varepsilon_{xx} & 0 \\ 0 & 0 & -\varepsilon_{zz} \end{pmatrix}\begin{pmatrix} E_x \\ E_y \\ E_z \end{pmatrix} = 0. \qquad (9.27)$$

The eigenvalues of this matrix (for which Eq. 9.27 is fulfilled with $\vec{E}(\vec{r}) \neq 0$), and the corresponding eigenvectors are

$$n^2 = \varepsilon_{xx} \pm j\varepsilon_{xy}, \quad \vec{E} = \begin{pmatrix} 1 \\ \pm j \\ 0 \end{pmatrix} E = \begin{pmatrix} 1 \\ e^{\pm j\frac{\pi}{2}} \\ 0 \end{pmatrix} E. \qquad (9.28)$$

The field defined by Eq. 9.28 is circularly polarized (either left-circularly polarized (LCP) with $+\pi/2$, or right-circularly polarized (RCP) with $-\pi/2$). Each of these circularly polarized waves travels at a different speed, i.e., with the refractive index $n_{\pm} = \sqrt{\varepsilon_{xx} \pm j\varepsilon_{xy}}$. Usually, the off-diagonal terms are small, leading to the following approximation:

$$n_{\pm} \approx n_o \pm \frac{\Delta n}{2}, \quad n_o = \sqrt{\varepsilon_{xx}}, \quad \Delta n = n_+ - n_- = j\frac{\varepsilon_{xy}}{\sqrt{\varepsilon_{xx}}}. \qquad (9.29)$$

Both diagonal and off-diagonal terms of the tensor contribute to the chiral effect.

When a linearly polarized beam impinges such a material, the effect of chirality is easily obtained by decomposing the incident field into LCP and RCP fields:

$$\vec{E} = \begin{pmatrix} 1 \\ 0 \\ 0 \end{pmatrix} E = \begin{pmatrix} 1 \\ e^{j\frac{\pi}{2}} \\ 0 \end{pmatrix}\frac{E}{2} + \begin{pmatrix} 1 \\ e^{-j\frac{\pi}{2}} \\ 0 \end{pmatrix}\frac{E}{2} = \vec{E}_{LCP} + \vec{E}_{RCP}. \qquad (9.30)$$

Each of these fields travels with a different speed. After propagation over a distance d in the chiral material, the output field is given by

$$\vec{E}\,(d) = \begin{pmatrix} 1 \\ e^{j\frac{\pi}{2}} \\ 0 \end{pmatrix} \frac{E}{2} e^{-j\frac{\omega}{c}n_{+}d} + \begin{pmatrix} 1 \\ e^{-j\frac{\pi}{2}} \\ 0 \end{pmatrix} \frac{E}{2} e^{-j\frac{\omega}{c}n_{-}d}$$

$$= \begin{pmatrix} \cos\varphi \\ -\sin\varphi \\ 0 \end{pmatrix} E\, e^{-j\left(\frac{\omega}{c}n_{-}d-\varphi\right)}. \qquad (9.31)$$

The output is still linearly polarized, but its polarization has been rotated by an angle $\varphi = \omega\,\Delta n\, d/2c$. A chiral material is characterized by its rotatory power ρ defined by

$$\rho = \frac{\omega}{2c}\Delta n \quad \Rightarrow \quad \varphi = \rho\, d, \quad n_{\pm} \approx n_{\mathrm{o}} \pm \frac{\rho}{c\,\omega}. \qquad (9.32)$$

The rotatory power depends on the frequency of the light beam. Thus, for natural crystals, it is in the far infrared about three orders of magnitude larger than in visible.

Let us point out the important question of left/right handedness convention. If the observer looks at the circularly polarized beam from the source or the receiver, he/she will see the polarization rotating clockwise or anti-clockwise. In physics and optics, one generally uses the convention of watching at the beam from the receiver. Thus, levorotary (dextrorotary) substances make the plane of polarization rotating to the left (right) (see Fig. 9.1), where the levorotary case is represented. The reason for this convention is purely historical: the French scientist Biot, who, together with Arago, was the first to study optical activity, used his eye as a receiver to observe the rotation of polarization through different substances [6].

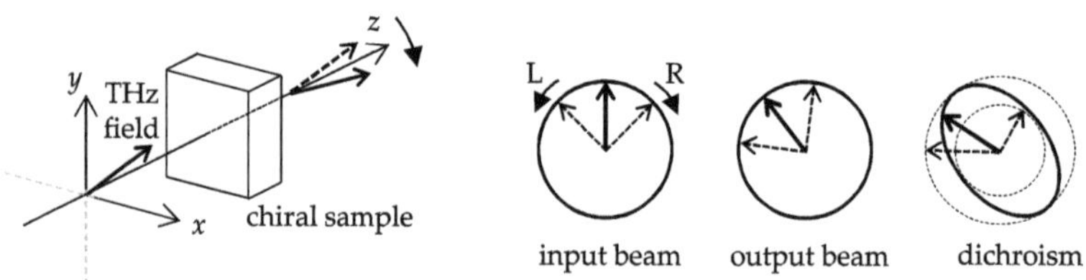

Figure 9.1 Rotation of the plane of polarization after propagation through a chiral sample (here the material is levorotary). The left drawings are as seen from the receiver.

9.3 Experimental Schemes

Polarimetric studies require us to monitor and measure the polarization of the emitted and detected THz beams. Unfortunately, simple waveplates that are broadband and handful in the THz domain do not exist. Thus, manipulating the THz polarization is not easy. On the other hand, grid polarizers exhibit great performance, but they do limit bandwidth.

When dealing with a THz-TDS setup based on photo-conducting antennas, such antennas deliver rather well linearly polarized beams. For example, using Grischkowsky's dipole switches, Duvillaret et al. measured an eccentricity of the order of 99.5% [7]. THz waves generated by optical rectification in EO crystals are also linearly polarized when excited by a linearly polarized laser beam [8]. Thanks to the reciprocity theorem, the polarization sensitivity of such THz antennas used as receivers is as good as in emission. Any misalignment in the set-up makes the polarization of the THz beam elliptical can be corrected using grid polarizers. Typically, commercial wire-grid polarizers exhibit an extinction ratio of about 10^{-3} [9]. The corresponding relative precision in the determination of the THz polarization is about 0.5 mrad [10].

Rotating the polarization of THz beams generated by photo-conducting antennas is feasible: the best way is to rotate the antenna at 45° from the horizontal axis (see Fig. 9.2). Then a polarizer selects the desired THz polarization orientation. The drawback of this scheme is the loss of signal (30% maximum in field strength). A similar scheme may be used in detection. Also, one may mount the receiving antenna on a rotating holder. The laser excitation of the antenna, which requests a tight focusing onto the switch gap, could be performed through a fiber attached to the antenna [11, 12].

Using both emitting and receiving EO antennas facilitates the control and detection of the polarization of the THz beam. The basic idea is to employ a cubic crystal with a <111> cut [13, 14]. The electro-optic tensor of cubic crystals, which is related to both optical rectification (emission) and Pockels effect (detection), writes (see

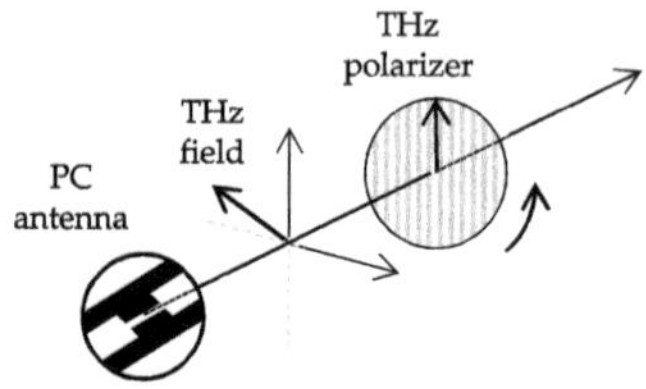

Figure 9.2 Scheme to select the polarization of a THz beam emitted by a PC antenna.

Annex, Fig. A.4):

$$\overset{\leftrightarrow}{d} = \begin{pmatrix} 0 & 0 & 0 \\ 0 & 0 & 0 \\ 0 & 0 & 0 \\ d_{41} & 0 & 0 \\ 0 & d_{41} & 0 \\ 0 & 0 & d_{41} \end{pmatrix} \tag{9.33}$$

Let us first treat THz emission. In the laboratory reference frame (x, y, z), z is the direction of the laser and THz beams that propagate perpendicularly to the crystal. ψ is the polarization angle of the laser beam as regards to the x direction. In the crystal reference frame (X, Y, Z), the photo-induced THz polarization is

$$\vec{E}_{cryst} = 2d_{14}E_{laser}^2 \times \begin{pmatrix} -(\cos\theta\sin\varphi\cos\psi + \cos\varphi\sin\psi)\sin\theta\cos\psi \\ -(\cos\theta\cos\varphi\cos\psi - \sin\varphi\sin\psi)\sin\theta\cos\psi \\ (\cos\theta\cos\varphi\cos\psi - \sin\varphi\sin\psi)(\cos\theta\sin\varphi\cos\psi + \cos\varphi\sin\psi) \end{pmatrix}$$

The crystal reference frame is oriented with an azimuthal angle θ around y, and a polar angle φ around z. In the laboratory, the THz generated field is [15]

$$\vec{E}_{lab,THz} \propto \begin{pmatrix} E_{cryst,X}\cos\varphi\cos\theta + E_{cryst,Y}\sin\varphi\cos\theta + E_{cryst,Z}\sin\theta \\ -E_{cryst,X}\sin\varphi + E_{cryst,Y}\cos\varphi \\ -E_{cryst,X}\sin\theta\cos\varphi - E_{cryst,Y}\sin\theta\sin\varphi + E_{cryst,Z}\cos\theta \end{pmatrix}. \tag{9.34}$$

It is easy to show that the THz field amplitude does not depend on the laser polarization angle ψ if $\cos\theta = 1/\sqrt{3}$ and $\varphi = \pi/4$, i.e., if

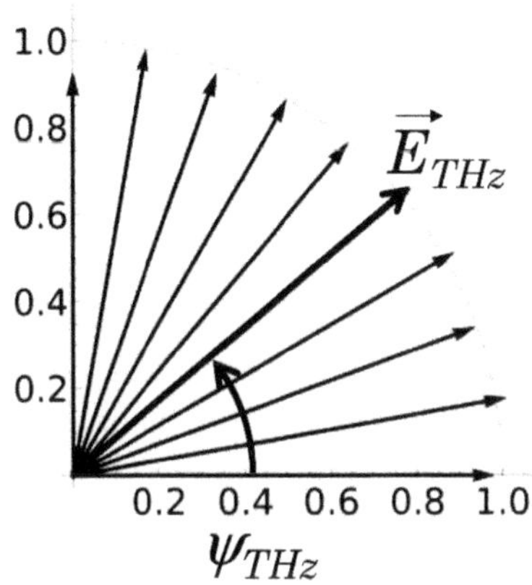

Figure 9.3 THz field emitted by a <111> cubic EO crystal and detected by a PC antenna (adapted from Gaborit et al. [15]). The polarization angle ψ_{THz} is tuned by rotating the polarization ψ of the pumping laser beam ($\psi_{\text{THz}} = -2\psi$).

the crystal cut is <111>. In this case, this field is written as

$$\vec{E}_{lab,THz} \propto 2d_{14}E^2_{laser}\begin{pmatrix} -\cos 2\psi \\ \sin 2\psi \\ -1/\sqrt{2} \end{pmatrix}. \tag{9.35}$$

The longitudinal component (along z) of the THz field does not propagate outside the crystal (it is totally reflected at the crystal surface). Thus, the THz wave delivered by the EO crystal is linearly polarized with a polarization angle ψ_{THz} equal to -2ψ, i.e., two times the polarization angle ψ of the exciting laser beam. This constitutes the great advantage of this technique: one can choose the THz polarization by a proper selection of the laser beam polarization, which is easier to perform.

This is illustrated in Fig. 9.3, which shows the THz E-field vector radiated by a ZnTe <111> EO crystal and measured with a photo-conducting antenna. The THz field direction is selected by simply rotating the EO crystal. Here the effect is almost frequency-independent on the experimental THz bandwidth.

Detection of the THz field through the Pockels effect in the same crystal (<111> cubic crystal) is independent of the incoming THz polarization [8, 15, 16]. In the crystal reference frame, the index ellipsoid is

$$\left(\frac{X}{n_o}\right)^2 + \left(\frac{Y}{n_o}\right)^2 + \left(\frac{Z}{n_o}\right)^2 + 2r_{41}\left(E_{cryst,X}YZ + E_{cryst,Y}XZ + E_{cryst,Z}XY\right) = 1. \quad (9.36)$$

n_o is the refractive index of the crystal at the laser wavelength, and $E_{cryst,X}$, $E_{cryst,Y}$, $E_{cryst,Z}$ are the components of the incoming THz field. When projecting the incoming THz field vector of polarization angle ψ along the X, Y, Z directions, inputting them in Eq. 9.36 and solving this equation yields the refractive index seen by the laser probe beam, i.e., the eigenvalues of Eq. 9.36:

$$n_{\pm} = n_o + \frac{n_o^3 r_{41}}{2\sqrt{3}}\left(\cos\psi \pm \sqrt{2}\sin\psi\right)\left|\vec{E}_{THz}\right|. \quad (9.37)$$

With a classical polarimetric detection scheme using a balanced photodetector, the obtained signal is [16]

$$S_{detection} \propto \left|\vec{E}_{THz}\right| \sin\left(-3\gamma + 4\alpha + \psi\right). \quad (9.38)$$

α is the angle of the principal axis of the $\lambda/2$ plate located in front of the polarimetric detector, and γ is the angle between the <11$\bar{2}$> and x directions (it corresponds to the orientation of the crystal). As desired, the recorded signal depends on the magnitude of the THz field. Moreover, for any orientation of the <111> crystal and of the incoming THz field polarization, there exists an adjustment of the $\lambda/2$ plate that maximizes the recorded signal. Let us notice that modulating the polarization of the laser probe beam can be used to improve the THz field detection efficiency [8, 17]. Figure 9.4 presents the measured polarization direction of a THz versus the polarizer angle that defines the THz polarization. The agreement is excellent, even at the very small mrad scale.

When characterising the anisotropic properties of a sample with THz-TDS techniques, the sample is illuminated by a linearly polarized THz pulse. In most experiments, this is performed under a normal incidence and there is no deviation of the beam because of anisotropy effects. Therefore, the point is to measure the polarization state of the transmitted or reflected beam. It could be useful to record simultaneously two orthogonal components of the transmitted/reflected THz signal, instead of

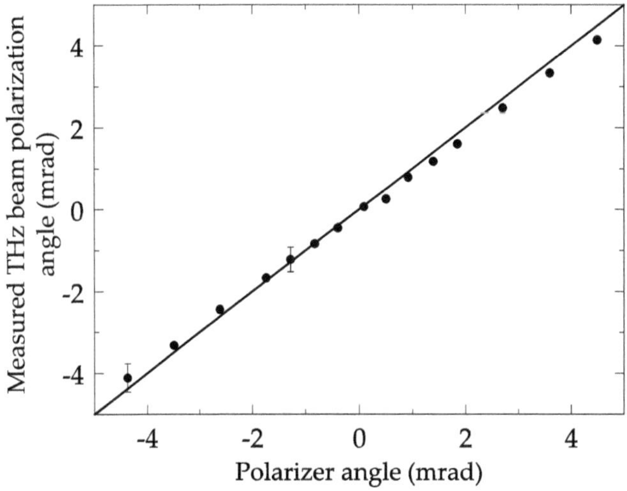

Figure 9.4 THz polarization angle ψ_{THz} measured by EO polarimetric detection versus the angle of the polarizer that defines the THz beam polarization. Adapted from Nemoto et al. [8].

rotating either emitter, sample or receiver to access some changes in the polarization state. To that purpose, several groups [18–22] have studied PC THz antennas with three or four electrodes that show good performance in delivering simultaneously and independently the THz field orthogonal components.

Comprehensive analyses of the extraction of the polarization-dependent properties of THz pulses and materials in THz-TDS have been published in [21–23]. Here, we base the extraction analysis on the paper by Castro-Camus and Johnson [23]. First, we write the relations between the different possible states of polarization, i.e., linear polarization (either "horizontal" $\vec{E}_H$ or "vertical" $\vec{E}_V$), and circular (either "left" $\vec{E}_L$ or "right" $\vec{E}_R$):

$$\vec{E}_H = \frac{\vec{E}_R + \vec{E}_L}{\sqrt{2}}, \quad \vec{E}_V = j\frac{\vec{E}_R - \vec{E}_L}{\sqrt{2}},$$
$$\vec{E}_R = \frac{\vec{E}_H - j\vec{E}_V}{\sqrt{2}}, \quad \vec{E}_L = \frac{\vec{E}_H + j\vec{E}_V}{\sqrt{2}}. \tag{9.39}$$

The reference frame of the laboratory is (x, y, z), where z is the direction of propagation of the THz beam, x is the "horizontal" axis and y the "vertical" one. In the Fourier space, each of the fields in

Eq. 9.39 is obtained with an amplitude and a phase:

$$\vec{E}_i = E_{ix}\, e^{j\varphi_{ix}} \vec{u}_x + E_{iy}\, e^{j\varphi_{iy}} \vec{u}_y. \tag{9.40}$$

In Eq. 9.40 and from now on, the ω dependence of all the values is omitted for the sake of simplicity. The main direction of polarization of the field $\vec{E}_i$, defined by the angle ψ of this direction with respect to x, and its ellipticity ξ_i are given by

$$\psi_i = \frac{1}{2} a\tan\left[2\frac{E_{ix}E_{iy}}{E_{ix}^2 - E_{iy}^2}\cos\left(\varphi_{iy} - \varphi_{ix}\right)\right], \tag{9.41}$$

$$\xi_i = \frac{b_i}{a_i} = \tan\left[\frac{1}{2} a\tan\left(2\frac{E_{ix}E_{iy}}{E_{ix}^2 + E_{iy}^2}\sin\left(\varphi_{iy} - \varphi_{ix}\right)\right)\right]. \tag{9.42}$$

b_i and a_i are the major and minor axes of the ellipse. In common THz-TDS set-ups, the fields are received either along the x or the y directions. Combining Eqs. 9.39 and 9.40 allows one to derive the fields in other bases, if necessary. For example, in a circular basis, one gets

$$\begin{cases} \vec{E}_R \equiv E_R\, e^{j\varphi_R} \vec{u}_R = \dfrac{E_H\, e^{j\varphi_H} \vec{u}_x - j\, E_V\, e^{j\varphi_V} \vec{u}_y}{\sqrt{2}}, \\ \vec{E}_L \equiv E_L\, e^{j\varphi_L} \vec{u}_L = \dfrac{E_H\, e^{j\varphi_H} \vec{u}_x + j\, E_V\, e^{j\varphi_V} \vec{u}_y}{\sqrt{2}}. \end{cases} \tag{9.43}$$

In a state of polarization i ($i = H, V, R, L$), the complex transmittance of a sample is equal to the ratio of the transmitted and incident fields in the frequency domain:

$$t_i = \frac{E_{i,sample}}{E_{i,ref}} e^{j\left(\varphi_{i,sample} - \varphi_{i,ref}\right)}. \tag{9.44}$$

If internal reflections in the sample can be neglected or rejected by a proper time-windowing, real and imaginary parts of the refractive index in the i polarization state are simply deduced from Eq. 9.44:

$$n_i = 1 + \frac{c}{\omega\, d}\left(\varphi_{i,sample} - \varphi_{i,ref}\right), \quad \kappa_i = \frac{c}{2\omega\, d}\ln\frac{E_{i,sample}}{E_{i,ref} t_i^{12} t_i^{21}}, \tag{9.45}$$

with the Fresnel coefficients at the air/samples interfaces:

$$t_i^{12} = \frac{2}{n_i + 1}, \quad t_i^{21} = \frac{2\, n_i}{n_i + 1}.$$

The main cases of anisotropy are easily treated using this formalism [22], namely

$$\text{birefingence: } \Delta n = n_E - n_O = \frac{c}{\omega d}\left[\varphi_{E,sample} - \varphi_{O,sample} - \varphi_{E,ref} + \varphi_{O,ref}\right], \quad (9.46)$$

$$\text{activity: } \rho = \frac{\omega}{2c}(n_L - n_R) = \frac{1}{2d}\left[\varphi_{L,sample} - \varphi_{R,sample} - \varphi_{L,ref} + \varphi_{R,ref}\right], \quad (9.47)$$

$$\text{linear dichroism: } LD = \alpha_E - \alpha_O = \frac{1}{d}\ln\left[\frac{t_O^{12}t_O^{21}}{t_E^{12}t_E^{21}}\frac{E_{E,sample}E_{O,ref}}{E_{O,sample}E_{E,ref}}\right], \quad (9.48)$$

$$\text{circular dichroism: } CD = \alpha_R - \alpha_L = \frac{1}{d}\ln\left[\frac{t_L^{12}t_L^{21}}{t_R^{12}t_R^{21}}\frac{E_{R,sample}E_{L,ref}}{E_{L,sample}E_{R,ref}}\right]. \quad (9.49)$$

In these equations, light whose polarization is perpendicular to the optic axis is governed by a refractive index n_O (for "ordinary") and light whose polarization is in the direction of the optic axis sees an index n_E (for "extraordinary"). Dong et al. [21] have shown that measuring the two orthogonal components of the received THz beam using a polarizer directed at 0°, 45°, and 90° (with respect to x) in front of the THz receiver, permits the determination of the whole anisotropic response of the sample, i.e., its Mueller matrix.

The different experimental arrangements of a polarimetric THz-TDS set up are plotted in Fig. 9.5. The emitting and receiving antennas are either PC or EO, and are supposed to be ideally sensitive to a single linear polarization. In the left scheme, the emitted beam is vertically polarized (see arrow). After passing through the sample, it becomes elliptic. Then it reaches the receiver that is sensitive to the vertical component of the elliptical THz field. To measure the horizontal one, the receiving antenna must be rotated by 90°. The sample itself can be rotated to get all the information about its anisotropic parameters.

In the right scheme, both emitting and receiving antenna radiates and detects a THz field that is oriented at 45° with respect to the x axis. In front of the sample, a polarizer attached to a rotation stage

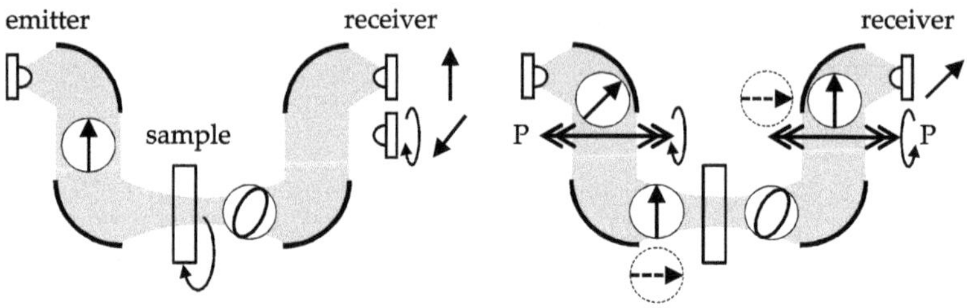

Figure 9.5 Two different polarimetric THz-TDS setups. The circles with arrow or ellipse indicate the state of polarization of the THz beam. Single arrows give the direction of polarization sensitivity of the receiver. P is for polarizer.

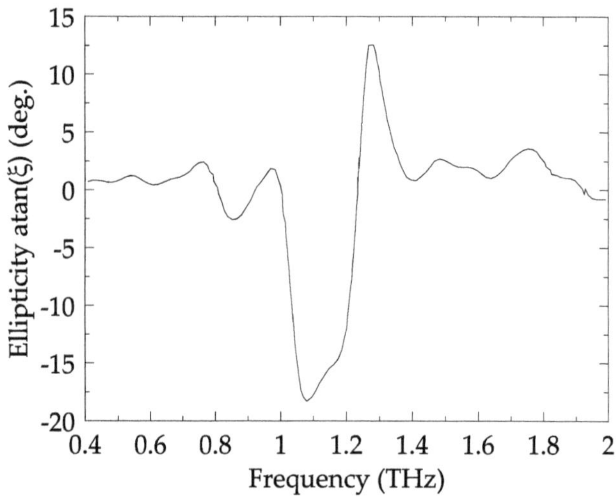

Figure 9.6 Ellipticity angle (atan ξ) spectrum of a 2D chiral metamaterial made of Archimedean spirals. Here the spirals are right-handed. Adapted from Aschaffenburg et al. [24].

allows to get either the vertical or horizontal polarization. After the sample, a second polarizer can be rotated to select the vertical or horizontal polarization component of the transmitted THz field. Regarding geometrical adjustment, this latter scheme is preferable, but 75% of the THz field strength is lost because of the polarizers.

Figure 9.6 presents an example of THz-TDS polarimetry data using the schemes depicted in Fig. 9.5. It shows the ellipticity spectrum of the THz beam that is transmitted through a chiral sample. This sample is a high resistivity Si substrate on which 250 nm thick aluminum Archimedean spirals (arranged in a 2D rhombohedral lattice) have been deposited using electron-beam lithography.

9.4 THz-TDS Study of Anisotropic Dielectric Crystals

Here the relative permeability of the material is equal to 1, thus $\mu = \mu_0$, but the permittivity exhibits a tensorial behavior leading birefringence. In the THz range, the literature on anisotropic dielectrics can be arranged in three topics: (i) determination of the refractive index and absorption spectra, (ii) observation of anisotropy induced by an external cause (pressure, etc.), by the shape of the sample, or by the sub-wavelength geometry (metamaterials), and (iii) anisotropic devices.

In uniaxial crystals, the refractive index for one polarization (let us say along Z) is different from the two other ones that are equal: $n_Z \equiv n_E \neq n_X = n_Y \equiv n_O$. n_E is the extraordinary index, while n_O is the ordinary one (when $n_E > n_O$, the crystal is said uniaxial positive). Z is the optical axis of the crystal, which is also an axis of symmetry. When the optical axis is oriented perpendicular to the surface of a flat sample, a wave impinging at normal incidence will "see" the ordinary refractive index n_O. If the axis is in the surface plane of the sample, it will see the following refractive index:

$$\frac{1}{n^2(\psi)} = \frac{\cos^2\psi}{n_O^2} + \frac{\sin^2\psi}{n_E^2}. \tag{9.50}$$

ψ is the angle between the optical axis and the polarization vector. Therefore, if the polarization of the THz beam is aligned with the optical axis or is perpendicular to it, the crystal will behave as an isotropic medium of refractive index n_E and n_O, respectively. The study of the THz properties of anisotropic crystals (sapphire and quartz) was initially published in one of the first pioneering papers in the field by Grischkowsky and coworkers [25] (see Fig. 9.7

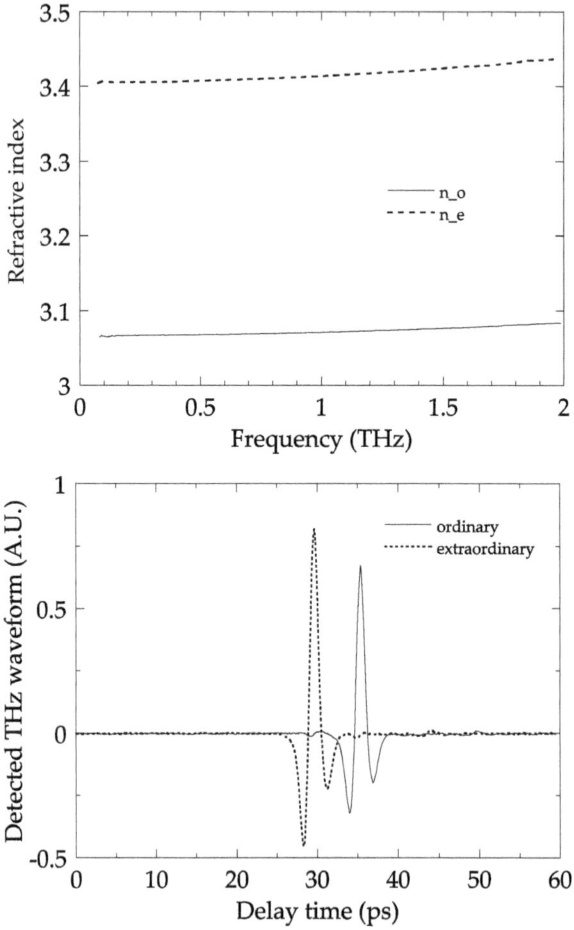

Figure 9.7 Top: Ordinary and extraordinary indices of crystalline sapphire versus frequency (adapted from Grischkowsky et al. [25]). Bottom: THz waveforms, with ordinary and extraordinary polarizations, transmitted through a 1.1 mm thick $LiNbO_3$ crystal. Typically, $n_E \approx 5.2$ and $n_O \approx 6.7$ (extracted from F. Garet's Ph.D. thesis, University of Grenoble, 1997, unpublished).

adapted from [25]). Optical anisotropy is a macroscopic outcome of crystallography: while cubic crystals and amorphous materials are isotropic, tetragonal, hexagonal, and trigonal crystals are uniaxial.

The other crystal systems (orthorhombic, monoclinic, and triclinic) are biaxial.

In general, ferroelectric crystals, like $LiNbO_3$ or $LiTaO_3$, show a very large birefringence up to $n_E - n_O \approx 1.5$, while absorption is also anisotropic [26]. In the time domain, such a big birefringence serves in selecting the polarization of an incoming beam. Indeed, the component of the THz wave polarized in the extraordinary direction travels in the crystal at a speed c/n_E, which is different from the ordinary polarization (c/n_O). If we can neglect dispersion, the extraordinary and ordinary pulses will pass through a sample of thickness d with a time delay $\tau_{E\text{-}O} = d(n_E - n_O)/c$. They can be easily separated as soon as the THz pulse duration is shorter $\tau_{E\text{-}O} > \tau_{THz}$ (see the example given in Fig. 9.7 bottom). Typically, with $\tau_{THz} \approx 1$ ps and $n_E - n_O = 1.5$, d should be larger than a few tens of millimeters.

In materials made with fibers, polymeric chains or long molecules, anisotropy results from the alignment of most of these microscopic structures along one direction. This preferred alignment is due either to geometrical constraints (e.g., thin films deposited over a surface [27]) or to the fabrication process (growth of nanowires normal to the substrate surface, polymeric chains aligned by an applied electrical field during the polymerization process). Uniaxial THz anisotropy has been observed in silicon nanowire films [28, 29], organic crystals [30], oriented polymeric films [31], as well as in polymers reinforced by fibers [32] or carbon nanotubes [33]. Polymer films deformed under pressure exhibit a stress-induced anisotropic behavior [34]. Even more exotic natural materials like wood [35–37] or plant leaves [37] exhibit a strong uniaxial anisotropy due to the natural orientation of the cells and fibers. Liquid crystals have also been widely studied since the first reports by the teams of M. Koch [38] and C. L. Pan [39], because they can be used to manufacture electrically-active THz components.

Biaxial anisotropy is more challenging to study. Only a few reports have been published so far. For example, Liang and coworkers [40] have measured the THz spectrum of the three refractive indices of $LiInSe_2$, an orthorhombic crystal, and have determined the frequency and strength of several phonon modes in the crystal. As well, biaxial anisotropy has been observed in slanted metallic columnar thin films [41].

9.5 THz Study of Magneto-optical Effects

9.5.1 Introduction

As explained previously, magneto-optical effects lead to chirality and also to several nonlinear phenomena [42]. The first magneto-optical effect ever reported was by Faraday in 1846 [43]. A magnetic material transmits differently right and left polarized waves, leading to rotating the polarization of the transmitted beam, or more generally to transform a linearly polarized beam into an elliptical transmitted one. The angle of the major axis of the ellipse is a linear function of magnetization. The effect also occurs in reflection as discovered by Kerr, namely the magneto-optical Kerr effect (MOKE). The Voigt effect is similar to the Faraday effect, but the ellipse angle varies as the square of the magnetization. Magnetic circular dichroism corresponds to a different absorption of left and right circularly polarized waves in magnetic materials. Magnetic linear dichroism involves the differential absorption of linearly polarized fields, and it is proportional to the square of magnetization. Here magnetization can be natural or induced by an applied magnetic field. Understanding the physical origin of magneto-optics was initially performed in the framework of classical physics (Lorentz and Drude models). Quantum mechanics permits a better description of the phenomenon: it appears that magneto-optical effects involve the spin of electrons through a spin-orbit coupling.

9.5.2 THz-TDS MOKE

Let us focus on MOKE as it is well studied because many magnetic or magneto-optical materials are opaque in the THz range. The goal of MOKE study is to obtain, from polarimetric THz-TDS reflectometry, the components of the $\overleftrightarrow{\varepsilon}(\vec{M})$ tensor. There are three geometries for MOKE experiments, depicted in Fig. 9.8, namely polar ($\vec{M}$ is normal

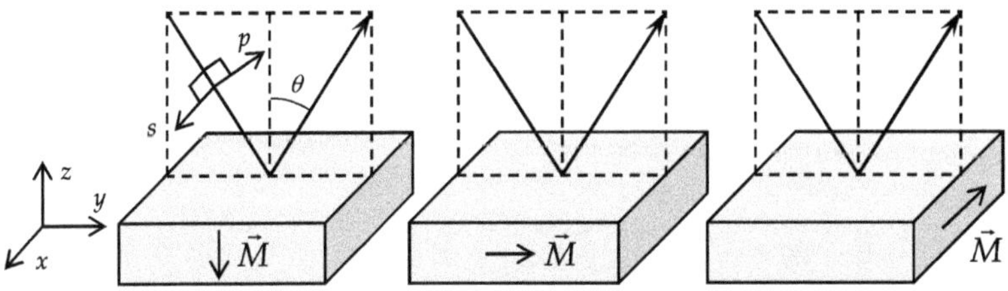

Figure 9.8 Different geometries of MOKE experiments.

to the sample surface), longitudinal and transversal MOKE. In the case of polar MOKE, the dielectric tensor is

$$\overleftrightarrow{\varepsilon}\left(\vec{M}\right) = \begin{pmatrix} \tilde{\varepsilon}_{xx} & \tilde{\varepsilon}_{xy} & 0 \\ -\tilde{\varepsilon}_{xy} & \tilde{\varepsilon}_{xx} & 0 \\ 0 & 0 & \tilde{\varepsilon}_{zz} \end{pmatrix}. \tag{9.51}$$

Because there are three unknown complex values, namely $\tilde{\varepsilon}_{xx}$, $\tilde{\varepsilon}_{zz}$, $\tilde{\varepsilon}_{xy}$, different measurements are requested in view of determining these coefficients. This is achieved by recording the reflected signal under oblique incidence and varying the polarization of the incident beam and the received signal. The incident polarization (see Fig. 9.8) is chosen either TE (or s polarization: the electrical field is transverse, i.e., the field perpendicular to the plane of incidence) or TM (or p polarization: the magnetic field is transverse). As well, the polarization sensitivity of the receiver is chosen either s or p. Then one measures by THz-TDS the $\tilde{r}_{lm}$ complex coefficient of reflection, where l and m denote the polarization state (s or p) of respectively the source and receiver. The calculation of $\tilde{r}_{lm}$ is long and tedious. In a first approximation, one gets

$$\begin{cases} \tilde{r}_{ss} = \dfrac{\cos\theta - \tilde{n}\cos\tilde{\theta}_s}{\cos\theta + \tilde{n}\cos\tilde{\theta}_s}, \quad \tilde{r}_{pp} = \dfrac{\tilde{n}\cos\theta - \cos\tilde{\theta}_s}{\tilde{n}\cos\theta + \cos\tilde{\theta}_s}, \\ \tilde{r}_{sp} = \tilde{r}_{ps} = \dfrac{jn_o\left(n_+ - n_-\right)\cos\theta}{\left(\tilde{n}\cos\tilde{\theta}_s + \cos\theta\right)\left(\tilde{n}\cos\theta + \cos\tilde{\theta}_s\right)\cos\tilde{\theta}_s}. \end{cases} \tag{9.52}$$

θ is the angle of incidence and θ_s the angle of refraction in the nonmagnetized material (whose refractive index is n_o), $n_\pm^2 = \varepsilon_{xx} \mp j\varepsilon_{xy}\cos\theta_s$, and $\tilde{n} = \sqrt{\varepsilon_{xx}}$. Therefore, the diagonal and off-diagonal elements of the $\overleftrightarrow{\varepsilon}(\vec{M})$ matrix can be determined by measuring the

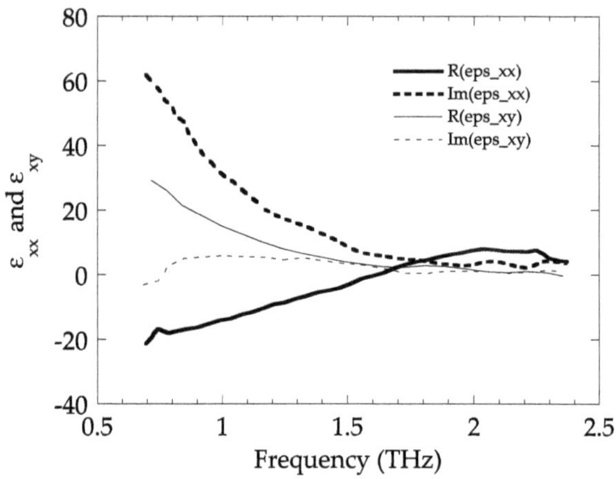

Figure 9.9 Spectra of the real (continuous curves) and imaginary (dashed curves) parts of ε_{xx} (thick curves) and ε_{xy} (thin curves) of n-doped InAs under a 0.48 T magnetic field. Adapted from Shimano et al. [44].

$\tilde{r}_{lm}$ coefficients and inverting Eqs. 9.52. Figure 9.9 presents the real and imaginary parts of ε_{xx} and ε_{xy} of n-doped InAs, under a magnetic field of 0.48 T, determined from a polar THz-TDS MOKE experiment by Shimano et al. [44].

In the case of longitudinal MOKE, one gets

$$\overleftrightarrow{\varepsilon}\left(\vec{M}\right) = \begin{pmatrix} \tilde{\varepsilon}_{xx} & 0 & \tilde{\varepsilon}_{xy} \\ 0 & \tilde{\varepsilon}_{yy} & 0 \\ -\tilde{\varepsilon}_{xy} & 0 & \tilde{\varepsilon}_{zz} \end{pmatrix}. \tag{9.53}$$

Here we have made the substitution $\tilde{\varepsilon}_{xz} = \tilde{\varepsilon}_{xy}$. This leads to the same formulae (9.52) than for polar MOKE, except that $\tilde{r}_{sp}$ has to be multiplied by -1.

In a way, the case of transversal MOKE is simpler. There is no $s-p$ coupling ($\tilde{r}_{sp} = \tilde{r}_{ps} = 0$) and only the p polarization is sensible to the magnetization:

$$\tilde{r}_{pp} = \tilde{r}_{pp,o}\left[1 + \frac{\varepsilon_o \tilde{\varepsilon}_{xy} \sin 2\theta}{\tilde{\varepsilon}_{xx}^2 \cos^2\theta - \varepsilon_o \tilde{\varepsilon}_{xx} + \varepsilon_o^2 \sin^2\theta_s}\right]. \tag{9.54}$$

$\tilde{r}_{pp,o}$ is the coefficient of reflection of the nonmagnetized material. An example of THz-TDS transversal MOKE measurement is given by Chen et al. [45].

9.5.3 Faraday Effect

The Faraday effect can also be studied by treating separately the longitudinal and polar cases. In polar geometry, the incident linearly polarized beam is transformed into an elliptically polarized one. The angle $\phi_{Faraday}$ of the major axis of the ellipse (Faraday rotation) and the Faraday ellipticity $\varepsilon_{Faraday}$ ($\varepsilon_{Faraday} = \text{atan}(b/a)$, where a and b are respectively the short and long axes of the ellipse), are given by

$$\phi_{Faraday} - j\,\varepsilon_{Faraday} = -j\frac{\omega}{c}d\frac{\tilde{\varepsilon}_{xy}}{2\sqrt{\tilde{\varepsilon}_{xx}}}, \tag{9.55}$$

d is the sample thickness. Note that this expression does not depend on the angle of incidence. In the longitudinal geometry, one has

$$\phi_{Faraday} - j\,\varepsilon_{Faraday} = -j\frac{\omega}{c}d\frac{\tilde{\varepsilon}_{xy}}{2\sqrt{\tilde{\varepsilon}_{xx}}}\tan\theta_s\,. \tag{9.56}$$

Here the expression depends on the angle of incidence. Several measurements at different angles of incidence may be performed to derive all the magneto-optical parameters of the tested material. Because of the tangent term in Eq. 9.56, the Faraday effect is more pronounced at large incident angles, i.e., close to the grazing incidence.

Figure 9.10 [46] shows the spectrum of the Faraday rotation angle $\phi_{Faraday}$, measured by THz-TDS, of a commercial ferrofluid (10 nm-sized nanoparticles of Fe_3O_4 dispersed in a hydrocarbon carrier fluid). The sample is 10 mm thick, and a magnetic field (amplitude given in the figure) is applied in the longitudinal direction.

Faraday effects are also observed at THz frequencies in magnets. Figure 9.11 adapted from [47] presents the real part n and imaginary part κ of the refractive index of cobalt chromite ($CoCr_2O_4$). $CoCr_2O_4$ is a cubic normal spinel ferrimagnet. The top plot is obtained for left-circularly polarized (LCP) light, while the bottom one is for right-circularly polarized (RCP) light. Dashed and continuous lines are achieved respectively without and with an applied magnetic field (3 T). Without applied field (dashed lines), one clearly sees a difference of absorption between the two circular polarization (dichroism). With the applied magnetic field, the difference becomes more pronounced around 0.6 THz, due to a resonant absorption

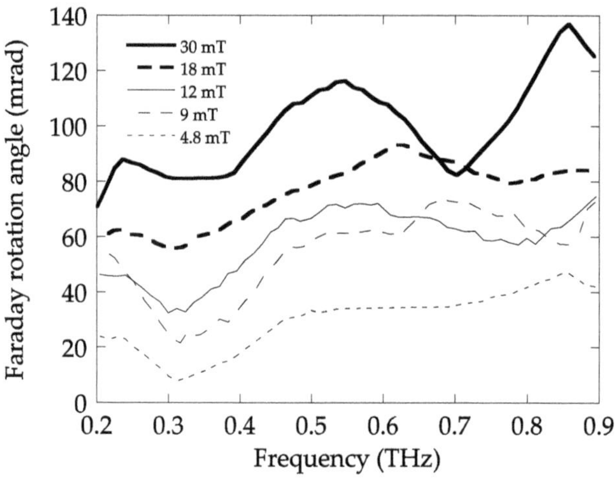

Figure 9.10 The spectrum of the Faraday rotation angle of a ferrofluid determined at different applied magnetic fields from THz-TDS data. Adapted from Shalaby et al. [46].

induced by the magnetic dipole transition from inter-sublattice exchange mode. The small difference of n for RCP and LCP lights gives rise to a Faraday rotation.

9.5.4 Voigt Effect

The Voigt effect results from a variation of the refractive index of a material, which proportional to the square of applied magnetic field amplitude. Generally, the effect is observed in the transversal geometry under normal incidence. For example, let us suppose that $\vec{M}$ is aligned along the x direction (Fig. 9.6) and the incident THz field is polarized at 45° ($E_x(0) = E_y(0) = E_i/\sqrt{2}$). After travelling over a distance d in the material, the two components of the E field exhibit a complex phase difference:

$$\frac{E_x(d)}{E_y(d)} = e^{-j\varphi} \quad \text{with } \varphi = \delta_V - j\eta_V \approx \frac{\omega}{2\tilde{n}c} d\left(\varepsilon_{xx} - \varepsilon_{yy} - \frac{\varepsilon_{yz}^2}{\varepsilon_{yy}}\right). \tag{9.57}$$

δ_V is the linear magnetic birefringence, and η_V is the linear magnetic dichroism. As for all the magneto-optical effects, a physical model

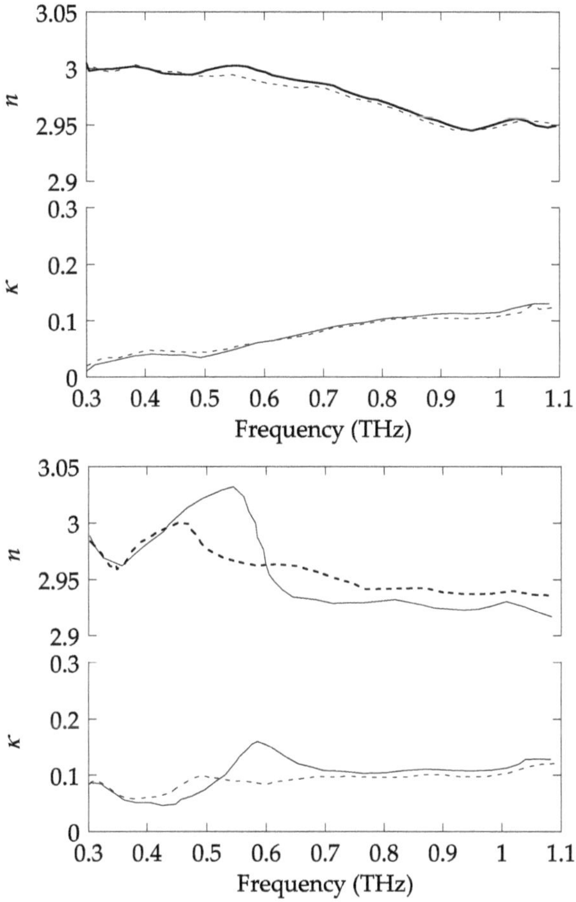

Figure 9.11 The spectrum of the real n and imaginary κ parts of the refractive index of $CoCr_2O_4$ for a left-circular polarized THz light (top plots) and a right-circular one (bottom plots) (adapted from Zhang et al. [47]). The continuous lines are obtained without any applied magnetic field, while the dashed ones correspond to a -3 T magnetic field.

is necessary to know the dependence on $\vec{B}$ of the dielectric matrix elements ε_{ij}. It could be a classical model (Lorentz, Drude) but with a limited accuracy, or a more sophisticated one based on quantum mechanics (here the calculation starts with the linear response of a many-electron system as initially established by Kubo

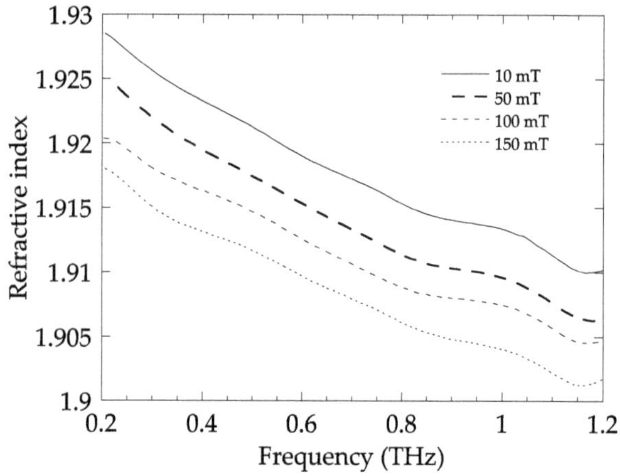

Figure 9.12 Refractive index of a ferrofluid versus frequency measured under different applied magnetic field (adapted from Chen et al. [45]). Here the concentration of Fe_3O_4 nanoparticles is 17.7%.

[48]). For a good review of these theories, the reader is referred to Oppeneer [1].

Voigt effect was observed, using THz-TDS, in a ferrofluid (Fe_3O_4 nanoparticles dispersed in a hydrocarbon oil) [45]. Transmission THz-TDS was performed with the magnetic field applied parallel to the sample surface (transversal magnetization). The variation of the refractive index with the applied magnetic field (see Fig. 9.12) is fitted by Chen and coauthors [45] with a dielectric model including the plasma and cyclotron frequencies of the metallic nanoparticles, as well as their magnetization. This permits to check that the observed phenomenon is actually due to the Voigt effect.

9.6 Anisotropic Metamaterials

9.6.1 Introduction

Chirality in nonmagnetic natural dielectric crystals at THz frequencies is rarely studied and reported because the effect is very weak in natural materials: we can only cite related works on chiral and racemic molecules [49, 50] and crystals [51]. Moreover, as a general rule, magneto-optical effects are limited. This thorny problem is nicely smoothed out with metamaterials.

Metamaterials are artificial materials made of a periodic arrangement of sub-wavelength structures. Strictly speaking, the period of the structures is also sub-wavelength, and consequently, the macroscopic electromagnetic response of the metamaterial can be described by the bulk parameters (ε, μ) of a homogeneous material. However, these macroscopic parameters are governed by the collective response of the sub-wavelength structures. This is the basic advantage of metamaterials: one can design the sub-wavelength structure to optimize the macroscopic properties of the metamaterial. Especially, one can make resonant micro-structures in view to enhance the interaction with an impinging THz beam. This is mostly achieved with metallic micro-structures. The illuminating wave induces currents in the metallic features and accumulation of charges at edges, resulting in tailored dielectric and magnetic properties of the metamaterials. As compared to natural materials, these properties can be so greatly enhanced that 3D metamaterials are not necessary to observe the desired dielectric/magnetic properties: even 2D metamaterials show an amazing electromagnetic response. On the other hand, the resonant character of the micro-structure response leads to a rather limited operating bandwidth. There is an additional advantage in the THz domain: as the wavelengths are typically sub-millimeter, the size of the microstructures and their geometrical details are of the order of a few microns up to tens of microns. This makes the fabrication of the metamaterials quite easy. Due to these two advantages, i.e., optimization of the desired electromagnetic response and simple fabrication process, THz metamaterials have been widely studied, and the related literature is extensive [52–54].

In the next section, we will focus on the anisotropic properties of metamaterials in the THz range, and on their characterization by THz-TDS. Because the effects are similar to those occurring in natural materials, the measurement techniques are the same (see above). We will present only a very limited number of typical examples extracted from the abundant literature.

9.6.2 Birefringence with Metamaterials

The anisotropic dielectric properties of metamaterials originate in the shape of the microstructures, which must show a pronounced dissymmetry in a given direction. Because the effect can be very strong, most metamaterials with enhanced birefringence have metallic metasurfaces, i.e., either metallic structures deposited on a dielectric substrate, or sheets of metal perforated with a holes' array.

One of the first and most popular metamaterials, which exhibits a strong anisotropy of absorption, is the famous wire-grid polarizer, conceived from pioneering ideas from J. Bose [55]. Metallic and parallel wires are stretched in a frame or are deposited over a thin transparent substrate. Both wire diameter and distance between two wires are sub-wavelength. When the incident E field is parallel to the wire, it induces a current in the wire as it would be induced in a metallic planar mirror. Because of the sub-wavelength dimensions, the grid behaves as a mirror and reflects the wave. On the opposite, when the field is perpendicular to the main direction of the wires, the induced current cannot flow over a distance longer than the wire diameter. The interaction between the wave and the wires is weak, and thus the wave is transmitted. Figure 9.13 shows the excellent TE wave rejection level over the whole THz range of a grid polarizer

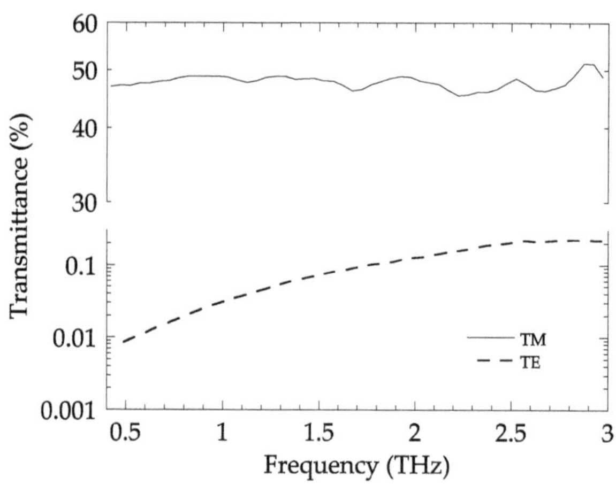

Figure 9.13 The transmittance of a micrometer-pitch Al grating polarizer. Adapted from Grady et al. [57].

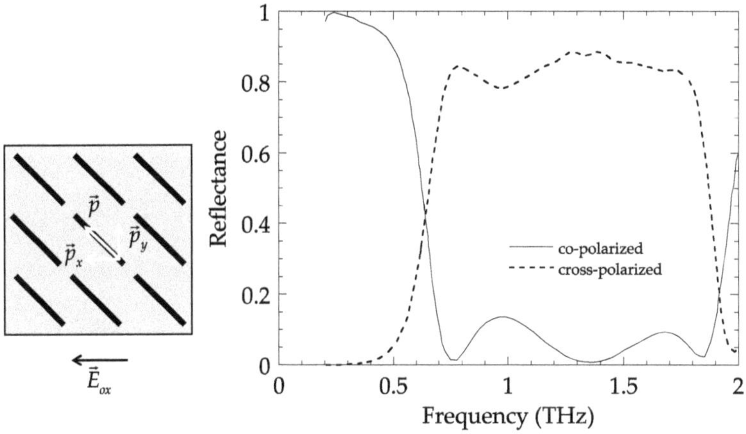

Figure 9.14 The reflectance spectrum of the 2D metamaterial shown on the left. Adapted from Grady et al. [57].

made of micrometer-pitch Al strips deposited on a Si wafer [56]. The curve was measured with a standard THz-TDS setup.

More complicated 2D patterns could lead to other birefringence effects. For example, Grady and coauthors [57] have designed a 2D metamaterial made of a periodic assembly of gold cut-wires deposited on a polyimide substrate (Fig. 9.14). The opposite face of the substrate is covered by a gold film and the device is studied in a reflection scheme. When the device is illuminated with an E field polarized in the x direction at 45° from the cut-wire axis, it induces a dipole in the cut wire, which is aligned along the axis of the wire. This dipolar vector has components parallel and perpendicular to the incoming E field.

The field reflected by the device is polarized according to the dipole direction, and thus it includes a component with the same polarization as the incoming field (co-polarized) and another one which is perpendicular (cross-polarized). The reflectance of the device (Fig. 9.14), measured by THz-TDS, shows that, between 0.7 and 1.9 THz, this device acts as a very efficient polarization converter. Moreover, around 0.6 THz, the amplitudes of the co- and cross-polarized reflected fields are equal while their phase difference is close to 90°, which means that the reflected beam is almost perfectly circularly polarized.

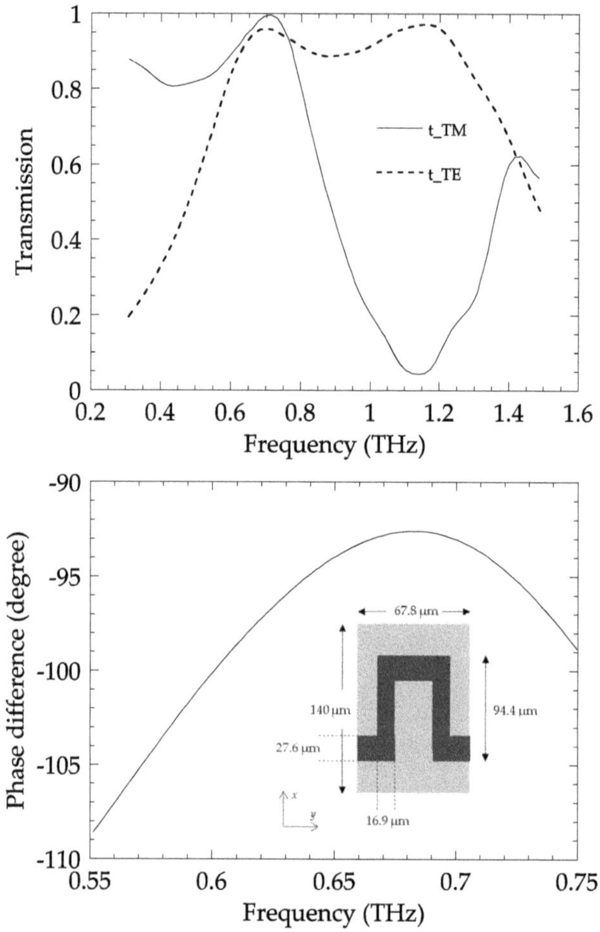

Figure 9.15 Top: Transmission spectrum and right phase difference spectrum of the 2D metamaterial shown in the inset (adapted from Strikwerda et al. [58]). TM means a polarization oriented along *x*. Bottom: Phase difference between the *x* and *y* polarized transmitted beams.

The last example given here is a meander structure [58], whose unit cell is shown in the inset of Fig. 9.15. Its electromagnetic response can be easily understood in terms of lumped element circuit. When the incoming light is polarized along *y*, the two vertical arms (along *x*) of the meander behave like a capacitor, while there is an inductance effect in the horizontal arms when the light is polarized along x. Therefore, the device shows a strong anisotropy; this is observed in the transmission spectra (Fig. 9.15), measured by THz-TDS. Transmission for the two orthogonal polarizations differ strongly; however, around 0.7 THz, the transmission values are however more or less equal. Moreover, the phase difference between the transmitted x and y polarized beams is close to 90°: hence, when illuminated with a linearly polarized beam (polarization direction at 45° from the x and y axes), the transmitted beam is circular around 0.7 THz and the transmission efficiency is close to 100%. At this frequency, this device serves as an efficient $\lambda/4$ waveplate.

It has to be noted that such a huge effect is achieved with a single metasurface, with an extremely small thickness compared to the THz wavelength. Normalized to the device thickness, these polarization phenomena in metamaterials are several orders of magnitude stronger than in natural materials.

9.6.3 Magnetic Effects with Metamaterials

Magnetism is achievable in metamaterials made of nonmagnetic materials. The microstructure must exhibit a magnetic dipolar moment, which could be easily obtained if the microstructure geometry supports current loops when illuminated. The first metamaterial with a magnetic response in the THz range was demonstrated in 2004 by Yen et al. [59]

Following this pioneering work, nearly all metamaterial schemes have been designed and modelled, and their electric/magnetic responses (effective permittivity and permeability) have been characterized. Figure 9.16 presents the unit cell of some popular 2D metallic patterns, which exhibit strong electric and magnetic responses. By an adequate selection of the dimensions, electric and magnetic resonances can be excited with an impinging THz wave. The pattern on the left in the upper row is the famous split ring resonator (SRR) that shows both an electrical resonance (dipolar electrical effect at the gap) and a magnetic one (dipolar magnetic momentum by the current induced in the ring). The drawing on the right side is a fractal pattern in which resonances at different frequencies can be excited [60].

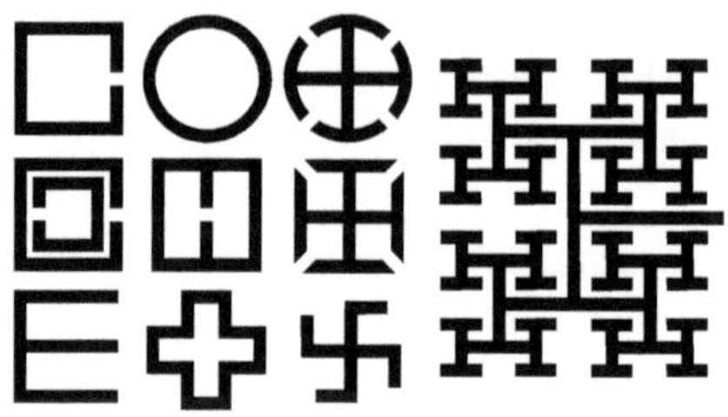

Figure 9.16 The unit cell of several popular 2D metallic metamaterials. The black lines represent the metal strips that are deposited over a substrate.

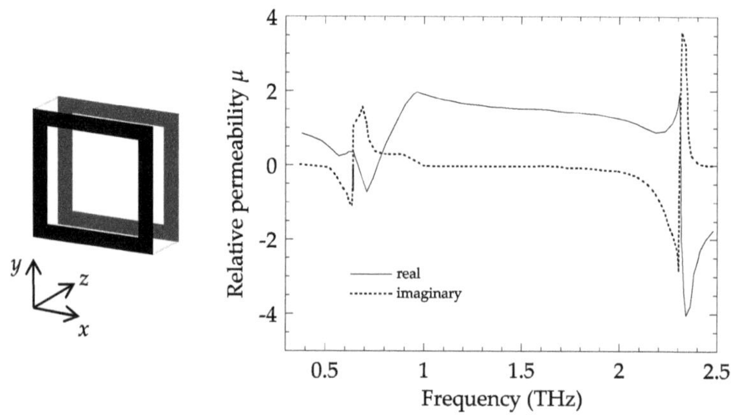

Figure 9.17 The spectrum of the relative permeability of a 2D metallic metamaterial, whose unit cell is drawn on the right side. Adapted from Cai and Shalaev [61].

Here, we select only two representative examples from the many seen in the literature. Figure 9.17 depicts the relative effective permittivity of a metamaterial made of patterns like the one shown on the left side of the figure [61]. Two metallic square rings are deposited face to face on each side of a thin Mylar film. An impinging THz beam, polarized along x, induces a current in the horizontal arms of the first ring. By coupling, a current flows in the opposite direction in the horizontal arms of the second ring. These two currents behave as a kind of horizontal loops, which give rise to the magnetic response of the metamaterial. The permeability of such a device has been determined from THz-TDS measurements. Both real and imaginary parts of the relative effective permeability are plotted versus frequency in Fig. 9.17 (right). In the range, $1 \sim 2.3$ THz, μ is larger than 1. Resonances around 0.75 and 2.3 THz are clearly seen.

9.6.4 Left-Handed Metamaterials

Maybe the most spectacular property of metamaterials is their left-handedness character, which leads to a negative refractive index (let us notice that such a negative index is not observed in natural materials, with the very rare exception being some multiferroics [62, 63]). Let us first remember some basic ideas about left-handed optical materials. For a plane wave propagating in a homogeneous medium (for the sake of simplicity, we suppose the medium being isotropic and nonconductive, i.e., $\sigma = 0$), the Maxwell equations (9.6) become

$$\vec{k} \times \vec{E} = \omega\,\mu_0\mu\vec{H}, \qquad \vec{k} \times \vec{H} = -\,\omega\,\varepsilon_0\varepsilon\vec{E} \tag{9.58}$$

$$\Rightarrow \quad \vec{k} \times \vec{k} \times \vec{E} = \vec{k}\left(\vec{k}\cdot\vec{E}\right) - k^2\vec{E} = -\frac{\omega^2}{c^2}\,\varepsilon\mu\vec{E} \tag{9.59}$$

As usual, ε and μ are the relative permittivity and permeability. Because $\vec{E}$ is perpendicular to $\vec{k}$, one gets

$$k^2 = \frac{\omega^2}{c^2}\,\varepsilon\mu \equiv \frac{\omega^2}{c^2}n^2 \quad \Leftrightarrow \quad n = \sqrt{\varepsilon\mu}. \tag{9.60}$$

Depending on the signs of ε and μ, there exist four different regimes of propagation defined in Fig. 9.18 (this figure is drawn for media without absorption).

When both ε and μ are negative, the material is said left handed. Indeed, the $\vec{E}, \vec{H}, \vec{k}$ set is direct for a plane wave travelling in common isotropic materials (Fig. 9.19), while it is indirect for left-handed materials through relations (9.58) in which ε and μ are negative.

In common right-handed materials, the wavevector $\vec{k}$ and the Poynting vector $\vec{P}$ (which gives the direction of electromagnetic energy flow) are directed along the same axis. In left-handed metamaterials, $\vec{k}$ and $\vec{P}$ are anti-parallel, which can be only described by making the refractive index n negative. This is mathematically possible because n is equal to $\sqrt{\varepsilon\mu}$ and thus can be either positive or negative.

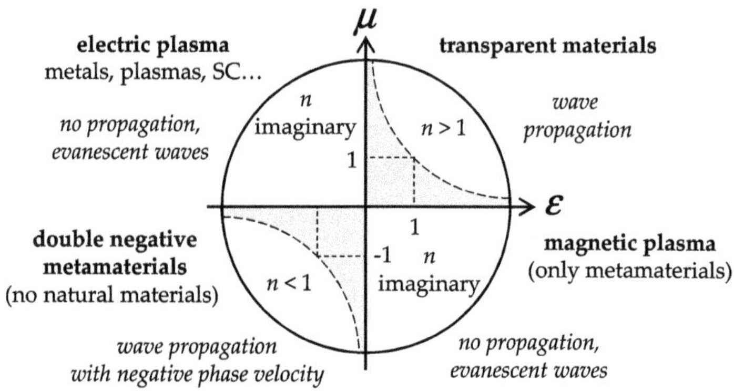

Figure 9.18 Chart of the four propagation regimes in the $\varepsilon - \mu$ space. For the sake of simplicity, the media are supposed to be lossless, leading to a refractive index n that is purely either real or imaginary. For absorbing materials, n is complex but mainly real or imaginary as indicated on the chart.

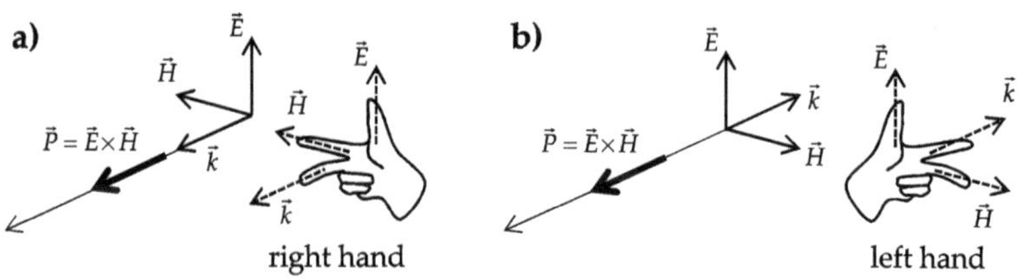

Figure 9.19 Propagation of a plane wave in (a) a common isotropic material (right-handed material) and (b) a left-handed metamaterial. $\vec{P}$ is the Poynting vector.

Left-handed optical materials exhibit amazing properties like negative refraction, inverse Doppler shift or Cherenkov radiation, evanescent waves with a "growing" amplitude, etc. These properties lead to astonishing applications: invisibility and cloaking, sub-wavelength focusing, super-lensing, etc. (see, e.g., [61]). Here, we will only treat left-handed metamaterials for the THz range, and how to characterize them by THz-TDS. Additional details on the physics and applications of these metamaterials is given in [52–54].

Obtaining simultaneously, i.e., at the same frequency, both ε and μ negative, can be realized only through resonances. Let us consider a dielectric resonance that obeys the classical Lorentz model:

$$\varepsilon = \varepsilon_r - j\varepsilon_i = \varepsilon_\infty + \frac{\omega_{pbe}^2}{\omega_o^2 - \omega^2 + j\omega\Gamma}, \tag{9.61}$$

$$\Rightarrow \quad \varepsilon_r = \varepsilon_\infty + \frac{\omega_{pbe}^2\left(\omega_o^2 - \omega^2\right)}{\left(\omega_o^2 - \omega^2\right)^2 + \omega^2\Gamma^2}, \qquad \varepsilon_i = \frac{\omega_{pbe}^2\omega\,\Gamma}{\left(\omega_o^2 - \omega^2\right)^2 + \omega^2\Gamma^2}. \tag{9.62}$$

ω_o is the resonance frequency, Γ a damping constant, ω_{pbe} a plasma-like frequency which accounts for the bound electron density, and ε_∞ is the value of ε at higher frequencies. Figure 9.20 presents the spectra of both real and imaginary parts (Eq. 9.62) of ε with $\omega_o = 5$ (normalized value), $\varepsilon_\infty = 10$, $\omega_{pbe} = 6$, and Γ varying from 0.2 up to 1. When Γ is small, the resonance feature of ε_r is not strong enough to make ε_r smaller than 0. It is what happens in natural materials. On the opposite, when $\Gamma = 1$, ε_r is negative in a given band with frequencies slightly higher than ω_o, and whose bandwidth depends on Γ, ε_∞ and ω_{pbe}: this is what could be observed in metamaterials. Similarly, magnetic permeability shows resonances that could be described by a Lorentz model [64], and thus a negative μ can be observed in metamaterials just after the resonance peak.

Negative permittivity or permeability could be observed nearby a resonance in some natural compounds, like ferroelectrics or ferromagnetics [63]. However, the magnetic effects occur preferentially up to the GHz range, and very rarely at THz frequencies (MnF_2, FeF_2, etc.). This is the great advantage of metamaterials: a properly designed unit cell pattern may exhibit a strong dipolar magnetic moment, for example when its shape is similar to a loop, as proposed initially by Pendry [65]. Moreover, the same metamaterial could be electrically sensitive, by inserting in the pattern an electric dipole-like structure. By choosing the right dimensions of the pattern details, both electric and magnetic effects could be resonant, i.e., very effective, at nearly the same frequency, and thus the metamaterial shows left-handed properties.

THz-TDS is an excellent technique to characterize left-handed metamaterials because it gives simultaneously the amplitude and the phase of the transmitted/reflected fields. The effect of a negative refractive index appears directly in the phase spectrum (we recall

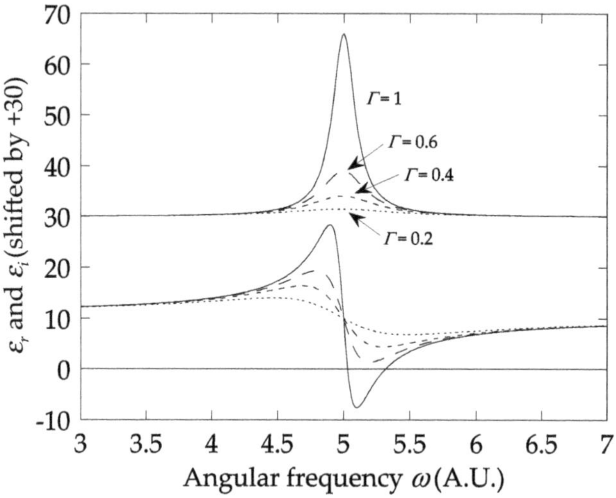

Figure 9.20 Real (lower curve) and imaginary (upper curve) parts of the relative permittivity versus frequency, calculated using a Lorentz model with $\omega_o = 5$ (normalized value), $\varepsilon_\infty = 10$, $\omega_{pbe} = 6$, and Γ varying from 0.2 up to 1.

the reader that in a left-handed material, the phase seems to oscillate in the direction opposite to the flow of energy). For example, Fig. 9.21 depicts both amplitude and phase of an SRR 2D metamaterial, which has been measured using THz-TDS [66]. Here transmission is defined as the ratio of the Fourier-transformed spectra of the sample to the reference, which is a blank slab identical to the substrate of the SRRs. The resonances appear as dips in the amplitude spectrum, from which the left-handed character cannot be determined. On the other hand, the regions of the negative index are clearly identified in the phase spectrum. Generally, the refractive index and the coefficient of absorption are retrieved either with similar methods as the ones for common materials (see Chapter 4) [67] or by a fit of the transmission curve (amplitude and phase) with a numerical model.

Another direct way of experimentally show the negative index is to record the angular deflection behavior of a metamaterial prism. The angle of deviation D_p by a prism (see Fig. 9.22, left column), whose angle is α_p, is given by (2.26). Here we suppose that the THz

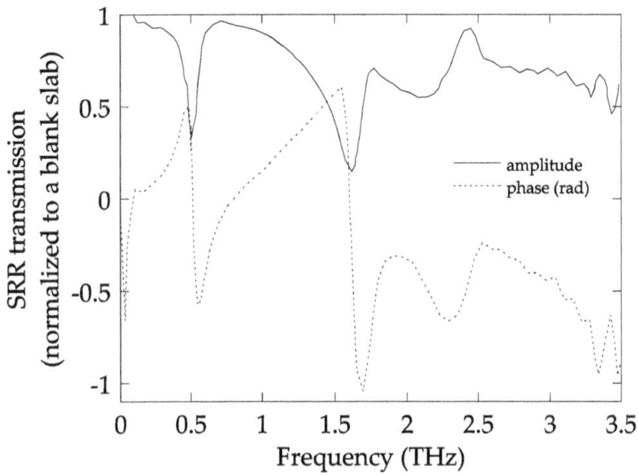

Figure 9.21 Transmission spectrum (amplitude and phase) of an SRR metamaterial measured by THz-TDS. Adapted from Gu et al. [66].

impinges normally to the entrance face of the prism and we get

$$D_p = \arcsin\left(n \sin \alpha_p\right) \tag{9.63}$$

Thus, depending on the sign of n, the deviation will be positive or negative. This is illustrated in Fig. 9.22 adapted from [68]. The prism is made of a stack of hole arrays drilled in metal sheets. The THz beam delivered by a THz-TDS set up is almost Gaussian and is deviated by the prism. The plot in the right column shows the measured beam profile (full circles) at different frequencies versus the angle between the optical axis and the direction of detection. The vertical line gives the "zero" angle. For 454 and 475 GHz, this angle is negative, which is the signature of the left-handed behavior, while the angle is positive for frequencies above 496 GHz. The open circles correspond to the incident beam profile whose deviation has been calculated using Eq. 9.63. The only unknown parameter in Eq. 9.63 is the refractive index n, which can be retrieved through the best fitting of the experimental data.

Figure 9.23 (top) shows the refractive index (real n and imaginary κ parts) of the SRR metamaterial pictured in Fig. 9.17. Here the left-handed character arises from a dipolar resonance along one side of the square rings, and from an inductive (magnetic)

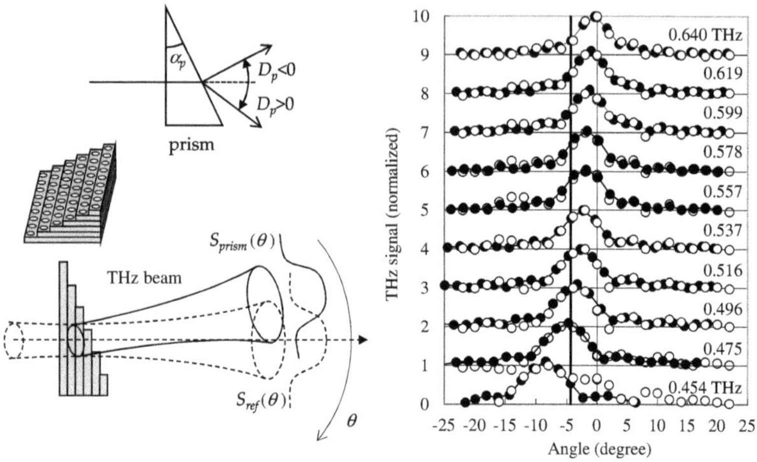

Figure 9.22 Metamaterial prism made of a stack of fishnet layers. The left column shows the device, the positive and negative deviation angles, and the deviation of a Gaussian beam. The experimental angular behavior of the deflected beam is plotted on the right side for different frequencies.

resonance between the two rings. The dimensions of the rings are optimized to observe the two resonances at almost the same frequency. It should be noted that (i) the left-handed character is achieved with only one single layer of sub-wavelength structures, (ii) because of causality, the loss in the metamaterial increases nearby the resonance, thus the spectral range of negative index is absorbing, (iii) the left-handed character, due to its resonant origin, is narrow band. Besides SRR devices, another popular and efficient design to achieve a left-handed behavior is a stack of metallic sheets perforated with an array of holes [69]. The incoming THz field induces a polarization of each hole (electric resonance) of the first sheet, i.e., opposite free charges of the metal are gathered at opposite edges of the hole. The movement of these charges around the hole creates a current. By induction, another current flows in the opposite direction in the second sheet. These two currents could be seen as the one circulating in a loop perpendicular to the sheets, giving a magnetic response to the device. By optimizing the size of the holes and the gap between two sheets, the magnetic response could also be resonant at the same frequency as the dipolar one. Figure 9.23

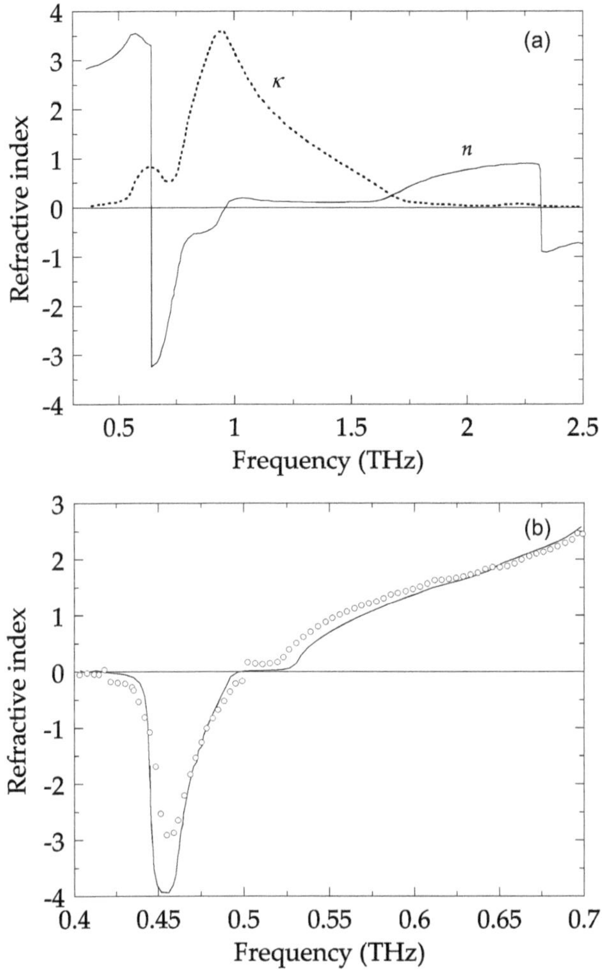

Figure 9.23 Top: n and κ of the SRR metamaterial pictured in Fig. 9.17 measured versus frequency (adapted from Gu et al. [66]). Bottom: n versus frequency (open circles: experimental data, continuous line: simulation) for a fishnet-like metamaterial (adapted from Wang et al. [70]).

(bottom) presents the refractive spectrum of such a metamaterial [70], in which a polymer material helps in making shorter and constant the gap between sheets. Here again, the effect is quite pronounced ($n = -3$ at 0.46 THz) and narrow band ($\sim$50 GHz).

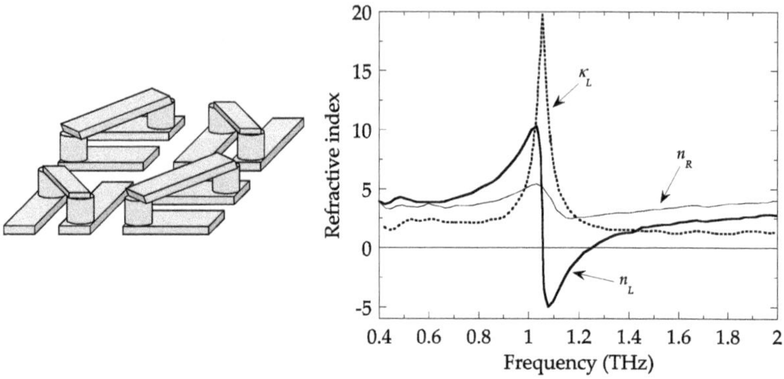

Figure 9.24 Chiral 2D metamaterial adapted from Zhang et al. [71]. Left: Schematic drawing of a unit cell. Right: Refractive index versus frequency, extracted from THz-TDS data, for a left-circularly polarization (n_L and κ_L in bold lines) and a right-circularly one (only n_R, narrow line).

Agreement between experimental data, extracted from THz-TDS measurements, and the numerical modeling is excellent.

As already explained, metamaterials with unit cells that do not exhibit mirror symmetries have a chiral behavior. Many chiral structures have been proposed and demonstrated, for example see [71, 75]. In short, there are two possibilities: either each unit cell is chiral, i.e., their pattern does not show mirror symmetries, or the cell is achiral, but is coupled with a second stack of cells that are twisted as compared to the first stack. A specimen [71] of the chiral pattern is presented in Fig. 9.24 (left). Thanks to air bridges, the upper part of the pattern is rotated by 45° as compared to the main directions of the lower part. Fishnet structures are an example an achiral cell: chirality of the fishnet metamaterial is obtained by rotating the pattern of the second stack as compared to the first one. This is illustrated by the drawing in Fig. 9.24 (left) [75]: here the fishnet is made of a stack of metallic sheets perforated by elliptical holes. The ellipses are rotated by 45° from one layer to the next one.

The great advantage of metamaterials over natural materials is their superior polarization response. In the case of chirality, the rotatory power ρ (9.32) could be very large, in such a way $\Delta n > n_o$ (9.29). Therefore, the refractive index $n_R = n_-$ of the right-circularly

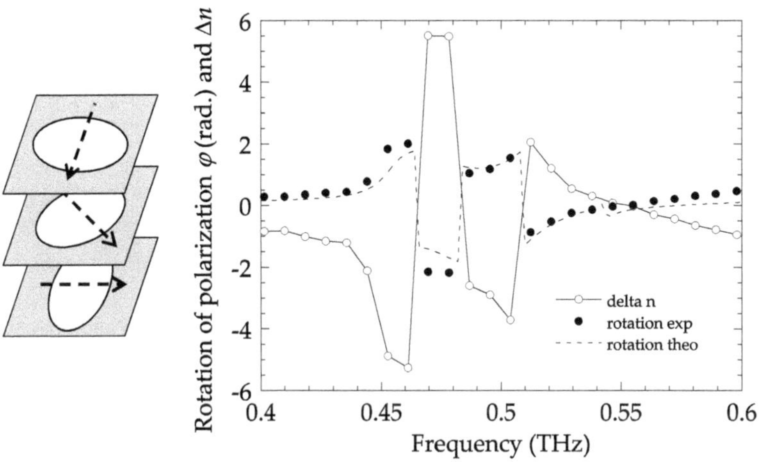

Figure 9.25 Fishnet chiral metamaterial adapted from Coutaz et al. [68] and Wang et al. [75]. Left: Schematic drawing of a unit cell. Right: Angle of rotation ϕ of the polarization (full circles: experimental data, dashed line: simulation) and Δn extracted from the experimental THz-TDS data (open circles: the continuous line is a guide to the eye) versus frequency.

polarized wave could be negative, while the left one $n_L = n_+$ is positive. Thus chirality leads to a left-handed character [75–76]. This is shown in Fig. 9.24, where the spectra of n_L, κ_L and n_R are plotted [71]. The data have been derived from THz-TDS measurements. Around 1.1 THz, n_L is strongly negative while n_R remains positive. However, due to the resonance-based chirality, losses for the right-circularly polarized wave are quite large at the same frequencies.

Figure 9.25 shows the angle of rotation ϕ of a linearly polarized beam crossing the fishnet material whose unit cell is depicted on the left side. Full circles are data measured by THz-TDS, while the dashed line is simulated. In the vicinity of resonance, φ is as big as $\pm\pi/2$, which corresponds to a rotational power of $\sim$1000° per wavelength, i.e., about 2.5×10^4 times the one of natural quartz at THz frequencies. Δn, extracted from these polarization measurements, is also plotted in the same figure. Its maximum value is 6, which explained the left-handed character of this chiral metamaterial.

References

1. Oppeneer, P. M. (2001). Magneto-optical Kerr spectra, in *Handbook of Magnetic Materials*, Vol. **13**, Buschow, K. H. J., ed. (Elsevier), pp. 229–422.
2. Landry, G. D., and Maldonado, T. A. (1995). Complete method to determine transmission and reflection characteristics at a planar interface between arbitrarily oriented biaxial media, *J. Opt. Soc. Am. A*, **12**, p. 2048.
3. Roberts, D. A. (1992). Simplified characterization of uniaxial and biaxial nonlinear optical crystals: a plea for standardization of nomenclature and conventions, *IEEE J. Quantum Electron.*, **28**, pp. 2057–2074.
4. Flack, H. D. (2003). Chiral and achiral crystal structures, *Helv. Chem. Acta*, **86**, p. 905.
5. Landau, L., and Lifshitz, E. (1960). *Electrodynamics of Continuous Media* (Pergamon Press, London, UK).
6. Biot, J.-B. (1812). Expériences sur les plaques de cristal de roche taillées perpendiculairement à l'axe de cristallisation (Experiments with slabs of rock crystal whose optical axis is normal to the slab surface), *Memoires de l'Institut des sciences, lettres et arts*, **13**, p. 218.
7. Duvillaret, L., Garet, F., and Coutaz, J.-L. (1999). Evidence of frequency dependent THz beam polarization in time-domain spectroscopy, in *THz Spectroscopy and Applications*, Sherwin, M. S., ed., *Proc. SPIE*, **3617**, pp. 30–37.
8. Nemoto, N., Higuchi, T., Kanda, N., Konishi, K., and Kuwata-Gonokami, M. (2014). Highly precise and accurate terahertz polarization measurements based on electro-optic sampling with polarization modulation of probe pulses, *Opt. Express*, **22**, p. 17915.
9. Castro-Camus, E. (2012). Polarization-resolved terahertz time-domain spectroscopy, *J. Infrared Millimeter THz Waves*, **33**, pp. 418–430.
10. Ikebe, Y., and Shimano, R. (2008). Characterization of doped silicon in low carrier density region by terahertz frequency Faraday effect, *Appl. Phys. Lett.*, **92**, p. 012111.
11. Sartorius, B., Roehle, H., Künzel, H., Böttcher, J., Schlak, M., Stanze, D., Venghaus, H., and Schell, M. (2008). All-fiber terahertz time-domain spectrometer operating at 1.5 μm telecom wavelengths, *Opt. Express*, **16**, pp. 9565–9570.

12. Goy, C., Scheller, M., Scherger, B., Wallace, V. P., and Koch. M. (2013). Terahertz waveguide prism, *Opt. Express*, **21**, p. 19292.

13. Neshat, M., and Armitage, N. P. (2012). Terahertz time-domain spectroscopic ellipsometry: instrumentation and calibration, *Opt. Express*, **20**, p. 29063.

14. Sato, M., Higuchi, T., Kanda, N., Konishi, K., Yoshioka, K., Suzuki, T., Misawa, K., and Kuwata-Gonokami, M. (2013). Terahertz polarization pulse shaping with arbitrary field control, *Nat. Photonics*, **7**, pp. 724–731.

15. Gaborit, G., Biciunas, A., Bernier, M., and Coutaz, J.-L. (2105). Emitting and receiving terahertz vectorial antennas based on cubic electro-optic crystals, *IEEE THz Sci. Technol.*, **5**, p. 828.

16. van der Valk, N. C. J., van der Marel, W. A., and Planken, P. C. M. (2005). THz polarization imaging, *Opt. Lett.*, **30**, pp. 2802–2804.

17. Yasumatsu, N., Kasatani, A., Oguchi, K., and Watanabe, S. (2014). High-speed terahertz time-domain polarimeter based on an electro-optic modulation technique, *Appl. Phys. Express*, **7**, p. 092401.

18. Hussain, A., and Andrews, S. R. (2008). Ultrabroadband polarization analysis of terahertz pulses, *Opt. Express*, **16**, p. 7251.

19. Makabe, H., Hirota, Y., Tani, M., and Hangyo, M. (2007). Polarization state measurement of terahertz electromagnetic radiation by three-contact photoconductive antenna, *Opt. Express*, **15**, p. 11650.

20. Castro-Camus, E., Lloyd-Hughes, J., Johnston, M. B., Fraser, M. D., Tan, H. H., and Jagadish, C. (2005). Polarization-sensitive terahertz detection by multicontact photoconductive receivers, *Appl. Phys. Lett.*, **86**, p. 254102.

21. Dong, H., Gong, Y. D., Paulose, V., and Hong, M. H. (2009). Polarization state and Mueller matrix measurements in terahertz-time domain spectroscopy, *Opt. Commun.*, **282**, p. 3671.

22. Shan, J., Dadap, J. I., and Heinz, T. F. (2009). Circularly polarized light in the single-cycle limit: the nature of highly polychromatic radiation of defined polarization, *Opt. Express*, **17**, p. 7431.

23. Castro-Camus, E., and Johnston, M. B. (2009). Extraction of the anisotropic dielectric properties of materials from polarization-resolved terahertz time-domain spectra, *J. Opt. A: Pure Appl. Opt.*, **11**, p. 105206.

24. Aschaffenburg, D. J., Williams, M. R. C., Talbayev, D., Santavica, D. F., Prober, D. E., and Schmuttenmaer, C. A. (2012). Efficient measurement of broadband terahertz optical activity, *Appl. Phys. Lett.*, **100**, p. 241114.

25. Grischkowsky, D., Keiding, S., Van Exter, M., and Fattinger, C. (1990). Far infrared time_domain spectroscopy with terahertz beams of dielectrics and semiconductors, *J. Opt. Soc. Am. B*, **7**, pp. 2006–2015.
26. Schall, M., Helm, H., and Keiding, S. R. (1999). Far infrared properties of electro-optic crystals measured by THz time-domain spectroscopy, *Int. J. Infrared Millimeter Waves*, **20**, pp. 595–604.
27. Baxter, J. B., and Schmuttenmaer, C. A. (2006). Conductivity of ZnO nanowires, nanoparticles, and thin films using time-resolved terahertz spectroscopy, *J. Phys. Chem. B*, **110**, p. 25229.
28. Ree, M., Chen, K.-J., Kirby, D. P., Katzenellenbogen, N., and Grischkowsky, D. (1992). Anisotropic properties of high-temperature polyimide thin films: Dielectric and thermal-expansion behaviors, *J. Appl. Phys.*, **72**, pp. 2014–2021.
29. Lim, M., Choi, S.-J., Lee, G.-S., Seol, M.-L., Do, Y., Choi, Y.-K., and Han, H. (2012). Terahertz time-domain spectroscopy of anisotropic complex conductivity tensors in silicon nanowire films, *Appl. Phys. Lett.*, **100**, p. 211102.
30. Singh, R., George, D. K., Benedict, J. B., Korter, T. M., and Markelz, A. G. (2012). Improved mode assignment for molecular crystals through anisotropic terahertz spectroscopy, *J. Phys. Chem. A*, **116**, pp. 10359–10364.
31. Unuma, T., Umemoto, A., and Kishida, H. (2013). Anisotropic terahertz complex conductivities in oriented polythiophene films, *Appl. Phys. Lett.*, **103**, p. 213305.
32. Jördens, C., Scheller, M., Wietzke, S., Romeike, D., Jansen, C., Zentgraf, T., Wiesauer, K., Reisecker, V., and Koch, M. (2010). Terahertz spectroscopy to study the orientation of glass fibres in reinforced plastics, *Compos. Sci. Technol.*, **70**, pp. 472–477.
33. Tanabe, T., Watanabe, K., Oyama, Y., and Seo, K. (2010). Polarization sensitive THz absorption spectroscopy for the evaluation of uniaxially deformed ultra-high molecular weight polyethylene, *NDT Int.*, **43**, pp. 329–333.
34. Bychanok, D. S., Kanygin, M. A., Okotrub, A. V., Shuba, M. V., Paddubskaya, A. G., Pliushch, A. O., Kuzhir, P. P., and Maksimenko, S. A. (2011). Anisotropy of the electromagnetic properties of polymer composites based on multiwall carbon nanotubes in the gigahertz frequency range, *JETP Lett.*, **93**, p. 607.
35. Reid, M., and Fedosejevs, R. (2006). Terahertz birefringence and attenuation properties of wood and paper, *Appl. Opt.*, **45**, p. 2766.

36. Tanaka, S., Shiraga, K., Ogawa, Y., Fujii, Y., and Okumura, S. (2014). Effect of pore conformation on dielectric anisotropy of oven-dry wood evaluated using terahertz time-domain spectroscopy and eigenvalue problems for two-dimensional photonic crystals, *J. Wood Sci.*, **60**, pp. 194–200.
37. Jördens, C., Scheller, M., Wichmann, M., Mikulics, M., Wiesauer, K., and Koch, M. (2009). Terahertz birefringence for orientation analysis, *Appl. Opt.*, **48**, pp. 2037–2044.
38. Turchinovich, D., Knobloch, P., Luessem, G., and Koch, M. (2001). THz time-domain spectroscopy on 4-trans-4-pentylcyclohexyl-benzonitril, in *Liquid Crystals V*, Khoo, I. C., ed., *Proc. SPIE*, **4463**, pp. 65–70.
39. Tsai, T.-R., Chen, C.-Y., Pan, C.-L., Pan, R.-P., and Zhang, X.-C. (2003). Terahertz time-domain spectroscopy studies of the optical constants of the nematic liquid crystal 5CB, *Appl. Opt.*, **42**, pp. 2372–2376.
40. Liang, Q., Wang, S., Tao, X., and Dekorsy, T. (2014). Optical properties of $LiInSe_2$ in the THz frequency regime, *Opt. Mater. Express*, **4**, p. 1336.
41. Hofmann, T., Schmidt, D., Boosalis, A., Kühne, P., Skomski, R., Herzinger, C. M., Woollam, J. A., Schubert, M., and Schubert, E. (2011). THz dielectric anisotropy of metal slanted columnar thin films, *Appl. Phys. Lett.*, **99**, p. 081903.
42. Kleemann, W. (2007). Magneto-optical materials, in *Handbook of Magnetism and Advanced Magnetic Materials*, Vol. 4: Novel Materials, Kronmuller, H., and Parkin, S., eds. (John Wiley & Sons, Chichester, UK).
43. Faraday, M. (1846). On the magnetization of light and the illumination of magnetic lines of force, *Phil. Trans. R. Soc.*, **126**, pp. 1–20.
44. Shimano, R., Ino, Y., Svirko, Y. P., and Kuwata-Gonokami, M. (2002). Terahertz frequency Hall measurement by magneto-optical Kerr spectroscopy in InAs, *Appl. Phys. Lett.*, **81**, p. 199.
45. Chen, S., Fan, F.,Chang, S., Miao, Y., Chen, M., Li, J., Wang, X., and Lin, L. (2014). Tunable optical and magneto-optical properties of ferrofluid in the terahertz regime, *Opt. Express*, **22**, p. 6313.
46. Shalaby, M., Peccianti, M., Ozturk, Y., Clerici, M., Al-Naib, I., Luca Razzari, L., Ozaki, T., Mazhorova, A., Skorobogatiy, M., and Morandotti, R. (2012). Terahertz Faraday rotation in a magnetic liquid: High magneto-optical figure of merit and broadband operation in a ferrofluid, *Appl. Phys. Lett.*, **100**, p. 241107.
47. Zhang, P., Su, F., Zhang, S., Mei, H., Zhang, C., Luo, X., Dai, J., and Pi, L. (2015). Terahertz magnetic circular dichroism induced by exchange resonance in $CoCr_2O_4$ single crystal, *Opt. Express*, **23**, p. 17805.

48. Kubo, R. (1957). Statistical-mechanical theory of irreversible processes. I. General theory and simple applications to magnetic and conduction problems, *J. Phys. Soc. Jpn.*, **12**, pp. 570–586.
49. Parrott, E. P. J., Zeitler, J. A., Friscic, T., Pepper, M., Jones, W., Day, G. M., and Gladden, L. F. (2009). Testing the sensitivity of terahertz spectroscopy to changes in molecular and supramolecular structure: a study of structurally similar cocrystals, *Cryst. Growth Des.*, **9**, pp. 1452–1460.
50. Du, Y., Liu, J., and Hong, Z. (2013). Characteristics of chiral and racemic ketoprofen drugs using terahertz time-domain spectroscopy, *Proc. SPIE, Intern. Symp. Photoelectronic Detection and Imaging 2013: THz Technol. and Appl.*, **8909**, 89090M.
51. Franz, M., Fischer, B. M., and Walther, M. (2011). Probing structure and phase-transitions in molecular crystals by terahertz time-domain spectroscopy, *J. Mol. Struct.*, **1006**, pp. 34–40.
52. Averitt, R. D., Padilla, W. J., Chen, H. T., O'Hara, J. F., Taylor, A. J., Highstrete, C., Lee, M., Zide, J. M. O., Bank, S. R., and Gossard, A. C. (2007). Terahertz metamaterial devices, *Proc. of SPIE, Terahertz Physics, Devices, and Systems II*, Anwar, M., DeMaria, A. J., and Shur, M. S., eds., **6772**, p. 677209.
53. Withayachumnankul, W., and Abbott, D. (2009). Metamaterials in the terahertz regime, *IEEE Photonics J.*, **1**, pp. 99–118.
54. Reinhard, B., Paul, O., and Rahm, M. (2013). Metamaterial-based photonic devices for terahertz technology, *IEEE J. Sel. Top. Quantum Electron.*, **19**, pp. 8500912–8500912.
55. Emerson, D. T. (1997). The work of Jagadis Chandra Bose: 100 years of millimeter-wave research, *IEEE Trans. Microwave Theory Tech.*, **45**, p. 2267.
56. Yamada, I., Takano, K., Hangyo, M., Saito, M., and Watanabe, W. (2009). Terahertz wire-grid polarizers with micrometer-pitch Al gratings, *Opt. Lett.*, **34**, p. 274.
57. Grady, N. K., Heyes, J. E., Chowdhury, D. R., Zeng, Y., Reiten, M. T., Azad, A. K., Taylor, A. J., Dalvit, D. A. R., and Chen, H.-T. (2013). Terahertz metamaterials for linear polarization conversion and anomalous refraction, *Science*, **340**, pp. 1304–1307.
58. Strikwerda, A. C., Fan, K., Tao, H., Pilon, D. V., Zhang, X., and Averitt, R. D. (2009). Comparison of birefringent electric split-ring resonator and meanderline structures as quarter-wave plates at terahertz frequencies, *Opt. Express*, **17**, pp. 136–149.

59. Yen, T. J., Padilla, W. J., Fang, N., Vier, D. C., Smith, D. R., Pendry, J. B., Basov, D. N., and Zhang, X. (2004). Terahertz magnetic response from artificial materials, *Science*, **303**, pp. 1494–1496.

60. Miyamaru, F., Saito, Y., Takeda, M. W., Hou, B., Liu, L., Wen, W., and Sheng, P. (2008). Terahertz electric response of fractal metamaterial structures, *Phys. Rev. B*, **77**, p. 045124.

61. Cai, W., and Shalaev, V. (2010). *Optical Metamaterials: Fundamentals and Applications* (Springer Science+Business Media, New York).

62. Pimenov, A., Loidl, A., Przyslupski, P., and Dabrowski, B. (2007). Negative refraction in ferromagnet-superconductor superlattices, *Phys. Rev. Lett.*, **98**, p. 197401.

63. Gevorkian, S. (2009). *Ferroelectrics in Microwave Devices, Circuits and Systems: Physics, Modeling, Circuits and Systems* (Springer-Verlag, London, UK).

64. Boardman, A. D., and Marinov, K. (2006). Electromagnetic energy in a dispersive metamaterial, *Phys. Rev. B*, **73**, p. 165110.

65. Pendry, J. B., Holden, A. J., Robbins, D. J., and Stewart, W. J. (1999). Magnetism from conductors and enhanced non-linear phenomena, *IEEE Trans. Microwave Theory Tech.*, **47**, p. 2075.

66. Gu, J., Han, J., Lu, X., Singh, R., Tian, Z., Xing, Q., and Zhang, W. (2009). A close-ring pair terahertz metamaterial resonating at normal incidence, *Opt. Express*, **17**, pp. 20307–20312.

67. Han, J. (2008). Probing negative refractive index of metamaterials by terahertz time-domain spectroscopy, *Opt. Express*, **16**, p. 1354.

68. Coutaz, J.-L., Bernier, M., Garet, F., Joly, S., Miyake, Y., Minamide, H., Lheurette, E., and Lippens, D. (2014). Terahertz time-domain spectroscopy of absorbing materials and of metamaterials, *Int. J. Terahertz Sci. Technol.*, **7**, pp. 53–69.

69. Beruete, M., Sorolla, M., and Campillo, I. (2006). Left-handed extraordinary optical transmission through a photonic crystal of subwavelength hole arrays, *Opt. Express.*, **14**, pp. 5445–5455.

70. Wang, S., Garet, F., Blary, K., Croënne, C., Lheurette, E., Coutaz, J.-L., and Lippens, D. (2011). Composite left/right-handed stacked hole arrays at submillimeter wavelengths, *J. Appl. Phys.*, **107**, p. 074510.

71. Zhang, S., Park, Y.-S., Li, J., Lu, X., Zhang, W., and Zhang, X. (2009). Chiral negative-index metamaterials in terahertz, *Phys. Rev. Lett.*, **102**, p. 02390.

72. Yogesh, N., Fu, T., Lan, F., and Ouyang, Z. (2015). Far-infrared circular polarization and polarization filtering based on Fermat's spiral chiral metamaterial, *IEEE Photonics J.*, **7**, p. 4600212.
73. Wu, J., Ng, B., Turaga, S. P., Breese, M. B. H., Maier, S. A., Hong, M., Bettiol, A. A., and Moser, H. O. (2013). Free-standing terahertz chiral meta-foils exhibiting strong optical activity and negative refractive index, *Appl. Phys. Lett.*, **103**, p. 141106.
74. Li, Z., Alici, K. B., Caglayan, H., Kafesaki, M., Soukoulis, C. M., and Ozbay, E. (2012). Composite chiral metamaterials with negative refractive index and high values of the figure of merit, *Opt. Express*, **20**, p. 6146.
75. Wang, S., Garet, F., Lheurette, E., Astic, M., Coutaz J.-L., and Lippens, D. (2013). Giant rotary power of a fishnet-like metamaterial, *APL Mater.*, **1**, p. 032116.
76. Tretyakov, S., Nefedov, I., Sihvola, A., Maslovski, S., and Simovski, C. (2003). Waves and energy in chiral nihility, *J. Electromagn. Waves Appl.*, **17**, p. 695.
77. Pendry, J. B. (2004). A chiral route to negative refraction, *Science*, **306**, p. 1353.

Chapter 10

THz Characterization of Heterogeneous Materials

10.1 Introduction

Many intended and practical applications of the THz waves involve the study of inhomogeneous materials like powder mixtures (explosives, drugs, medical tablets, etc.), multilayered samples (paints), micro-structured layers or surfaces (fabrics, carbon composites, etc.), and so on. In these materials, characterization in terms of determining the refractive index, absorption, or complex permittivity, should be viewed with caution. Indeed, depending on the size of the constituent parts in such composites materials, different phenomena can occur like scattering or diffraction of THz waves, which makes the characterization challenging. In such cases, losses can be greatly modified (most of the time increased), leading to an overestimation of the absorption if retrieved using methods like Kramer–Kronig analysis (see Chapter 13) or common numerical THz-TDS algorithms as discussed in Chapters 5 and 6, as those methods do not take into account these effects. More precisely, most of the characterization experiments in the THz regime are obtained through THz-TDS or FTIR transmission (or

Principles of Terahertz Time-Domain Spectroscopy
Jean-Louis Coutaz, Frédéric Garet, and Vincent Wallace

ISBN 978-981-4774-56-7 (Hardcover), 978-0-203-70969-6 (eBook)
www.panstanford.com

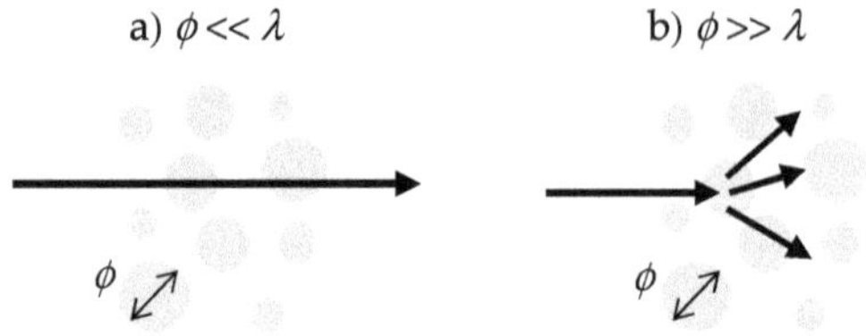

Figure 10.1 Propagation of an electromagnetic wave in a mixture of materials: (a) case of the homogeneous mixture (effective medium) and (b) case of inhomogeneous material (scattering).

reflection) measurements (see Chapter 2) that only measure the beam transmitted by the sample in the direction of the incoming beam (or in the specular direction). Any part of the THz beam that deviates will not reach the point detector and thus is not measured and is wrongly considered as lost due to absorption. Thus, in the case of scattering, for example, losses do not account only for the absorption of the material but also for the scattering "losses." The total attenuation of the signal is the sum of absorption and scattering. The extracted loss (or attenuation) coefficient is not the absorption coefficient but the so-called extinction coefficient, α_{ext}:

$$\alpha_{ext} = \alpha_{abs} + \alpha_{scat}. \tag{10.1}$$

The refractive index calculation is also affected by this scattering and diffraction, but to a much lower extent. Moreover, due to the wide range of the THz wavelengths, from $\lambda = 30$ μm to 3 mm, the behavior of mixtures or composites should be considered in different ways, depending on the size of the inhomogeneities compared with the wavelength. In mixtures, depending on the size ϕ and concentration of the particles, the material response can either be modeled using effective medium approaches (EMA) if, on the macroscopic level, the material can be assumed to be homogeneous, typically when $\phi \ll \lambda$, or using scattering theories, for $\phi \sim\lambda$ or $\phi \gg \lambda$. In the first case, the propagation of an electromagnetic wave in the material is not impacted by scattering ($\alpha_{scat} = 0$), whereas it is no more the case ($\alpha_{scat} \neq 0$) when the size of the particles increases and thus the material behaves as inhomogeneous (see Fig. 10.1).

In this chapter, we will illustrate the different behaviors encountered in samples made of different materials (mixtures, micro-structured or composite materials) for two common families: pressed pellets of powder and fabrics. Let us notice that we mainly focus on materials presenting a random repartition of the homogeneities. Thereby, depending on the size of the powder grains, pressed powders can either be considered as homogeneous or scattering materials, whereas the materials such as fabrics can also produce diffraction phenomena due to the periodicity of the arrangement of the thread.

10.2 Homogeneous Mixtures of Dielectric Materials

10.2.1 Effective Medium Approaches (EMA)

10.2.1.1 Maxwell–Garnett EMA

In a composite material, if the size of the inclusions is much smaller than the wavelength of interest (Fig. 10.1a), scattering effects are negligible and quasi-static modeling perfectly predicts the electromagnetic behavior of such mixtures. Several EMAs have been used in the THz domain to describe the response of materials like particles embedded in a host material, mixtures of different powders, natural inhomogeneous materials, and so on [1–3]. EMA models are used to bypass the rigorous problem that considers each particle of a mixture.

One of the most popular EMA is the Maxwell–Garnett (MG) theory [4] that expresses the effective complex permittivity $\tilde{\varepsilon}_{\text{eff}}$ of a mixture composed of a volume fraction k_{p} of particles, whose complex permittivity is $\tilde{\varepsilon}_{\text{p}}$, embedded in a host matrix of permittivity $\tilde{\varepsilon}_{\text{m}}$. $\tilde{\varepsilon}_{\text{eff}}$ obeys the following relation:

$$\frac{\tilde{\varepsilon}_{\text{eff}} - \tilde{\varepsilon}_{\text{m}}}{\tilde{\varepsilon}_{\text{eff}} + 2\tilde{\varepsilon}_{\text{m}}} = k_{\text{p}} \frac{\tilde{\varepsilon}_{\text{p}} - \tilde{\varepsilon}_{\text{m}}}{\tilde{\varepsilon}_{\text{p}} + \tilde{\varepsilon}\varepsilon_{\text{m}}}, \tag{10.2}$$

which leads to

$$\tilde{\varepsilon}_{\text{eff}} = \tilde{\varepsilon}_{\text{m}} \frac{\tilde{\varepsilon}_{\text{p}} \left(1 + 2k_{\text{p}}\right) + 2\tilde{\varepsilon}_{\text{m}} \left(1 - k_{\text{p}}\right)}{\tilde{\varepsilon}_{\text{p}} \left(1 - k_{\text{p}}\right) + \tilde{\varepsilon}_{\text{m}} \left(2 + k_{\text{p}}\right)}. \tag{10.3}$$

Relation 10.2 is derived from the Clausius–Mossotti equation that links the relative permittivity $\tilde{\varepsilon}_{\text{eff}}$ of a material composed of D_{p}

spherical particles p of radius a_p per unit volume (D_p is the particle number density), whose polarizability is $\tilde{\alpha}_\mathrm{p}$, embedded in a host material of permittivity $\tilde{\varepsilon}_\mathrm{m}$ [5, 6]:

$$\frac{\tilde{\varepsilon}_\mathrm{eff}-\tilde{\varepsilon}_\mathrm{m}}{\tilde{\varepsilon}_\mathrm{eff}+2\tilde{\varepsilon}_\mathrm{m}}=\frac{1}{3\varepsilon_0}D_\mathrm{p}\tilde{\alpha}_\mathrm{p},\ \text{with}\ \tilde{\alpha}_\mathrm{p}=4\pi\varepsilon_0\frac{\tilde{\varepsilon}_\mathrm{p}-\tilde{\varepsilon}_\mathrm{m}}{\tilde{\varepsilon}_\mathrm{p}+2\tilde{\varepsilon}_\mathrm{m}}a_\mathrm{p}^3. \quad (10.4)$$

Equation 10.2 is easily deduced from Eqs. 10.3 and 10.4 because the volume fraction of all the particles is written as

$$k_\mathrm{p}=\frac{4\pi}{3}D_\mathrm{p}\,a_\mathrm{p}^3. \quad (10.5)$$

Let us remark that the MG model can be generalized to mixtures of p different particles:

$$\frac{\tilde{\varepsilon}_\mathrm{eff}-\tilde{\varepsilon}_\mathrm{m}}{\tilde{\varepsilon}_\mathrm{eff}+2\tilde{\varepsilon}_\mathrm{m}}=\sum_p k_\mathrm{p}\frac{\tilde{\varepsilon}_\mathrm{p}-\tilde{\varepsilon}_\mathrm{m}}{\tilde{\varepsilon}_\mathrm{p}+2\tilde{\varepsilon}_\mathrm{m}}. \quad (10.6)$$

All the permittivities are complex numbers, which permits to take into account absorption in the material. Nevertheless, the MG model only considers inclusions of non-interacting spherical particles and is limited to a small concentration of particles embedded in a host matrix, i.e., a large dilution. When the relative volume of particles increases, the MG model is no more valid [7].

A way to extend the MG model to non-spherical particles is to introduce in the model a depolarization factor N_p that accounts for the aspect ratio of the particles p [8]. This factor takes different values N_{px}, N_{py}, N_{pz} in the three directions x_p, y_p, z_p in the case of ellipsoidal particles [9], always with $N_{px}+N_{py}+N_{pz}=1$. In direction x, the depolarization factor N_{px} is written as

$$N_{px}=\frac{x_\mathrm{p}y_\mathrm{p}z_\mathrm{p}}{2}\int_0^\infty\frac{du}{\left(x_\mathrm{p}^2+u\right)\sqrt{\left(x_\mathrm{p}^2+u\right)\left(y_\mathrm{p}^2+u\right)\left(z_\mathrm{p}^2+u\right)}}. \quad (10.7)$$

In the case of only one type of particles and taking into account the depolarization factor N_p, the MG equation becomes

$$\varepsilon_\mathrm{eff}=\varepsilon_\mathrm{m}+\frac{k_\mathrm{p}\left(\varepsilon_\mathrm{p}-\varepsilon_\mathrm{m}\right)\sum\limits_{j=x,y,z}\frac{\varepsilon_\mathrm{m}}{\varepsilon_\mathrm{m}+\left(\varepsilon_\mathrm{p}-\varepsilon_\mathrm{m}\right)N_{pj}}}{3-k_\mathrm{p}\left(\varepsilon_\mathrm{p}-\varepsilon_\mathrm{m}\right)\sum\limits_{j=x,y,z}\frac{N_{pj}}{\varepsilon_\mathrm{m}+\left(\varepsilon_\mathrm{p}-\varepsilon_\mathrm{m}\right)N_{pj}}}. \quad (10.8)$$

In the case of spherical particles, i.e., for $N_{pj}=1/3$ (see Table 10.1), relation 10.8 takes the same expression as the classical

Table 10.1 Values of the depolarization factor for ellipsoids. For disks and cylinders, x is the direction of the axis of rotation

Shape of the inclusions	Depolarization factor			
	N_{px}	N_{py}	N_{pz}	
Disk	0	1	0	Depending on the field direction
Cylinder	~1/2	~1/2	~0	
Sphere	1/3	1/3	1/3	

MG one (10.3). Even if MG theory, combined with the depolarization factor approach, can render for mixtures behavior of any shape of inclusions, it remains limited to rather low concentration and low contrast of permittivity between inclusions and host matrix.

10.2.1.2 Polder and van Santen EMA

To go over the limitations of the MG model, Scheller et al. [10] used, in the THz domain, the Polder and van Santen (PvS) model: it is based on the MG theory, but it also takes into account the change of permittivity, in the particle environment, due to the presence of the other particles. This model is thus adapted to materials with a larger concentration of particles. Thereby relation 10.2 is slightly modified by replacing the permittivity of the host matrix ε_{m} in Eq. 10.2 by the effective one $\varepsilon_{\mathrm{eff}}$. The new relation is

$$\frac{\varepsilon_{\mathrm{eff}}-\varepsilon_{\mathrm{m}}}{3\varepsilon_{\mathrm{eff}}}=k_{\mathrm{p}}\frac{\varepsilon_{\mathrm{p}}-\varepsilon_{\mathrm{m}}}{\varepsilon_{\mathrm{p}}+2\varepsilon_{\mathrm{eff}}}. \tag{10.9}$$

Such modeling, which is also limited to spherical particles, can be associated with the depolarization factor N_p [1–3, 10], to give

$$\varepsilon_{\mathrm{eff}}=\frac{\varepsilon_{\mathrm{m}}}{1-\frac{1}{3}k_{\mathrm{p}}\left(\varepsilon_{\mathrm{p}}-\varepsilon_{\mathrm{m}}\right)\sum\limits_{j=x,y,z}\frac{1}{\varepsilon_{\mathrm{eff}}+\left(\varepsilon_{\mathrm{p}}-\varepsilon_{\mathrm{eff}}\right)N_{pj}}}. \tag{10.10}$$

The simplest particle shapes lead to quite simple formulas [1–3, 11]:

$$\text{Spheres:}\quad \frac{\varepsilon_{\mathrm{eff}}-\varepsilon_{\mathrm{m}}}{3\varepsilon_{\mathrm{eff}}}=k_{\mathrm{p}}\frac{\varepsilon_{\mathrm{p}}-\varepsilon_{\mathrm{m}}}{\varepsilon_{\mathrm{p}}+2\varepsilon_{\mathrm{eff}}}, \tag{10.11}$$

$$\text{Disks:}\quad \frac{\varepsilon_{\mathrm{eff}}-\varepsilon_{\mathrm{m}}}{\varepsilon_{\mathrm{eff}}+2\varepsilon_{\mathrm{p}}}=k_{\mathrm{p}}\frac{\varepsilon_{\mathrm{p}}-\varepsilon_{\mathrm{m}}}{3\varepsilon_{\mathrm{p}}}, \tag{10.12}$$

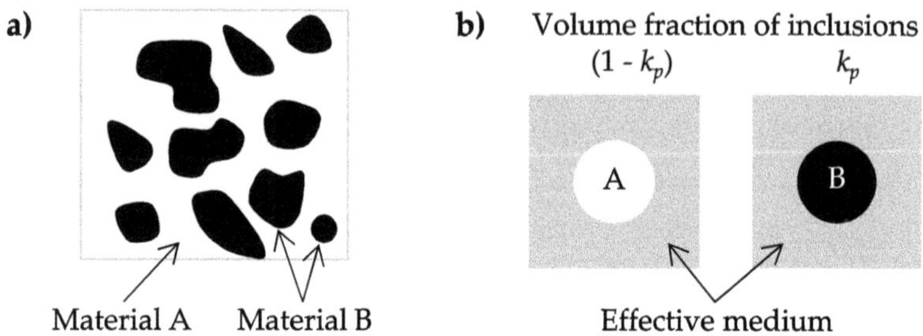

Figure 10.2 Scheme of the principle of the Bruggeman EMA model: (a) real material; (b) model.

$$\text{Rods:}\quad \frac{\varepsilon_{\text{eff}}-\varepsilon_{\text{m}}}{5\varepsilon_{\text{eff}}+\varepsilon_{\text{p}}}=k_{\text{p}}\frac{\varepsilon_{\text{p}}-\varepsilon_{\text{m}}}{3\left(\varepsilon_{\text{p}}+\varepsilon_{\text{eff}}\right)}. \tag{10.13}$$

Even if this PvS model is expected to describe many cases, it remains limited to low concentrations of particles. Moreover, Scheller et al. [10] have shown that some experimental results cannot be precisely retrieved using this model, as it is very dependent on the type of mixtures studied.

10.2.1.3 Bruggeman EMA

The theory proposed by Bruggeman [12] can also be applied to more various samples than the MG one. Instead of considering inclusions of particles embedded in a host matrix, the Bruggeman model (BG) considers the embedding of inclusions for the two materials of the mixture (A and B) in an effective material host matrix (Fig. 10.2). This model permits to study samples with a high particle concentration.

For spherical particles, the BG model is written as

$$k_A\frac{\varepsilon_A-\varepsilon_{\text{eff}}}{\varepsilon_A+2\varepsilon_{\text{eff}}}+k_B\frac{\varepsilon_B-\varepsilon_{\text{eff}}}{\varepsilon_B+2\varepsilon_{\text{eff}}}=0, \tag{10.14}$$

where ε_A, ε_B are respectively the permittivity of the two components A and B of the mixture, and k_A and k_B are their respective volume fractions ($k_A+k_B=1$). ε_{eff} is the permittivity of the effective medium. This model can be generalized whatever is the number N

of components in the mixture:

$$\sum_{n=1}^{N} k_i \frac{\varepsilon_n - \varepsilon_{\text{eff}}}{\varepsilon_n + 2\varepsilon_{\text{eff}}} = 0. \tag{10.15}$$

The BG model well describes high contrast permittivity mixtures; nevertheless, it is limited to spherical particles.

10.2.1.4 Improvement of the classical EMA models

The classical EMA models have been improved to deal with particularly material cases. Scheller et al. [10] take into account the interaction between ellipsoidal-shaped particles by combining PvS with the depolarization factor formalism and a differential approach. The following substitutions should be performed in Eq. 10.10:

$$k_{\text{p}} \to \frac{dk_{\text{p}}}{1 - k_{\text{p}}}, \qquad \varepsilon_{\text{m}} \to \varepsilon_{\text{eff}}, \qquad \varepsilon_{\text{eff}} \to \varepsilon_{\text{eff}} + d\varepsilon_{\text{eff}}. \tag{10.16}$$

The particle volume fraction is then given by the following differential equation:

$$\frac{dk_{\text{p}}}{1 - k_{\text{p}}} = \frac{3\,d\varepsilon_{\text{eff}}}{\varepsilon_{\text{eff}}} \frac{\left(\varepsilon_{\text{p}}^2 - \varepsilon_{\text{eff}}^2\right) N\left(N-1\right) - \varepsilon_{\text{p}}\varepsilon_{\text{eff}}\left(2N^2 - N + 1\right)}{-\varepsilon_{\text{p}}^2\left(3N^2+1\right) + 2\varepsilon_{\text{p}}\varepsilon_{\text{eff}}\left(3N-2\right) + \varepsilon_{\text{eff}}\left(-3N+5\right)}. \tag{10.17}$$

Solving Eq. 10.17 leads to [10]

$$k_{\text{p}} = 1 - \frac{\left(\frac{\varepsilon_{\text{p}} - \varepsilon_{\text{eff}}}{\varepsilon_{\text{p}} - \varepsilon_h}\right)}{\left(\frac{\varepsilon_h}{\varepsilon_{\text{eff}}}\right)^{3N\frac{N-1}{3N+1}} \left(\frac{(1+3N)\varepsilon_{\text{p}} + (5-3N)\varepsilon_h}{(1+3N)\varepsilon_{\text{p}} + (5-3N)\varepsilon_{\text{eff}}}\right)^{\frac{2}{9}\frac{(N-1)^2}{9N^2-12N-5}}}. \tag{10.18}$$

EMA models are also modified to take into account non-uniform inclusions (particles of different types, shapes and/or sizes): the particle set is then treated as distributions. Considering N classes of particles of the same material, the total inclusion volume fraction becomes

$$k_{\text{p}} = \sum_{n=1}^{N} k_{pn}, \tag{10.19}$$

k_{pn} being the relative volume fraction of each class n of particles p. Equation 10.19 can be written down for particles of M different materials, each with their permittivity value ε_{pni}. Moreover, each class of material can also divided into N groups of different shapes [10] and so on. The general expression of k_{p} becomes

$$k_{\mathrm{p}} = \sum_{n=1}^{N} \sum_{m=1}^{M} \sum_{r=1}^{R} \sum_{\ldots}^{\ldots} k_{pnmr\ldots} \tag{10.20}$$

Such extension of classical EMA models is very efficient to treat some specific cases. However, they are generally more complicated to implement and require inputting in the calculation some geometrical parameters that could be unknown: in this case, these unknown parameters serve as adjustable variables in the modeling. One must be vigilant to the use of a given model and the choice of the adjustable parameters, at the risk of losing any physical significance of the achieved modeling results!

10.2.1.5 Power-law approaches

Power-law models can be used for higher concentrations of particles: they are based on empirical studies. However, some physical insight can be attributed to the parameters involved in the power-law expressions. For example, Zhou et al. [13] define the parameter α in 10.21 as the mean depolarization factor along the external field, which can be attributed to the particle shape. In this case,

$$\varepsilon_{\mathrm{eff}}^{\alpha} = \sum_{n=1}^{N} k_{pn} \varepsilon_{n}^{\alpha}. \tag{10.21}$$

$\alpha \in [-1, 1]$ is a dimensionless parameter, whose value depends on the micro-geometry of the material. For example, the Landau–Lifshitz–Looyenda (LLL) model [14] is obtained for $\alpha = 1/3$ and, for two components A and B, is written as

$$\varepsilon_{\mathrm{eff}}^{1/3} = k_A \varepsilon_A^{1/3} + (1 - k_A)\, \varepsilon_B^{1/3}. \tag{10.22}$$

The Lichtenecker and Rother (LR) [15, 16] model is based on the Lichtenecker formula [17, 18], i.e., on relation 10.21, when $\alpha \to 0$:

$$\ln \varepsilon_{\mathrm{eff}} = \sum_{n=1}^{N} k_{pn} \ln \varepsilon_{n}. \tag{10.23}$$

Table 10.2 Elements of comparison of the main EMA usable in the THz domain (from Scheller et al. [1])

Predicting model	Concentration of inclusions	Particle shape	Restrictions
Maxwell–Garnett (MG)	Low	Spherical	Small permittivity contrast
MG + depolarization factor	Low	Ellipsoidal	
Polder and van Santen (PvS)	Low	Ellipsoidal	Anisotropic materials
Bruggeman (BG)	High	Spherical	Any permittivity contrast
BG + depolarization factor	High	Ellipsoidal	
Landau–Lifshitz–Looyenda (LLL)	Middle	Not taken into account	Small permittivity contrast
Lichteneker and Rother (LR)	Middle	Not taken into account	Small permittivity contrast

Therefore, we get

$$\varepsilon_{\text{eff}} = \prod_{n=1}^{N} \varepsilon_n^{k_{pn}}. \tag{10.24}$$

This model is mainly valid when the difference between the permittivities of the different materials of the mixture remains small [19].

Table 10.2 summarizes the range of validity of the above-described EMA models. A list of other effective medium theories can be found in [20] as well as their range of applications and limits.

10.2.2 Example of Powder Pellets

Pellets are often used to measure the complex refractive index or permittivity of powders because samples are easy to fabricate using a press. Moreover, powders of absorbing materials, or of materials with strong absorption bands, can be diluted in a transparent powder (like polyethylene), which permits transmission measurement. Such dilution leads to heterogeneous compounds that contain the different materials of the mixtures. Even in the case of pure powder, the pellet will contain a fraction of air, and thus the sample cannot be considered as bulky.

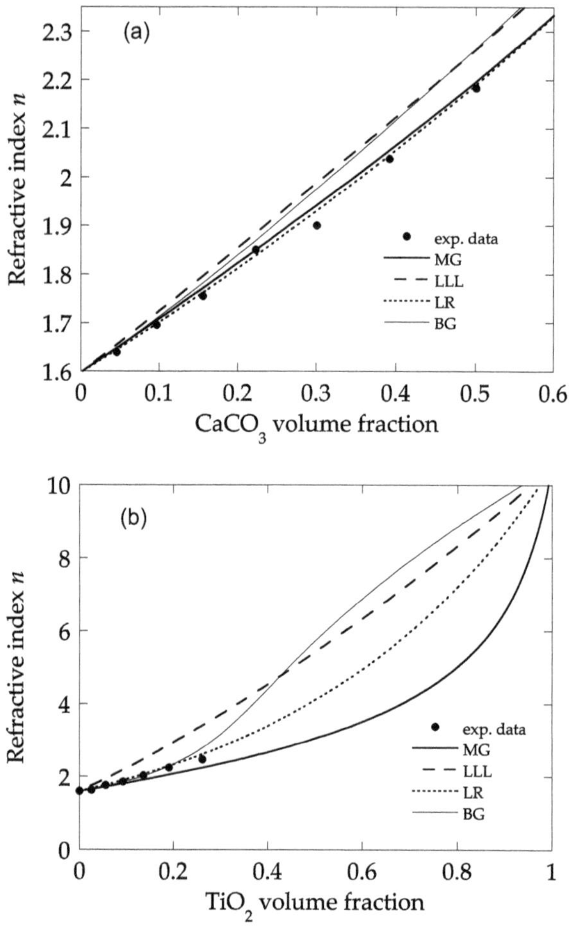

Figure 10.3 Refractive index spectra of two mixtures materials at 0.5 THz: $CaCO_3$ (a) and TiO_2 (b) particles embedded in latex. Full circles are experimental data, while lines are the best fits using different EMA models [21].

Figure 10.3 presents the refractive index measured at 0.5 THz of $CaCO_3$ (10.3a) or TiO_2 (10.3b) powders embedded in latex, versus the volume concentration [21]. The pellets were fabricated with a molding process. Experimental results (dots) have been obtained using a classical transmission THz-TDS setup. Different EMA models are used to fit (lines) at best the experimental data. For these samples, a rather good agreement is obtained with the LR model. For the calculation, the refractive indixes of $CaCO_3$, TiO_2, and latex are respectively 3, 10.5, and 1.6. It has to be noticed that these values have been measured with a molded sample of pure latex and with bulk $CaCO_3$ and TiO_2 wafers. If one uses pellets of $CaCO_3$ or TiO_2 powders to determine their refractive index, this value could be erroneous due to the presence of air in the pellets. For example, a remaining proportion of air of about 12–15% [10, 22] lowers the refractive index of $CaCO_3$ from 3 to 2.5. However, for some materials, this procedure is not perfectly correct, because the refractive index of the particles could differ from the bulk value, due to contamination, oxidation or the presence of moisture. In the present example, the decrease from 3 to 2.5 is perfectly calculated using the MG model. Here the TiO_2 powder is made from a rutile crystal, which is uniaxial with slow and fast refractive indices of about 12 and 9.3, respectively [1]. In a random powder, the average value of the refractive index is about 10.5. In the case of $CaCO_3$ inclusions in latex, the refraction index difference between inclusion and host matrix is not large (1.6 for latex and 3 for $CaCO_3$). Moreover, since the volume concentration of $CaCO_3$ remains low, all the EMA models predict similar behaviors, in good agreement with the experimental results. Here, the LR and MG models fit nicely the experimental data.

In the case of TiO_2 inclusions, the permittivity contrast between inclusions and host material is very large: this makes the models leading to different results, typically over a volume ratio of about 10%. The LLL model does not describe the THz response of the material well, even with a concentration below 10%. Let us notice that, because of the density of TiO_2 and mechanical cohesion issues, it is not possible to fabricate samples with a volume ratio of TiO_2:latex above 25%.

Figure 10.4 presents the extinction coefficient of the same samples versus the inclusion volume ratio. No EMA models can fit the experimental data. This can be explained by scattering because the inclusion sizes are not very small compared to the wavelength in the material ($\lambda \sim 600$ μm). Scattering leads to an overestimation of the losses, while it does not affect so much the refractive index (Fig. 10.3). Such scattering effects are not taken into account by EMA models.

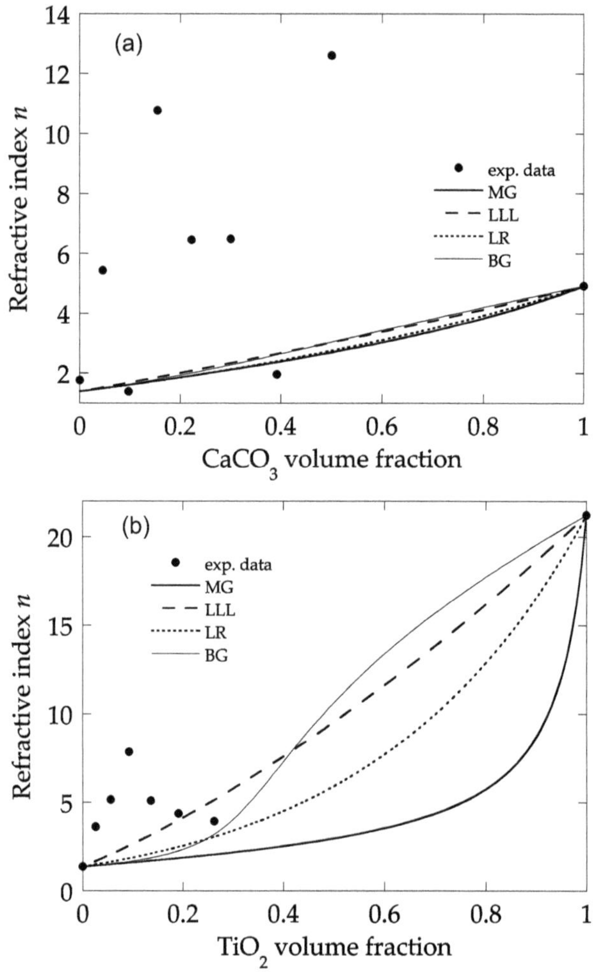

Figure 10.4 Extinction coefficient of two mixtures materials at 0.5 THz: $CaCO_3$ (a) and TiO_2 (b) embedded in latex. Full circles are experimental data, while lines are the best fits using different EMA models [21].

10.3 Scattering Materials

10.3.1 Overview of Scattering Phenomena

We consider the size of particles or inclusions in relation to the THz wavelength. These particles are assumed to be randomly arranged

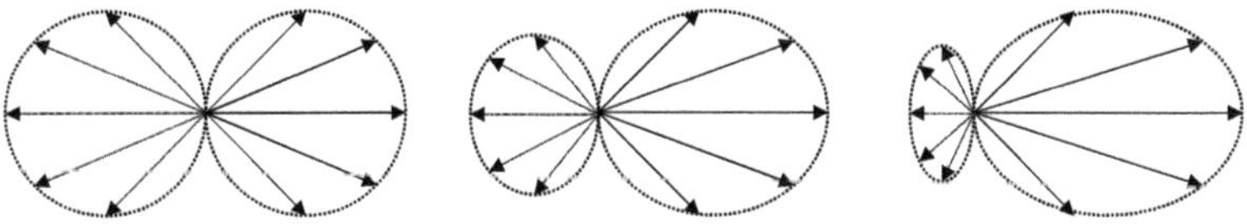

Figure 10.5 Schematic patterns of the THz light scattered by a particle: (a) small, (b) medium, and (c) large particles (compared to the wavelength).

in the material. Because of the particles size, THz waves interact resonantly with the particles, and scattering effects are observed. Scattering is then mainly governed by two parameters:

- the size parameter $x = 2\pi\,\phi/\lambda$, where ϕ is the diameter of the particle and λ is the wavelength,
- the particle properties relative to the surrounding medium, more precisely the refractive index contrast $\Delta n = n_{\text{particle}} - n_{\text{surrouding}}$.

We distinguish three cases of scattering (Fig. 10.5), in which the wavelengths of the exciting and scattered light are equal (elastic scattering):

- Rayleigh scattering for "small" particles, i.e., when $x \ll 1$ that corresponds typically to $\lambda \gg 10\ \phi$: the THz wave is isotropically scattered in the whole space (i.e., over 4π sr). In this regime, the shape of the scattering particle is not very significant and is often treated as a sphere.
- Mie scattering for medium-size and large particles, i.e., when $x \geq 1$ $(\phi/10 \leq \lambda \leq 10\ \phi)$: the wave is mainly scattered along the beam direction. In this wavelength regime, the particle shape becomes more significant.
- For larger particles ($x \gg 1$), geometric ray optics is sufficient to describe the particle-wave interaction.

Inelastic interactions like Raman scattering are described in Chapter 13.

10.3.1.1 Rayleigh scattering

Rayleigh scattering may be modeled as the interaction of the exciting beam with spherical particles much smaller than the wavelength. Nevertheless, the particles are bigger than when EMA models apply. The scattered wave intensity varies with the inverse of the wavelength to the power four. It is simply written [23] as

$$I = I_0 \left(\frac{1+\cos^2\theta}{2R^2}\right)\left(\frac{2\pi}{\lambda}\right)^4\left(\frac{n^2-1}{n^2+2}\right)^2\left(\frac{\phi}{2}\right)^6, \tag{10.25}$$

$$\Leftrightarrow \begin{cases} \dfrac{I}{I_0} = \left(\dfrac{1+\cos^2\theta}{2R^2}\right)\left(\dfrac{n^2-1}{n^2+2}\right)^2\left(\dfrac{x^2}{8}\right)^2\phi^2, \\ \dfrac{I}{I_0} = \left(\dfrac{1+\cos^2\theta}{2R^2}\right)\left(\dfrac{n^2-1}{n^2+2}\right)^2\left(\dfrac{x^3}{16\pi}\right)^2\lambda^2. \end{cases} \tag{10.26}$$

I_0 is the incident wave intensity, R the distance from the particle to the detector, θ the scattering angle, n the refractive index of the particle, and ϕ its diameter. The hosting material is supposed to be air.

10.3.1.2 Mie scattering

Mie scattering is probably the most difficult case to solve as it deals with particles whose size is comparable or greater than the incident wavelength. In such a case, there is no simple analytical solution to derive the scattered wave intensity. Indeed, one needs to solve a more complex problem based on the solution to Maxwell's equations, to evaluate the electromagnetic field inside and outside the scattering particle. Nevertheless, the numerical resolution of such problems no longer causes major difficulties with modern computers. It is useful, when dealing with Mie scattering, to define the following values:

- The scattering cross section σ is the ratio of the power of the light scattered by the particle and the power density that illuminates the particle. Thus, the unit of σ is m^2.
- The extinction efficiency Q_{ext} is the ratio of the scattering cross section σ and the area of the particle profile. Q_{ext} is without unit.

Mie theory remains restricted to scattering of plane electromagnetic waves by spherical particles, but it can be modified to give a good approach of scattering by nonspherical centers and real beams. Initially, Mie solutions were based on Riccati–Bessel functions [24], but this approach is computationally intensive. Other methods are based on spherical Bessel functions [25–27] and permit to obtain the desired parameters: particle cross section, scattering efficiency factor, extinction coefficient, etc. Thus, a wide variety of different cases have been addressed, for example, simple homogeneous spheres [28], stratified spheres [29], cluster of spheres [30], coated [28], distorted, magnetic spheres [31], etc. Originally, Mie theory was also limited to non-absorbing materials but was extended to absorbing spheres and/or spheres embedded in an absorbing matrix [32], to describe most of the practical cases.

As an example, Fig. 10.6 presents the extinction efficiency Q_{ext}, versus the size parameter, of spheres of latex ($n_{\text{Latex}} = 1.59$) hosted in air or in a dielectric medium ($n_{\text{host}} = 1.33$). The effect of absorption in latex is treated. As previously explained, Q_{ext} represents the diminution rate of the transmitted light because of scattering and absorption through the scattering medium. For small size parameter, Q_{ext} varies as x^6 as expected from Rayleigh scattering theory (Eq. 10.26). For larger values of x, the scattering efficiency oscillates to reach a quasi-stationary and damped regime at very large size parameters. Both losses in the sphere material and in the surrounding matrix greatly affect the extinction efficiency. Losses increase the damping of the oscillating phenomena, whereas changing the refractive index of the host matrix shifts the maximum of the curve to higher size parameter values, broadens the oscillation and slightly decreases the scattering phenomena intensity.

10.3.1.3 Scattering in the THz regime

Many studies have been performed on scattering phenomena induced by the propagation of a THz wave in random and/or heterogeneous media, based on far-field THz-TDS measurements in almost all the cases. We can, for example, cite studies of scattering by one single sphere [33] or cylinder [34], or by a dense medium made of large Teflon spheres (800 μm diameter, the volume fraction

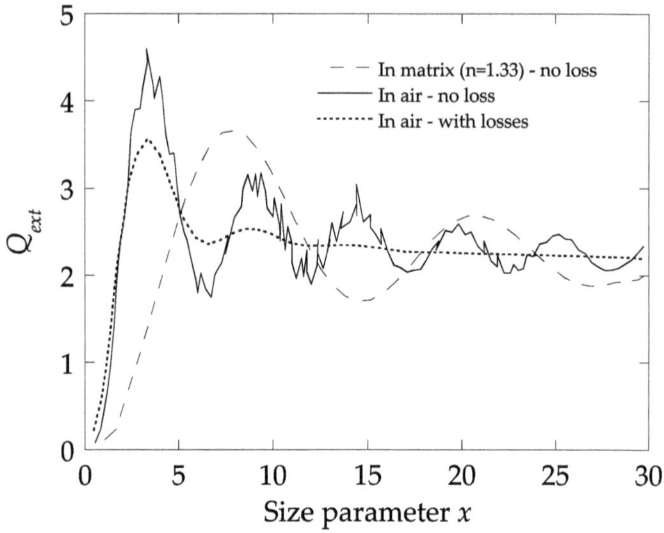

Figure 10.6 Extinction efficiency of a sphere of latex with $n_{\text{latex}} = 1.59$ and $\kappa_{\text{latex}} = 10^{-4}$ (roughly speaking without loss) and $\kappa_{\text{latex}} = 10^{-1}$ (with losses), embedded either in air or in a material $n = 1.33$. The calculation is performed in the optical domain, i.e. $\lambda = 1$ μm. However, as parameter x is dimensionless, similar patterns would be seen at THz frequencies. Adapted from Bohren and Gilra [32].

of about 56%) [35]. In view of detecting or studying materials with characteristic fingerprint spectral features, powders of these materials, like sucrose [36] or explosives [37, 38], were mixed with a transparent powder, like polyethylene (PE) or Teflon.

The main effect of scattering on THz spectra is a significant distortion of the spectral signature of a material, mostly through an increase in the losses. As with all electromagnetic waves, scattering depends on the particles size and is stronger when the refractive index contrast between the components of the mixture is large. However, in the THz domain, this contrast could be 10 times larger [39–41] than at the optical wavelengths, because the wide bandwidth can give a wide range of values for the refractive indices, from less than 1.5 for plastics, glass, etc., up to more than 3 for semiconductors, and even higher in ferroelectric materials.

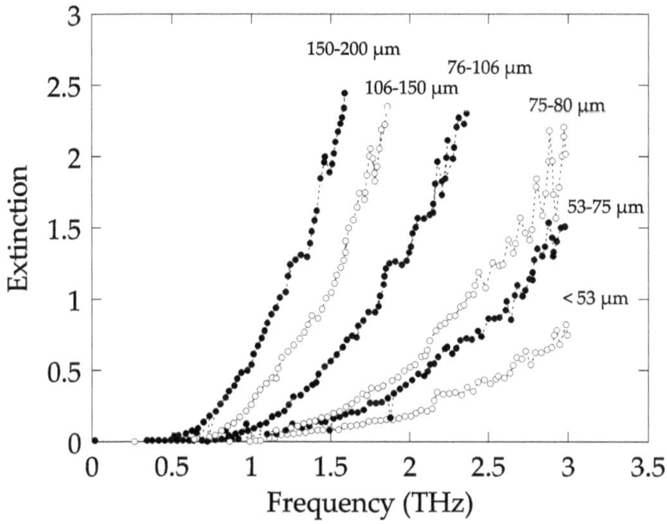

Figure 10.7 Measured extinction versus size parameter measured of a polyethylene powder for different grain sizes. Adapted from Shen et al. [36].

More generally, scattering depends on several geometrical parameters of the studied mixture: some of them can be experimentally and directly determined from the pure materials before mixing (grain size, volume fraction, etc.), whereas others (homogeneity, density, etc.) are more difficult to evaluate and depend on the sample preparation.

Most of the time, people are interested in determining the absorption of the scattering particles and try to sort absorption and scattering contribution from the total loss measurement. For example, Shen et al. [36] proposed a simple model to get rid of the scattering effect and to evaluate the optical parameters of the scattering material. They empirically showed, by studying uncompressed samples made of PE grains, that extinction roughly varies as the power of the frequency to the third (see Fig. 10.7):

$$\alpha \approx 1.5 \times 10^{-3} \phi^{2.8} f^{3.3}. \tag{10.27}$$

In this study, the problem was quite simple to solve because PE is transparent in the THz domain: thus, the measured losses are only due to scattering. The results by Shen et al. [36] are in rather good agreement with the theory by Bohren and Huffmann [28], or in the limit of small size parameter, with the Rayleigh theory (Eq. 10.25), which predicts a variation of the scattering efficiency with the frequency to the fourth power. However, in the Rayleigh theory, this efficiency depends on the sixth power of the particle size, whereas Shen et al. measured a variation approximately to the power three. Such a disagreement is explained by a decrease of the particle number density when the particle size increases (the decrease is proportional to the third power of the particle size), keeping the mass of PE in the sample constant.

10.3.2 Scattering in Powder Mixtures

Materials made from powders must be considered as heterogeneous if the grain size is not much smaller than the wavelength. Depending on this relative grain size, scattering will appear more or less intensely, whether these materials are pure or made from mixtures of different powders. Even when these powder samples are packaged in pressed pellets, scattering remains present, but may slightly vary with the pressure applied during the fabrication of the samples. Such conditioning is usually necessary to evaluate precisely the THz parameters (mostly the absorption) of the materials that constitute the samples, given established databases [42].

To study some materials, a mixture of powders is necessary. When these materials exhibit a high absorption over the whole THz range or over spectral bands, performing transmission THz-TDS is not possible unless the samples are very thin. Manufacturing thin films is not always technically possible; for example, if the material is produced as a powder, very thin pressed pellets could be too brittle. The solution to this problem is to mix the material with a binding powder which is inert and without any strong absorption lines in the THz range (i.e. transparent), like polyethylene (PE). This technique is widely employed in THz spectroscopy to study drugs, explosives or other finger print materials, where often it is only the position of spectral absorption lines that is of interest, rather than their intensity. In the case of explosives, it also diminishes the risk of explosion. Here, we will give the example of fructose-HDPE (high-density polyethylene) powder pellets.

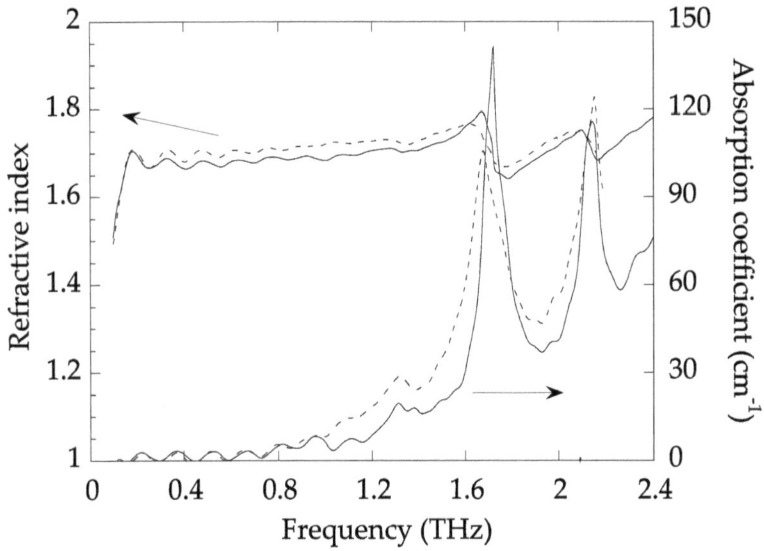

Figure 10.8 The frequency signatures of HDPE-fructose pellets mixture for two different fructose grain size: 50 μm (continuous line) and 150–200 μm (dashed line). Fructose-HDPE volume ratio is 36–74%.

These pellets have been characterized using THz-TDS spectroscopy. Two different pellets are studied with fructose grain size of ~50 μm and 150 ~ 200 μm. Other parameters are similar: the pellets have been formed under 10 ton of pressure, the host material is HDPE, and the volume fraction of fructose is 40%. Figure 10.8 presents the refractive index and extinction coefficient spectra of the two pellets (continuous lines, 50 μm grain size; and dashed line, 150 ~ 200 μm grain size). As expected, depending on the grain size, the spectral signature is substantially changed because of scattering: the absorption base line increases with the fructose grain size whereas the peaks are broadened and are less intense. The refractive index is less impacted. However, at the position of the absorption peaks, the index jump is softened due to the peak broadening. It is concluded from these observations that deriving the parameters of a constituent of the powder pellet is not trivial, because the spectral response depends on the granularity.

10.3.2.1 Single-photon scattering

As previously explained, scattering may be described as an elastic deviation of the incoming photons by the scattering particles. In the limit of a low volume density of particles and of the 1D model, the probability η of a photon to be scattered, when travelling through a slab of thickness *dx*, is proportional to *dx* and the particle concentration C. It follows immediately that the decay of the photon density $N(x)$ of the beam through the sample is

$$\eta = \sigma C\, dx \ \Rightarrow\ dN(x) = -\sigma\, C\, N(x)\, dx$$
$$\Leftrightarrow\ N(x) = N_0 e^{-\sigma C x} = N_0 e^{-\alpha x}. \qquad (10.28)$$

σ is the scattering cross-section of the particles and N_0 the photon density at the entrance of the sample. Such a simple model implies that each photon is scattered no more than once when crossing the sample. Relation (Eq. 10.28) is nothing but the Beer–Lambert law. Because most scattered photons move away from the beam, only a few of them reach the detector and their contribution to the received signal is negligible. Thus, one cannot separate losses due to absorption in the particles and scattering. Therefore [28, 43], one can define the extinction decay coefficient α_{ext} and the extinction cross-section σ_{ext}:

$$\alpha_{\text{ext}} = \sigma_{\text{ext}} C = \alpha_{\text{abs}} + \alpha_{\text{scat}}. \qquad (10.29)$$

In this relation, α_{abs} and α_{scat} correspond to the loss of energy respectively by absorption and by scattering. When no multiscattering takes place, α_{ext} does not depend on the sample thickness.

Figure 10.9 presents the measured extinction coefficient of similar pressed pellets (a similar mixture of HDPE and fructose powders: the volume fraction and the grain size of fructose are respectively 36% and 150 $\sim$ 200 μm) versus the sample thickness. In the thickness range 0.3–1.7 mm, α_{ext} remains roughly constant whatever the frequency. Let us notice that the uncertainty $\Delta\alpha_{\text{ext}}$ is about 5% and is due to the reproducibility of the samples (composition and homogeneity): the fructose grain shape and size may vary because of the mechanical method used to sort the size (set of sieves). Density (mainly the air ratio) may also change from a sample to another one with the amount of pressed material.

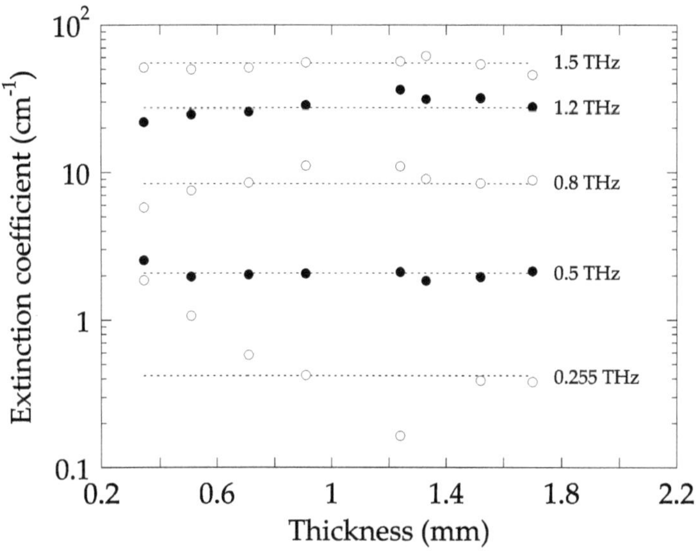

Figure 10.9 Extinction coefficient measured on PE-fructose pellets of different thickness (fructose particles diameter ~150–200 μm) and plotted for five different frequencies.

Thus, we can conclude that multiple scattering is negligible in such samples or at least is a second order effect.

σ_{ext} can be evaluated in case of dielectric spheres using numerical codes like the one available in [44]. A comparison of measured and calculated values of σ_{ext} for fructose grains (size 150 μm) is given in Table 10.3 [39].

Measurements and calculation give the same tendency for the variation of σ_{ext} with frequency. This agreement is tolerable if we take into account the practical limits of such analysis: in the

Table 10.3 Comparisons of the extension cross section C_{ext} of the fructose grains obtained from THz-TDS measurements and calculated using a numerical code [39]

Frequency (GHz)	255	500	800	1200	1500
C_{ext} - measured (μm^2)	245	982	4663	13990	26998
C_{ext} - calculated (μm^2)	381	3046	10976	29670	46124
Mean free path λ_D(mm)	12–20	1.6–5	0.4–1.25	0.2–0.4	0.1–0.2

calculation, all the fructose particles are identical and spherical, with the same concentration in all the samples and no inclusion of air. Also, due to the numerical aperture of the parabolic mirrors used in the experiment, only a part of the scattered light is collected, leading to a sub-evaluation of the scattering cross section. σ_{ext} varies as the third power of the frequency f, in agreement with Shen et al. observation [36]: this behavior is transitional between Rayleigh scattering ($\sigma_{\text{ext}} \propto f^4$) and Mie scattering ($\sigma_{\text{ext}} \propto f^2$). In the present case, σ_{abs} is typically smaller (about 10 times less) than σ_{scat} and cannot be invoked to explain the difference between calculated and experimental results.

Some other simple models have also been successfully used in the THz range. For example, the phenomenological expression, adapted by Franz et al. [41] from previous works by Raman [45], introduces the effect of scattering in the optical parameters of the mixture:

$$\alpha_{\text{ext}} = \alpha_{\text{abs}} + \left(K\frac{\omega}{c}\left(n_{\text{particles}} - n_{\text{matrix}}\right)\right)^2 \Delta, \tag{10.30}$$

$$n_{\text{ext}} = n_{\text{mixture}} + \varphi\frac{c}{\omega\phi}. \tag{10.31}$$

In the model by Raman, scattering is treated by calculating the reflected and transmitted waves in a stack of layers made of the different materials of the mixture (2D model). Here K is a numerical parameter depending on the shape, the concentration and the distribution of the particles, and Δ is the thickness of the layers. ϕ is an additional frequency dependent phase term to account for the scattering effect on the refractive index n_{ext} and φ the effective path length of the THz wave through the pellet.

10.3.2.2 Multiple-photon scattering

When multiple scattering occurs, not only photons disappear from the beam because of scattering (relation 10.28), but they could be re-injected in the beam through the multiple scattering processes:

$$dN(x) = -N(x)\sigma C\, dx + \sigma C dx \int_0^x \xi\left(x'\right) N\left(x'\right) dx' \tag{10.32}$$

Here the second term (integral) represents the contribution of the scattered photons. $\xi(x')$ renders for photons scattered in x' that are re-injected in the beam at x. In three dimensions, Eq. 10.32 is even more complex, and its derivation as well as its resolution have been the subject of many studies [43, 46, 47] that are outside the scope of this book.

The main result related to THz-TDS is that the Beer–Lambert law must be corrected and that multiple scattering is supposed to contribute efficiently when

$$\lambda_D = \frac{1}{\alpha_{\text{ext}}} = \frac{1}{C\sigma_{\text{ext}}} = \frac{1}{C\left(\sigma_{\text{abs}} + \sigma_{\text{scat}}\right)} < d. \tag{10.33}$$

d is the sample thickness and λ_D is the mean free path of the photons between two interactions with scattering centers. In the previous experimental example (Fig. 10.9), relation 10.33 implies that multiple scattering can be neglected for frequencies below 0.8 THz, but not at 1.2 and 1.5 THz. Samples are at the limit between single and multiple scatterings, which may explain the discrepancy between the calculated and the experimental results of Table 10.3. Finally, looking more precisely at Fig. 10.9, the slight increase of the extinction coefficient with the thickness of the sample probably finds its origin in the presence of the multiple scattering.

Multiple scattering of THz waves in powder materials has been quite rarely treated. In the beginning of the 2000s, the team of D. Mittleman studied the propagation of the THz pulses in random and dense medium [48, 49]. They experimentally reported and quantified the dispersion induced by multiple scattering on both the extinction coefficient and the effective refractive index. Moreover, they evaluated the mean free path λ_D of the THz pulse in the material and more precisely the additional path induced by diffusion depending on the frequency of the THz wave.

10.3.2.3 Effect of the inclusion size

To understand how the scattering is impacted by the size of the inclusions (grain size), we plot in Fig. 10.10 both the refractive index and the extinction coefficient measured for HDPE-fructose powder mixtures (pellets), versus the fructose grain size for the different frequencies chosen out of the absorption bands of fructose. The

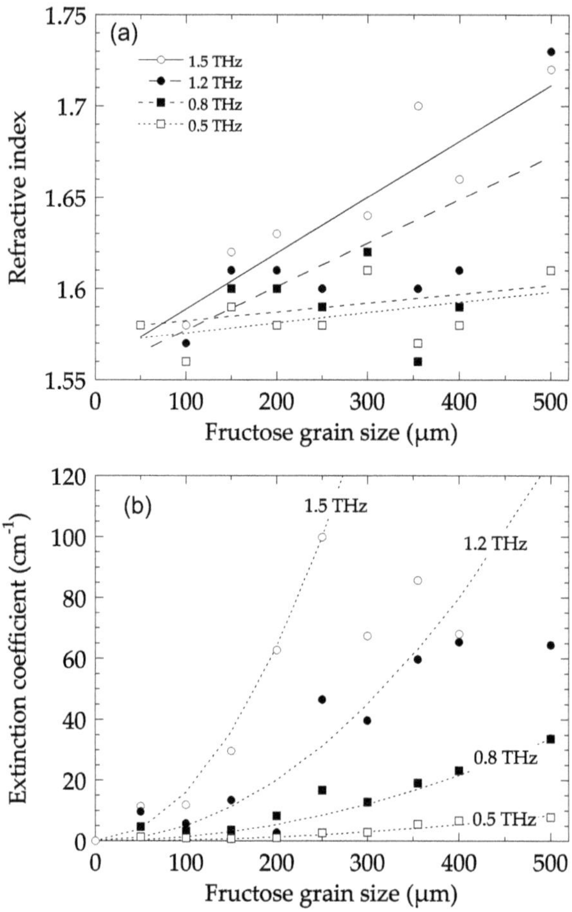

Figure 10.10 Refractive index (a) and extinction coefficient (b) measured on HDPE-fructose pellets versus the fructose grains diameter (volume fraction HDPE-fructose 64–36%).

fructose volume fraction remains constant and equal to 36%, as well as the sample thickness around 600 μm. The grain size range spreads from about 50 to 500 μm. The refractive index variation with the grain size (Fig. 10.10a) greatly fluctuates depending on the considered frequency. The general tendency is a linear increase with the grain size. However, let us notice that the refractive index is not strongly impacted by the grain size, since its highest variation, observed at 1.5 THz, remains smaller than 6%.

On the other hand, the extinction coefficient (Fig. 10.10b) variation is large even at the lowest frequency and can reach 1000%. It is fitted by a power–law curve, whose power coefficient is between 2 to 3 (see dashed lines in Fig. 10.10b), depending on the frequency. Such a result is consistent with the ones by Shen et al. [36], who reported that the extinction coefficient depends on the third power of the grain size for mixtures of low and medium density. At first order and for such a type of mixture materials, we can consider that the extinction coefficient is greatly impacted by scattering that becomes the main source of losses, whereas it is not so important on the refractive index. For a given grain size, this tendency is especially true when the frequency is high (short wavelengths).

10.3.2.4 Mie scattering

Apart from the cases of multiple scattering or for large nonspherical particles, Mie theory can be successfully used [35, 50] to predict the size dependent effect of scattering in the THz region. Here, again, we give the example of fructose-HDPE pellets [50]. Figure 10.11 presents the extinction coefficient versus the fructose grain size: experimental results are plotted (dots), as well as Mie theory results (lines) calculated using a free software [44], which is based on the formalism by Bohren and Huffman [51]. The optical parameters of the pure materials are required in such a calculation and are given in Table 10.4 for fructose. For HDPE, we considered two cases for the refractive index: $n_{\mathrm{HDPE}} = 1.41$ corresponds to the measured value of a pressed pellet of pure HDPE (HDPE + ~16% of residual air), and $n_{\mathrm{HDPE}} = 1.51$, which is the value of the bulk material.

Table 10.4 Refractive index and absorption of fructose measured using THz-TDS on pure fructose pressed powder pellets

f (THz)	0.5	0.8	1.2	1.5
n	1.97	1.98	2.01	2.04
$\alpha(\mathrm{cm}^{-1})$	2.5	6	11	19

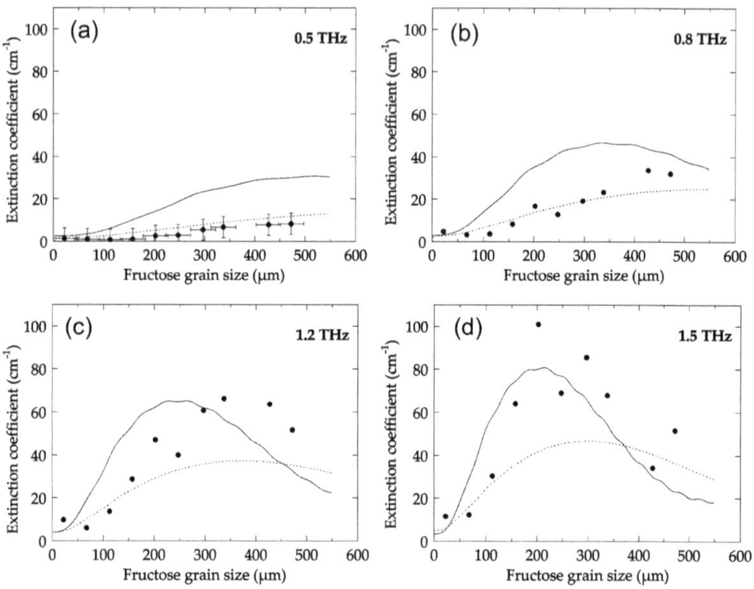

Figure 10.11 Extinction coefficient of a fructose-PEHD mixture versus fructose grains size: measured (dots) and calculation using Mie theory with $n_{\mathrm{HDPE}} = 1.51$ (dashed line) and with $n_{\mathrm{HDPE}} = 1.41$ (continuous line). Adapted from Garet et al. [50].

For the four studied frequencies, the measured data distribution is bell-shaped (truncated for the lowest ones) with a maximum of extinction occurring when the grain diameter is roughly equal to twice the wavelength in the fructose grain. This overall shape is in good agreement with the theoretical results predicted by the Mie theory (continuous and dashed lines), which are calculated without any adjusting parameter.

Here, we can fit the experimental results by setting $n_{\mathrm{HDPE}} = 1.41$ at the highest frequencies and $n_{\mathrm{HDPE}} = 1.51$ for the lowest frequencies. This can be explained by the fact that the pellets still include a given percentage of air, and are a three components mixture, i.e. of air, HDPE and fructose. Also, data for pure fructose, entered into the calculation, have been measured with a pressed pellet of fructose powder, and thus they differ slightly from the bulk value, because of residual scattering. This hypothesis was validated by simulations by Garet et al. [50].

On the other hand, the size and shape of the scattering particles also modify the scattering efficiency. In the case of nonspherical particles with a specified distribution of sizes, we can wonder about the Mie theory to finely describing the scattering phenomena. In fact, for actual materials, the microscopic structure is more intricate than the simple ones used in the model. Moreover, this structure, for example, the shape of the inclusions, is sometimes not known with any great precision. Therefore, in any case, results of simulation should be considered with a critical mind.

10.3.2.5 Absorption or scattering?

To quantitatively estimate the relative contribution of scattering and absorption to the total losses of powder mixtures, we define the relative scattering contribution to the whole losses by the ratio η:

$$\eta = \frac{\alpha_{\text{scat}}}{\alpha_{\text{ext}}} = \frac{\alpha_{\text{ext}} - \alpha_{\text{abs}}}{\alpha_{\text{ext}}}. \tag{10.34}$$

Here we give the example of the HDPE-fructose pellets depicted in the previous paragraphs. Figure 10.12 presents η versus the particles size, which has been calculated using the Mie theory software for four frequencies (lines) and the experimental values extracted from Fig. 10.10 at 1.5 THz (dots). As explained above, α_{ext} is determined directly from THz-TDS data, while pure absorption α_{abs} of the mixture is derived from data measured for a pellet of fructose powder with very small particles (extrapolated at small grain sizes from Fig. 10.10). As for the extinction coefficient in Fig. 10.11, η exhibits a maximum when the fructose grain size is of the order of the wavelength. At this maximum, losses induced by scattering represent 80 $\sim$ 90% of the total losses. Therefore, the intrinsic absorption of the sample is difficult to evaluate as it contributes to only 10% of the total losses.

10.3.2.6 Effect of the concentration of inclusions

Varying the concentration of volume fraction of the constituents allows one to evaluate the limits of validity of the models. Indeed,

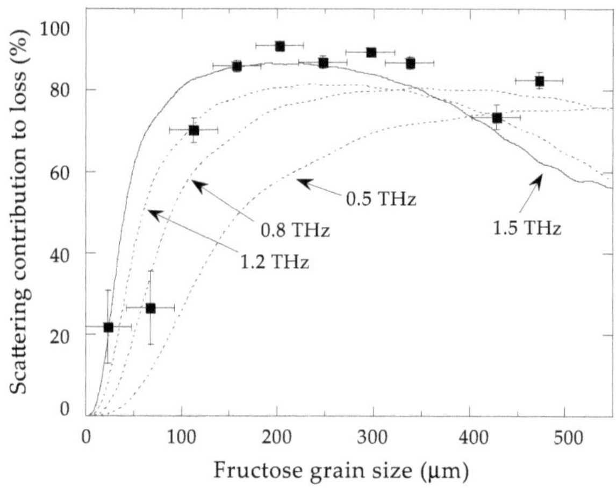

Figure 10.12 Relative scattering contribution η versus the particle diameter and for the four frequencies, calculated using $n_{\text{PEHD}} = 1.41$ (lines) and directly evaluated from the experimental data of Fig. 10.10b at 1.5 THz (dots). Redrawn from Garet et al. [50].

several models are based on the hypothesis of weak inclusion concentration. We still give here the example of HDPE-fructose powder pellets. Figure 10.13 presents both the refractive index and the extinction coefficient of the sample (150 $\sim$ 200 μm fructose grain size) versus the volume fraction of fructose. The studied frequencies are chosen outside the absorption bands. The refractive index increases roughly linearly with the fructose fraction. This behavior is rather well predicted by the MG theory: this proves again that scattering weakly influences the refractive index. In other terms, the refractive index is mostly determined by ballistic photons, i.e., photons that are not scattered when passing through the sample.

On the contrary, the behavior of the extinction coefficient over the whole volume fraction range (0–100%) does not fit any of the EMA models. Indeed, the extinction coefficient increases up to a volume fraction of about 35% then it decreases. Let us note that it should have instead increased up to the value of pure fructose if only the absorption of fructose and EMA modeling are considered. Moreover, the extinction values are larger than the one obtained for

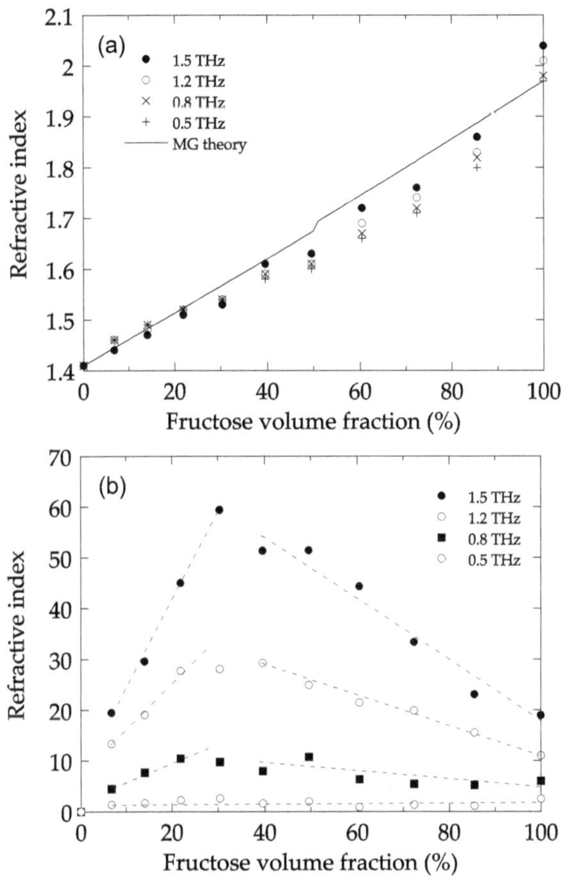

Figure 10.13 The refractive index (a) and extinction coefficient (b) of a fructose-PEHD mixture versus the fructose volume fraction (grain size 150 ~ 200 μm). On the bottom-side plot, dashed lines are just a guide to the eye).

pure fructose (see Table 10.4). These observations are explained by scattering, which is the foremost contribution to the extinction coefficient for frequencies above 500 GHz. Indeed, the scattering is high when the concentration of fructose is large. However, over a given limit, the material should be considered as made of inclusions of HDPE in a fructose matrix (and not the reverse). Here again, scattering will increase with the HDPE inclusion volume fraction. The trend of both scattering curves cross for a given fructose concentration, as shown by the straight lines in Fig. 10.13b. This volume fraction limit is not at 50% because the extinction coefficient depends on both the volume fraction and cross sections of the components. This phenomenon is even more pronounced at higher frequencies, for which scattering is more effective.

10.3.3 Scattering in Fabrics

Within the 10–15 years, several publications have been devoted to the study of the THz response of fabric, which is probably the most widespread material behind which it is sought to identify substances or objects, primarily for security screening applications. Fabrics are complex materials, with most having a weaved structure that leads to scattering when meshes are small and tight, and to other phenomena like diffraction for larger meshes. Thereby, few authors venture to present the optical parameters (index of refraction and absorption) of the fabrics, because they are very difficult to determine and, even more, to define their real thickness. Thus, most of the results found in the literature, simply show the transmission or the absorbance of samples of a given thickness: these results are difficult to compare to other samples with different thicknesses [52–55].

Figure 10.14 shows the transmission of different fabrics commonly used, like linen, denim, silk, etc. For all these materials, the transmission varies roughly as the frequency to the square, or sometimes the third, i.e., it does not really correspond to the Rayleigh regime. More precisely, Rayleigh regime cannot be invoked for all the frequency bandwidth. Although the meshes of some fabrics are small, they cannot be really considered as much smaller than the THz wavelengths. Furthermore, most these published studies do not aim to evaluate the contribution of scattering to total losses. A reason for this is probably the complexity of the material itself (mesh consisting of alternating air and woven threads, themselves constituted by twisted and entangled fibers) and whose geometrical parameters are difficult to know. Moreover, the refractive index contrast between air and yarns remains small. Because of the very variable size of the inhomogeneous patterns relative to the THz frequencies, we are sometimes at the limit of

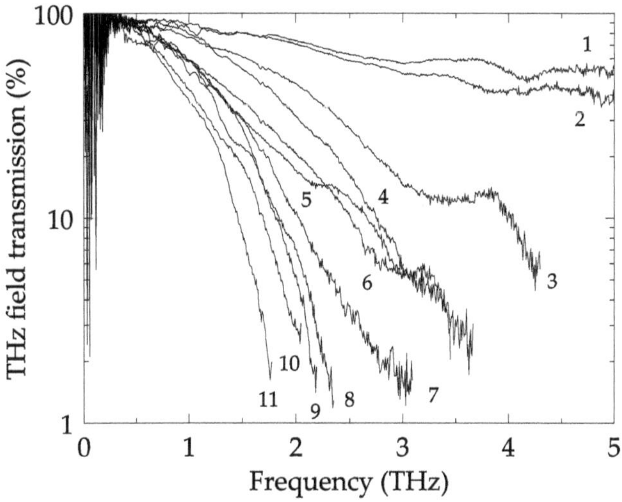

Figure 10.14 Transmission of common fabrics: dainty shining synthetic (1), clean room overall (2), natural silk (3), shirt (4), T-shirt (5), table top (6), leatherette (7), fine sport fabrics (8), denim (9), Lacoste shirt (10), linen (11). Redrawn from Hérault et al. [55].

the Rayleigh scattering, sometimes in full Mie diffusion, or even straddling both, because of the broad bandwidth of the THz signal and of the different sizes of scattering centers in the same fabric.

Few studies attempt to quantitatively analyze scattering in fabrics. Fletcher et al. [56] proposed a modeling for loss to describe the propagation of THz waves through a randomly structured material like powders or fabrics. This model shows a consistent behavior compared to the rigorous Mie model in the case of spherical diffusing centers. Also, it leads to a good agreement with the experimental results obtained with different fabrics like tweed, fleece, shirt (see Fig. 10.15; solid lines are theoretical fits).

However, modeling the THz response of all the fabrics requires different algorithms. In shirt (typically made from cotton and/or polyester) and thin fabrics, yarns are considered as an array of uniform cylinders whose refractive index values have been adjusted to give the best fit of the experimental data. On the contrary, in tweed (made from wool), for example, the fibers exhibit a larger diameter: they are considered as randomly arranged and occupying 50% of

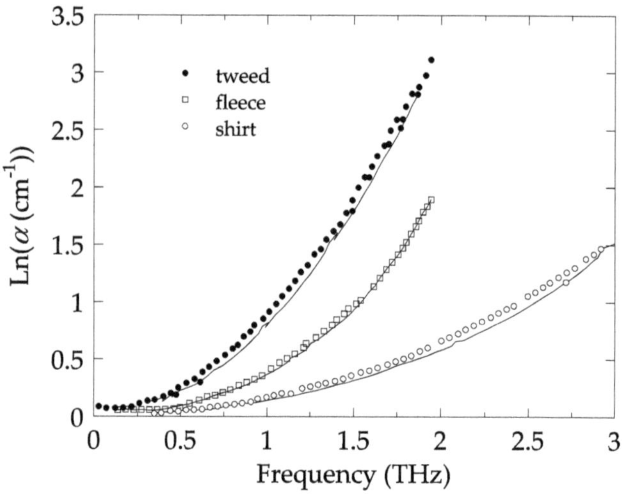

Figure 10.15 Attenuation of various fabrics. Solid lines are theoretical fits. Adapted from Fletcher et al. [56].

the volume of the fabric. An empirical adjustment of the refractive index is also used, to fit as well as possible the experimental results obtained with wool. Even if the agreement is good, such a model is then difficult to apply since it requires a particular adjustment to each fabric under study. The prediction is in good agreement with the experimental results as long as there is no absorption in the materials.

10.3.4 Diffuse Reflection

Most of the time, the sample to be analyzed cannot permit transmission measurements, simply because it is too opaque due to it being too thick, or high absorption, or a combination of both. In such a case, reflection measurements are needed, which is not a major problem even in the case of smooth samples, if we are only interested in the modulus (or intensity) of the reflected THz field and not in its phase. Indeed, in TDS reflection experiments with flexible materials like fabrics, it is practically impossible to precisely measure the phase variation of the signal induced by reflection, due to the problem of positioning reference and sample at the the same location. Moreover, defining what is the surface of the fabric is difficult, when the yarn diameter is of the order of the wavelength, and thus the surface exhibits strong reliefs or roughness.

For rough or/and heterogeneous materials, like powders or fabrics, scattering greatly complicates the problem of the signal analysis in reflection, even more than in transmission, because

- the THz field reflected in the specular direction is lower than expected in the case of smooth/homogeneous samples, and the signal-to-noise ratio could be weak;
- energy scattered in all the other directions will make difficult a precise tracking of the specular direction;
- any attempt to measure the phase will always be doomed to failure. As explained in Chapter 6, a precise phase measurement is necessary to determine precisely the absorption in reflection. Therefore, TDS spectroscopy of rough/heterogeneous samples in reflection is almost impossible and, at least, a Kramers–Kronig transform must be used.

Researches on the reflection of THz waves by rough surfaces have been carried out by X.-C. Zhang and his team [57], in view of dealing with actual issues like detection of drugs, explosives, etc. Figure 10.16 adapted from [57] shows typical results obtained with fingerprint materials that exhibit diffuse reflection. The sample is a pellet made of RDX, an explosive exhibiting specific absorption peaks between 0.5 and 2 THz that permit an easy identification of the material. The experimental setup collects the THz signal in a $\pi/2$ solid angle around the specularly reflected direction, and a Kramers–Kronig transform has been used to retrieve the absorbance of the RDX from the modulus of the reflected signal.

These results clearly show that all the peaks of RDX, measured from diffuse reflection THz-TDS in the range 0.2–1.8 THz, are smoothed as compared to the ones obtained in transmission, especially the very characteristic one at 0.82 THz. The baseline of the absorbance is also less accurately determined and, even if some of these peaks remain visible, the contrast of the measurement is

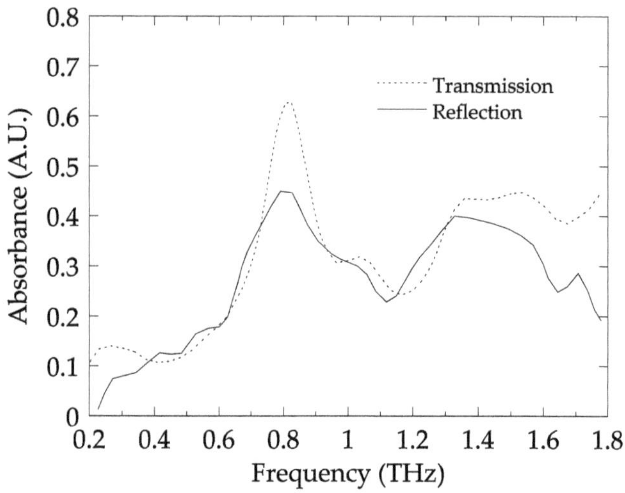

Figure 10.16 Absorbance of RDX versus frequency obtained from the transmission and diffuse reflection THz-TDS measurements. Adapted from Liu et al. [57].

drastically reduced. Practically, if the material to be detected cannot be packaged in the form of a tablet, which is a rather favorable case, but only as a cluster of powder, this contrast will be even more degraded, making challenging the identification of the explosive.

In practice, prohibited materials are hidden or protected by a covering material, for example, paper (letter), fabrics (cloth, luggage, etc., which also attenuates and scatters the signal THz to be detected, complicating the problem even more. The effect of different common coverings has been addressed by X.-C. Zhang [57]. The same RDX pellet as above is covered by either white paper (0.05 mm thick), polyethylene (0.1 mm), leather (0.3 mm), or polyester (0.4 mm). The covered sample was characterized by reflection (Fig. 10.17) under a relative humidity of 20%, and not in nitrogen as for Fig. 10.16. The covers lead to a distortion and a slight shift of the peaks. Secondly, humidity, even at a moderate level, significantly affects the measurements, especially at the water vapor lines around 1 THz, and even more around 1.7 THz. The combined impact of these two phenomena causes a very significant change and variability of the RDX signature. Most of the lines become difficult to identify and

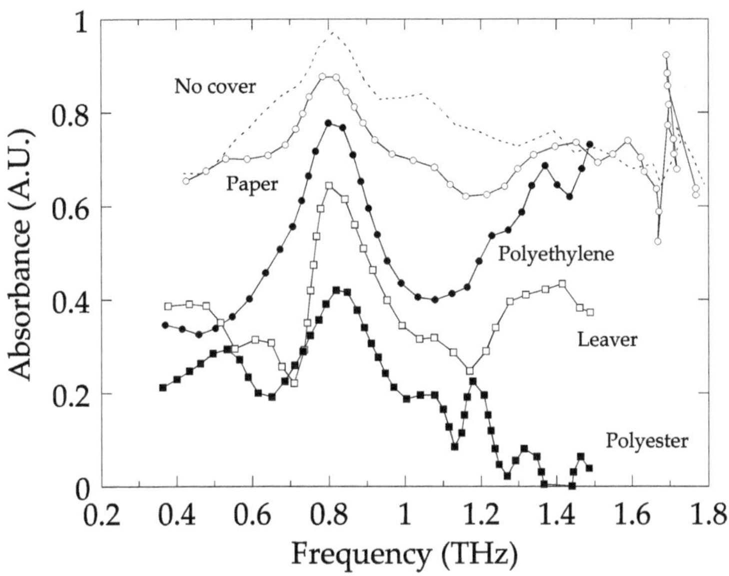

Figure 10.17 The absorption of RDX (redrawn from [57]) obtained from diffuse reflection measurements on RDX pellets covers by a different material (curves are vertically shifted for the sake of clarity).

any interpretation of such a diffuse signatures should be done with caution.

10.4 Materials with a Periodic Microstructure: Example of Fabrics

Fabrics are among the most common materials used to conceal prohibited substances. Therefore, studying their properties in the THz range is of prime importance for security and defense. Fabrics are complex inhomogeneous materials: most of them exhibit an intrinsic and periodic sub-millimeter structuration originating in their manufacturing. Fabrics can either be considered as a scattering material, for very small weaving pattern or as a diffractive device when the mesh size is of the order of the wavelength. Scattering phenomena have been discussed in Section 10.3.3. Here we will

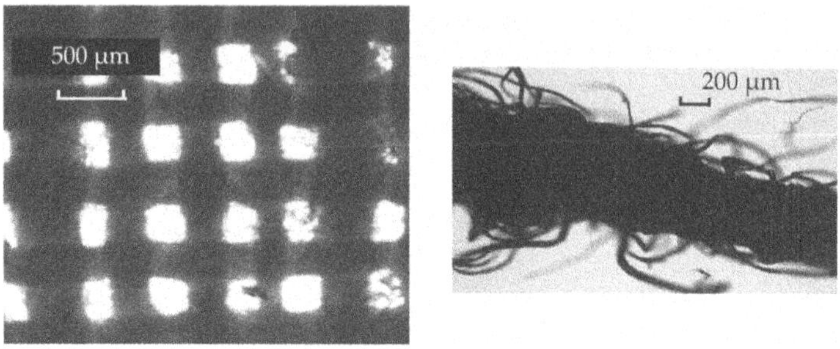

Figure 10.18 Pictures of the mesh of a linen fabric (left) and of a linen yarn (right).

only treat diffraction effects. Diffracted signals may disrupt the THz signature of fabrics and of the underneath hidden materials.

10.4.1 Signature of Diffraction by Fabrics

Most of the fabrics are woven. Thus, they look like a periodic mesh made by the interlaced yarns. Of course, yarns are not ideal cylinders, and the mesh is not perfectly periodic, but both show strong random disorders and irregularities. Therefore, as already explained, fabrics scatter THz waves. Moreover, when the period of the weave is comparable to the THz wavelength, the weave acts as a 2D grid (see Fig. 10.18) that diffracts the THz waves, as a 2D diffraction grating does.

This diffraction phenomenon has been recently reported by Hérault et al. [58], who recorded the geometrical shape of a THz beam transmitted through a piece of linen. The linen threads (Fig. 10.18) exhibit a mean diameter of 375 µm ($\pm$30%) and the stitches are almost rectangular, with average dimensions $d_x = 785$ µm and $d_y = 585$ µm. The authors used a THz-TDS system, in which the receiving antenna is attached to a rotating arm, while the sample is located at the center of rotation (THz-TDS goniometer setup). The receiving antenna is triggered by the laser beam that is delivered through an optical fiber, which permits the rotation while keeping constant the optical excitation of the antenna. The incident THz beam is almost Gaussian with a waist size that depends on frequency: in the experiment, the THz beam is collimated, and its

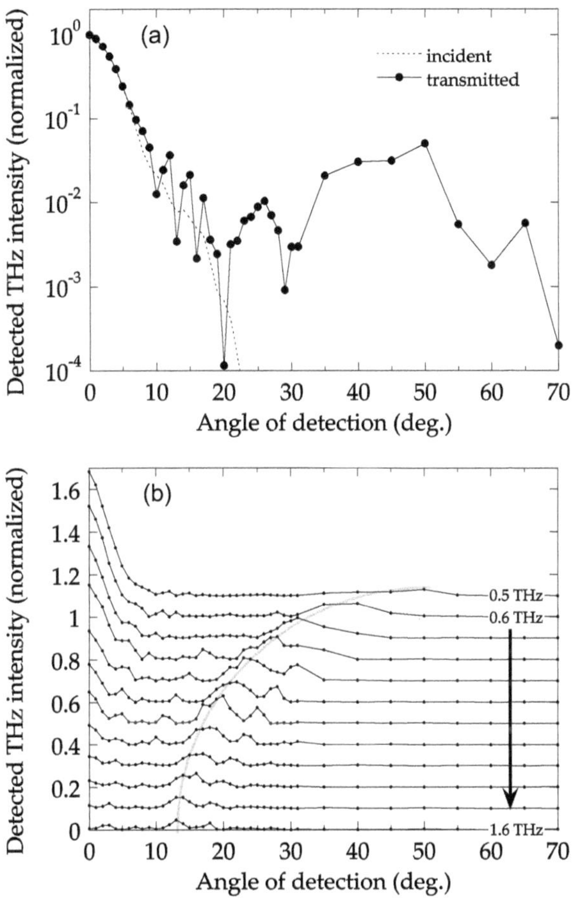

Figure 10.19 Angle-resolved intensity transmitted through a piece of linen with the THz field polarized along the y axis. (a) Continuous and dashed curves represented respectively the incident and transmitted THz field intensities (normalized) at 0.5 THz. (b) Normalized transmission between 0.5 THz and 1.6 THz by step of 0.1 THz. For the sake of legibility, each plot is vertically shifted by 0.1.

diameter at 0.5 THz is $\sim$15 mm. The angular resolution of the setup is limited to 5° by the size of the silicon lens attached to the receiving antenna. The THz beam is linearly polarized, which permits the characterization of the transmitted which is oriented along one of the two directions x and y of the woven yarns.

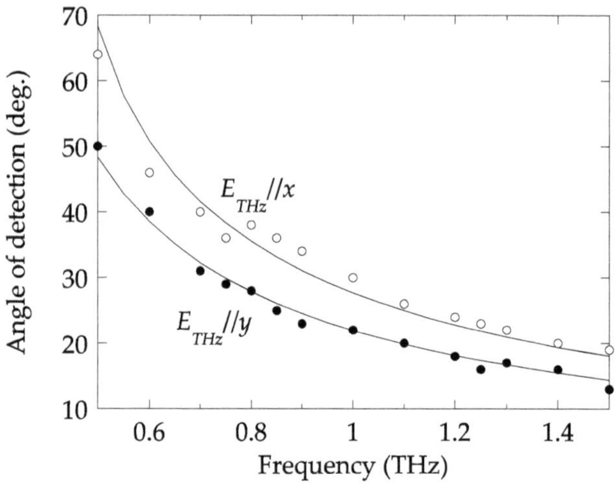

Figure 10.20 The angular position of the diffracted peak versus frequency. Circles are experimental data, while the lines are calculated with the diffraction grating Eq. 10.35.

The transmitted THz intensity of the linen sample is plotted in Fig. 10.19 versus the detection angle: at 0.5 THz on the top (logarithm scale) and for several frequencies from 0.5 to 1.6 THz on the bottom (linear scale); these latter curves have been vertically shifted for the sake of clarity. The transmission curves are normalized to the results obtained at 0°. The dashed line in Fig. 10.19a represents the incident field, i.e., measured without the sample.

In Fig. 10.19, the bump at 0° is attributed to the signal directly transmitted through the sample. Its amplitude is weakened by both absorption and scattering, and it decreases with frequency (Fig. 10.19a), which is in agreement with published results [10]. For angles typically larger than 10°, a second bump is seen on each curve. Its angular position varies with frequency from 15° up to 50° (a trend curve is shown in grey in Fig. 10.19b). For the lower frequencies, its intensity is rather weak, around 10% of the 0° peak, whereas for frequencies above 1 THz, the amplitudes of the two peaks are comparable. This second peak is due to diffraction, as it will be demonstrated below.

Figure 10.20 presents the angular position of the second peak versus frequency. The experimental data are well fitted by the well-known grating diffraction law [59, 60]:

$$d\left(\sin\theta_i + \sin\theta_d\right) = m\frac{c}{f}, \tag{10.35}$$

where θ_i and θ_d are the incident and diffracted angles, d is the grating period, and m is an integer that labels the order of diffraction. Here $\theta_i = 0$ and $m = 1$. The best fit is obtained for $d_x = 800$ μm and $d_y = 545$ μm, which is very close to $d_x = 785$ μm and $d_y = 585$ μm deduced from the photo (Fig. 10.18).

A simulation of the diffraction by the linen sample is obtained using numerical software like ANSYS HFSS [61], which is based on a finite element theory. The parameters needed for the simulation are the geometrical of the thread and the complex optical index of the materials. The validity of the simulation is limited because the sample is described as a perfect 2D periodic structure, whose pattern dimensions are constant. In addition, the determination of the refractive index of the thread is a non-trivial problem. Indeed, fabric is composed of threads and air, and only the optical index of this effective medium can be measured using classical THz-TDS. For linen, $n_{\mathrm{eff}} \approx 1.25$ over the THz frequency range. Taking into account the ratio of air in the sample and using an effective medium approach like Maxwell–Garnett theory (see Section 10.2), the optical index of the linen thread is evaluated to be $n_{\mathrm{thread}} \approx 1.33$. This low value can be explained by the composite structure of the yarn itself, which is also composed of interspersed fibers and air.

From these parameters, it is possible to simulate the transmission by the linen fabric. Figure 10.21 presents the theoretical (continuous lines) and experimental (open circles) results for different polarizations of the THz beam. We can observe a good overall agreement between the calculated and measured data whatever the frequency or the orientations of the sample, especially if we take into account that the theoretical results are obtained without any adjusting parameter. The slight differences between simulated and measured values arise from the poor description of the fabric: actual threads are not cylindrical, their sizes are different, the period of the knit is not constant, etc. Moreover, scattering in each yarn by the interlocked fibers and absorption in the yarns are

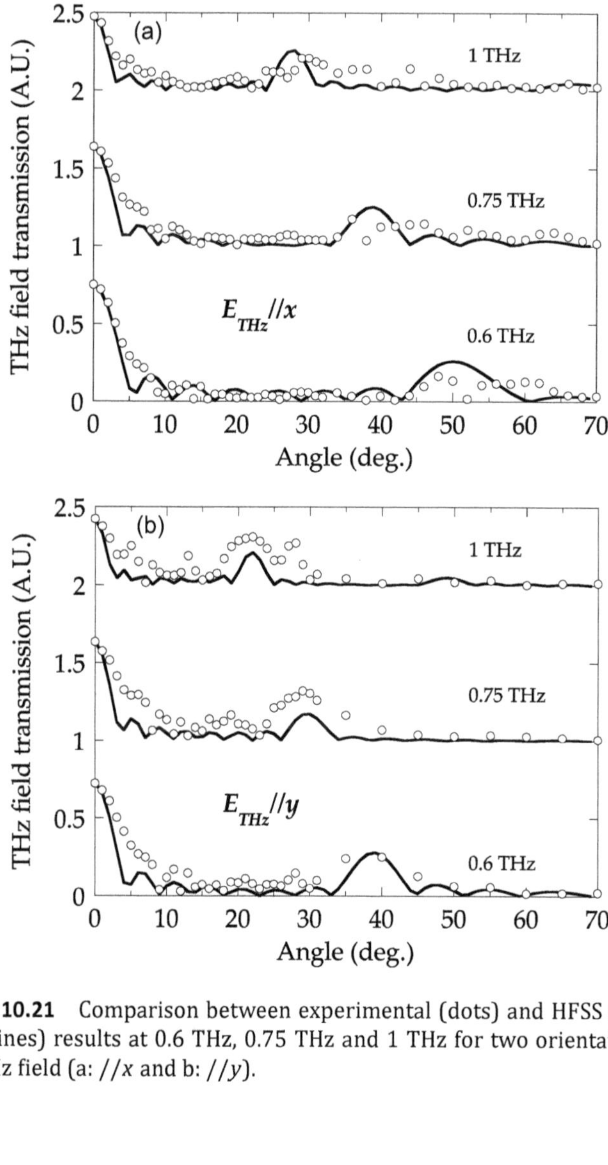

Figure 10.21 Comparison between experimental (dots) and HFSS simulation (lines) results at 0.6 THz, 0.75 THz and 1 THz for two orientations of the THz field (a: $//x$ and b: $//y$).

neglected. Despite these approximations, this simple model of the mesh is efficient enough to demonstrate the diffractive behavior of linen and more generally of fabrics, whose mesh size of the order of the wavelength.

10.4.2 Diffraction Contribution

Diffraction is a phenomenon, which must be taken into account when dealing with fabrics in the THz domain. Nevertheless, its contribution is not the same whatever is the frequency or the size of mesh. To understand the impact of diffraction in the total losses of a piece of fabrics, let us take into account the particular case of the previously described linen sample. In classical THz-TDS measurements, only the directly transmitted light is recorded since the diffracted, or scattered light is not received by the detector. In fabrics, because of their heterogeneous structure and their thinness, echoes can be neglected in first approximation, leading to a simplified expression of transmission at normal incidence ($\theta = 0°$):

$$T(f) = \frac{I_{\text{sample}}(f, \theta = 0)}{(1 - R(f))^2 I_{\text{reference}}(f, \theta = 0)} = e^{-\alpha_{\text{ext}} L}, \qquad (10.36)$$

where $T(f)$ is the transmittance of the sample, I_{sample} and $I_{\text{reference}}$ are the intensities detected respectively with and without the sample in the THz beam, θ is the detection angle, $R(f)$ the reflection coefficient at the air/sample interfaces, α_{ext} the extinction coefficient and L the sample thickness. In our case, R is weak enough ($\sim$1 %) to be neglected.

The extinction coefficient α_{ext}, deduced from Eq. 10.36, is overestimated because this expression does not include the signal that does not reach the detector. For a more accurate evaluation, the transmitted signal must be integrated over the 2π sr solid angle. Thanks to the symmetry of the system, such integration can be limited between 0 and $\pi/2$:

$$T(f) = \frac{\int_0^{\pi/2} I_{\text{sample}}(f, \theta)\, d\theta}{\int_0^{\pi/2} I_{\text{reference}}(f, \theta)\, d\theta} = e^{-\alpha_{\text{ext}} L}. \qquad (10.37)$$

Let us notice that the depolarized signal is neglected in Eq. 10.37: it has been measured to be smaller than 0.5% for the linen fabric. Therefore, the extinction coefficient, calculated with Eq. 10.37, mainly reports on the absorption of linen: it is plotted in Fig. 10.22, as well as its value directly obtained from TDS measurements, i.e., without taking into account the scattered and

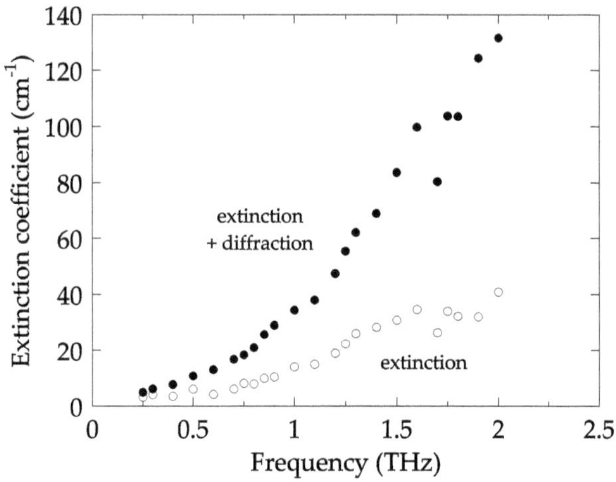

Figure 10.22 Extinction coefficient obtained from classical TDS procedure (triangles) or by taking into account diffraction effect (dots).

diffracted signal (Eq. 10.36). In the latter case, α_{ext} includes losses due to both absorption and scattering/diffraction: thus, it is strongly overestimated by a factor of up to three at the highest frequencies, around 2 THz. Although the contribution of diffraction has been removed in Eq. 10.37, α_{ext} cannot be considered as pure absorption; indeed, because of the intrinsic structure of the fabric, it is the sum of absorption and scattering.

For some other fabrics, such as fleece, denim or T-shirt, diffraction phenomena are also observed since their weaving presents a stitch size more or less similar to the linen, i.e., comparable to the THz wavelength, but this is not always the case. Indeed, the strength of the diffraction effect differs for each type of fabric, depending on the more or less fine weaving, or of the diameter of the threads used. For example, silk and some sport fabrics exhibit almost no diffraction as their stitch size is sub-wavelength (i.e., below several hundreds of micrometers) and their thinner threads are almost in contact.

Of course, such results can be generalized to any 2D periodic structure with the size of the order of the wavelength, like for example composite materials (carbon or glass fiber cloths impregnated with resin). In a practical point of view, and although this phenomenon is less common than scattering, diffraction effect could be a serious snag in sensing or imaging as it can drastically change the THz signature of any material to be detected. More precisely, unlike scattered signals that are almost isotropic, a diffracted signal is concentrated at over specific angles. If the detector is placed in one of these directions, the additional diffracted signal could be interpreted as the signature of a substance, which is not present, or which could blind the sensor. Moreover, practical conditions will make very complex to predict and process the interaction between the THz light and several cloths layers or fabrics in contact with the medium to be analyzed.

Nevertheless, since diffraction is due to interference phenomena, a way to overcome this diffraction is to perform imaging with incoherent THz light, like in passive imaging, or like active imaging with the use of an incoherent source.

References

1. Scheller, M., Jansen, C., and Koch, M. (2010). Applications of effective medium theories in the terahertz regime, in *Recent Optical and Photonic Technologies*, Kim, K. Y., ed. (InTech), pp. 231–250.
2. Tanaka, S., Shiraga, K., Ogawa, Y., Fuji, Y., and Okumura, S. (2014). Applicability of effective medium theory to wood density measurements using terahertz time-domain spectroscopy, *J. Wood Sci.*, **60**, pp. 111–116.
3. Kuzel, P., and Nemec, H. (2014). Terahertz conductivity in nanoscaled system: effective medium theory aspects, *J. Phys. D: Appl. Phys.*, **47**, pp. 374005–374014.
4. Maxwell Garnett, J. C. (1904). Colours in metal glasses and in metallic films, *Phil. Trans. R. Soc. Lond. A*, **203**, pp. 385–420.
5. Kittel, C. (2004). *Introduction to Solid State Physics*, 8th edn. (John Wiley and Sons, New Jersey).
6. Jackson, J. D. (1999). *Classical Electrodynamics* (Wiley, New York).
7. Niklasson, G. A., Granqvist, C. G., and Hunderi, O. (1981). Effective medium models for the optical properties of inhomogeneous materials, *Appl. Opt.*, **20**, pp. 26–30.
8. Sihvola, A. (1988). Effective permittivity of dielectric mixtures, *IEEE Trans. Geosci. Remote Sens.*, **26**, pp. 420–429.

9. Koledintsseva, M. Y., Wu, J., Zhang, J., Drewniak, J. L., and Rozanov, K. N. (2004). Representation of permittivity for multi-phase dielectric mixtures in FDTD modeling, *Proc. IEEE Symp. Electromagn. Compat.*, **1**, pp. 309–314.
10. Scheller, M., Wietzke, S., Jansen, C., and Koch, M. (2009). Modelling heterogeneous dielectric mixtures in the terahertz regime: a quasi-static effective medium theory, *J. Phys. D: Appl. Phys.*, **42**, pp. 065415–065419.
11. Hale, D. K. (1976). The physical properties of composite materials, *J. Mater. Sci.*, **11**, pp. 2105–2141.
12. Bruggeman, D. A. G. (1935). Berechnung verschiedener physikalischer Konstanten von heterogenen Substanzen, *Ann. Phys. (Leipzig)*, **24**, pp. 636–679.
13. Zhou, P. H., Deng, J. H., Wu, B.-I., and Kong, J. A. (2008). Influence of scatterer's geometry on power-law formula in random mixing composite, *Prog. Electromagnet. Res.*, **85**, pp. 69–82.
14. Looyenda, H. (1965). Dielectric constant of heterogeneous mixtures, *Physica*, **31**, pp. 401–406.
15. Lichtenecker, K., and Rother, K. (1931). Die Herleitung des logarithmischen Mischungs-gesetzes aus allegemeinen Prinzipien der station äaren Strömung, *Physik Zeitschr.*, **32**, pp. 255–260.
16. Buchner, A. (1939). Das Mischkorperproblem in der Kondensa torentechnik, *Wissenschoftliche Veroffentlichungen Aus Der Siemens-Werken*, **18**, pp. 84–96.
17. Zarki, T., Laurent, J.-P., and Vauclin, M. (1998). Theoretical evidence for "Lichtenecker's mixture formulae" based on the effective medium theory, *J. Phys. D: Appl. Phys.*, **31**, pp. 1589–1594.
18. Lichteneker, K. (1926). Die dielektrizitaskonstante naturlicher und kuntlicher mischkoper, *Physik Zeitschr.*, **27**, p. 115.
19. Goncharenko, A. V., Lozovski, V. Z., and Venger, E. F. (2000). Lichteneker's equation: applicability and limitations, *Opt. Commun.*, **174**, pp. 19–32.
20. Kiley, E. M., and Yakolev, V. V. (2012). Applicability study of classical and contemporary models for effective complex permittivity of metal powder, *J. Microwave Power Electromagnet. Energy*, **46**, pp. 26–38.
21. Hamdi, M. (2014). Design and realization of chipless RFID TAGs in the THz domain, PhD thesis, University of Savoy (unpublished).

22. Kaushik, M., Ng, B. W.-H., Fisher, B. M., and Abbott, D. (2012). Terahertz scattering by granular composite materials: an effective medium approach, *App. Phys. Lett.*, **100**, p. 011107.
23. Rayleigh, J. W. (1899). On the transmission of light through an atmosphere containing small particles in suspension, and on the origin of the blue of the sky, *Philos. Mag.*, **47**, pp. 375–384.
24. Dave, J. V. (1968). Subroutines for computing the parameters of the electromagnetic radiation scattered by a sphere, *Technical Rep. No. 320–3237*, IBM Scientific Center, Palo Alto, Calif.
25. Giese, R. H. (1961). Streuung elektromagnetischer Wellen an absorbierenden und dielektrischen kugelförmigen Einzelteilchen und an Gemischen solcher Teilchen, *Zeitschrift für Astrophysik*, **51**, p. 119.
26. Deirmendjann, D. (1969). *The Electromagnetic Scattering on Spherical Polydispersion* (Elsevier, New York).
27. Deirmendjann, D., Clasen, R., and Viezee, W. (1961). Mie scattering with complex index of refraction, *J. Opt. Soc. Am.*, **51**, pp. 620–633.
28. Bohren, C. F., and Huffman, D. R. (1998). *Absorption and Scattering of Light by Small Particles* (Wiley, New York).
29. Toon, O. B., and Ackerman, T. P. (1981). Algorithms for the calculation of scattering by stratified spheres, *Appl. Opt.*, **20**, pp. 3657–3660.
30. Xu, Y.-L., and Gustafson, B. A. S. (2001). A generalized multiparticle Mie-solution: further experimental verification, *J. Quant. Spectrosc. Radiat. Transfer*, **70**, pp. 395–419.
31. Hergert, W., and Wriedt, T. (2012). *The Mie Theory: Basics and Applications* (Springer Series in Optical Sciences, Vol. 169).
32. Bohren, C. F., and Gilra, D. P. (1979). Extinction by a spherical particle in an absorbing medium, *J. Colloid Interface Sci.*, **72**, p. 215.
33. Cheville, R. A., McGowan, R. W., and Grischkowsky, D. (1998). Time resolved measurements which isolate the mechanisms responsible for terahertz glory scattering from dielectric spheres, *Phys. Rev. Lett.*, **80**, pp. 269–272.
34. Png, G. M., Fumeaux, C., Stringer, M., Miles, R., and Abbott, D. (2001). Terahertz scattering by subwavelength cylindrical arrays, *Opt. Express.*, **19**, pp. 10138–10152.
35. Jian, Z., Pearce, J., and Mittleman, D. M. (2003). Characterizing individual scattering events by measuring the amplitude and phase of the electric field diffusing through a random medium, *Phys. Rev. Lett.*, **91**, p. 033903.

36. Shen, Y. C., Taday, P. F., and Pepper M. (2008). Elimination of scattering effects in spectral measurement of granulated materials using terahertz pulsed spectroscopy, *App. Phys. Lett.*, **92**, p. 051103.
37. Chen, Y., Liu, H., Deng, Y., Schauki, D., Fitch, M. J., Osiander, R., Dobson, C., Spicer, J. B., Shur, M., and Zhang X.-C. (2004). THz spectroscopic investigation of 2,4-dinitrotoluene, *Chem. Phys. Lett.*, **400**, pp. 357–361.
38. Liu, H.-B., Chen, Y., Bastians, G. J., and Zhang, X.-C. (2006). Detection and identification of explosive RDX by THz diffuse reflection spectroscopy, *Opt. Express*, **14**, pp. 415–423.
39. Zurk, L. M., Orlowski, B., Winebrenner, D. P., Thorsos, E. I., Leahy-Hoppa, M. R., and Hayden, L. M. (2007). Terahertz scattering from granular material, *J. Opt. Soc. Am. B*, **24**, pp. 2238–2243.
40. Bandyopadhyay, A., Sengupta, A., Barat, R. B., Gary, D. E., Federici, J. F., Chen, M., and Tanner, D. B. (2007). Effects of scattering on THz spectra of granular solids, *Int. J. Infrared Millimeter Waves*, **28**, pp. 969–978.
41. Franz, M., Fisher, B. M., and Walther, M. (2008). The Christiansen effect in terahertz time-domain spectra of coarse-grained powders, *App. Phys. Lett.*, **92**, p. 021107.
42. http://www.riken.jp/THzdatabase/
43. Bohren, C. F. (1987). Multiple scattering of light and some of its observable consequences, *Am. J. Phys.*, **55**, pp. 524–533.
44. http://omlc.ogi.edu/software/mie/
45. Raman, C. V. (1949). The theory of the Christiansen experiment, *Proc. Indian Acad. Sci.*, **A29**, pp. 380–381.
46. Kokhanovsky, A. A. (2002). Analytical solutions of multiple light scattering problems: a review, *Meas. Sci. Technol.*, **13**, pp. 233–240.
47. Pine, D. J., Weitz, D. A., Zhu, J. X., and Herbolzheimer, E. (1990). Diffusing-wave spectroscopy: dynamic light scattering in the multiple scattering limit, *J. Phys.*, **51**, pp. 2101–2127.
48. Pearce, J., Jian, Z., and Mittleman, D. (2003). Statistics of multiply scattered broadband terahertz pulses, *Phys. Rev. Lett.*, **91**, p. 043903.
49. Pearce, J., and Mittleman, D. (2001). Propagation of single-cycle terahertz pulses in random media, *Opt. Lett.*, **26**, pp. 2002–2004.
50. Garet, F., Hofman, M., Meilhan, J., Simoens, F., and Coutaz, J.-L. (2014). Evidence of Mie Scattering at terahertz frequencies in powder materials, *App. Phys. Lett.*, **105**, p. 031106.
51. Bohren, C. F., and Huffman, D. R. (1983). *Absorption and Scattering of Light by Small Particles* (John Wiley & Sons, New York), p. 531.

52. Bjarnason, J. E., Chang, T. L. J., Lee, A. W. M., Celis, M. A., and Brown, E. R. (2004). Millimeter-waves, terahertz and mid-infrared transmission through common clothing, *Appl. Phys. Lett.*, **85**, pp. 519–521.

53. Puc, U., Abina, A., Rutar, M., Zidansek, A., Jeglic, A., and Valusis, G. (May 2015). Terahertz spectroscopic identification of explosive and drug simulants concealed by various hiding techniques, *Appl. Opt.*, **54**, pp. 4495–4502.

54. Naftaly, M., Molloy, J. F., Lanskii, G. V., Kokh, K. A., and Andreev, Y. M. (2013). Terahertz time-domain spectroscopy for textile identification, *Appl. Opt.*, **52**(19), pp. 4433–4437.

55. Hérault, E., Garet, F., and Coutaz, J.-L. (2016). On the possibility of identifying substances by remote active THz spectroscopy, *IEEE Trans. THz Sci. Technol.*, **6**, pp. 12–19.

56. Fletcher, J. R., Swift, G. P., Dai, D. C., Levitt, J. A., and Chamberlain, J. M. (2007). Propagation of terahertz radiation through random structures: an alternative theoretical approach and experimental validation, *J. Appl. Phys.*, **101**, p. 013102.

57. Liu, H.-B., Zhong, H., Karpowicz, N. Y., Chen, and Zhang, X.-C. (2007). Terahertz spectroscopy and imaging for defense and security applications, *IEEE Proc.*, **95**, pp. 1514–1527.

58. Herault, E., Hofman, M., Garet, F., and Coutaz, J.-L. (2013). Observation of terahertz beam diffraction by fabrics, *Opt. Lett.*, **38**, pp. 2708–2710.

59. Vincent, P. (1980). *Theory of Grating*, Petit, R., ed. (Springer Verlag, Berlin).

60. Nevière, M., Popov, E., Reinisch, R., and Vitrant, G. (2000). *Electromagnetic Resonances in Nonlinear Optics* (Gordon and Breach Science Publisher, Amsterdam).

61. http://www.ansys.com.

Chapter 11

Noise in THz-TDS Systems

11.1 Errors and Precision in THz-TDS

As any value determined from experiments, material properties calculated from THz-TDS data are affected by uncertainties and errors from the measurement. In THz-TDS, measured data are electrical signals $S_{\text{meas}}(x)$ delivered by the detectors and recorded as a function of the position, x, of a delay line. These raw electrical signals are altered by random and systematic errors [1] and thus differ from the "true" value, defined as $S(x)$.

Random errors are related to the unpredictable variation (fluctuations) of the recorded value over the course of several measurements. They are due to noise, poor equipment sensitivity, and sometimes to imprecise definitions. Thus, any record i of the signal is written as

$$S_{\text{meas},i}(x) = S(x) + \Delta S_{\text{i}}(x). \tag{11.1}$$

$\Delta S_{\text{i}}(x)$ is the difference between the recorded and true values. One can perform a statistical treatment of the random error: the mean value $\overline{S_{\text{meas}}(x)}$ of a set of records $S_{\text{meas},i}(x)$ corresponds to the best

Principles of Terahertz Time-Domain Spectroscopy
Jean-Louis Coutaz, Frédéric Garet, and Vincent Wallace

ISBN 978-981-4774-56-7 (Hardcover), 978-0-203-70969-6 (eBook)
www.panstanford.com

experimental estimate of the signal:

$$\overline{S_{\text{meas}}(x)} = \frac{1}{I}\sum_{i=1}^{I} S_{\text{meas},i}(x) = S(x) + \frac{1}{I}\sum_{i=1}^{I} \Delta S_{\text{i}}(x) = S(x) + \overline{\Delta S(x)} \tag{11.2}$$

Here, I is the total number of recordings of the signal. Terms with a bar over the top correspond to mean values. Due to the randomness of the measurements, the mean value of the uncertainty tends to zero while increasing I (due to both positive and negative contributions). Thus, when $I \rightarrow \infty$,

$$\overline{\Delta S(x)}_{I\rightarrow\infty} \rightarrow 0 \quad \Rightarrow \quad \overline{S_{\text{meas}}(x)} \underset{I\rightarrow\infty}{=} S(x). \tag{11.3}$$

While its mean value is null (this is not the general case, but this is commonly observed in common THz-TDS setups), the noise is better described by its standard deviation $\sigma(x)$:

$$\sigma^2(x) = \frac{1}{I}\sum_{i=1}^{I} \left(S_{\text{meas},i}(x) - \overline{S_{\text{meas}}(x)}\right)^2 = \frac{1}{I}\sum_{i=1}^{I} (\Delta S_{\text{i}}(x))^2 = \overline{\Delta S^2(x)}. \tag{11.4}$$

If $\sigma(x) << S(x)$, it follows that

$$S_{\text{meas}}(x) = S(x) \pm \sigma(x). \tag{11.5}$$

Systematic errors make the best estimate of the measured signal shifted compared to the expected signal. They are due to biases inherent to the experimental setup or/and to the experimental procedure. For example, a poorly calibrated delay line could give erroneous values of the distance x. Systematic errors are often very difficult to identify and sometimes to correct. Contrary to random errors, they are not reduced by averaging the signal over a large number of recordings. It follows that the relation between the measured and the actual signals is

$$S_{\text{meas}}(x) = S(x) \pm \Delta S(x) = S(x) + \Delta S_{\text{shift}}(x) \pm \sigma(x). \tag{11.6}$$

Both random and systematic errors are schematically presented in Fig. 11.1.

From the recorded electrical signals $\overline{S_{\text{meas}}(x)}$, one can determine the complex refractive index $\tilde{n}(\omega) = n(\omega) - j\kappa(\omega)$ of the sample material under study as explained in Chapter 4. Extraction methods are based on a model for transmission or reflection of the THz beam

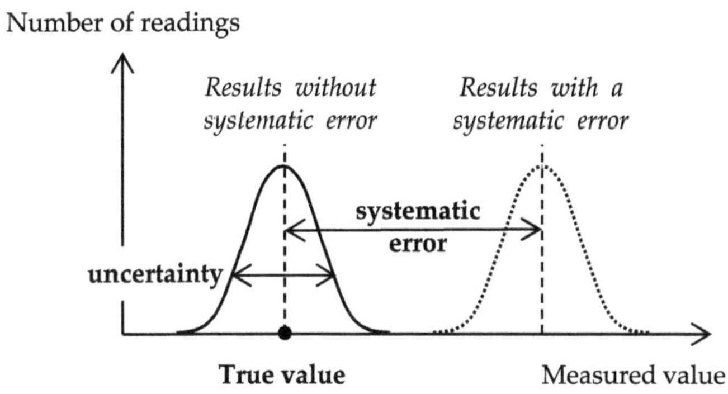

Figure 11.1 A simplified representation of random and systematic errors.

through or by the sample, and the numerical solution of the inverse electromagnetic problem. This determination is affected by several errors:

- Propagation of the uncertainty $\Delta S(x)$ in $S(x)$, which means that entering $S(x) \pm \Delta S(x)$ instead of $S(x)$ in the calculation induces an error in $\tilde{n}(\omega)$.
- An inaccurate model or poor approximation of the transmission/reflection of the actual THz beam through/by the sample. For example, most of the extraction procedures are based on a plane wave analysis, which does not correspond rigorously to actual THz beam shape in THz-TDS systems.
- For the sample itself: an erroneous knowledge of its geometry and/or of its structure could lead a large uncertainty. For example, if the sample is heterogeneous at the sub-wavelength scale, scattering may be misinterpreted as absorption.
- Numerical rounding in computation.

These different errors that can occur during the measurement and the subsequent extraction process are summarized in Fig. 11.2.

These sources of error have been clearly classified as random errors, systematic errors, or both, by Withayachumnankul et al. [2], as presented in the following table, adapted from this reference. The noise and uncertainties in THz-TDS experiments and their impact

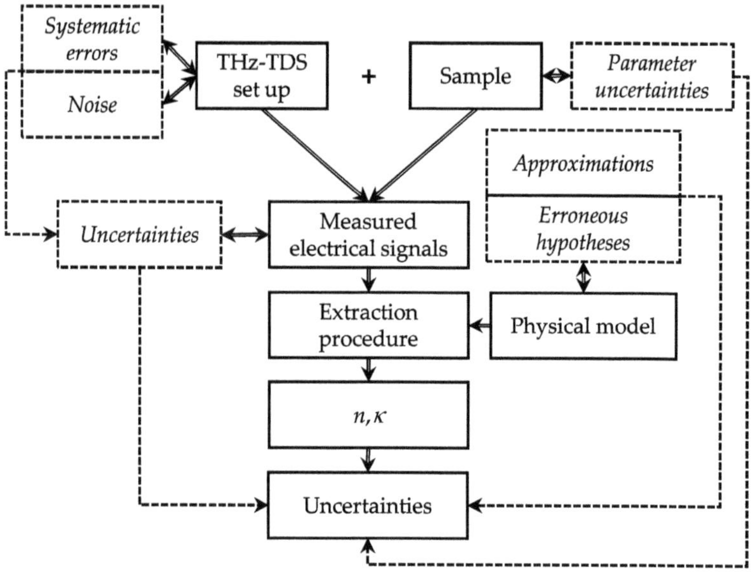

Figure 11.2 Sources of error in THz-TDS measurements.

on the extraction of material parameters are reviewed by the same author in a comprehensive reference [3].

Table 11.1 Random and systematic errors in THz-TDS measurements from Withayachumnankul et al. [2]

Electronic and optic noise	Random and systematic
Sample alignment	Random or systematic
Sample thickness measurement	Random and systematic
Approximate transmission/reflection transfer function	Systematic
Refractive index of air	Systematic
Fabry-Pérot reflections in the sample	Systematic

In this chapter, we will treat the noise in THz-TDS measurements. The next chapter is devoted to the errors, including those that result from noise.

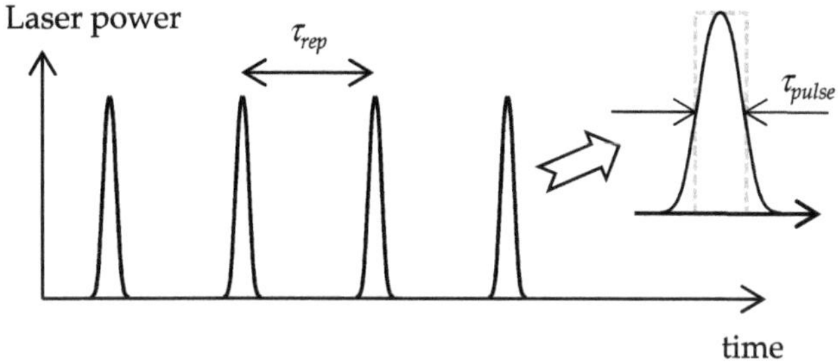

Figure 11.3 Parameters of the laser pulse train.

11.2 Origin of Noise in THz-TDS Systems

11.2.1 Contribution of the Laser Shot Noise to Noise in Emission

In common THz-TDS systems, emission of the THz pulses is prompted by the excitation of either photoconductive antennas or electro-optic crystals with a train of femtosecond laser pulses delivered by self mode-locked lasers. The noise of the laser beam is related to the particle aspect of light (photon noise) and the fluctuations of the laser.

11.2.1.1 Photo-conducting antennas

The photon noise can be addressed as follows. Let us first consider the laser pulse train of repetition frequency $f_{\text{rep}} = 1/\tau_{\text{rep}}$ (Fig. 11.3) and of pulse duration τ_{pulse}. In a simple model, each pulse could be substituted by a rectangular pulse, of time width τ_{pulse}, during which the power is constant (see inset on the right of Fig. 11.3). The mean number N_{pulse} of photons per pulse is

$$N_{\text{pulse}} = \frac{1}{f_{\text{rep}}} \frac{P_{\text{laser}}}{h\, f_{\text{laser}}}. \tag{11.7}$$

P_{laser} is the mean laser power, and f_{laser} is the central optical frequency of the pulsed laser beam. The laser photon statistics obeys a Poisson distribution [4], even when dealing with femtosecond laser pulses [5], thus N_{pulse} is known with a precision ΔN_{pulse}:

$$\left(\Delta N_{\text{pulse}}\right)^2 = N_{\text{pulse}} = \frac{1}{f_{\text{rep}}} \frac{P_{\text{laser}}}{h\, f_{\text{laser}}}. \tag{11.8}$$

In photoconductive antennas, the incident train of laser pulses generates bursts of electrons (here we neglect the contribution of the holes). As the photo-generation of electrons is also a stochastic process, the uncertainty $\Delta N_{e/p}$ is as well governed by the Poisson statistics. Therefore, the number of electrons per pulse and its noise are

$$N_{e/p} = \eta N_{pulse} = \frac{\eta}{f_{rep}} \frac{P_{laser}}{h f_{laser}}, \quad (\Delta N_{e/p})^2 = N_{e/p}. \tag{11.9}$$

η is the quantum efficiency of the photo-excitation process. The peak current i_{pulse} expresses the number $N_{e/p}$ of collected electrons per pulse times the electron charge q divided by the pulse duration τ_{pulse}:

$$i_{pulse} = \frac{N_{e/p} q}{\tau_{pulse}} = \frac{\eta}{f_{rep}} \frac{q}{\tau_{pulse}} \frac{P_{laser}}{h f_{laser}},$$

$$(\Delta i_{pulse})^2 = \left(\frac{q}{\tau_{pulse}}\right)^2 N_{e/p} = \frac{q}{\tau_{pulse}} i_{pulse}. \tag{11.10}$$

Here, we assume that the pulse duration does not exhibit noise; and the mean current i is derived from i_{pulse} by taking into account the comb rate:

$$i = f_{rep} \tau_{pulse} i_{pulse} = q\, f_{rep} N_{e/p} = \eta\, q \frac{P_{laser}}{h f_{laser}}, \quad (\Delta i)^2 = (q\, f_{rep})\, i. \tag{11.11}$$

We have supposed that there is no current noise between the current pulses. These expressions 11.11 must be corrected because, in ultrafast semiconductors, most photoelectrons are trapped before reaching the electrodes. Using a rough estimate, the percentage of photoelectrons that are collected by the electrode is the ratio of the displacement x induced the bias field ($x = v\, \tau_{fc}$, where v is the transient carrier velocity and τ_{fc} the carrier lifetime) and the inter-electrode gap g. It follows that

$$i = \frac{v\, \tau_{fc}}{g} \eta\, q \frac{P_{laser}}{h f_{laser}}, \quad (\Delta i)^2 = \left(\frac{v\, \tau_{fc}}{g} q\, f_{rep}\right) i. \tag{11.12}$$

For typical THz-TDS LT-GaAs antennas (P_{laser} =10 mW, $\tau_{pulse} = 0.5$ ps, $g = 10$ µm, $\lambda_{laser} = 0.8$ µm, $\eta = 0.5$), a 10-V bias gives the electrons a velocity of the order of $v = 2 \times 10^5$ m/s. Therefore, only $v\, \tau_{fc}/g \approx 1\%$ of these electrons participate to the current flowing in the connection wires. This current is typically $i = 32$ µA and $\Delta i = 2$ nA ($\Delta i / i \approx 6 \times 10^{-5}$).

The performance of a noisy system is well assessed by defining the signal-to-noise ratio (*SNR*), which is the ratio of the signal power (here proportional to the current to the square) and the noise power, i.e., proportional to $(\Delta i)^2$:

$$SNR = \frac{i^2}{(\Delta i)^2} = \frac{i}{\frac{v\,\tau_{fc}}{g} q\, f_{rep}} = \frac{\eta}{f_{rep}} \frac{P_{laser}}{h\, f_{laser}}. \tag{11.13}$$

With the above parameters of THz-TDS LT-GaAs antennas, the current *SNR* is of the order of 90 dB. Even if this estimation has been crudely calculated, the orders of magnitude are close to that measured experimentally. At the moment, the dark current, i.e. the random current flowing even without laser excitation, and the Johnson noise, due to the thermal agitation of the charge carriers, have not been considered.

The THz radiated far field is proportional to the time derivative of the current pulses (Eq. 11.10). Thus, we need to take into account the temporal shape $h\,(t)$ of the pulses. Let us write the number $N_e\,(t)$ of photo-excited carriers at time t as:

$$N_e\,(t) = N_{e,p} h\,(t), \quad \text{with} \int_{-\infty}^{+\infty} h\,(t)\, dt = 1. \tag{11.14}$$

The current at time t is written as

$$i_{pulse}\,(t) = q N_e\,(t) = q\, h\,(t)\, N_{e,p}. \tag{11.15}$$

Because of the Poisson statistics, it follows immediately that

$$(\Delta N_e\,(t))^2 = N_{e,p} h\,(t), \quad \left(\Delta i_{pulse}\,(t)\right)^2 = q^2\, h\,(t)^2\, N_{e,p}. \tag{11.16}$$

The temporal shape $h\,(t)$ of the pulses is supposed not to vary from pulse to pulse. The THz field radiated far from the antenna is proportional to the time derivative of the current:

$$E_{THz,pulse}\,(t) = a \frac{d\, i_{pulse}\,(t)}{dt} = a N_{e,p} q \frac{dh\,(t)}{dt}.$$

$$\Rightarrow (\Delta E_{THz}\,(t))^2 = \left(aq \frac{dh\,(t)}{dt}\right)^2 N_{e,p},$$

$$SNR = \left(\frac{E_{THz,\delta t}\,(t)}{\Delta E_{THz,\delta t}\,(t)}\right)^2 = N_{e,p} \tag{11.17}$$

a is a proportionality constant. Note that here we use i_{pulse}, i.e., the current flowing in the switch gap and not the one reaching the electrodes, because it is i_{pulse} that radiates the THz waves. For the same parameters as previously, i.e., $P_{\text{laser}} = 10$ mW, $\lambda_{\text{laser}} = 0.8$ μm, $\eta = 0.5$, $f_{\text{rep}} = 100$ MHz, the shot noise-limited *SNR* of the THz signal is ~ 73 dB.

11.2.1.2 Electro-optic antennas

When generating THz pulses by optical rectification in electro-optic (EO) crystals, the THz field waveform is given by expression 4.20:

$$E_{\text{THz,pulse}}(t) = -\frac{\mu_o}{2n_{\omega_{\text{opt}}}n_\omega}\chi^{(2)}(\omega, \omega_{\text{opt}})\, z\frac{d\,P_{\text{laser}}(t)}{dt} \equiv a'\frac{d\,P_{\text{laser}}(t)}{dt}. \tag{11.18}$$

z is the crystal thickness. Employing the same procedure as used above, we get

$$E_{\text{THz,pulse}}(t) = a' P_{\text{laser,peak}}\frac{d\,h(t)}{dt} = a'\frac{P_{\text{laser}}}{f_{\text{rep}}\tau}\frac{d\,h(t)}{dt}, \tag{11.19}$$

$$SNR(\text{THz}) = \left(\frac{P_{\text{laser,peak}}}{\Delta P_{\text{laser,peak}}}\right)^2 = \frac{P_{\text{laser}}}{h\,f_{\text{laser}}\,f_{\text{rep}}}. \tag{11.20}$$

Here, for a 10 mW laser beam at a 100 MHz repetition rate, $SNR(\text{THz}) \approx 86$ dB. Roughly speaking, the THz field *SNR* is of the same order of magnitude for both photoconductive and electro-optic antennas.

11.2.2 Contribution of the Laser Shot Noise to Noise in Reception

11.2.2.1 Photo-conducting antennas

In photo-conducting antennas, the signal of interest is the current arising from photo-generated carriers that are accelerated by the incoming THz field. We treat the case of similar antennas in emission and reception. With the same notation as previously, we write

$$i_{\text{pulse,det}}(t, \tau) = \frac{v\,\tau_{\text{fc}}}{g}E_{\text{THz}}(t+\tau)\,N_{\text{e}}(t)\,q\mu_{\text{e}}, \tag{11.21}$$

with $N_{\text{e}}(t) = \frac{\eta P_{\text{laser}} h(t)}{f_{\text{rep}} h\, f_{\text{laser}}} = N_{\text{e,p}} h(t)$.

μ_{e} is the carrier mobility and τ is the time delay between THz and laser pulses at the receiver. Usually, the electronics is slow as compared to the involved picosecond waveforms. The pulsed current $i_{\text{pulse,det}}(t, \tau)$ is thus integrated over time t. When the laser pulse repetition rate is f_{rep}, the recorded current is

$$i_{\text{det}}(\tau) = f_{\text{rep}} \int_{\text{pulse}} i_{\text{pulse,det}}(t, \tau)\, dt$$
$$\Rightarrow i_{\text{det}}(\tau) = a\mu_{\text{e}} f_{\text{rep}} \frac{v\,\tau_{\text{fc}}}{g} \left(qN_{\text{e,p}}\right)^2 \int_{\text{pulse}} \frac{dh(t+\tau)}{dt} h(t)\, dt. \tag{11.22}$$

Here we consider only the shot noise. Therefore, in Eq. 11.22, only $N_{\text{e,p}}$ is affected by this noise. Therefore, we get

$$(\Delta i_{\text{det}}(\tau))^2 = \left(\frac{v\,\tau_{\text{fc}}}{g} a f_{\text{rep}} q^2 \mu_{\text{e}}\right)^2 \left(\int_{\text{pulse}} h(t) \frac{dh(t+\tau)}{dt} dt\right)^2 \left(\Delta\left(N_{\text{e,p}}^2\right)\right)^2$$

Assuming $\Delta N_{\text{e,p}} << N_{\text{e,p}}$, i.e. $\Delta\left(N_{\text{e,p}}^2\right) \approx 2N_{\text{e,p}}\Delta N_{\text{e,p}}$, leads to

$$(\Delta i_{\text{det}}(\tau))^2 = (i_{\text{det}}(\tau))^2 \left(\frac{2\Delta\left(N_{\text{e,p}}\right)}{N_{\text{e,p}}}\right)^2 = \frac{4}{N_{\text{e,p}}}(i_{\text{det}}(\tau))^2$$
$$\Rightarrow SNR = \left(\frac{i_{\text{det}}(\tau)}{\Delta i_{\text{det}}(\tau)}\right)^2 = \frac{N_{\text{e,p}}}{4}. \tag{11.23}$$

For $f_{\text{rep}} = 100$ MHz, $P_{\text{laser}} = 10$ mW, $\lambda = 0.8$ μm and $\eta = 0.5$, $SNR \approx 67$ dB. Therefore, it is preferable to use the highest possible laser power to trigger both emitting and receiving antennas to decrease the shot noise. For example, having $P_{\text{laser}} = 100$ mW instead of 10 mW increases the SNR by 10 dB.

It is interesting to separate the contribution of the shot noise in the emitter and in the receiver. We write again the current in the receiver, but now as a function of the carriers per pulse in the emitter, $N_{e,p,E}$, and in the receiver, $N_{e,p,R}$:

$$i_{\text{det}} \propto N_{e,p,E} N_{e,p,R},$$

$$\Rightarrow \left(\frac{\Delta i_{\text{det}}}{i_{\text{det}}}\right)^2 = \left(\frac{\Delta N_{e,p,E}}{N_{e,p,E}}\right)^2 + \left(\frac{\Delta N_{e,p,R}}{N_{e,p,R}}\right)^2 + 2\left(\frac{\Delta N_{e,p,E}}{N_{e,p,E}}\right)\left(\frac{\Delta N_{e,p,R}}{N_{e,p,R}}\right),$$

$$\Rightarrow \frac{1}{SNR_{(i_{\text{det}})}} = \frac{1}{SNR_{(N_{e,p,E})}} + \frac{1}{SNR_{(N_{e,p,R})}} + \frac{2}{\sqrt{SNR_{(N_{e,p,R})} SNR_{(N_{e,p,D})}}}. \tag{11.24}$$

If $N_{e,p,E} \approx N_{e,p,R}$ and if both values follow the same uncertainty law, for example, a Poisson like one, the noise is multiplied by 4 (6 dB) when one takes into account the shot noise in emission ($1/SNR_{N_{e,p,E}} \neq 0$ in (11.24)).

11.2.2.2 Electro-optic antennas

When using electro-optic detection to record the THz pulsed field, the expression of the recorded signal is given by (Eq. 4.34)

$$S\,(\text{THz}) = \frac{i_1 - i_2}{i_1 + i_2} = \frac{I_1 - I_2}{I_1 + I_2} \approx \frac{\delta}{2} = \frac{1}{4}\frac{\omega_{\text{opt}}}{c} n_{\text{las}}^3 r\, L\, E_{\text{THz}}. \tag{11.25}$$

i_{p} and I_{p} ($p = 1{,}2$) are the electrical currents delivered by the photodiode labeled p and the laser intensity that excites this photodiode, respectively. For the sake of simplicity, we do not use the matrix form of Eq. 4.34, and thus r is the order of magnitude of the EO coefficient, n_{las} is the refractive index of the EO crystal at an optical frequency, and L is the crystal length. It follows that the measured THz field is given by

$$E_{\text{THz}} = \frac{4\,c}{\omega_{\text{opt}} n_{\text{las}}^3 r\, L} \times \frac{i_1 - i_2}{i_1 + i_2}. \tag{11.26}$$

The uncertainty in E_{THz} is due to noise from the thermally induced variations of the crystal properties (L, r and n_{las}), and the shot noise of the recorded current. It follows that

$$\frac{\Delta E_{\text{THz}}}{E_{\text{THz}}} = \frac{\Delta L}{L} + \frac{\Delta n_{\text{las}}}{n_{\text{las}}} + \frac{\Delta r}{r} + 2\frac{i_2 \Delta i_1 + i_1 \Delta i_2}{i_1^2 - i_2^2}. \tag{11.27}$$

Assuming the uncertainty of current i due to shot noise is $\sigma_{\text{i}}^2 = 2Bqi$, the THz field SNR obeys the following relation:

$$\frac{1}{SNR} = \frac{\sigma_{E_{\text{THz}}}^2}{E_{\text{THz}}^2} = \frac{\sigma_L^2}{L^2} + \frac{\sigma_{n_{\text{las}}}^2}{n_{\text{las}}^2} + \frac{\sigma_r^2}{r^2} + \frac{8Bq\, i_1 i_2}{(i_1^2 - i_2^2)\,(i_1 - i_2)}. \tag{11.28}$$

B is the receiver bandwidth, and q is the charge of the electron. The last term of (11.28) originates in the shot noise, while the first three terms are related to the thermal induced changes of the crystal. They depend on the variation ΔT of the crystal temperature:

$$\frac{\Delta L}{L} = \alpha_{\text{therm}} \Delta T, \quad \frac{\Delta n_{\text{las}}}{n_{\text{las}}} = \frac{dn_{\text{las}}}{dT} \Delta T, \quad \frac{\Delta r}{r} = \alpha_{\text{EO}} \Delta T. \quad (11.29)$$

α_{therm} is the thermal expansion coefficient of the crystal material, dn_{las}/dT is the temperature derivative of its refractive index and α_{EO} is temperature coefficient of its linear EO coefficient. Therefore, we get

$$\frac{1}{SNR} = \left(\alpha_{\text{therm}}^2 + \left(\frac{dn_{\text{las}}}{dT} \right)^2 + \alpha_{\text{EO}}^2 \right) \Delta T^2 + \frac{8Bq\, i_1 i_2}{\left(i_1^2 - i_2^2\right)(i_1 - i_2)}. \quad (11.30)$$

For ZnTe, $\alpha_{\text{therm}} \approx 8.2 \times 10^{-6}$ K^{-1} [6], $dn_{\text{las}}/dT \approx 6 \times 10^{-5}$ K^{-1} [7] and $\alpha_{\text{EO}} \approx 15 \times 10^{-4}$ K^{-1} (r_{41}) [8]. The change of the EO coefficient value is by far the most important. Typical temperature fluctuations within the crystal of the order of $\Delta T \approx 10$ mK lead to $\alpha_{\text{EO}}^2 \Delta T^2 \approx 2 \times 10^{-10}$, i.e., a *SNR* of 97 dB. On the other hand, the shot noise contribution in (11.30) can be rewritten as follows:

$$\frac{1}{SNR_{\text{shot}}} = \frac{8Bq\, i_1 i_2}{\left(i_1^2 - i_2^2\right)(i_1 - i_2)}$$

$$\Rightarrow \quad \sigma_{\text{shot}}^2 = \frac{8Bq\, i_1 i_2}{(i_1 + i_2)^3} \left(\frac{4c}{\omega_{\text{opt}} n_{\text{las}}^3 r L} \right)^2.$$

Because the two currents i_1 and i_2 are almost equal to i (the THz induced difference $i_1 - i_2 = 2\Delta i$ is small), we obtain

$$\sigma_{\text{shot}}^2 \approx \frac{Bq}{i} \left(\frac{4c}{\omega_{\text{opt}} n_{\text{las}}^3 r L} \right)^2 = \frac{Bq}{i} \left(\frac{2\lambda_{\text{opt}}}{\pi n_{\text{las}}^3 r L} \right)^2. \quad (11.31)$$

In the weak signal limit, the standard deviation of the THz current does not depend on the THz field, but only on the recorded current, i.e., on the laser intensity that reads the EO signal. Therefore, the shot-noise limited THz current *SNR* is

$$SNR_{\text{shot}} \approx \frac{\Delta i^2}{Bqi} = \frac{i}{Bq} \left(\frac{\pi n_{\text{las}}^3 r L}{\lambda_{\text{opt}}} E_{\text{THz}} \right)^2. \quad (11.32)$$

For a typical (non-amplified laser) THz-TDS system, $i = 1$ mA, $\lambda = 0.8$ μm, $B = 10$ Hz (100 ms integration time), and with a $L = 1$ mm thick

ZnTe crystal ($n \sim 3, r \sim 4$ pm/V), we obtain $SNR_{\text{shot}} \approx 12 \times E_{\text{THz}}^2$. With a typical $10 \sim 100$ V/cm THz field, $SNR_{\text{shot}} \approx 70 \sim 90$ dB.

11.2.3 Flicker Noise and Other Laser Instabilities

Fluctuations of the laser beam intensity, in addition to the photon noise, are generated by different phenomena, leading to the well-known flicker noise, which varies typically as $1/f$, where f is the frequency of the Fourier transform of the noise power. Its main origin is the amplification of the noise in the pump laser. The use of solid-state diode-pumped (SSDP) lasers has led to a great gain in performance as compared to the Ar-ion pump laser that was originally used. Minor differences are observed when using single-mode or multi-mode SSDP lasers, as shown in Fig. 11.4a adapted from [9].

The total noise integrated over the whole spectrum is very low (Fig. 11.4b) [10]. Typically, the flicker noise is reduced significantly when the signals are recorded with a lock-in amplifier (LIA) running at an acquisition frequency higher than 1 kHz. For example, the typical relative intensity noise (RIN) around 1 kHz is $\sim 10^{-10}\,\text{Hz}^{-1}$. If the signal is read with lock-in amplifier of 100 ms integration time (bandwidth 10 Hz), the effective RIN-limited *SNR* is 90 dB. When taking into account noise in emission and reception (Eq. 11.22), one should add 6 dB of noise, i.e. then total RIN induced noise is around 84 dB. Here the measurements are impacted more by the shot noise than by RIN. However, if the recordings are performed at a smaller LIA frequency, for example, 10 Hz instead of 1 kHz, the RIN-limited *SNR* is 70 dB, i.e., smaller than the shot-noise limited *SNR*.

More important is the drift of the laser power or other instabilities due to a bad adjustment of the laser cavity or worn out laser components (pumping diodes, chiller pump, etc.). The related effects are difficult to quantify, and even, to identify. This can be only achieved by regularly comparing the recorded signals (waveform amplitude and shape, spectrum bandwidth, noise) with the ones obtained when the laser is perfectly regulated. The drift of the laser power occurs mostly when the laser is heating up or if the room temperature is varying. It could lead to a decrease or increase of the signal during recording the waveform or from a record to

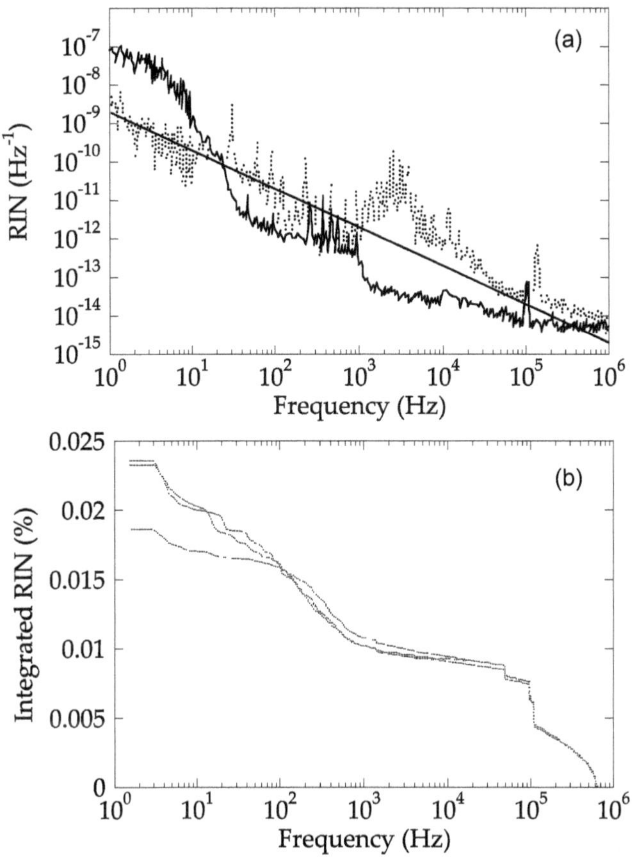

Figure 11.4 (a) Relative intensity noise of a Coherent Verdi-V6 laser (continuous curve) and a Spectra-Physics Millennia Xs laser (dotted curve) (adapted from Matos et al. [9]). The straight line corresponds to a $1/f$ noise law for the Spectra-Physics laser. (b) Integrated RIN of a Ti:Sapph femtosecond oscillator (Femtolasers Femtosource 20) pumped by a Coherent-Verdi V10 laser (adapted from Vernaleken et al. [10]).

another one, if several records are performed in view of averaging. Taking a regular record of the reference signal helps in detecting and removing drift effects. Moreover, when the noise in the THz records suddenly increases, it could be the signature of a deterioration of the laser functioning. Let us also notice that in the earlier literature on THz-TDS, most of the systems used, at first, a dye CPM laser, and subsequently a Ti:Sapph laser pumped by an Ar-ion laser. Both these lasers were much noisier than present diode-pumped Ti:Sapph or doped fiber lasers. Thus, the reported contribution of the laser instability was quite huge [11, 12].

11.2.4 Contribution of the Dark Current in the Emitter

In photo-conducting antennas, the switch exhibits a finite resistivity even when not illuminated. This dark conductivity originates mostly in ion defects for materials with a rather large bandgap, like LT-GaAs. In small gap semiconductors, like LT-InGaAs, the thermal excitation of carriers through the gap, contributes mainly to the dark conductivity. The dark resistivity of LT-GaAs is of the order of 10^6 Ω·cm [13], while, for operation at 1.5 μm, it is only about 1000 Ω·cm in the best homogeneous materials (ion-implanted InGaAsP [14]), as well as in layered structures [15]. Typically, the resulting dark resistance of switches with a 5–10 μm gap is about 20 ~ 30 MΩ and 0.2 ~ 0.5 MΩ respectively for LT-GaAs and 1.5 μm material switches. Under a 10 V DC bias, the dark currents are 0.2 ~ 0.3 μA and 20 ~ 50 μA, respectively. It is essential to note that only the temporal fluctuations of the dark current, i.e., its noise, generates a THz signal. The noise of this current (Johnson noise) is given by $\sigma_i^2 = 4\, k_B T\, B/R_{dark}$, where k_B is the Boltzmann constant, R_{dark} is the dark resistance, and B is the bandwidth over which the noise is measured. If $B = 10$ Hz, we get $\sigma_i \approx 0.1$ pA and $\sigma_i \approx 1$ pA, respectively. This has to be compared with the peak value i_{peak} of the photo-excited pulse: here, in this comparison, we have to take into account the ratio (typically 10^{-4}) between excitation duration (typically 1 ps) and repetition period of the laser pulses (typically 10 ns), because the dark current signal is generated continuously. For a typical $R_{on} \approx 100$ Ω and a 10 V DC voltage, $i_{peak} \times 10^{-4} \approx 10$ μA. The corresponding *SNR* is 90 and 80 dB respectively for 0.8 and 1.5 μm materials.

In EO emitting antenna, there is no dark signal because the EO crystal does not radiate a THz signal when not illuminated (the thermal signal from the crystal contributes to the overall thermal radiation of the room environment, which is treated below).

11.2.5 Dark Current in the Receiver

In photo-conducting antennas, the recorded current arises from the acceleration of the carriers (photo-generated by the laser probe beam) by the electrical field of the incoming THz pulse. When not polarizing the switch with the incoming THz signal, there still exist photo-generated carriers in the semiconductor and a current may flow due to some intrinsic voltage at the switch. This voltage could be caused by a Schottky contact at the switch electrodes and some asymmetry of the electrodes geometry. It can be easily removed by adjusting the lock-in amplifier that reads the current. A second source of noise is the presence of free carriers in the semiconductor material even if not excited by the laser probe beam. They will lead to a current whatever the delay between the THz and laser pulse. Even the lighting in the room could excite these free carriers. Typically, under common lighting conditions, about 10^{12} photons impinge a 1 cm^2 surface per second [4]. Assuming the average wavelength of the visible lighting is 550 nm, the corresponding power received by a 10×10 μm^2 switch area is 0.4 pW, which is negligible compared to the typical 10 mW laser power. However, the lightning illuminates not only the switch gap area but also the access connections to the photo-switch. The preceding 0.4 pW power is thus underestimated. Performing the measurement in a dark room may reduce this effect.

In EO detection system, a dark current flows in the photodiodes. It corresponds to the saturation current in the reverse-biased diodes. Typically, in Si PIN diodes, the dark current i_d is of the order of 10 nA, while the responsivity of the diodes is about $0.6 \sim 0.8$ A/W, i.e., a $0.6 \sim 0.8$ mA current for a typical 1 mW laser probe beam power. Here again, the dark current contribution can be properly removed by adjusting the lock-in amplifier, while the corresponding noise is negligible.

11.2.6 Fluctuations of the Laser Beam Pointing

The laser beam is not only fluctuating in intensity, but its direction may also vary randomly and is known as pointing stability. This is mostly due to mechanical vibrations of the optical components within the laser cavity, especially the output mirror, but also due to thermal effects, for example in the optical gain material [16]. Typically, the beam pointing fluctuation $\delta\theta$ is smaller than a tenth of a mrad [17]. Fluctuations of the air density along the beam path between the laser and the THz antennas lead to smaller pointing variations. If the laser beam is focused by a convergent lens (of focal distance F) onto the THz antenna (photo-conducting or electro-optic), the focused beam spot location varies laterally as $\delta x \approx F\,\delta\theta$. Typically, with $F = 100$ mm, δx is of the order of a few μm. This corresponds to about 1/20 of the focused laser spot diameter (assuming a 5 mm laser beam waist at the lens). When using a photo-conducting antenna with a 20 μm inter-electrode gap, this induces a 0.1 % variation of the exciting laser power. For narrower gaps, the effect is even weaker.

Along with beam pointing fluctuations, mechanical vibrations within the setup make the laser beam illuminating the THz antennas at random positions. The effect of such vibrations is difficult to quantify. The vibrations could be reduced using good quality dampers to hold the optical table, as well as by tightly binding the optical components and devices to the table.

11.2.7 Thermal Background

When measurements are performed at room temperature, the thermal radiation from the environment acts the THz receiver. The random (incoherent) character of the thermal radiation makes null the average induced signal. Nevertheless, this radiation contributes to the noise. An estimation of this contribution may be performed starting from Planck's law, assuming the receiver is at thermal equilibrium with the room. The blackbody electromagnetic energy dU' in a volume dV and a df bandwidth is

$$dU' = 8\pi h \frac{f^3}{c^3} \frac{1}{e^{hf/k_B T} - 1} dV df \approx \frac{8\pi f^2}{c^3} k_B T\, dV df. \qquad (11.33)$$

Let us consider a photo-conducting receiving antenna. The switch area dimensions are gap width L, electrode width W, active layer thickness d. Thus, the volume of the active detecting region is $dV = LWd$. Only the thermal radiation with an electrical field oriented towards the switch electrodes contributes to the current flowing through the switch. Taking into account the corresponding solid angles leads to the effective thermal radiation dU within the active region:

$$dU \approx dU' \times \frac{2d}{2\pi L} \times \frac{2W}{2\pi L} = dU' \times \frac{Wd}{\pi^2 L^2} = \frac{8\pi f^2}{c^3} k_B T \frac{Wd}{\pi^2 L^2} dV df. \tag{11.34}$$

This electromagnetic energy can also be derived from electromagnetic theory:

$$dU = \frac{1}{2}\varepsilon_o \varepsilon c E_{\mathrm{THz}}^2 dV. \tag{11.35}$$

ε is the relative dielectric constant of the switch material and E_{THz} is the corresponding electrical field. We get its magnitude by comparing Eqs. 11.34 and 11.35:

$$E_{\mathrm{THz}}^2 = \left(\frac{4f}{c^2 L}\right)^2 \frac{k_B T}{\pi \varepsilon_o \varepsilon} Wd\, df. \tag{11.36}$$

Without integrating this expression over the available THz range, we crudely obtain $E_{\mathrm{THz}}^2 = df \times 10^{-24}\ \mathrm{V^2/m^2}$. For a 1 THz bandwidth, the field is about $E_{\mathrm{THz}} = 10^{-6}$ V/m. In common THz-TDS setups, the field amplitude at the THz pulse peak is about kV/cm [18], i.e., 10^4 V/m. Thus, the peak signal is about 100 dB over thermal radiation noise.

11.2.8 Johnson Noise and Noise in the Electronics

As already explained, the receiving device, in a THz-TDS setup, delivers an electrical current whose intensity is proportional to the THz field. Due to the random instantaneous motion of electrons in conductors, any electrical current is altered by noise, known as the Johnson noise (or as Johnson–Nyquist noise, or simply speaking as thermal noise). Statistical physics gives the standard deviation of the current i, flowing in a resistor R, due to Johnson noise:

$$\sigma_i^2 = 4\frac{k_B T}{R} B. \tag{11.37}$$

B is the bandwidth of the recording electronics, and T is the temperature of the resistor. The Johnson noise depends on the laser probe power (through R) but not on the THz signal! In common photo-conducting antennas, the switch is open between two laser pulses and is closed during τ_{pulse} (typically $\tau_{\mathrm{pulse}} = 1$ ps). The resistance R appearing in Eq. 11.37 is the mean value of the switch resistance because the lock-in amplifier integrates the signal over many laser pulses (this hypothesis is less valid for systems based on 1 kHz repetition rate amplified lasers):

$$\frac{T_{\mathrm{rep}}}{R} = \frac{T_{\mathrm{rep}} - \tau_{\mathrm{pulse}}}{R_{\mathrm{dark}}} + \frac{\tau_{\mathrm{pulse}}}{R_{\mathrm{on}}} \quad \Rightarrow \quad \frac{1}{R} \approx \frac{1}{R_{\mathrm{dark}}} + \frac{\tau_{\mathrm{pulse}}}{R_{\mathrm{on}} \times T_{\mathrm{rep}}}. \tag{11.38}$$

Typically, $T_{\mathrm{rep}} = 10$ ns, $R_{\mathrm{dark}} = 20$ MΩ, $R_{\mathrm{on}} = 50\Omega$. Thus $R =$ 500 kΩ. For a 100 ms integration time of the lock-in amplifier ($B =$ 10 Hz), $\sigma_{\mathrm{i}} \approx 0.6$ pA, while the maximum mean value of the THz-induced current is about $4 \sim 5$ nA. Thus, the Johnson-limited *SNR* is about 80 dB.

Noise in the active components of the electronics is more difficult to evaluate. Current/voltage preamplifiers of modern lock-in amplifiers exhibit an input noise of a few fA/$\sqrt{\mathrm{Hz}}$ and a gain of 10^9 V/A.

11.2.9 Quantization and Numerical Errors [19]

In addition to the experimental errors and uncertainties, the numerical treatment of the recorded signal is also noisy.

11.2.9.1 Round-off noise

Let us first treat the rounding (or round-off) noise, i.e., the effect of rounding a numerical value. We set x the actual value and x_{round} the value that will be treated in the calculations. x and x_{round} share M digits. We write

$$x = x_{\mathrm{round}} + \Delta x \times 10^{-M} \quad \Rightarrow \quad x \times 10^{M} = x_{\mathrm{round}} \times 10^{M} + \Delta x. \tag{11.39}$$

In Eq. 11.36, $x_{\mathrm{round}} \times 10^M$ is an integer and Δx is a real belonging to the interval $[-\Delta/2, +\Delta/2]$. When dealing with rounding effects, $\Delta = 1$ (if $|\Delta x| > 0.5$, $x \times 10^M$ is rounded to the neighbor integer $x_{\mathrm{round}} \times 10^M \pm 1$), but the same formalism could be applied to analog-to-digital conversion, for which Δ is different from 1. Thus, we save Δ in the next equations for the sake of generality. From one measure to another one, Δx varies randomly. We suppose that the probability of having any value Δx is the same over the available interval $[-\Delta/2, +\Delta/2]$. Therefore, the mean square value of Δx is

$$\overline{(\Delta x)^2} = \frac{1}{\Delta} \int_{-\Delta/2}^{\Delta/2} (\Delta x)^2 \, d(\Delta x) = \frac{\Delta^2}{12}. \tag{11.40}$$

The variance of x is given by

$$\sigma_x^2 = \overline{(x - x_{\text{round}})^2} = \overline{(\Delta x)^2} \times 10^{-2M} = \frac{\Delta^2}{12} \times 10^{-2M}. \tag{11.41}$$

Within a calculation process in which only rounding occurs, i.e. $\Delta = 1$, *SNR* (here named signal-to-quantization-noise ratio, i.e., *SQNR*) is

$$SQNR \ (\text{dB}) = 20 \log \frac{x}{\sigma_x} \approx 10.79 + 20 \times M + 20 \log x. \tag{11.42}$$

Typically, for an $M = 9$-digit calculation, with $x \approx 1$, $SQNR \approx 191$ dB, while it is only 131 dB for a 6-digit calculation. In fact, computations are performed on a base of binary digits (bits). In such a base, each number x is represented by

$$x = 10^{-M} \left(-b_{\text{o}} + \sum_{i=1}^{\infty} \frac{b_{\text{i}}}{2^i} \right), \quad b_{\text{i}} = 0 \text{ or } 1. \tag{11.43}$$

b_{o} defines the sign of x (positive or negative). x_{round} is defined by the same expression in which the sum is limited to B, which is the number of bits that encodes x:

$$x_{\text{round}} = 10^{-M} \left(-b_{\text{o}} + \sum_{i=1}^{B} \frac{b_{\text{i}}}{2^i} \right). \tag{11.44}$$

Combining Eqs. 11.42 and 11.44 leads to the expression of Δx:

$$\Delta x = 10^{-M} \sum_{i=B}^{\infty} \frac{b_{\text{i}}}{2^i} \quad \Rightarrow \quad \Delta x > \ 10^{-M} \times 2^{-B}. \tag{11.45}$$

Thus, the minimum value of Δx is $10^{-M} \times 2^{-B}$. The variance of x is obtained as previously (see 11.41):

$$\sigma_x^2 = \overline{(x - x_{\text{round}})^2} = \frac{2^{-2B} \times 10^{-2M}}{12}. \tag{11.46}$$

The influence of the quantization error on the output of the Fourier transform is trickier to evaluate, and its evaluation is out of the scope of this book. A clear demonstration may be read in reference [19], which ends up with the following result:

$$SQNR\,(k) = SQNR = \frac{2^{2B}}{4N} \Rightarrow SQNR\,(\text{dB}) \approx 6\,(B-1) - 10\log N. \tag{11.47}$$

$SQNR\ (k)$ is the $SQNR$ of the spectral component k, N is the number of samples and B is the number of bits that encode each temporal sample. It appears immediately that $SQNR\ (k)$ is the same for any spectral component. Typically, for modern PC computers, $B = 32$ or 64, which leads to $SQNR$ larger than 150 dB, as soon as the number of recorded time samples is larger than 512. Such a noise appears in Fig. 11.7. The peak value of the spectrum is ~0.5, while the noise level is $\sim 10^{-8}$, and the corresponding $SQNR$ is ~154 dB. This is the value given by expression (11.42), for $N = 1064$ and $B = 32$ bits.

11.2.9.2 Noise from analog-to-digital conversion

More significant is the quantization effect due to the analog-to-digital converter (ADC). Each sample of the waveform is transformed into a digital number coded over B bits. For ideal ADC, the numerical voltage resolution V_{res} is equal to the effective number of bits (ENOB), i.e.,

$$V_{\text{res}} = \frac{V_{\text{FS}}}{2^B}. \tag{11.48}$$

V_{FS} is the full scale voltage range that can address the ADC. Typically, for $V_{\text{FS}} = 10$ V and an 8-bit ADC, $V_{\text{res}} = 10/2^8 = 10/256 \approx 39$ mV. With the same formalism as above, we obtain

$$SQNR = \frac{12}{8}\left(\frac{2^B V_{\text{signal}}}{V_{\text{FS}}}\right)^2,$$

$$\Rightarrow SQNR\,(\text{dB}) \approx 1.76 + 6.02 \times B + 20\log\left(\frac{V_{\text{signal}}}{V_{\text{FS}}}\right) \tag{11.49}$$

In the upper limit of the signal to be equal to V_{FS}, we get the maximum $SQNR$ delivered by the ADC:

$$SQNR_{\text{max}}\,(\text{dB}) \approx 1.76 + 6.02 \times B. \tag{11.50}$$

For a 16-bit ADC, $SQNR_{\text{max}} \approx 97.8$ dB. The $SQNR$ of real ADC's is smaller than this ideal value, typically around 80 dB.

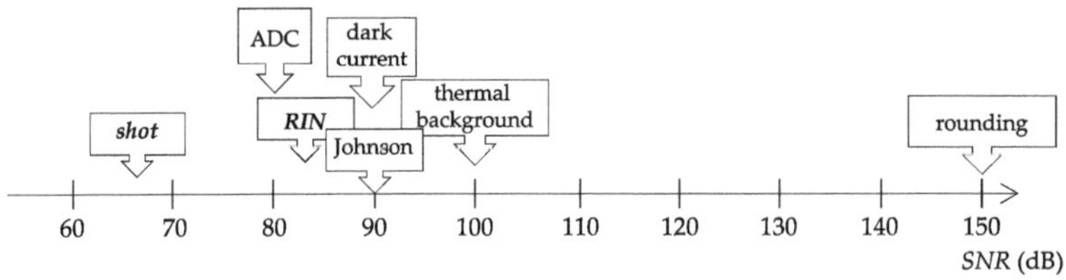

Figure 11.5 Schematic plot of the noise contributions in the time domain.

11.2.10 Summary

The order of magnitude of the different noise contributions is shown in the following figure, where the different *SNR*s that impact the THz waveforms are listed (noises in bold italic are THz signal-dependent). This figure has been established for common modern THz-TDS setups. Some contributions, like laser drift and instabilities, or noise in the electronics, are not given here, even if they could be the noisiest ones, because they depend strongly on each system, and no general rule can be given. Here, the main source of noise is the shot noise that affects simultaneously both emitter and receiver. Then ADC, dark current, RIN, and Johnson noises contribute also, but at about a 15–25 dB lower value. Other noise sources, including the ambient thermal background, are negligible.

11.3 Phase Noise

11.3.1 Phase Noise in Mode-Locked Lasers

In addition to the intensity fluctuations of the laser and the shot noise, mode-locked lasers exhibit specific noise due to their pulsed operation [20]. Either there is a jitter in the periodicity of the pulses, or/and a shift of the phase of the optical wave (compared to the pulse envelope), which vary from pulse to pulse (namely the carrier envelope offset – CEO). Femtosecond mode-locked Ti:Sapph lasers offer exceptional performance, with a pulse time jitter of about 0.05 fs. Femtosecond fiber lasers present a longer time jitter, typically 2 fs [21]. The mean value of the phase jitter $\Delta\varphi_{\mathrm{CEO}}$ is related to the CEO frequency defined as $f_{\mathrm{CEO}} = \Delta\varphi_{\mathrm{CEO}}/2\pi\, T_{\mathrm{rep}}$, where T_{rep}

is the pulse repetition period. f_{CEO} spectra of Ti:Sapph and solid-state lasers exhibit a maximum around $0.5 \sim 1$ kHz [22-23], leading to $\Delta\varphi_{\mathrm{CEO}} \approx 30$ mrad with $T_{\mathrm{rep}} \approx 10$ ns. Because of the pump-and-probe scheme of any THz-TDS systems, phase fluctuation effects in THz emission are normally compensated in detection, providing the optical setup arrangements in both emitting and receiving arms show equivalent optical dispersions.

11.3.2 Phase Noise in TDS Setups

The phase of the experimental spectrum is related to the delay time τ between the arrival of the THz pulse and the probe laser pulse at the receiver. More precisely, because both emitter and receiver are excited by the same pulsed laser beam that has been divided into two parts by a beam splitter, τ depends on

- the optical path from the beam splitter to the emitting antenna,
- the time of flight of the THz pulse from the emitter to the receiver, and
- the optical path from the beam splitter to the receiving antenna (this last path includes the delay line).

The random variations of τ are due to mechanical vibrations of the optical components in the setup and to fluctuation of the optical/THz path in air due to turbulence. The contribution of mechanical vibration is difficult to determine because it depends on the specific setup. It can be reduced by strongly binding good quality optical components and making the paths as short as possible. The influence of air turbulence, resulting from both airflow and thermal fluctuations, has been estimated in several papers related to optical interferometry. Typically, the thermal fluctuations in a room are of the order of 0.01°C, resulting in relative optical path changes of about a few 10^{-8} [24]. For a 3 m optical path, this leads to a time error of about $\Delta\tau = \Delta L/c = 0.1$ fs, which is negligible. The fluctuations of the path between the emitter and receiver are avoided by performing the measurements in a vacuum. If not, because the refractive index of air in the THz domain

($n_{\text{air,THz}} - 1 \approx 6 \times 10^{-5}$ [25]) is slightly smaller than in the visible domain ($n_{\text{air,vis}} - 1 \approx 8.3 \times 10^{-5}$ [26]), THz path length fluctuations are also expected to be smaller than in the visible, and thus are insignificant.

In fiber-coupled THz-TDS systems, the situation is different if the fibers exceed several meters in length. Soltani et al. [27] have observed that, in a THz-TDS system, whose antennas are excited by the femtosecond laser after 6–7 m of propagation in fibers, $\Delta\tau$ is of the order of 140 fs.

11.4 Effect of Noise: From Time to Frequency Domains

11.4.1 Basic Relations

The signal corresponding to the detection of the THz waveform and recorded by the receiving detector, and associated electronics, is slightly noisy, as explained previously. Let us write it, at any delay time τ, as the sum of its expected "true" value $S(\tau)$ and the noise $\Delta S(\tau)$:

$$S_{\text{meas}}(\tau) = S(\tau) + \Delta S(\tau). \tag{11.51}$$

The Fourier transform of the waveform is consequently the sum of the spectra of the expected waveform and noise:

$$S_{\text{meas}}(\omega) = \frac{1}{\sqrt{2\pi}} \int_{-\infty}^{+\infty} (S(\tau) + \Delta S(\tau))\, e^{-j\omega\tau} d\tau \equiv S(\omega) + \Delta S(\omega). \tag{11.52}$$

Therefore, the variance of the noisy spectrum is also the variance of the noise:

$$\sigma_S^2(\omega) = \overline{|S_{\text{meas}}(\omega) - S(\omega)|^2} = \overline{|\Delta S(\omega)|^2}. \tag{11.53}$$

In Eq. 11.53, $\bar{x}$ denotes the mean value of x, calculated from I measurements x_i:

$$\bar{x} = \frac{1}{I} \sum_{i=1}^{I} x_i. \tag{11.54}$$

We would like to emphasize the fact that the time appearing in Eqs. 11.51 and 11.52 is the delay time τ in the THz-TDS experiment, and that $S_{\text{meas}}(\tau)$ is the recorded waveform: its mean value $S(\tau)$ is obtained by averaging the signal over a given duration at a given time delay τ, or by averaging several independent records.

Actual experiments are not performed over an infinite recording time. Thus, the limits of the Fourier transform integral in Eq. 11.52 are not infinity, but equal to 0 and T_{record}, where T_{record} is the equivalent total duration of the record:

$$S_{\text{meas}}(\omega) = \frac{1}{\sqrt{2\pi}} \int_{0}^{T_{\text{record}}} (S(\tau) + \Delta S(\tau))\, e^{-j\omega\tau} d\tau,$$

$$= \frac{1}{\sqrt{2\pi}} \int_{-\infty}^{+\infty} (S(\tau) + \Delta S(\tau))\, W_{T_{\text{record}}}(\tau)\, e^{-j\omega\tau} d\tau. \quad (11.55)$$

$W_{T_{\text{record}}}(\tau)$ is a rectangle function corresponding to the recording window:

$$W_{T_{\text{record}}}(\tau \in U) = 1, \quad W_{T_{\text{record}}}(\tau \notin U) = 0, \quad U = [0, T_{\text{record}}]. \quad (11.56)$$

In an optimized experiment, the value of T_{record} should be chosen to have the signal $S(\tau)$ null before and after the time window. Thus the integral of $S(\tau)$ over the actual recording window T_{record} is equal to its integral over an infinite recording time, i.e. equal to the rigorous Fourier transform of $S(\tau)$. However, there is noise outside the recording time window and thus we must write

$$S_{\text{meas}}(\omega) = \frac{1}{\sqrt{2\pi}} \int_{-\infty}^{+\infty} S(\tau)\, e^{-j\omega\tau} d\tau$$
$$+ \frac{1}{\sqrt{2\pi}} \int_{-\infty}^{+\infty} \Delta S(\tau)\, W_{T_{\text{record}}}(\tau)\, e^{-j\omega\tau} d\tau. \quad (11.57)$$

Because the Fourier transform of a product is the convolution of the Fourier transforms, we obtain

$$S_{\text{meas}}(\omega) = S(\omega) + \Delta S(\omega) \otimes W_{T_{\text{record}}}(\omega), \quad (11.58)$$

$$\Rightarrow \quad \sigma_S^2(\omega) = \overline{|S_{\text{meas}}(\omega) - S(\omega)|^2} = \overline{\left|\Delta S(\omega) \otimes W_{T_{\text{record}}}(\omega)\right|^2}. \tag{11.59}$$

Usually, the recording window is much longer than the THz waveform. The cardinal sine function, which is the Fourier transform of the time window, could be approximated by a rectangle of amplitude 1 and width $2\pi/T_{\text{record}}$, in which $\Delta S(\omega)$ does not vary notably. Therefore,

$$\Delta S(\omega) \otimes W_{T_{\text{record}}}(\omega) \approx \frac{2\pi}{T_{\text{record}}} \Delta S(\omega),$$

$$\Rightarrow \sigma_S^2(\omega) \approx \left(\frac{2\pi}{T_{\text{record}}}\right)^2 \overline{|\Delta S(\omega)|^2}, \tag{11.60}$$

$$SNR(\omega) = \left(\frac{T_{\text{record}}}{2\pi}\right)^2 \frac{|S(\omega)|^2}{\overline{|\Delta S(\omega)|^2}}. \tag{11.61}$$

$SNR(\omega)$ varies as the square of the duration T_{record} of the waveform recording. We now estimate first the term $\Delta S(\omega)^2$and then the variance:

$$\begin{aligned}
|\Delta S(\omega)|^2 &= \frac{1}{2\pi} \left| \int_{-\infty}^{+\infty} \Delta S(\tau)\, e^{-j\omega\tau}\, d\tau \right|^2 \\
&= \frac{1}{2\pi} \left| \int_{-\infty}^{+\infty} \int_{-\infty}^{+\infty} \Delta S(\tau)\, \Delta S(\tau')\, e^{-j\omega(\tau-\tau')}\, d\tau\, d\tau' \right|, \\
&= \frac{1}{2\pi} \left| \int_{-\infty}^{+\infty} \left(\int_{-\infty}^{+\infty} \Delta S(\tau)\, \Delta S(\tau - \tau'')\, d\tau \right) e^{-j\omega\tau''}\, d\tau'' \right|, \\
&\text{with} \quad \tau'' = \tau - \tau'.
\end{aligned} \tag{11.62}$$

In Eq. 11.62, the integral over τ is the autocorrelation (see A.99) of the noise:

$$\int_{-\infty}^{+\infty} \Delta S(\tau)\, \Delta S(\tau - \tau'')\, d\tau = \Delta S(\tau'') \otimes \Delta S(\tau''),$$

$$\Rightarrow |\Delta S(\omega)|^2 = \frac{1}{2\pi} \left| \int_{-\infty}^{+\infty} \Delta S(\tau) \otimes \Delta S(\tau)\, e^{-j\omega t}\, dt \right|$$

$$= \frac{|\text{TF}[\Delta S(\tau) \otimes \Delta S(\tau)]|}{\sqrt{2\pi}}. \tag{11.63}$$

$|\Delta S(\omega)|^2$ is named the noise power spectral density (PSD). For white noise, *PSD* is constant and does not vary with frequency.

11.4.2 Temporal Samples and Related Noise

In fact, one records discrete values of the temporal waveform by moving the delay line step by step in view of sampling. Thus, one does not perform a continuous Fourier transform, but a discrete Fourier transform (DFT). A pedagogical and complete description of the processing of discrete signals can be found in the book by Oppenheim and Schaffer [28]. Let N be the number of recorded points and $\Delta\tau$ the time interval between two successive points (thus $T_{\text{record}} = (N-1) \times \Delta\tau$). At any discrete time $\tau_n = n \times \Delta\tau$ (n is an integer, $0 \leq n \leq N-1$), the recorded signal is equal to the integral of the actual THz field $E_{\text{THz}}(t)$ times the impulse response $I(t)$ of the receiver:

$$S_{\text{meas}}(n\Delta\tau) \equiv S_{\text{meas},n} \propto \int\limits_{t=-\infty}^{+\infty} \left(E_{\text{THz}}(t) + \Delta E_{\text{THz}}(t)\right) I(t - n\Delta\tau)\, dt. \tag{11.64}$$

In a first approximation, the impulse response could be substituted by a rectangle function of amplitude 1 and temporal width δt. We get

$$\begin{aligned} S_{\text{meas},n} &\propto \int\limits_{t=n\Delta\tau-\frac{\delta t}{2}}^{n\Delta\tau+\frac{\delta t}{2}} \left(E_{\text{THz}}(t) + \Delta E_{\text{THz}}(t)\right) dt \\ &\approx \delta t \times E_{\text{THz}}(n\Delta t) + \delta t \times \Delta E_{\text{THz}}(n\Delta t). \end{aligned}$$

Thus, we can simply write

$$S_{\text{meas},n} = S_n + \Delta S_n, \tag{11.65}$$

with

$$S_n \propto \delta t \times E_{\text{THz}}(n\Delta t), \qquad \Delta S_n \propto \delta t \times \Delta E_{\text{THz}}(n\Delta t). \tag{11.66}$$

The recorded signal to be Fourier-transformed is thus a set of N data.

Therefore, the recorded signal (Fig. 11.6) looks like a series of square pulses of different amplitudes, equally separated in time (period $\Delta\tau$).

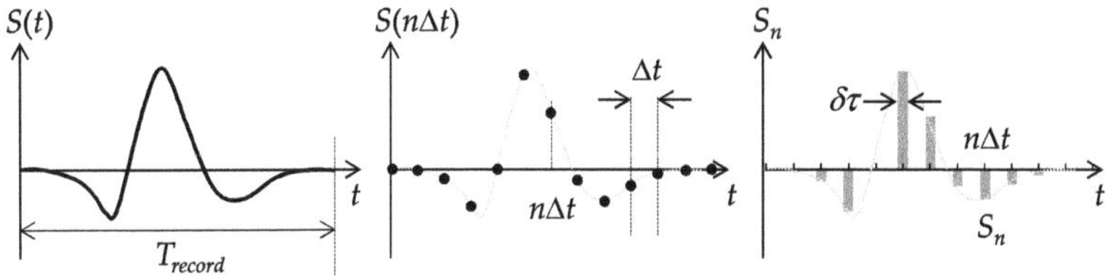

Figure 11.6 Typical THz pulse waveform (left) and sampled waveform (right).

11.4.3 Spectral Samples

The recorded signal to be Fourier-transformed is a set of N data points, whose values are S_n $(0 \leq n \leq N-1)$. The Fourier transform of this signal is also a set of N discrete spectral data, at frequencies ω_k defined by $\omega_k = 2\pi k/N\Delta\tau$ $(0 \leq k \leq N-1)$:

$$S_{\text{rec}}\left(\omega_k\right) \equiv S_k = \frac{1}{\sqrt{N}}\sum_{n=0}^{N-1} S_n e^{-j\omega_k \Delta\tau n} = \frac{1}{\sqrt{N}}\sum_{n=0}^{N-1} S_n e^{-j2\pi k\frac{n}{N}} = \sum_{n=0}^{N-1} S_{n,k}. \tag{11.67}$$

From here, the n and k subscripts refer to the time and frequency domains, respectively. Definition 11.64 may be found under different forms in the literature: the present form makes the reciprocal definition of S_n related to S_k similar to Eq. 11.67: if the factor $1/\sqrt{N}$ is omitted in Eq. 11.67, a factor $1/N$ appears in the series expression of S_n.

To illustrate all the preceding definitions, we compute a waveform that resembles an experimental one. The curve is the sum of three Gaussian functions:

$$S\left(\tau\right) = \exp\left(-\left(\frac{\tau-4}{0.2}\right)^2\right) - 0.1 \times \exp\left(-\left(\frac{\tau-3.5}{0.16}\right)^2\right)$$
$$-0.2 \times \exp\left(-\left(\frac{\tau-4.5}{0.25}\right)^2\right)$$

By using an analytical function, we can easily choose the number of samples, and add white noise or signal-dependent noise, in view of exemplifying the preceding relations as well as coming ones that deal with the noise question. The computed waveform is plotted in Fig. 11.7. The total "recording" window is $T_{\text{record}} = 20.46$ ps. We

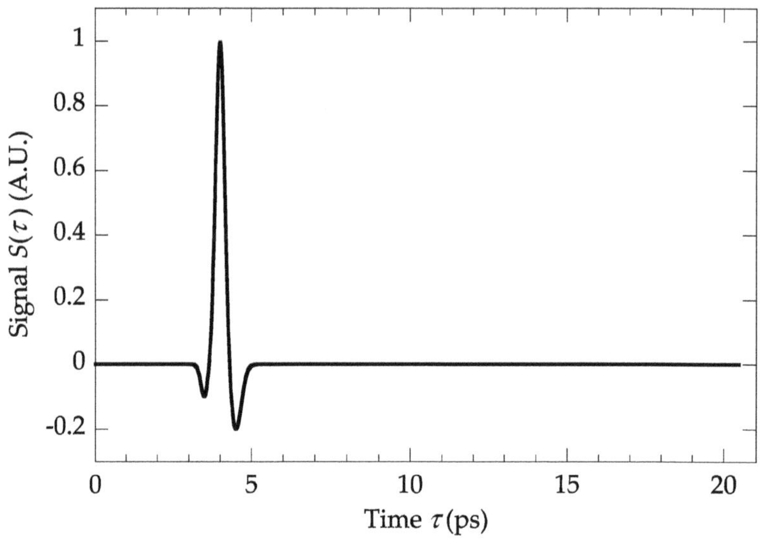

Figure 11.7 Computed THz waveform.

first compute 1024 temporal samples, therefore with a time interval, between two successive samples, equal to $\Delta\tau = 20$ fs. Next we compute only 512 points over the same T_{record}, i.e., the time interval is doubled, $\Delta\tau = 40$ fs.

Then we perform a DFT of the computed waveform, either with 1024 or 512 temporal samples. The Fourier transform moduli are shown in Fig. 11.8. Because of the very short time interval $\Delta\tau$, the spectra spread up to a large maximum frequency, $f_{\max} = 1/\Delta\tau$. As expected they are symmetric around $k \sim N/2$, i.e., $f = f_{\max}/2$. Thus, the calculated 512-point spectrum is twice narrower than the 1064 one. The curves minima are limited to $10^{-8} \sim 10^{-9}$ by numerical noise when performing the Fourier transform calculation (see below). The zoom at smaller frequencies (Fig. 11.8b) shows the suitable part of the spectra, i.e., corresponding to the THz signal that emerges from noise.

The magnitude of the spectral points varies approximately as $\sqrt{N}$: the 1024-point spectral data are thus $\sqrt{2}$ larger than the 512-point ones.

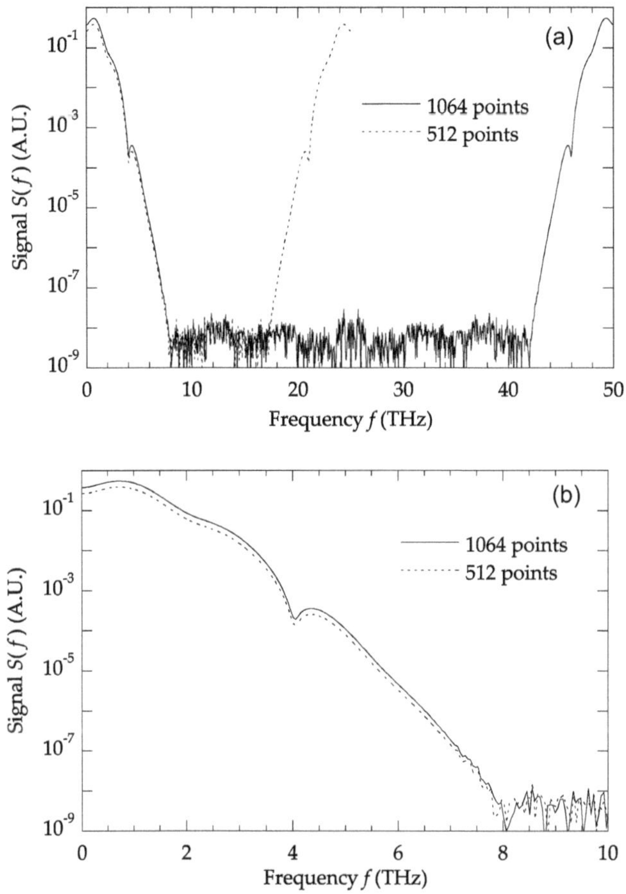

Figure 11.8 (a) The modulus of the discrete Fourier transform of the waveform shown in Fig. 11.6. The dashed curve is calculated with 512 samples, while the continuous one is obtained with 1064 samples; (b) Zoom of the spectra at low frequencies.

11.4.4 Noise of Temporal Samples

Let us now enter the random noise in the preceding expressions:

$$S_{\text{meas},n} = S_n + \Delta S_n$$

$$\Rightarrow S_n = \frac{1}{I}\sum_{i=1}^{I} S_{\text{meas},n,i},\ \sigma_n^2 = \overline{\Delta S_n^2} = \frac{1}{I}\sum_{i=1}^{I} \Delta S_{n,i}^2.$$

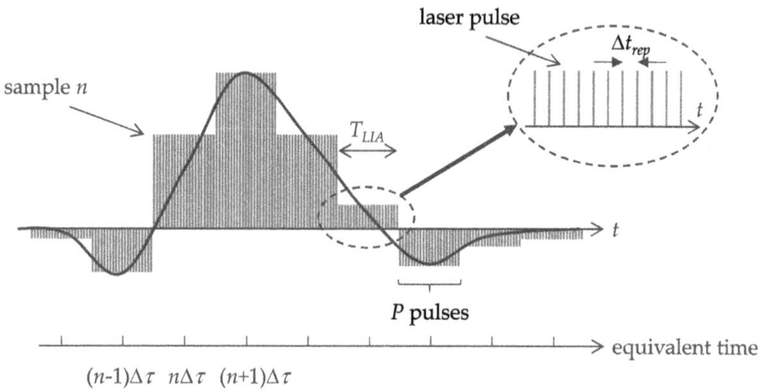

Figure 11.9 Actual recorded signal (inset: zoom of a time-equivalent sample).

I is the number of records $S_{n,i}$ performed at delay time $n\Delta\tau$. Here, each temporal sample S_n of a given THz waveform results from the integration of the detected signal over the impulse response time of the receiver. Because they are delivered by mode-locked lasers, the THz pulses are repetitive. Usually, mode-locked lasers in THz-TDS systems run at high repetition rates, typically 80 $\sim$ 100 MHz, and the signals are recorded with a lock-in amplifier (LIA) at a lock-in frequency $f_{\text{LIA}} = 1/\Delta t_{\text{LIA}}$. The THz signal is modulated at f_{LIA} using a mechanical or electronic chopper. The LIA integration time T_{LIA} is much longer than both Δt_{LIA} and the delay Δt_{rep} between two consecutive laser pulses. Therefore (see Fig. 11.9), the LIA signal is the sum of the recorded signal over P laser pulses$\left(P = T_{\text{LIA}}/\left(2 \times \Delta t_{\text{rep}}\right)\right)$; the chopper saves only half of the laser beam, which explains the factor of $1/2$ in the expression of P:

$$S_{\text{LIA,meas},n} = \sum_{p=1}^{P} S_{\text{meas},n,p},$$

$$\Rightarrow \quad S_{\text{LIA},n} = P \times S_n, \quad \Delta S_{\text{LIA},n} = \sum_{p=1}^{P} \Delta S_{n,p}. \tag{11.68}$$

If the noise of each pulse p of an LIA record is not correlated with one of the other pulses, the random behavior of the noise together

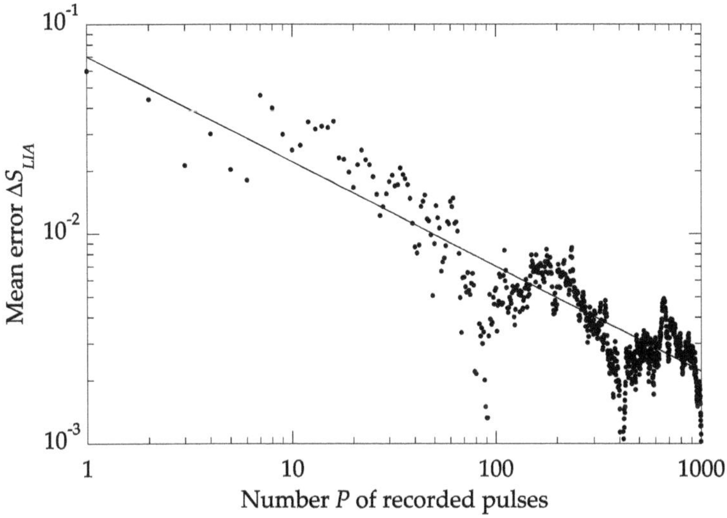

Figure 11.10 Mean error vs. the number P of averaged data for white noise. Dots are calculated with a random number generator, while the continuous line is the $1/\sqrt{P}$ curve.

with the central limit theorem leads to

$$\overline{\Delta S_{\text{LIA},n}} = \frac{1}{\sqrt{P}}\overline{\Delta S_n}, \quad SNR_{\text{LIA},n} = \left(\frac{\overline{S_{\text{LIA},n}}}{\overline{\Delta S_{\text{LIA},n}}}\right)^2 = P. \qquad (11.69)$$

The random noise, at each temporal sample n, decreases as the inverse of the square root of recorded pulses, and thus it is equal to zero when P tends to infinity.

This is illustrated in Fig. 11.10, in which $\overline{\Delta S}$ is plotted versus the number P of involved data. Here $\Delta S_{\text{p}} = g(p)$, where $g(p)$ is a random real number, $-0.5 < g(p) < 0.5$, whose probability of occurrence is constant over the range [-0.5, 0.5]. In actual experiments, noise in successive pulses may be correlated. The correlation is related to the noise power spectral density.

When employing mode-locked lasers, the main noise with a characteristic time of the order of the LIA integration time is the flicker noise, i.e., the $1/f$ noise. In the general case of $1/f^{\alpha}$ noise ($\alpha$is a real), its variance varies as [29]:

$$|\Delta S(\omega)|^2 \propto \frac{A}{\omega^{\alpha}} \quad \Rightarrow \quad \sigma^2 = A\,C(\alpha)\,T_{\text{LIA}}^{1+\alpha},$$

with $C(\alpha) = \left(2^{2+\alpha} - 2^{1+2\alpha}\right)\pi^{\alpha-1}\Gamma(-1-\alpha)\sin\left(\frac{\alpha\pi}{2}\right)$. Here Γ is the gamma function. Thus, we get

$$\overline{\Delta S_{\mathrm{LIA},n}} = \frac{1}{P\sqrt{1+\alpha}}\overline{\Delta S_n}, \quad SNR_{\mathrm{LIA},n} = \left(\frac{S_{\mathrm{LIA},n}}{\overline{\Delta S_{\mathrm{LIA},n}}}\right)^2 = P^{(1-\alpha)}. \tag{11.70}$$

It follows that, for flicker noise when $\alpha > 1$, the *SNR* of the LIA signal decreases with P, i.e. with the LIA integration time T_{LIA}. It can be simply understood as follows: for a short T_{LIA}, the corresponding spectrum spreads to high frequencies, for which the noise is weak. The mean noise value, integrated over all the frequencies, is thus small, because of the high frequencies contribution. On the other hand, a longer T_{LIA} involves lower frequencies, and thus a larger noise.

Therefore, it is preferable several records over a short LIA time integration, and then to average them, than to perform recording over a long T_{LIA}.

From now on, for the sake of simplicity, we will write $S_n = S_{\mathrm{LIA},n}$ and $\sigma_n^2 = \sigma_{\mathrm{LIA},n}^2$, which are the temporal samples and their variances as measured with an LIA. Let us notice that, contrary to what occurs in the frequency domain, here all the temporal samples are real.

11.4.5 Noise of Spectral Samples

Thanks to the addition properties of the Fourier transform, the corresponding spectral data are

$$S_{\mathrm{meas},k} = \frac{1}{\sqrt{N}}\sum_{n=0}^{N-1} S_{\mathrm{meas},n}\, e^{-j\frac{2\pi k}{N}n} \tag{11.71}$$

$$\Rightarrow S_{\mathrm{meas},k} = \frac{1}{\sqrt{N}}\sum_{n=0}^{N-1} S_n e^{-j\frac{2\pi k}{N}n} + \frac{1}{\sqrt{N}}\sum_{n=0}^{N-1} \Delta S_n e^{-j\frac{2\pi k}{N}n} = S_k + \Delta S_k. \tag{11.72}$$

The noise ΔS_k is thus obtained from the sum of random values ΔS_n. If the noise of each temporal data n obeys some random law, and if the number N of data is sufficiently large, the central limit theorem states that ΔS_k is also a random variable with an asymptotic Gaussian distribution [30].

There are two ways of taking the average value of (11.72) when performing I measurements. The first one is to record I waveforms, to Fourier-transform each of them and then to take the average value of the Fourier transforms. We get

$$\overline{S_{\text{meas},k}} = S_k + \overline{\Delta S_k} = S_k + \frac{1}{I}\sum_{i=1}^{I}\left(\frac{1}{\sqrt{N}}\sum_{n=0}^{N-1}\Delta S_{n,i}e^{-j\frac{2\pi k}{N}n}\right). \quad (11.73)$$

In the second way, one averages the I waveforms and one performs the Fourier transform of the averaged waveform:

$$\overline{S_{\text{meas},k}} = S_k + \overline{\Delta S_k} = S_k + \frac{1}{\sqrt{N}}\sum_{n=0}^{N-1}\frac{1}{I}\left(\sum_{i=1}^{I}\Delta S_{n,i}\right)e^{-j\frac{2\pi k}{N}n}. \quad (11.74)$$

Because the i and n indices are involved simultaneously in only $\Delta S_{n,i}$, both expressions 11.73 and 11.74 are equal. For a random noise of null mean value, $\overline{\Delta S_k}$ tends towards zero for a large number I of recorded pulses, and $\overline{S_{\text{meas},k}}$ tends towards the actual S_k value.

The variance of each spectral component is named the "periodogram":

$$\begin{aligned}
\sigma_k^2 &= \overline{\left|S_{\text{meas},k} - S_k\right|^2} \\
&= \overline{|\Delta S_k|^2} = \frac{1}{N}\overline{\left|\sum_{n=0}^{N-1}\Delta S_n e^{-j\frac{2\pi k}{N}n}\right|^2} = \frac{1}{I}\sum_{i=0}^{I}\frac{1}{N}\left|\sum_{n=0}^{N-1}\Delta S_{n,i}e^{-j\frac{2\pi k}{N}n}\right|^2 \\
\Rightarrow \sigma_k^2 &= \frac{1}{N}\left|\sum_{n=0}^{N-1}\sum_{p=0}^{N-1}\overline{\Delta S_n \Delta S_p}e^{-j\frac{2\pi k}{N}n}e^{-j\frac{2\pi k}{N}p}\right|.
\end{aligned}$$

We set $n' = n + p$:

$$\sigma_k^2 = \frac{1}{N}\left|\sum_{n=0}^{N-1}\sum_{n'=n}^{n+N-1}\overline{\Delta S_n \Delta S_{n'-n}}e^{-j\frac{2\pi k}{N}n'}\right|. \quad (11.75)$$

This expression is the autocorrelation of the noise in the case of discrete values, i.e. the relation equivalent to Eq. 11.62. A simpler analytic expression cannot be derived in the general case. Schoukens and Renneboog [30] have studied white and colored noise sequences. In the case of a Gaussian white noise with $\overline{\Delta S_n^2} = \sigma_n^2$, they have shown that

$$\sigma_k^2 = \frac{N}{2}\sigma_n^2. \quad (11.76)$$

The variance of the discrete data in the frequency domain is equal to the one in the temporal domain divided by $N/2$. Thus, it is necessary to increase the number of recorded temporal data to make the noise-induced uncertainties as small as possible. This is illustrated by Fig. 11.11a. The ratio σ_k^2/σ_n^2 in dB (i.e., the difference σ_k^2 (dB) $-$ σ_n^2 (dB)) is plotted versus N (from $N = 2^8 = 256$ up to $N = 2^{16} = 65\,356$). For common recorded data size ($N = 2^{12} = 4096$), σ_k^2 is about 33 dB bigger than σ_n^2. Figure 11.11b shows a computer-simulated noisy signal. This is a Gaussian white noise: the dotted line corresponds to the ΔS_n data plotted versus n (here 1024 data), while the continuous line gives the corresponding ΔS_k data obtained by Fourier-transforming the ΔS_n's (here, for the sake of clarity, we plot the absolute value of the ΔS_n's and ΔS_k's). As expected, the mean value of the ΔS_k'sis about $\sqrt{N/2} = \sqrt{512} \approx 22.6$ larger than the ΔS_n's.

When dealing with relative noise (*SNR*), relation (11.76) should be amended [30], because the discrete Fourier transform usually introduces a scale factor $N/2$ for the Fourier coefficients S_k. Therefore, all the amplitudes have to be divided by $N/2$ and the variances by $(N/2)^2$. The relative (or scaled) variance is

$$\sigma^2_{\text{scaled},k} = \frac{2}{N}\sigma_n^2. \tag{11.77}$$

It is important to note potential confusion when using a commercial code to perform the Fourier transformation; some codes employ the same definition as here, while some others do not normalize to $N/2$, or some do to N instead of $N/2$.

11.4.6 Examples for Two Different THz-TDS Setups

11.4.6.1 Homemade system

The THz waveform delivered and recorded by LT-GaAs photo-conducting antennas in a home-built THz-TDS system has been recorded 20 times consecutively. The antennas have a 10-μm gap that is excited by a 10 mW Ti:Sapph pulsed beam (pulse duration 50 fs). The THz beam is focused by parabolic mirrors. The 1024-point waveform data are recorded each 57.2 fs (total time-equivalent duration 58.54 ps). Each waveform has been Fourier-transformed.

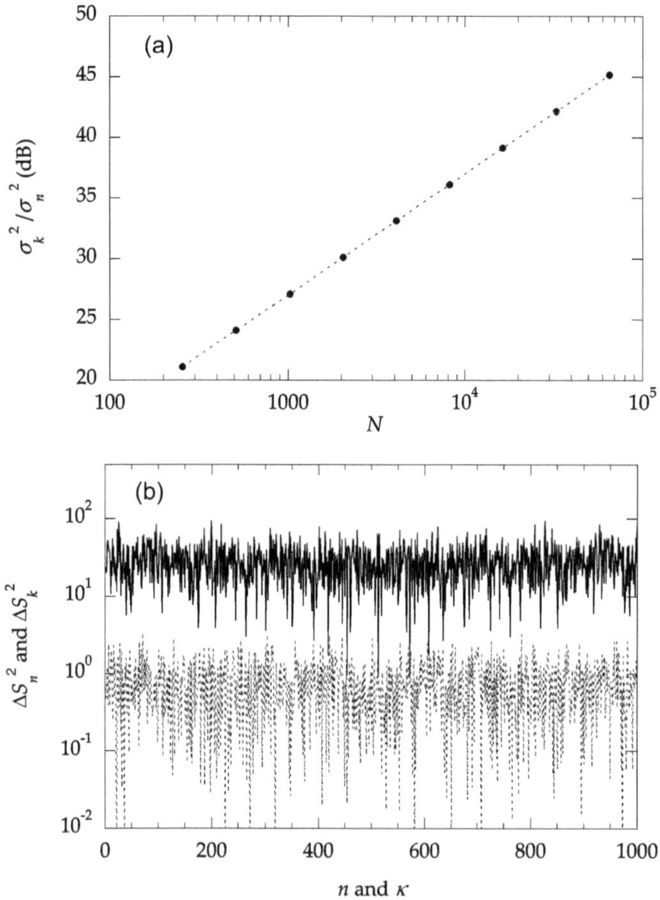

Figure 11.11 (a) σ_k^2/σ_n^2(in dB) versus the number N of recorded data. (b) $|\Delta S_n|$ versus n (logarithmic scale, lower plot) and $|\Delta S_k|$ versus k (upper plot) for a Gaussian white noise (1024 data points are used).

The variance of both temporal and frequency data (modulus) has been calculated from 20 recordings.

Figure 11.12 shows the averaged waveform and the standard deviation (multiplied by 100 for visibility). As expected, the noise presents a maximum simultaneously with the THz pulse, because part of the noise (mostly shot noise) is signal dependent.

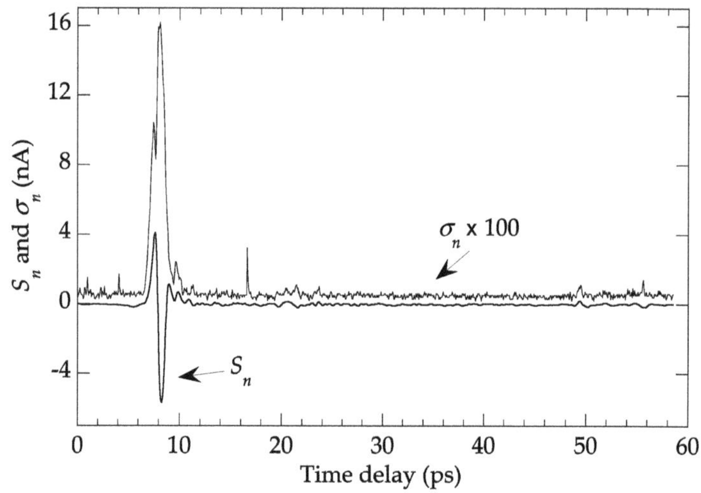

Figure 11.12 Experimental THz waveform (lower trace: the curve has been averaged over 20 recordings) and corresponding standard deviation (upper trace: values multiplied by 100).

The spectrum of the recorded waveform is plotted in Fig. 11.13a. The upper curve corresponds to the modulus of the spectral power density (i.e., S_k^2) given in dB relative to the maximum of the curve. The dashed curve shows σ_k^2 also given in dB. Finally, the lowest continuous curve depicts the best fit of the noise with a signal-dependent function, namely $\sigma_k^2 = a\,S_k^2 + b$ (here $a = 8 \times 10^{-6}$ and $b = 2.5 \times 10^{-5}$): The good agreement between the fit and the noise curve validates the signal-dependent noise behavior, to which a signal-independent noise is superposed. Because of this later noise, the spectrum is limited by the noise floor at higher frequency, which is about –60 dB. The mean value of σ_k^2, taken over 6 THz where the signal vanishes, is about –27 dB, while the value of σ_n^2 (Fig. 11.12) averaged over 30 ps is about –48 dB. The difference, 21 dB, is smaller than expected for white noise (ln (1024/2) = 27 dB), which could indicate that the noise is not purely white.

The setup dynamic range (DR) is defined as the maximum magnitude of the signal over the noise *rms* value, while *SNR* is the ratio of the mean magnitude and the standard deviation of the signal [31]. Here, in the absence of a THz signal, i.e., over 6 THz, S_k is

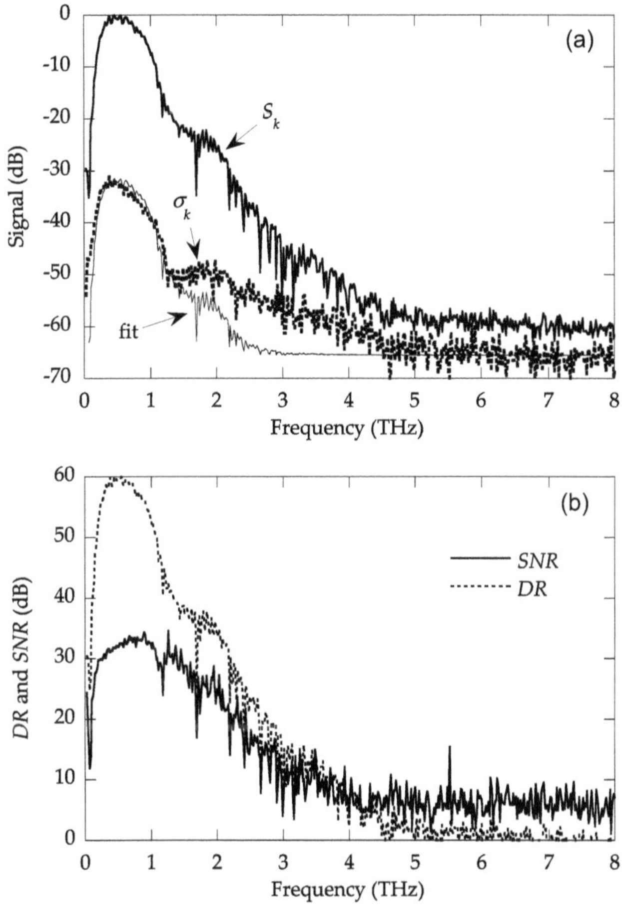

Figure 11.13 (a) THz-TDS spectrum (modulus) given in dB, i.e., 20 log $(S_k/S_{k,\max})$, of the waveform plotted in Fig. 11.12. The curve is obtained by averaging 20 spectra. The dashed line corresponds to the variance of the spectrum, while the lowest continuous line is the best fit of the noise using $\sigma_k^2 = (8\,S_k^2 + 25) \times 10^{-6}$. (b) *SNR* and DR of the spectrum.

about −60 dB. This noise floor value should be subtracted from the maximum THz signal to obtain the DR. In THz-TDS, the maximum signal is obtained without sample in the beam. Thus here DR $=$ $60 + S_k$ given in dB, where S_k is negative (Fig. 11.13a). Both DR and *SNR* are plotted in Fig. 11.13b. *SNR* spectrum resembles the THz one, and its maximum value is about 33 dB. Due to some residual water vapor in the setup, the DR and *SNR* are degraded at the water vapor absorption lines, for example at 1.1 THz.

11.4.6.2 Commercial systems

Using the same procedure as previously, we have characterized the random noise of the TAS 7500 TU model from Advantest (Fig. 11.14). Thanks to parametric generation in $LiNbO_3$, this system has a bandwidth of 7 THz. Moreover, the system does not make use of a classical mechanical delay line; the time sampling is performed by changing the pulse repetition rate of one of the two lasers that excite the emitting and receiving antennas. Here, we have limited the number of recorded data points to 4096 and the recording equivalent time window to 8.2 ps. The averaging has been performed over ten measurement runs.

As the measurements were made in a vacuum, the spectrum is not perturbed by water vapor and thus it reaches higher frequencies ($\sim$7 THz) and it appears smooth. DR is large ($>$80 dB), however *SNR* takes values similar to the home-built system, i.e., a maximum value around 32 dB. Similar values have been reported by [31] (we warn the reader that in [31], *SNR* and DR are defined regarding the THz field magnitude and not the THz power as we do here).

11.4.7 Effect of the Number N of Samples and of Zero Padding

To show the influence of the number N of recorded data, we start from the 4096 data points recorded with the Advantest system (Fig. 11.14). We arbitrary remove one data point over two in order to reduce the number of data points to 2048, and we repeat this procedure to the get 1024 data points. Figure 11.15a shows the three spectra (respectively calculated from 4096, 2048, and 1024 data points) and their variances. The modulus of the THz spectral power, i.e., $20\log(S_k)$, varies proportionally to N. The 4096-data spectrum is thus 6 dB ($= 20\log(2)$) larger than the 2048-data one. The variance varies as well. Therefore, *SNR* does not depend on the number of processed temporal data.

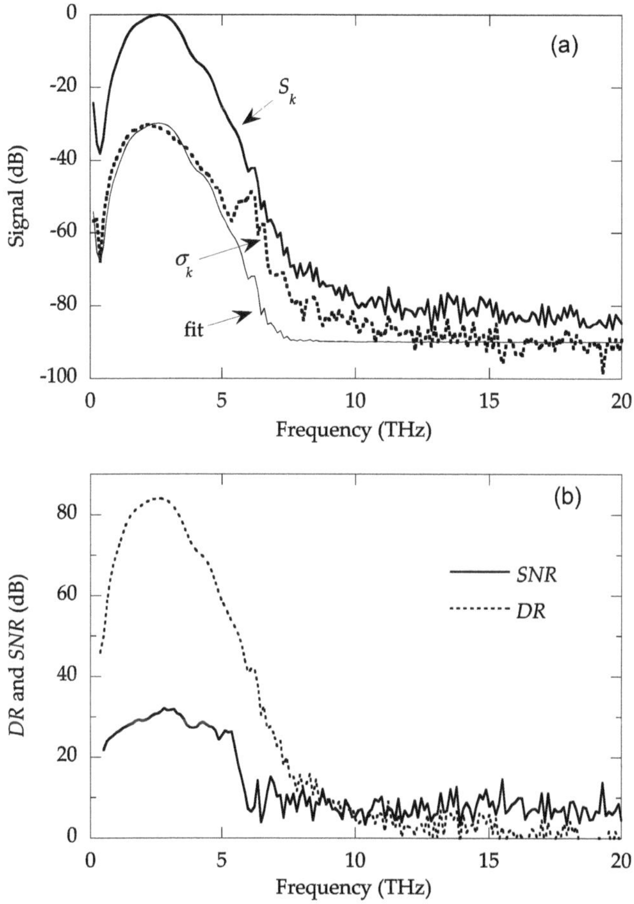

Figure 11.14 (a) THz-TDS spectrum (modulus) delivered by an Advantest TAS 7500 TU model. The curve is obtained by averaging 10 spectra. The dashed line corresponds to the variance of the spectrum, while the lowest continuous line is the best fit of the noise using $\sigma_k^2 = \left(120 \times S_k^2 + 1\right) \times 10^{-9}$. (b) DR and *SNR* of the spectrum.

Very often, to make the spectral curves softer, the number of time samples entering in the Fourier transform calculation is increased up to M, by adding $M-N$ zeros to the recorded set of N time samples (zero padding). Let us notice that these $M - N$ additional data are

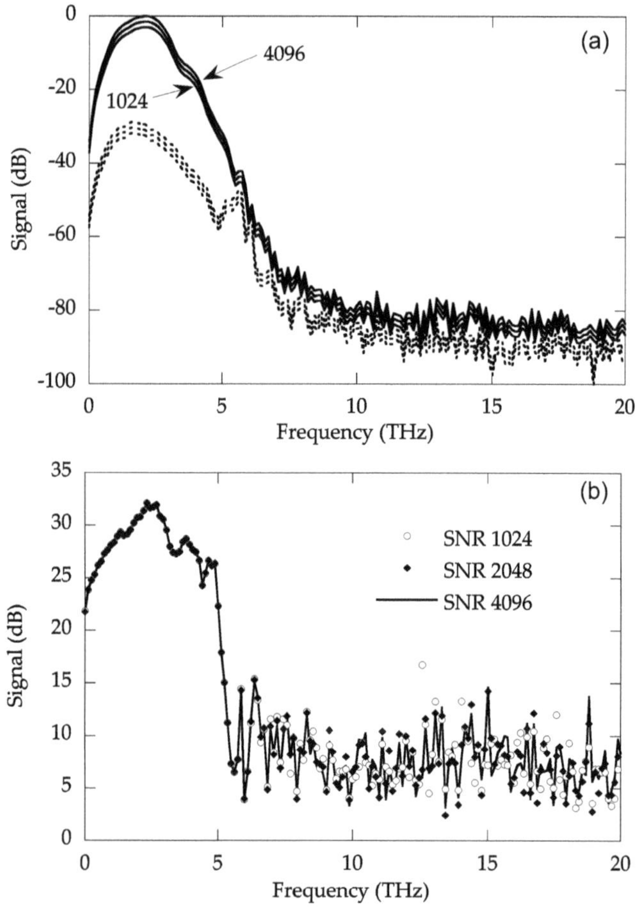

Figure 11.15 (a) THz-TDS spectrum (continuous lines) and its variance (dashed lines) delivered by the Advantest TAS 7500 TU model. The upper continuous or dashed lines are obtained with 4096 temporal data points, the mid ones with 2048 data points, and the lower ones with 1024 data points. (b) *SNR* spectra obtained from 4094, 2048, and 1024 temporal data points.

noiseless:

$$\begin{cases} S_{\text{noisy},m} = S_{\text{noisy},n}\ (0 \leq m \leq N-1),\ S_{\text{noisy},m} = 0\ (N-1 \leq m \leq M-1), \\ \Delta S_m = \Delta S_n\ (0 \leq m \leq N-1),\ \Delta S_m = 0\ (N-1 \leq m \leq M-1). \end{cases} \tag{11.78}$$

The spectral data $S'_{k'}$ of the signal with zero padding are written as

$$S'_{k'} = \frac{1}{\sqrt{M}} \sum_{n=0}^{M-1} S_n e^{-j\frac{2\pi k'}{M}n} = \frac{1}{\sqrt{M}} \sum_{n=0}^{N-1} S_n e^{-j\frac{2\pi k'}{M}n}, \qquad (11.79)$$

$$\sigma^2_{k'} = \frac{1}{M} \sum_{n=0}^{M-1} \sigma^2_n e^{-2j\frac{2\pi k'}{M}n} = \frac{1}{M} \sum_{n=0}^{N-1} \sigma^2_n e^{-2j\frac{2\pi k'}{M}n}. \qquad (11.80)$$

Let us compare these data with and without zero padding. The comparison should be done at the same frequency for both spectra, i.e., $N k' = M k$:

$$S'_{k'} = \sqrt{\frac{N}{M}} S_{k'}, \quad \sigma^2_{k'} = \frac{N}{M}\sigma^2_{k'} \quad \Rightarrow \quad SNR'_{k'} = SNR_{k'}. \qquad (11.81)$$

Thus, both magnitude and variance decrease when adding zeros. However, *SNR* remains unchanged. The zero padding procedure results only in extrapolating the spectral curve between consecutive spectral samples, but it does not improve the spectra in terms of information and *SNR*.

This is illustrated in Fig. 11.16. We add 1024 zeroes to the 1024 temporal data of the 10 measured Advantest waveforms. Figure 11.16a shows the spectra with and without zero padding, together with the related standard deviations. Here the curves differ only by 3 dB $\left(= 20 \log \sqrt{2048/1024}\right)$. Therefore, *SNR* is unchanged (see Fig. 11.16b). The zoom (inset of Fig. 11.16a: the signals are normalized to their maximum values) shows that the zero padding procedure only extrapolates spectral data between the raw values.

11.4.8 Noise of the Magnitude and Phase of the Spectral Samples

Finally, let us treat mathematically the spectral signals as complex values. The recorded waveform and thus the recorded temporal data S_n are real. From Eq. 11.67, we deduce

$$S_{\text{noisy},k} = \left|S_{\text{noisy},k}\right| e^{j\varphi_{\text{noisy},k}} = \text{Re}\left(S_{\text{noisy},k}\right) + j\,\text{Im}\left(S_{\text{noisy},k}\right)$$

$$= \frac{1}{\sqrt{N}} \sum_{n=0}^{N-1} S_{\text{noisy},n} \cos\left(\omega_k n \Delta t\right) + \frac{j}{\sqrt{N}} \sum_{n=0}^{N-1} S_{\text{noisy},n} \sin\left(\omega_k n \Delta t\right). \qquad (11.82)$$

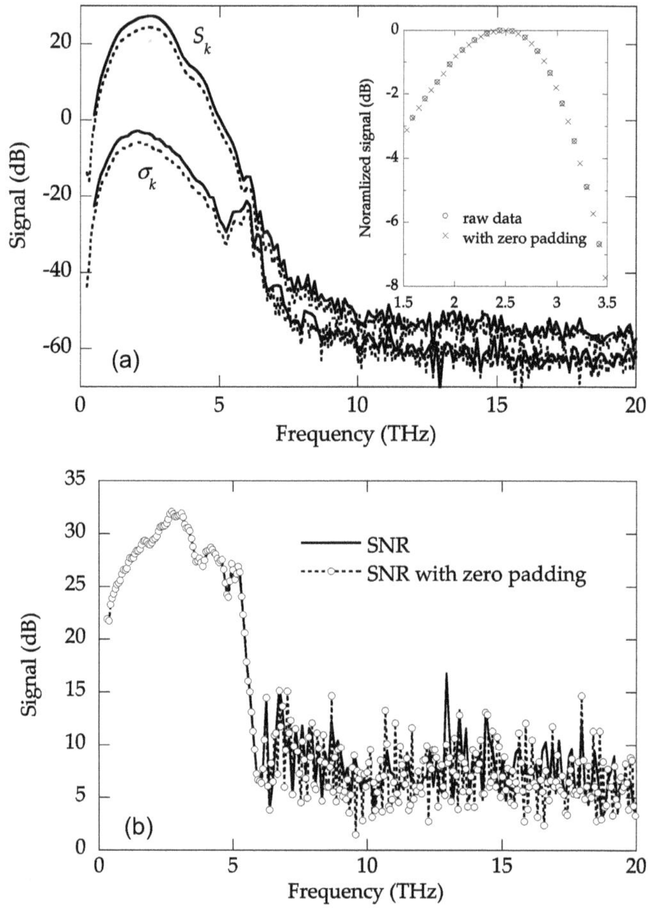

Figure 11.16 (a) Spectra, and their standard deviations, delivered by the Advantest TSA 7500 system, calculated from 1024 temporal samples (continuous line) and from the same 1024 data plus 1024 zeroes (dashed lines). The inset is a zoom at smaller frequencies of the normalized spectra. (b) Corresponding *SNR*s.

The modulus $|S_{\text{noisy},k}|$ and phase $\varphi_{\text{noisy},k}$ of the spectral component $S_{\text{noisy},k}$ are

$$\left|S_{\text{noisy},k}\right|^2 = \frac{1}{N}\sum_{n=0}^{N-1}\sum_{m=0}^{N-1} S_{\text{noisy},n}S_{\text{noisy},m}\cos\left(\omega_k\left(n-m\right)\Delta t\right), \quad (11.83)$$

$$\tan\left(\varphi_{\text{noisy},k}\right) = \frac{\sum_{n=0}^{N-1} S_{\text{noisy},n} \sin\left(\omega_k n\Delta t\right)}{\sum_{n=0}^{N-1} S_{\text{noisy},n} \cos\left(\omega_k n\Delta t\right)}. \tag{11.84}$$

Let us write the recorded spectral component $S_{\text{noisy},k}$ as the sum of the expected noiseless value S_k and a noise value ΔS_k ($\Delta S_k << S_k$)

$$S_{\text{noisy},k} = S_k + \Delta S_k = \left|S_{\text{noisy},k}\right| e^{j\varphi_{\text{noisy},k}} = |S_k|\, e^{j\varphi_{S,k}} + |\Delta S_k|\, e^{j\varphi_{\Delta S,k}}. \tag{11.85}$$

We easily get, within the hypothesis of a weak noise ($\Delta S_k << S_k$),

$$\left|S_{\text{noisy},k}\right| \approx |S_k| + |\Delta S_k| \cos\left(\varphi_{S,k} - \varphi_{\Delta S,k}\right), \tag{11.86}$$

$$\tan\varphi_{\text{noisy},k} \approx \tan\varphi_{S,k} + \frac{|\Delta S_k|}{|S_k|} \frac{\sin\left(\varphi_{\Delta S,k} - \varphi_{S,k}\right)}{\cos^2\varphi_{S,k}}, \tag{11.87}$$

$$\Rightarrow \tan\left(\varphi_{\text{noisy},k} - \varphi_{S,k}\right) = \tan\left(\Delta\varphi_k\right) \approx \Delta\varphi_k = \frac{|\Delta S_k|}{|S_k|} \sin\left(\varphi_{\Delta S,k} - \varphi_{S,k}\right). \tag{11.88}$$

The variances of modulus and phase are

$$\sigma^2_{|S_k|} = \overline{\left(\left|S_{\text{noisy},k}\right| - |S_k|\right)^2} = \overline{|\Delta S_k|^2 \cos^2\left(\varphi_{S,k} - \varphi_{\Delta S,k}\right)}, \tag{11.89}$$

$$\sigma^2_{\varphi_k} = \overline{\left(\Delta\varphi_k\right)^2} \approx \frac{1}{|S_k|^2} \overline{|\Delta S_k|^2 \sin^2\left(\varphi_{S,k} - \varphi_{\Delta S,k}\right)}. \tag{11.90}$$

If the noise of the magnitude and phase are not correlated, the mean value of the products in Eqs. 11.89 and 11.90 is equal to the product of the mean values:

$$\sigma^2_{|S_k|} = \overline{|\Delta S_k|^2}\, \overline{\cos^2\left(\varphi_{S,k} - \varphi_{\Delta S,k}\right)}, \quad \sigma^2_{\varphi_k} = \frac{\overline{|\Delta S_k|^2}}{|S_k|^2} \overline{\sin^2\left(\varphi_{S,k} - \varphi_{\Delta S,k}\right)}.$$

The mean value of the square of the sine/cosine function is calculated using an integral because the noise phase is uniformly spread over $[0, 2\pi]$:

$$\overline{\cos^2(\varphi_{S,k} - \varphi_{\Delta S,k})} = \frac{1}{2\pi}\int_0^{2\pi} \cos^2(\varphi_{S,k} - \varphi_{\Delta S,k})\, d\varphi_{\Delta S,k}$$

$$= \overline{\sin^2(\varphi_{S,k} - \varphi_{\Delta S,k})} = \frac{1}{2}, \quad (11.91)$$

$$\Rightarrow \quad \sigma^2_{|S_k|} = \frac{\overline{|\Delta S_k|^2}}{2} = \frac{\sigma_k^2}{2}, \quad \sigma^2_{\varphi_k} = \frac{\sigma_k^2}{2\,|S_k|^2} = \frac{1}{2\,SNR} = \frac{\sigma^2_{|S_k|}}{|S_k|^2}.$$

The variance of the phase of the k-th component is thus equal to the variance of the modulus divided by the square of the modulus.

The phase of the waveforms recorded by the home-built THz-TDS system (Fig. 11.12) is plotted in Fig. 11.17a. The average and the standard deviation are calculated over 10 measurements. Here we normalized the phase to ω/c (i.e., the plotted curve corresponds to a distance), in order to make the small spectral features not hidden by the strong ω/c dependency. In the spectral region (0.5 $\sim$ 2 THz) where the THz signal is strong, the phase is small and almost constant. At low frequencies, it strongly varies: as here the THz antennas are photo-conducting switches, this could be explained by the contribution of the current in the electrodes of the emitting switch, which is stronger at lower frequencies. The standard deviation (normalized to ω/c) is small up to 2.2 THz. At this frequency, because of noise, a phase jump artifact induces a step in the curve. Other jumps occur at higher frequencies. The validity of relation 11.91 is demonstrated in Fig. 11.17b. Measured and calculated σ_{φ_k} values (here not normalized) are plotted. Up to 2.2 THz, the agreement between experiment and modeling is very good. Over this frequency, the phase uncertainty is too large, and thus the hypothesis of a weak noise (see Eqs. 11.86 and 11.87) is not satisfied.

11.4.9 Noise of the Delay Line

Apart ASOP and related techniques, the delay line in most of common THz-TDS systems is a motorized translation stage on which a retro-reflector is attached. The temporal THz waveform is recorded by varying step by step the position of the retro-reflector.

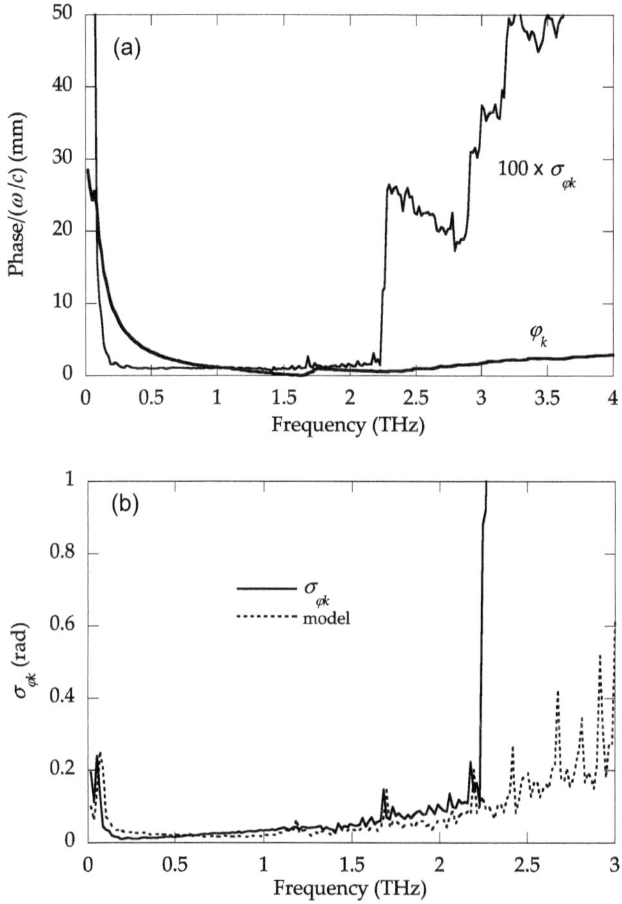

Figure 11.17 (a) Spectrum of the normalized phase ($\varphi_k/(\omega/c)$) and its standard deviation (multiplied by 100), of the waveform plotted in Fig 11.12 (homemade THz-TDS setup). (b) Spectra of the measured phase (σ_{φ_k}) and calculated one deduced from Eq. 11.90.

If Δx is the increment in position, the corresponding increment in time-delay is $\Delta\tau = 2\Delta x/c$. Thus, one records a discrete signal $S(m\,\Delta\tau)$ at given positions $x_m = m\,\Delta x$, i.e., at given time delays $m\,\Delta\tau$, where m is an integer that ranges from 0 to $M - 1$, where M is the total number of recorded data. As described in Chapter 4, a Fourier transform of these discrete temporal data is performed, which leads to a discrete spectrum (see relation 4.48, which we write again here):

$$S_{\text{rec}}(\omega) = \frac{\Delta T}{2\pi}\text{sinc}\left(\frac{\Delta T}{2}\omega\right) \otimes \sum_{m=0}^{M-1} S(m\,\Delta\tau)\,e^{-jm\omega\Delta\tau}.$$

ΔT is the time-equivalent duration of the recording window. For the sake of simplicity in the analysis of the translation stage noise, we suppose here that ΔT is very long, i.e., the sinc function tends towards a Dirac function. The expression of the spectrum simplifies into

$$S_{\text{rec}}(\omega) \simeq \sum_{m=0}^{M-1} S(m\,\Delta\tau)\,e^{-jm\omega\,\Delta\tau}. \tag{11.92}$$

The knowledge of the exact value of the positions x_m is affected by the precision of the translation stage as well as by errors that could be repeatable (drifts) or random. Repeatable errors will be treated in the next chapter. Here we look only at the noise. Noise in the position results from several causes, including variation in screw pitch or motor step angle, the accuracy of the position encoder, alignment of the encoder to the stage, interpolation of the encoder signal, and so on [32, 33]. Also, straightness, roll, flatness, or yaw may lead to errors in position. Therefore, each position x_m is erroneous by an amount δx_m ($\delta x_m \ll x_m$), and consequently $\tau_m = 2x_m/c$ is wrong by the amount $\delta\tau_m = 2\,\delta x_m/c$. The spectrum expression (11.92) becomes

$$S_{\text{rec,noisy}}(\omega) = \sum_{m=0}^{M-1} S\left(m\,\Delta\tau + \delta\tau_m\right)\,e^{-j(m\,\Delta\tau+\delta\tau_m)\,\omega}. \tag{11.93}$$

Assuming $\delta\tau_m << \tau_m = m\,\Delta\tau$, Eq. 11.93 leads to

$$\Delta S(\omega) = S_{\text{rec,noisy}}(\omega) - S_{\text{rec}}(\omega) \sim \sum_{m=0}^{M-1}\left(\left.\frac{dS}{d\tau}\right|_{m\,\Delta\tau} - j\,\omega S\left(m\,\Delta\tau\right)\right)\delta\tau_m\,e^{-jm\,\Delta\tau\,\omega}. \tag{11.94}$$

The related standard deviation is

$$\sigma^2(\omega) = \overline{\left(\Delta S(\omega)\right)^2} \sim \overline{\left(\sum_{m=0}^{M-1}\left(\left.\frac{dS}{d\tau}\right|_{m\,\Delta\tau} - j\,\omega S\left(m\,\Delta\tau\right)\right)\delta\tau_m\,e^{-jm\,\Delta\tau\,\omega}\right)^2},$$

$$\Rightarrow \sigma^2(\omega) \sim \sum_{m=0}^{M-1}\sum_{p=0}^{M-1}\left[\begin{array}{l}\left(\left.\frac{dS}{d\tau}\right|_{m\,\Delta\tau} - j\,\omega S\left(m\,\Delta\tau\right)\right)\left(\left.\frac{dS}{d\tau}\right|_{p\,\Delta\tau} - j\,\omega S\left(p\Delta\tau\right)\right)\\ \times\,\overline{\delta\tau_m\delta\tau_p}e^{-j(m+p)\,\Delta\tau\,\omega}\end{array}\right]. \tag{11.95}$$

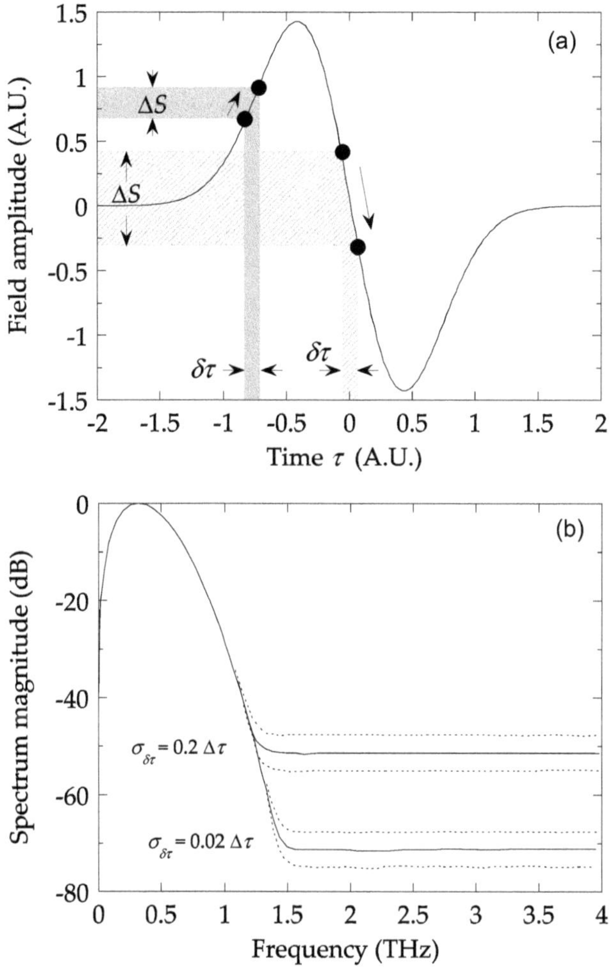

Figure 11.18 (a) Effect of a $\delta\tau$ shift on the discrete values taken in the THz waveform signal $S(\tau)$. (b) Computed spectrum (continuous lines) with two different time uncertainties $\sigma_{\delta\tau}$. Dashed lines correspond to the uncertainty due to the time-delay noise. Adapted from Jahn et al. [34].

Expression 11.95 cannot be calculated analytically: Jahn et al. [34] have computed the effect of such a noisy delay line through a Monte Carlo simulation. However, if the errors $\delta\tau_m$ are not correlated $\left(\overline{\delta\tau_m\delta\tau_{p\neq m}} = 0,\ \overline{\delta\tau_m^2} = \sigma_{\delta\tau}^2\right)$, this expression is greatly simplified:

$$\sigma^2(\omega) \propto \sigma_{\delta\tau}^2 \sum_{m=0}^{M-1} \left(\left.\frac{dS}{d\tau}\right|_{m\,\Delta\tau} - j\,\omega S\left(m\,\Delta\tau\right)\right)^2 e^{-j2m\,\Delta\tau\,\omega} = \sigma_{\delta\tau}^2\,\mathrm{TF}\left[\left(\frac{dS\left(\tau\right)}{d\tau}\right)^2\right]. \tag{11.96}$$

Thus, in the time domain, the error varies as the derivative of the waveform. This can be simply explained [34] by the effect of a small temporal shift that results in a change of signal proportional to the derivative of the curve (see Fig. 11.18 derived from [34]). In the frequency domain, $\sigma\left(\omega\right)$ is proportional to the standard deviation $\sigma_{\delta\tau}$ of the temporal steps. Jahn et al. [34] have observed, from a Monte Carlo simulation, that this noise impacts mostly the noise floor level of the spectrum of the signal. Therefore, when considering only the effect of the delay line, the following expression fits the *SNR* of the THz signal well:

$$SNR = \frac{S^2(\omega)}{\sigma^2(\omega)} = 20\ \log\left(C\ \sigma_{\delta\tau}\sqrt{\Delta\tau}\right). \tag{11.97}$$

They explain the $\sqrt{\Delta\tau}$ dependence as resulting from the central limit theorem. As seen in Fig. 11.18b, the change of *SNR* is mostly observed when $S\left(\omega\right)$ is below noise, i.e., at higher frequencies. Let us also notice that *SNR* increases while shortening $\Delta\tau$, i.e., while recording more temporal data (oversampling effect).

11.4.10 Summary

In the frequency domain, the magnitude of the THz spectrum is affected by two sources of noise: one that is proportional to the THz signal, and one that is independent. Both types have several origins, as explained in this chapter. The signal-dependent noise is dominated by the shot noise. Typically, this noise (rigorously speaking, its standard deviation) is about $30 \sim 40$ dB smaller than the THz spectral power density, i.e., $10^{-3} \sim 10^{-4}$ times smaller in terms of power and $30 \sim 100$ smaller regarding the field magnitude. Consequently, the *SNR* of THz-TDS systems is typically $30 \sim 40$ dB.

The signal-independent noise is 70 $\sim$ 90 dB below the maximum of the THz spectrum. This defines the DR of THz-TDS systems. The uncertainty on the phase is a few percent in the spectral region where the THz signal magnitude is several times bigger than the noise. Outside this region, the noise introduces random jumps of the phase, typical of one or a few π, together with an increase of the phase noise.

References

1. Taylor, J. R. (1997). *An Introduction to Error Analysis* (University Science Books, Sausalito).
2. Withayachumnankul, W., Fischer, B. M., Lin, H., and Abbott, D. (2008). Uncertainty in terahertz time-domain spectroscopy measurement, *J. Opt. Soc. Am. B*, **25**, pp. 1059–1072.
3. Withayachumnankul, W., and Naftaly, M. (2014). Fundamentals of measurement in terahertz time-domain spectroscopy, *J. Infrared Millimeter THz Waves*, **35**, pp. 610–637.
4. Saleh, B. E. A., and Teich, M. C. (2007). *Fundamentals of Photonics* (Wiley, New York).
5. Quinlan, F., Fortier, T. M., Jiang, H., and Diddams, S. A. (2013). Analysis of shot noise in the detection of ultrashort optical pulse trains, *J. Opt. Soc. Am. B*, **30**, pp. 1775–1785.
6. Berger, L. I. (1999). *Semiconductor Materials* (CRC Press, Boca Raton), p. 193.
7. Budenstein, P., ed. (1973). *Digest of Literature on Dielectrics*, Vol. 35 (National Academy of Sciences, Washington).
8. Okroashvili, G. (1979). Temperature dependence of linear electrooptical properties of some crystals of the sphalerite structure, *Opt. Spectrosc.*, **47**, pp. 442–443.
9. Matos, L., Mücke, O. D., Chen, J., and Kärtner, F. X. (2006). Carrier-envelope phase dynamics and noise analysis in octave-spanning Ti:sapphire lasers, *Opt. Express*, **14**, pp. 2497–2511.
10. Vernaleken, A., Schmidt, B., Wolferstetter, M., Hänsch, T. W., Holzwarth, R., and Hommelhoff, P. (2012). Carrier-envelope frequency stabilization of a Ti:sapphire oscillator using different pump lasers, *Opt. Express*, **20**, pp. 18387–18396.

11. van Exter, M., and Grischkowsky, D. (1990). Characterization of an optoelectronic terahertz beam system, *IEEE Trans. Microwave Theory Tech.*, **38**, pp. 1684–1691.
12. Duvillaret, L., Garet, F., and Coutaz, J.-L. (2000). Influence of noise on the characterization of materials by terahertz time-domain spectroscopy, *J. Opt. Soc. Am. B*, **17**, pp. 452–461
13. Look, D. C., Walters, D. C., Robinson, G. D., Sizelove, J. R., Mier, M. G., and Stutz, C. E. (1993). Annealing dynamics of molecular-beam epitaxial GaAs grown at 200°C, *J. Appl. Phys.*, **74**, pp. 306–310.
14. Fekecs, A., Bernier, M., Morris, D, Chicoine, M., Schiettekatte, F., Charette, P., and Arès, R. (2011). Fabrication of high resistivity cold-implanted InGaAsP photoconductors for efficient pulsed terahertz devices, *Opt. Mater. Express*, **1**, pp. 1165–1177.
15. Roehle, H., Dietz, R. J. B., Hensel, H. J., Böttcher, J., Künzel, H., Stanze, D., Schell, M., and Sartorius, B. (2010). Next generation 1.5 μm terahertz antennas: mesa-structuring of InGaAs/InAlAs photoconductive layers, *Opt. Express*, **18**, pp. 2296–2301.
16. Ivanov, E. N., Hollberg, L., and Diddams, S. A. (2002). Analysis of noise mechanisms limiting frequency stability of microwave signals generated with a femtosecond laser, *Proceedings of the IEEE International Frequency Control Symposium and PDA Exhibition*, pp. 435–451.
17. Paschotta, R. (2010). Noise in laser technology, Part 3: beam pointing fluctuations, *Opt. Photonik*, **5**, pp. 55–57.
18. Loeffler, T., Kress, M., Thomson, M., Hahn, T., Hasegawa, N., and Roskos, H., G. (2005). Comparative performance of terahertz emitters in amplifier-laser-based systems, *Semicond. Sci. Technol.*, **20**, pp. S134–S140.
19. Widrow, B., and Kollar, I. (2008). *Quantization Noise* (Cambridge University Press, Cambridge), see Chapter 15.
20. Paschotta, R., Schlatter, A., Zeller, S. C., Telle, H. R., and Keller, U. (2006). Optical phase noise and carrier-envelope offset noise of mode-locked lasers, *Appl. Phys. B*, **82**, pp. 265–273.
21. Cox, J. A., Benedick, A. J., and Kärtner, F. X. (2011). Characterization of timing jitter in ultrafast fiber and Ti:S lasers, *XXXth URSI General Assembly and Scientific Symposium*, manuscript: ABD.3.
22. Helbing, F. W., Steinmeyer, G., and Keller, U. (2003). Carrier-envelope offset phase-locking with attosecond timing jitter, *IEEE J. Sel. Top. Quantum Electron.*, **9**, pp. 1030–1040.

23. Schilt, S., and Südmeyer, T. (2015). Carrier-envelope offset stabilized ultrafast diode-pumped solid-state lasers, *Appl. Sci.*, **5**, pp. 787–816.

24. Bobroff, N. (1987). Residual errors in laser interferometry from air turbulence and nonlinearity, *Appl. Opt.*, **26**, pp. 2676–2682.

25. Yang, Y., Mandehgar, M., and Grischkowsky, D. (2012). Time domain measurement of the THz refractivity of water vapor, *Opt. Express*, **20**, pp. 26208–26218.

26. Edlen, B. (1966). The refractive index of air, *Metrologia*, **2**, p. 71.

27. Soltani, A., Probst, T., Busch, S. F., Schwerdtfeger, M., Castro-Camus, E., and Koch, M. (2014). Error from delay drift in terahertz attenuated total reflection spectroscopy, *J. Infrared Millimeter THz Waves*, **35**, pp. 468–477.

28. Oppenheim, A. V., and Schafer, R. W. (1999). *Discrete Time Signal Processing*, 2nd edn. (Prentice Hall, New Jersey).

29. McDowell, E. J., Ren, J., and Yang, C. (2008). Fundamental sensitivity limit imposed by dark 1/f noise in the low optical signal detection regime, *Opt. Express*, **16**, pp. 6822–6832.

30. Schoukens, J., and Renneboog, J. (1986). Modeling the noise influence on the Fourier coefficients after a discrete Fourier transform, *IEEE Trans. Instrum. Meas.*, **IM-35**, pp. 278–286.

31. Naftaly, M., and Dudley, R. (2009). Methodologies for determining the dynamic ranges and signal-to-noise ratios of terahertz time-domain spectrometer, *Opt. Lett.*, **34**, pp. 1213–1214.

32. See Technical note on *Motion Basics and Standards*, Newport Company.

33. Ramesh, R., Mannan, M. A., and Poo, A. N. (2000). Error compensation in machine tools—a review: Part I: geometric, cutting-force induced and fixture-dependent errors, *Int. J. Mach. Tools Manuf.*, **40**, pp. 1235–1256.

34. Jahn, D., Lippert, S., Bisi, M., Oberto, L., Balzer, J. C., and Koch, M. (2016). On the influence of delay line uncertainty in THz time-domain spectroscopy, *J. Infrared Millimeter THz Waves*, **37**, pp. 605–613.

Chapter 12

Precision of the Extraction of Material Properties from THz-TDS

12.1 Introduction

Measured data suffers from two sources of uncertainty: noise [1], which was described in the previous chapter, and repeatable errors [2, 3]. The latter originate from either the setup adjustment, the setup components, the extraction technique, or a bad knowledge of the sample parameters:

- Setup adjustment: the THz beam that is not parallel at the sample location, the sample is not perpendicular to the beam, etc.
- Setup components: delay line that does not move linearly, etc.
- Extraction technique: overly simplified models, numerical errors, etc.
- Sample parameters: erroneous knowledge of the sample thickness, sample faces that are neither flat nor parallel, roughness, inhomogeneity, unknown anisotropy, etc.

Principles of Terahertz Time-Domain Spectroscopy
Jean-Louis Coutaz, Frédéric Garet, and Vincent Wallace

ISBN 978-981-4774-56-7 (Hardcover), 978-0-203-70969-6 (eBook)
www.panstanford.com

12.2 Uncertainties on the Material Parameters due to Random Noise

12.2.1 General Relations

The influence of all the random noise that is described in the previous chapter, on the extraction of the material parameters, can be synthetized as follows. In both transmission and reflection THz-TDS, one measures the signal waveform as well as a reference waveform. Let us label S_{exp} the spectral signal obtained by Fourier-transforming any waveform; for the sake of simplicity, we omit to write the ω dependence of $S_{\text{exp}}(\omega)$. This experimental value is equal to the actual one S plus a noise term S_{N}:

$$S_{\text{exp}} \equiv s_{\text{exp}} e^{j\varphi_{\text{exp}}} = S + S_{\text{N}} = s\, e^{j\varphi} + s_{\text{N}} e^{j\varphi_{\text{N}}}. \tag{12.1}$$

s and φ are the amplitude and phase of the values. If $s_{\text{N}} \ll s$, we get

$$\begin{cases} s_{\text{exp}} \approx s + s_{\text{N}} \cos(\varphi_{\text{N}} - \varphi), \\ \varphi_{\text{exp}} \approx \arctan\left(\frac{s \sin\varphi + s_{\text{N}} \sin\varphi_{\text{N}}}{s\cos\varphi + s_{\text{N}}\cos\varphi_{\text{N}}}\right) \approx \varphi + \frac{s_{\text{N}}}{s}\sin(\varphi_{\text{N}} - \varphi). \end{cases} \tag{12.2}$$

The standard deviation σ_{s} of the amplitude s of the signal, which is equal to the precision Δs, is given by

$$\Delta s = \sigma_{\text{s}} = \sqrt{\overline{(s_{\text{exp}} - s)^2}} = \sqrt{\overline{s_{\text{N}}^2 \cos^2(\varphi_{\text{N}} - \varphi)}}. \tag{12.3}$$

Usually, the noise amplitude s_{N} and noise phase φ_{N} are not correlated. Also, φ_{N} randomly takes any value between 0 and 2π, so we have

$$\Delta s = \sqrt{\overline{s_{\text{N}}^2 \cos^2(\varphi_{\text{N}} - \varphi)}} = \sqrt{\frac{\overline{s_{\text{N}}^2}}{2}}. \tag{12.4}$$

As well, the precision on the phase φ is given by

$$\Delta\varphi = \sigma_{\varphi} = \sqrt{\overline{\frac{\sin^2(\varphi_{\text{N}} - \varphi)}{(s/s_{\text{N}})^2}}} = \frac{1}{s}\sqrt{\frac{\overline{s_{\text{N}}^2}}{2}} = \frac{\sigma_{\text{s}}}{s}. \tag{12.5}$$

The useful signal $\tilde{X}$, i.e., the coefficient of transmission ($\tilde{X} = \tilde{T}$) or of reflection ($\tilde{X} = \tilde{R}$) for the THz field, is equal to the ratio of the transmitted or reflected signal $\tilde{S}$ and of the reference signal, which is supposed to be equal to the incident signal $\tilde{S}_i$:

$$\tilde{X}_{\exp} = x_{\exp} e^{j\,\varphi_{x,\exp}} = \frac{\tilde{S}_{\exp}}{\tilde{S}_{i,\exp}} = \frac{\tilde{S} + \tilde{S}_{\mathrm{N}}}{\tilde{S}_i + \tilde{S}_{i,N}}$$
$$\approx \frac{\tilde{S}}{\tilde{S}_i}\left(1 + \frac{\tilde{S}_{\mathrm{N}}}{\tilde{S}} - \frac{\tilde{S}_{i,N}}{\tilde{S}_i}\right) = \tilde{X} + \frac{\tilde{S}_{\mathrm{N}} - \tilde{X}\,\tilde{S}_{i,N}}{\tilde{S}_i}. \quad (12.6)$$

It follows that the noisy part of $\tilde{X}_{\exp}$ is

$$\tilde{X}_{\mathrm{N}} = \tilde{X}_{\exp} - \tilde{X} = \frac{\tilde{S}_{\mathrm{N}} - \tilde{X}\,\tilde{S}_{i,N}}{\tilde{S}_i}. \quad (12.7)$$

12.2.2 ABC Noise Coefficients

From Eq. 12.7, assuming that the noise of the signal and the reference are not correlated, the standard deviation of the modulus x of $\tilde{X}$ is

$$\sigma_x^2 = \frac{\sigma_{\mathrm{s}}^2 + X^2\,\sigma_{s_i}^2}{s_i^2}. \quad (12.8)$$

The noise σ_{s}^2 of signal s (modulus) at the receiver is the sum of the noise σ_{e}^2 at the emitter times the transmission or reflection modulus x and the Johnson noise that is signal dependent ($\sigma_{\mathrm{J}}^2 = 2es\Delta f$), and other noises σ_{d}^2 in detection that are signal independent:

$$\sigma_{\mathrm{s}}^2 = x^2\sigma_{\mathrm{e}}^2 + 2es\Delta f + \sigma_{\mathrm{d}}^2. \quad (12.9)$$

Similarly, the reference signal noise is given by

$$\sigma_i^2 = \sigma_{\mathrm{e}}^2 + 2es_i\Delta f + \sigma_{\mathrm{d}}^2. \quad (12.10)$$

Therefore, the standard deviation of the transmission/reflection coefficient can be written as a second order polynomial expression of x:

$$\sigma_x^2 = \frac{2\,\sigma_{\mathrm{e}}^2 + 2es_i\Delta f + \sigma_{\mathrm{d}}^2}{s_i^2}x^2 + \frac{2e\Delta f}{s_i}x + \frac{\sigma_{\mathrm{d}}^2}{s_i^2}$$
$$\equiv A(\omega)x^2 + B\,(\omega)\,x + C\,(\omega). \quad (12.11)$$

The A, B, and C coefficients depend on frequency, and they are distinctive parameters of each set up [1, 4]. C is only determined by the signal-independent noise, while B originates in the Johnson noise. A includes all the noise sources. Practically, they are

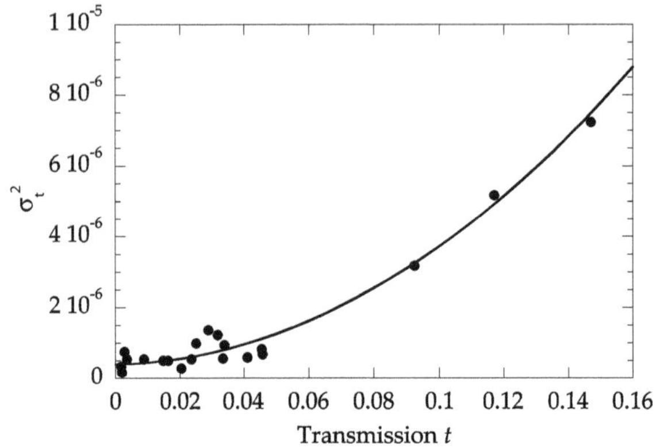

Figure 12.1 The standard deviation σ_t^2 versus t for the Advantest THz-TDS setup. The circles are measured data; the curve is the best fit calculated with Eq. 12.11.

determined from a set of measurements, mostly in transmission, using a rather transparent sample. Calculating the average and standard deviation of both transmitted and reference spectra leads easily to determine A, B, and C versus frequency [1, 4]. Figure 12.1 demonstrates the validity of relation 12.11 [4]. It shows the measured standard deviation σ_t^2 versus t for an Advantest THz-TDS system at 3.5 THz. Here the samples are pellets of bio-plastic (Biomer P226). Twenty-seven pellets with different thicknesses have been fabricated. Each of them exhibits a different transmission. Each point of Fig. 12.1 corresponds to one of these pellets. The statistical treatment has been performed over 20 measurements for each pellet.

The continuous curve is the best fit with expression 12.11. Even if, at low transmission, the signal is noisy, the calculated curve fits the experimental data well. The numerical fit leads directly to the coefficients A, B, and C. Getting the spectral dependence of these noise coefficients requires repeating this procedure at any frequency.

Figure 12.2 shows the A, B, and C coefficients versus frequency for our homemade TDS system (top) and the Advantest 7500TSA model (bottom). At the maximum of the THz spectra, A is about 10 times larger than B, which is also 10 times larger than C. This means

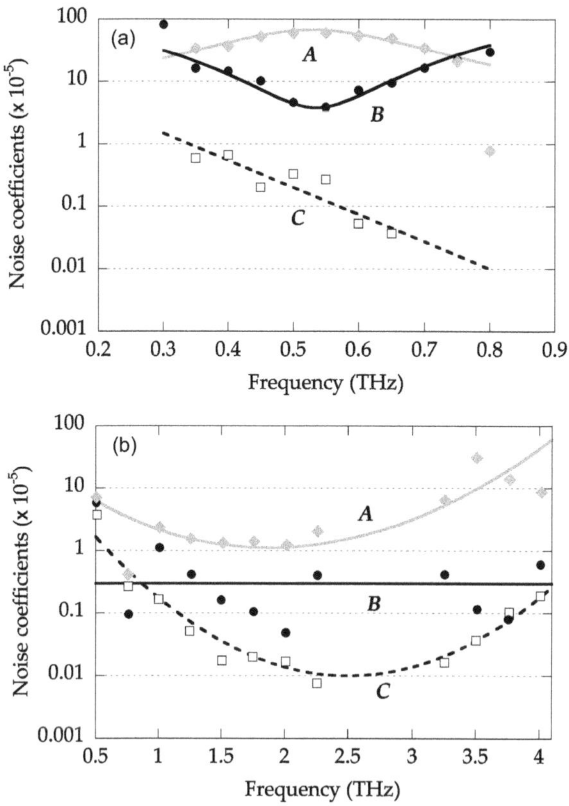

Figure 12.2 Measured noise parameters A, B, and C for a homemade TDS system (top) and the Advantest 7500TSA model (bottom). The continuous line is a guide to the eye.

that the Johnson noise is bigger than the signal independent noise in the receiver and that the noise in the emitter (mostly shot noise and laser instabilities) is the main contributor to the overall noise.

12.2.3 Reflection THz-TDS

12.2.3.1 Uncertainty on $r(\omega)$ due to noise

The related relative uncertainty on the coefficient of reflection is plotted in Fig. 12.3 for the homemade system. The uncertainty is around 2.5% at higher values of n and κ, but it increases strongly

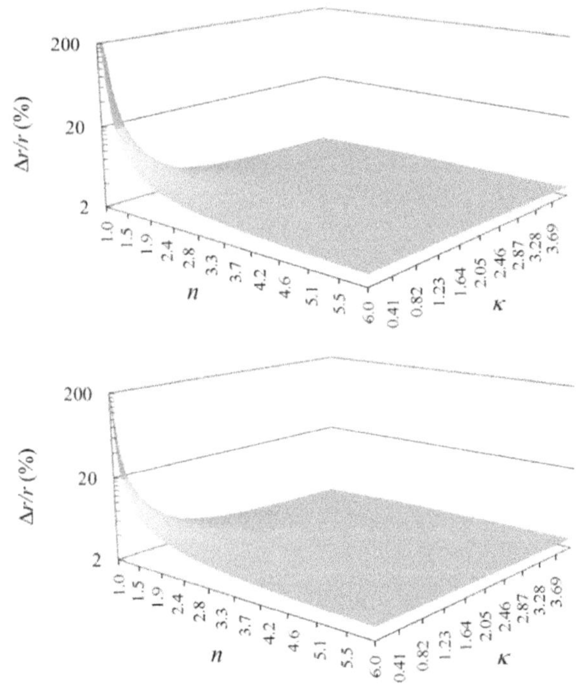

Figure 12.3 Calculated relative uncertainty $\Delta r/r$ versus n and κ for the homemade TDS system at 0.5 THz (top) and 0.75 THz (bottom).

when n and κ approach respectively 1 and 0 (at 0.5 THz, the uncertainty is larger than 0.5% for $n < 1.1$ and $\kappa < 0.08$ ($\alpha < 16$ cm^{-1} at 0.5 THz)). This was expected, as, for these values, r tends towards zero.

The relative phase uncertainty (Fig. 12.4) increases monotonously with n. Its variation with κ is more complex, even if, in any case, it diverges when κ tends to zero. Thus, reflection THz-TDS is not suitable for characterizing transparent samples. At higher values of n, it decreases with κ. At smaller values of n, it first strongly decreases with κ, and then it slightly increases with κ. Here the minimum value of $\Delta\varphi/\varphi$ ($\sim$3.4%) is obtained for $\kappa \sim 1.45$ (i.e., at $300 \sim 500$ cm^{-1} in the range 0.5–0.8 THz).

The same modulus and phase uncertainty maps, calculated for the Advantest 7500TSA system, are plotted in Figs. 12.5 and 12.6.

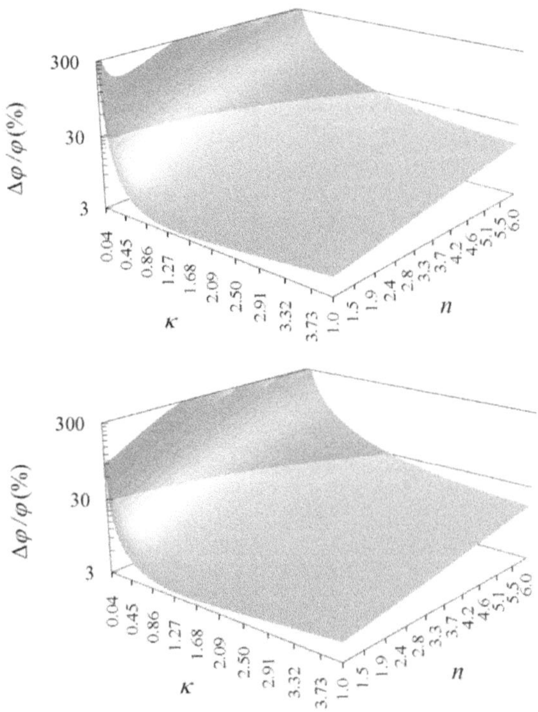

Figure 12.4 Calculated relative uncertainty $\Delta\varphi/\varphi$ versus n and κ for the homemade TDS system at 0.5 THz (top) and 0.75 THz (bottom).

The behavior of the $\Delta r/r$ map is similar to the one of the homemade system but with weaker values (at 1 THz). At 5 THz, the uncertainty is quite strong, of the order of 17 % for large values of n and κ.

The uncertainty on the phase at 1 THz is again quite similar to the one obtained with the homemade system, but with larger values, maybe due to the ASOPS-like triggering of the Advantest system. At 5 THz, the uncertainty is bigger than 10%.

Figure 12.7 shows the standard deviations Δr and $\Delta\varphi_r$ (dashed lines) determined from eight different measurements of the THz signal reflected by a high-resistivity Si wafer [5]. The dashed lines are calculated with the noise expression (Eq. 12.11), adapted to the reflection case (see relation Eq. 12.12 hereafter). In the range where the THz signal is intense (0.2 $\sim$ 1.4 THz), the agreement

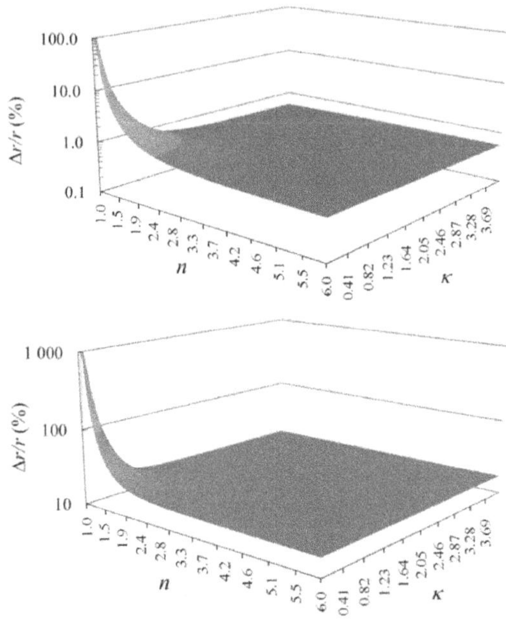

Figure 12.5 Calculated relative uncertainty $\Delta r/r$ versus n and κ for the Advantest TAS7500 system at 1 THz (top) and 5 THz (bottom).

between modeling and experiment is excellent. At lower or higher frequencies, the THz signal is weak, and the noise is rather large. Thus, the hypothesis $\sigma_s \ll s$ that serves to obtain Eq. 12.11 is not fulfilled: relation 12.11 is no more valid.

12.2.3.2 Uncertainty on *n* and κ

From now on, we still simplify the expressions by omitting the ω dependence. In the case of reflection THz-TDS, (12.11) becomes

$$\sigma_r^2 = Ar^2 + Br + C\,. \qquad (12.12)$$

The complex refractive index $\tilde{n}$ is simply deduced from the coefficient of reflection:

$$\tilde{n} = \frac{1 - r\,e^{j\varphi}}{1 + r\,e^{j\varphi}} = \frac{1 - r^2 - 2jr\sin\varphi}{1 + r^2 + 2r\cos\varphi}, \qquad (12.13)$$

$$\Rightarrow \quad n = \frac{1 - r^2}{1 + r^2 + 2r\cos\varphi}, \qquad \kappa = \frac{-2r\sin\varphi}{1 + r^2 + 2r\cos\varphi}. \qquad (12.14)$$

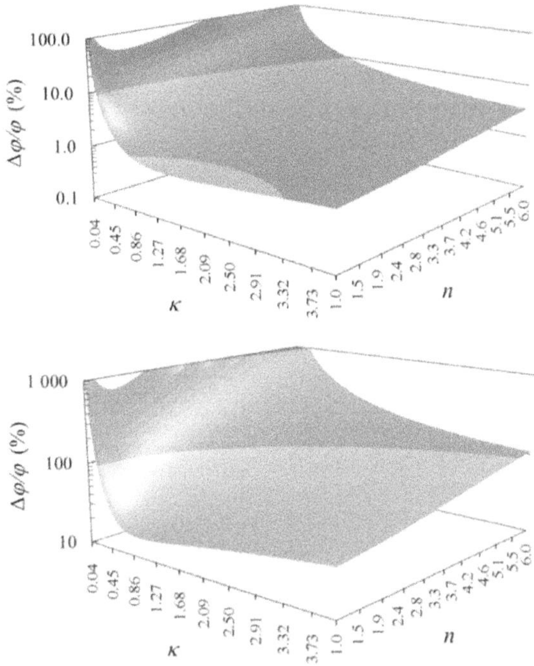

Figure 12.6 Calculated relative uncertainty $\Delta\varphi/\varphi$ versus n and κ for the Advantest TAS7500 system at 1 THz (top) and 5 THz (bottom).

r and φ are respectively the modulus and phase of the coefficient of reflection. The error Δn is obtained by deriving Eq. 12.14:

$$\Delta n = \left|\frac{\partial n}{\partial r}\right| \Delta r + \left|\frac{\partial n}{\partial \varphi}\right| \Delta\varphi = \left(r\left|\frac{\partial n}{\partial r}\right| + \left|\frac{\partial n}{\partial \varphi}\right|\right)\frac{\Delta r}{r}.$$

$$\Rightarrow \quad \Delta n = 2\frac{\left|2r + \left(1+r^2\right)\cos\varphi\right| + \left(1-r^2\right)\left|\sin\varphi\right|}{\left(1+r^2+2r\cos\varphi\right)^2}\Delta r. \quad (12.15)$$

Here, we made use of expression 12.5 that links $\Delta\varphi$ and Δr. In the same way, it is easy to show that

$$\Delta\kappa = \left(r\left|\frac{\partial \kappa}{\partial r}\right| + \left|\frac{\partial \kappa}{\partial \varphi}\right|\right)\frac{\Delta r}{r} = \Delta n. \quad (12.16)$$

In the reflection case, the error due to random noise is the same for n and κ. Writing expressions 12.15 and 12.16 versus only n and κ gives huge expressions that cannot be solved analytically.

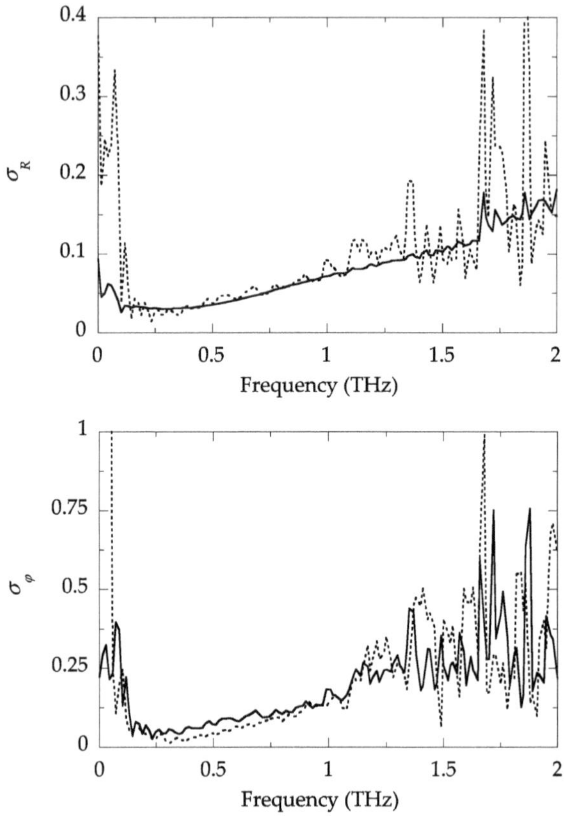

Figure 12.7 Standard deviations Δr and $\Delta\varphi_r$ vs. frequency for the THz signal reflected by an HR-Si wafer (adapted from Bernier et al. [5]). The dashed lines are obtained from eight different measurements, while the continuous lines are calculated with Eq. 12.12.

Figure 12.8 shows the calculated error on n and κ for the homemade system at 0.5 THz (at 0.75 THz, the curves are almost equal). Typically, $\Delta n/n$ is larger than 0.2% over the entire studied range. At low values of n, it increases up to 2%. $\Delta\kappa/\kappa$ increases exponentially when the sample becomes transparent. In any case, the uncertainty on κ is larger than 100%!

The uncertainty on n with the Advantest 7500TSA system at 1 THz (Fig. 12.9) is of the same order of magnitude than with the homemade system, typically less than 0.5%, while at 5 THz, it is one order of magnitude larger.

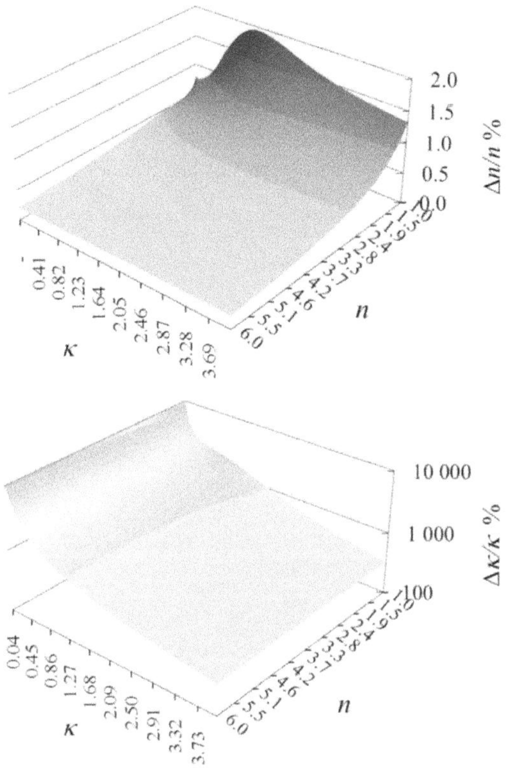

Figure 12.8 Uncertainty $\Delta n/n$ (top) and $\Delta\kappa/\kappa$ (bottom) versus n and κ, calculated from Eqs. 12.15 and 12.16, for the homemade system at 0.5 THz.

The relative error on κ increases inversely proportionally to κ (Fig. 12.10). It does not depend so much on n. At 1 THz, it is lower than 1% for $\kappa > 0.5 \sim 1$ (i.e., $\alpha > 200 \sim 400$ cm^{-1}). At 5 THz, the general behavior of the map is the same, but the uncertainty is one order of magnitude bigger.

Figure 12.11, adapted from Bernier et al. [5], shows the refractive index of high-resistivity Si extracted from THz-TDS measurements performed in reflection. The uncertainty, Δn, due to noise is also plotted: open circles have been determined from eight measurements, while the continuous line is calculated using the expression 12.15.

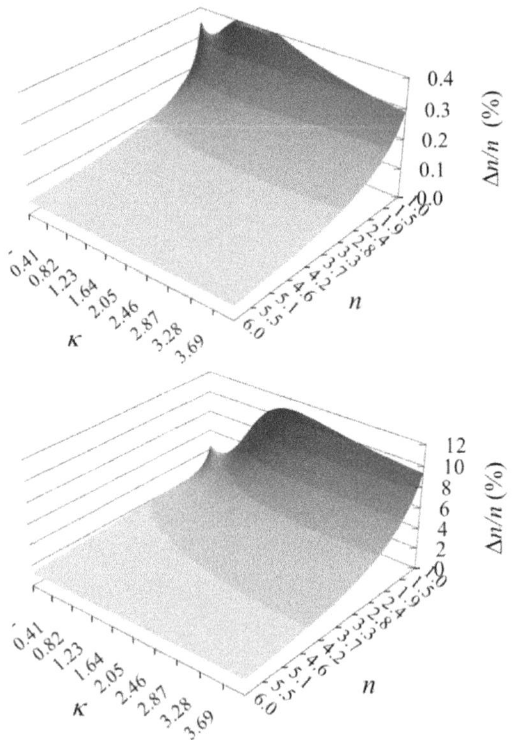

Figure 12.9 Uncertainty $\Delta n/n$ versus n and κ, calculated from Eq. 12.15, for the Advantest TAS7500 system at 1 THz (top) and 5 THz (bottom).

The agreement between modeling and experimental data is excellent up to 1.6 THz, over which the noise magnitude cannot be neglected as compared to the reflected signal. Here the uncertainty Δn is of the order of $15 \sim 30\%$. The large imprecision demonstrated, as already stated, that reflection THZ-TDS is not suited to characterize transparent materials like HR-Si.

12.2.3.3 Dynamic range

From the previous study, one can derive the dynamic range of THz-TDS in reflection, i.e., what are the minimum and maximum values of both n and κ one can determine [6]. Let us write again the expressions of the modulus r and the phase φ_r of the coefficient of

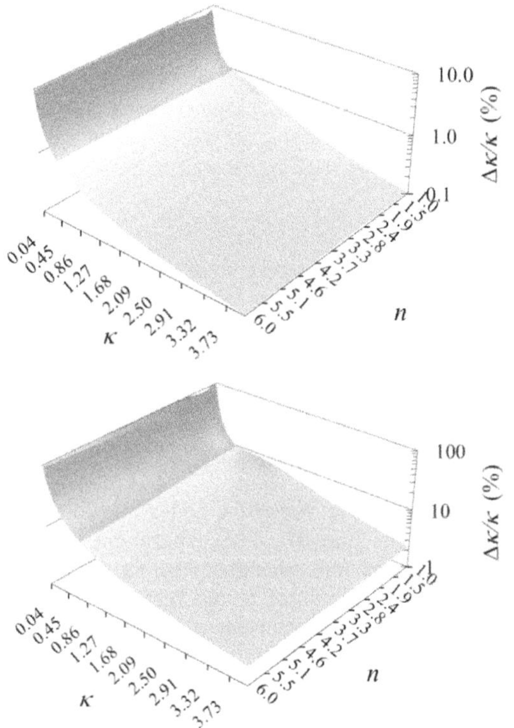

Figure 12.10 Uncertainty $\Delta\kappa/\kappa$ versus n and κ, calculated from Eq. 12.16, for the Advantest TAS7500 system at 1 THz (top) and 5 THz (bottom).

reflection:

$$r = \sqrt{\frac{(1-n)^2+\kappa^2}{(1+n)^2+\kappa^2}}, \quad \tan\varphi_r = \frac{2\kappa}{n^2+\kappa^2-1}. \tag{12.17}$$

When κ is large, r and φ_{r} tend respectively towards one and zero. However, one cannot measure exactly one and zero, but $1-\sigma_r$ and σ_φ because of the noise-induced uncertainty. Therefore, the maximum value of κ one can extract is calculated as follows:

$$r \approx 1-\sigma_{\mathrm{r}} = \sqrt{\frac{(1-n)^2+\kappa_{\max}^2}{(1+n)^2+\kappa_{\max}^2}}, \ \tan\varphi \approx \sigma_\varphi = \frac{\sigma_{\mathrm{r}}}{r} = \frac{2\kappa_{\max}^2}{n^2+\kappa_{\max}^2-1},$$

$$\Rightarrow \quad \kappa_{\max} = \frac{1-\sigma_{\mathrm{r}}}{\sigma_{\mathrm{r}}} + \sqrt{\left(\frac{1-\sigma_{\mathrm{r}}}{\sigma_{\mathrm{r}}}\right)^2 + n^2 - 1} \approx \frac{2}{\sigma_{\mathrm{r}}}. \tag{12.18}$$

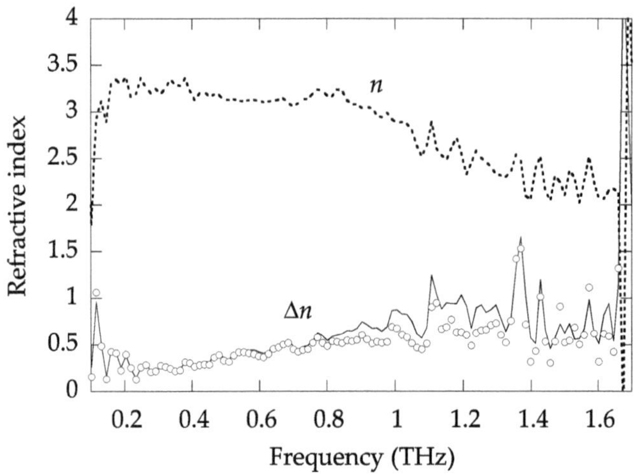

Figure 12.11 Spectra of the refractive index n (dotted line) of HR-Si determined from reflection THz-TDS experiment, and noise-induced uncertainty Δn (continuous line calculated with Eq. 12.15 and open circles obtained from eight measurements). Notice that the Δn values have been multiplied by 1000. Adapted from Bernier et al. [5].

The last simplified expression of $\kappa_{\max}$, which is valid when $\sigma_r \ll 1$, is the same as relation (8) in [6], but these authors do not simplify it with $\sigma_\varphi = \sigma_r/r$. Figure 12.12 displays $\alpha_{\max} = 2\kappa_{\max}\omega/c$ versus frequency for both homemade and Advantest systems. In the available frequency range, α can be measured up to $10^4 \sim 10^5$ cm^{-1}, which allows one to characterize practically any absorbing material.

The other limit that determines the dynamic range is when κ is small, i.e., when the samples are transparent: what is the minimum value of κ one can extract? In the limit of small κ, expressions 12.17 become

$$r \approx \frac{\sqrt{\left(1-n^2\right)^2 + 4n\kappa^2}}{(1+n)^2},\ \tan\varphi_r \approx \frac{2\kappa}{n^2-1+\kappa^2}. \qquad (12.19)$$

Two cases must be addressed: either $n \approx 1$ or $n - 1 \gg \kappa$. Let us start with the second case. We can simplify Eq. 12.19:

$$r \approx \frac{1-n}{1+n},\ \tan\varphi_r \approx \varphi_r = \sigma_\varphi = \frac{\sigma_r}{r} = \frac{2\kappa_{\min}}{n^2-1},$$

$$\Rightarrow \kappa_{\min} = \frac{\sigma_r}{r}\frac{n^2-1}{2} = \sigma_r\frac{(n+1)^2}{2}. \qquad (12.20)$$

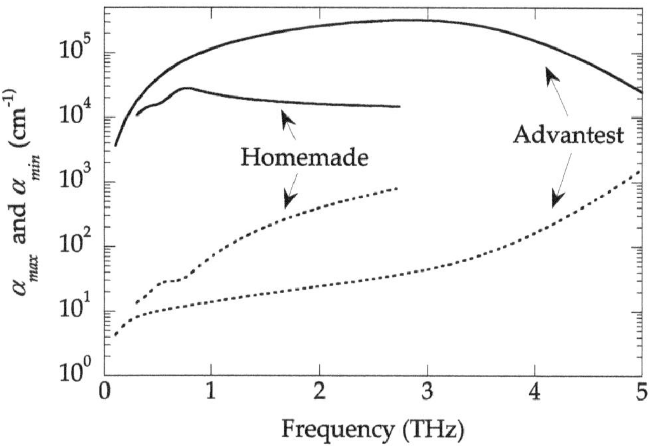

Figure 12.12 The maximum (continuous curves) and minimum (dotted curves) values of the absorption coefficient α, which can be extracted from reflection THz-TDS, versus frequency for the homemade and Advantest THz-TDS systems.

When $n \approx 1$ the problem is more difficult to treat because it depends on which is the largest among $n - 1$ and κ. The unrealistic case $n = 1$ is easier to solve:

$$r = \sigma_r \approx \frac{\kappa_{\min}}{2}, \Rightarrow \quad \kappa_{\min} \approx 2\sigma_r. \tag{12.21}$$

Figure 12.12 also shows the spectra of the minimum absorption coefficient one can extract from reflection measurements with the homemade and Advantest THz-TDS systems. Because $\alpha_{\min}$ depends on n (12.20), here the curves are calculated for $n = 2$. In the spectral range for which the THz signal is strongest (0.2 $\sim$ 1 THz for the homemade system, 0.5 $\sim$ 3 THz for the Advantest system), $\alpha_{\min} =$ 10 $\sim$ 20 cm^{-1}. Reflection THz-TDS is definitively not convenient for transparent samples. In the useful spectral range, the dynamic range is about 10^4 for both THz-TDS systems.

12.2.4 Transmission THz-TDS

12.2.4.1 Uncertainty on $t(\omega)$ due to noise

For the sake of simplicity, we treat the case of thick samples, for which the rebounds of the waveform can be rejected by a proper

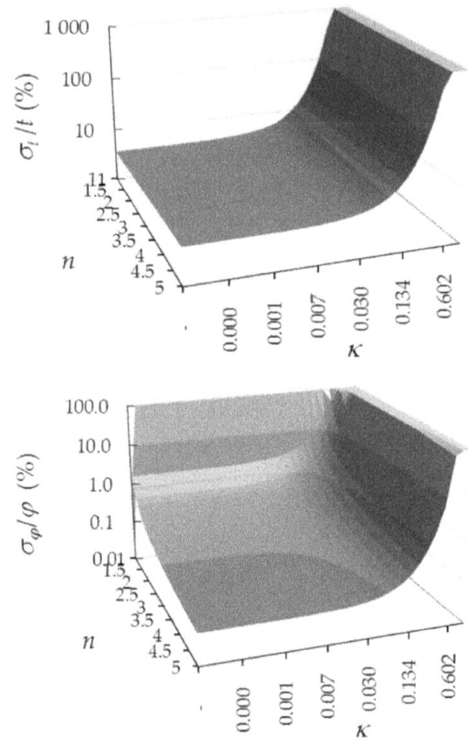

Figure 12.13 Maps of the relative uncertainties σ_t/t and σ_φ/φ versus n and κ for the homemade THz-TDS system. The sample is 1 mm thick, and the frequency is 1 THz.

time-window filtering. Therefore, the coefficient of transmission becomes

$$\tilde{t} = \frac{4\tilde{n}}{(\tilde{n}+1)^2} e^{-j\frac{\omega}{c}(\tilde{n}-1)d}. \tag{12.22}$$

Contrary to the reflection coefficient, $\tilde{t}$ depends on the sample thickness and directly on the frequency, because of the propagation exponent. Thus, $\frac{\omega}{c}d$ is used as a parameter in the following analysis. The error on t is derived from Eq. 12.11:

$$\sigma_t^2 = At^2 + Bt + C. \tag{12.23}$$

Figure 12.13 presents the relative uncertainties σ_t/t (12.13 top) and σ_φ/φ (12.13 bottom) versus n and κ for the homemade system. The values are calculated for $d = 1$ mm at $f = 1$ THz. For low values of κ (here $\kappa < 0.1$, i.e., $\alpha < 40$ cm^{-1}) the uncertainty σ_t/t originates mostly from the Fresnel coefficients in Eq. 12.22, and it is almost constant (here about 3.6%), whatever the values of n and κ. At higher values of κ, the main cause of uncertainty is the attenuation due to propagation through the sample. Thus, the uncertainty increases exponentially with κ. The map calculated for the Advantest system is very similar, but with lower values (for example, the 0.5% instead of 3.6% for transparent samples). As expected, the σ_φ/φ map exhibits the same behavior versus κ. However, because φ varies mostly as $(n-1)\,d\,\omega/c$, the relative uncertainty σ_φ/φ diverges when n approaches one.

These results confirm what can be deduced from good sense. Transmission THz-TDS is not convenient for low transmission and opaque samples. Moreover, if the refractive index is very small (gases, foams, etc.), the method is not precise. The highest absorption value, for a given sample, which permits measurement in transmission, is obtained when the transmitted signal is equal to noise, i.e., $t_{\mathrm{lim}} = \sigma_t$. Let us notice that $1/t_{\mathrm{lim}} = s_i/s_{\mathrm{N}} = DR$ is the dynamic range of the setup (see previous section 12.4.6.1). Entering this value of σ_t in Eq. 12.23 allows us to determine the expression of t_{lim} for a given setup, with characteristic noise parameters A, B, and C:

$$(A-1)\,t_{\mathrm{lim}}^2 + B t_{\mathrm{lim}} + C = 0 \;\Rightarrow\; t_{\mathrm{lim}} = \frac{B \pm \sqrt{B^2 + 4\,(1-A)\,C}}{2\,(1-A)}. \tag{12.24}$$

Because A, B, and C are very small, and $B \gg C$, $t_{\mathrm{lim}} = \sigma_{\mathrm{t}} \approx B$. Thus, the maximum value κ_{max} of κ, or α_{max} of α, that can be determined in transmission, is a solution of the following equation:

$$\sigma_t = \frac{4n}{(n+1)^2 + \kappa_{\mathrm{max}}^2}\, e^{-\frac{\omega}{c}\kappa_{\mathrm{max}} d} = \frac{4n}{(n+1)^2 + \left(\frac{c}{4\pi f}\alpha_{\mathrm{max}}\right)^2}\, e^{-\frac{\alpha_{\mathrm{max}} d}{2}}. \tag{12.25}$$

In the THz range, $4\pi f/c \approx 100 \sim 2000$ cm^{-1}. For absorption as smaller than a few tens of cm^{-1}, the denominator in Eq. 12.25 is almost equal to $(n+1)^2$. Hence, α_{max} is directly obtained from

Eq. 12.25 [6]:

$$\sigma_t \approx \frac{4n}{(n+1)^2} e^{-\frac{\alpha_{\max} d}{2}} \quad \Rightarrow \quad \alpha_{\max} = \frac{2}{d} \ln\left(\frac{4n}{(n+1)^2} DR\right). \tag{12.26}$$

On the other hand, the weakest absorption that could be determined is obtained when the phase φ equals the phase noise σ_φ:

$$\begin{aligned} \varphi = \sigma_\varphi &= \left| \arctan\frac{\kappa_{\min}}{n} - 2\arctan\frac{\kappa_{\min}}{n+1} - \frac{\omega}{c}(n-1)\,d \right| \\ &\approx \left| \frac{1-n}{1+n}\frac{\kappa_{\min}}{n} - \frac{\omega}{c}(n-1)\,d \right|, \\ \Rightarrow \kappa_{\min} &\approx \frac{n+1}{n-1} n \left(\sigma_\varphi - \frac{\omega}{c}(n-1)\,d\right). \end{aligned} \tag{12.27}$$

$\kappa_{\min}$ varies linearly with σ_φ. The larger the noise on the phase, the poorer the characterization of transparent materials. When the sample thickness d increases, $\kappa_{\min}$ decreases as well because the propagation through a thicker sample makes the absorption bigger. For a given thickness $(d_c = c\,\sigma_\varphi/\omega\,(n-1))$, $\kappa_{\min} = 0$, which means that for samples thicker than d_c, even transparent materials can be precisely characterized.

Figure 12.14 shows $\alpha_{\min}$ and $\alpha_{\max}$ versus the sample thickness d for three different frequencies, calculated using Eqs. 12.26 and 12.27

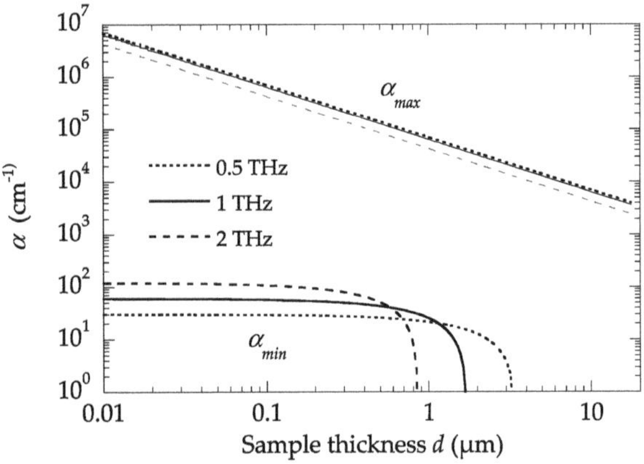

Figure 12.14 Minimum and maximum values of α that can be measured with the homemade THz-TDS system, versus the sample thickness. Here the curves are calculated with (12.27) for $n = 2$ and $\sigma_\varphi = 0.036$ rad.

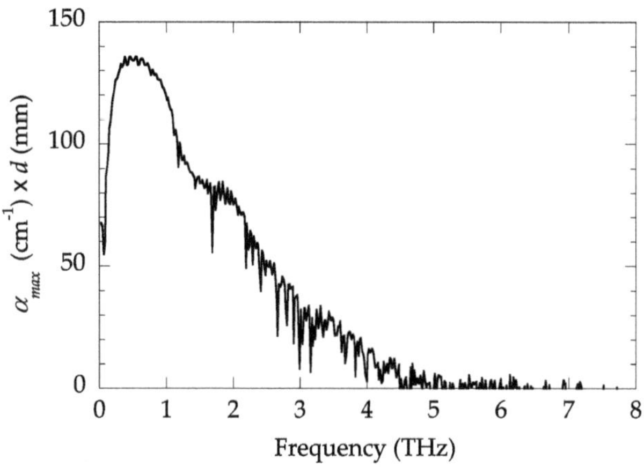

Figure 12.15 The maximum value of α that can be measured with the homemade THz-TDS system, versus the sample thickness. Here the curves are calculated with (12.27) for $n = 2$ and $\sigma_\varphi = 0.036$ rad. Dips are due to water vapor lines.

for the homemade system. The phase noise is $\sigma_\varphi = 0.036$ rad (see Fig. 12.13) and the refractive index is $n = 2$. The thickness threshold d_c over which α_{min} tends to zero is clearly seen. Below this value (2 μm at 1 THz), α_{min} is quite big (60 cm^{-1} at 1 THz). Only thin films with a large κ, like metals or water layers, can be characterized by THz-TDS in transmission. α_{max} is extremely large, especially when samples are thin.

It is also interesting to plot the dependence of α_{max} on frequency. We use here the *DR* spectrum of the homemade system (Fig. 11.13). We plot the product $\alpha_{max} \times d$ versus frequency for the homemade system (Fig. 12.15). Typically, at the highest THz spectral density, i.e., around 0.6 THz, 1 mm samples can be studied up to a maximum absorption $\alpha_{max} \sim 135$ cm^{-1}. To characterize materials that are more absorbing, thinner samples should be used. α_{max} decreases at higher frequencies, and becomes null at the higher spectral limit of the available THz beam (here around $4 \sim 5$ THz). This upper limit α_{max}may introduce artefacts in the characterization of samples. This has been clearly pointed out by Jepsen and Fischer [6], and this will be explained in the next paragraph.

12.2.4.2 Noise-induced uncertainty on $n(\omega)$ and κ (ω) [1–4, 7]

In the case of transmission, expressions are more complicated than in reflection, because they include both the Fresnel coefficients at the sample interface and a propagation term. The complex coefficient of transmission is given by (Eq. 5.3):

$$\tilde{t} = \frac{4\tilde{n}}{(\tilde{n}+1)^2} \times \frac{e^{-j\frac{\omega}{c}(\tilde{n}-1)d}}{1-\left(\frac{\tilde{n}-1}{\tilde{n}+1}e^{-j\frac{\omega}{c}\tilde{n}d}\right)^2}. \tag{12.28}$$

There exists no analytical expression of Δn and $\Delta\kappa$ versus Δt and $\Delta\varphi$, which can be derived from Eq. 12.28. Therefore, the errors Δn and $\Delta\kappa$ must be computed using the partial derivatives of $\tilde{t}$:

$$\tilde{t} = te^{-j\varphi} \quad \Rightarrow \quad dt = \frac{\partial t}{\partial n}dn + \frac{\partial t}{\partial \kappa}d\kappa, \quad d\varphi = \frac{\partial \varphi}{\partial n}dn + \frac{\partial \varphi}{\partial \kappa}d\kappa. \tag{12.29}$$

Hence,

$$dn = \frac{\frac{\partial\varphi}{\partial\kappa} - \frac{1}{t}\frac{\partial t}{\partial\kappa}}{\frac{\partial\varphi}{\partial\kappa}\frac{\partial t}{\partial n} - \frac{\partial t}{\partial\kappa}\frac{\partial\varphi}{\partial n}}dt, \quad d\kappa = \frac{\frac{1}{t}\frac{\partial t}{\partial n} - \frac{\partial\varphi}{\partial n}}{\frac{\partial\varphi}{\partial\kappa}\frac{\partial t}{\partial n} - \frac{\partial t}{\partial\kappa}\frac{\partial\varphi}{\partial n}}dt. \tag{12.30}$$

If the sample is thick enough to allow a time windowing of the waveform to save only the first transmitted pulse (temporal rebounds are filtered) and in the case of low absorption ($\kappa \ll 1$), the coefficient of transmission takes a simpler form:

$$t = \frac{4n}{(n+1)^2}e^{-\frac{\omega}{c}\kappa d}, \quad \varphi = (n-1)\left(\frac{\kappa}{n(1+n)} + \frac{\omega}{c}d\right). \tag{12.31}$$

Consequently, the errors can be analytically derived from Eqs. 12.30 and 12.31:

$$\Delta n = \frac{\left(n-1+\frac{\omega}{c}n(n+1)d\right)n(1+n)}{\left|(n-1)^2 + \left(\left(1+2n-n^2\right)\kappa + n^2(1+n)^2\frac{\omega}{c}d\right)\frac{\omega}{c}d\right|}\frac{\Delta t}{t}, \tag{12.32}$$

$$\Delta\kappa = \frac{\left|n\left(n^2-1\right) - \left(1+2n-n^2\right)\kappa - n^2(1+n)^2\frac{\omega}{c}d\right|}{\left|(n-1)^2 + \left(\left(1+2n-n^2\right)\kappa + n^2(1+n)^2\frac{\omega}{c}d\right)\frac{\omega}{c}d\right|}\frac{\Delta t}{t}. \tag{12.33}$$

As anticipated, it follows that the errors increase inversely with t, i.e., exponentially with $\kappa d\,\omega/c$: the weaker is the transmission, the larger are the uncertainties. The terms including $\omega d/c$ in Eqs. 12.32 and

12.33 are the contribution from the propagation through the sample, while other terms are due to the Fresnel coefficients at the sample interfaces. In the case of thick samples, i.e., $\frac{\omega}{c}d \gg \frac{1+2n-n^2}{n^2(1+n)^2}\kappa$ in Eqs. 12.32 and 12.33, the errors Δn and $\Delta\kappa$ are equal:

$$\Delta n \approx \frac{1+\zeta}{1+\zeta^2}\frac{(n+1)^3}{n-1}e^{\frac{\omega}{c}\kappa d}\frac{\Delta t}{4}, \qquad \Delta\kappa \approx \frac{\zeta-1}{\zeta+1}\Delta n, \tag{12.34}$$

with $\zeta = n\left(\frac{n+1}{n-1}\right)\frac{\omega}{c}d$. Additionally, one can simplify Eq. 12.34 when $\frac{\omega}{c}d \gg \frac{n-1}{n(n+1)}$ $(\zeta \gg 1)$, which is quite realistic in most cases since it corresponds to $f_{\text{THz}}d_{\text{mm}} \gg 8.5 \times 10^{-3}$:

$$\Delta n \approx \Delta\kappa \approx \frac{(n+1)^2}{n}\frac{e^{\frac{\omega}{c}\kappa d}}{\frac{\omega}{c}d}\frac{\Delta t}{4}, \tag{12.35}$$

Note that here $\Delta n \approx \Delta\kappa$. The dependence of the uncertainty (12.35) on the sample thickness d, for a given material of imaginary index κ, is determined by the ratio $\exp\left(\frac{\omega}{c}\kappa d\right)/\frac{\omega}{c}d$ $(\Delta t = \sigma_{\text{t}}$ is almost constant for the small values of κ; see Fig. 12.10). At small thicknesses, the uncertainty increases due to the weak phase induced by the propagation through the sample $(\frac{\omega}{c}d)^{-1}$. For large thicknesses, the uncertainty increases due to absorption in the material, namely $\exp(\frac{\omega}{c}\kappa d)$. The uncertainty is minimum when

$$\kappa\frac{\omega}{c}d = 1 \quad \Rightarrow \quad 2\,\alpha\,d = 1. \tag{12.36}$$

Figure 12.16 depicts the optimum thickness d for a precise extraction material parameters in transmission THz-TDS. Typically, for common absorption values $\left(\alpha \approx 1 \sim 10\ \text{cm}^{-1}\right)$, the samples should be 0.5 ~ 5 mm thick. For highly absorbing layers like metals, the thickness must be smaller than one μm. For water, $\alpha \approx 200\ \text{cm}^{-1}$ at 1 THz, and the water film thickness should be of the order or smaller than 100 μm.

On the other hand, for transparent media like gases, the optimum thickness should be more than tens of cm. This is extremely difficult to manage in THz-TDS, because the THz beam should be parallel over such a long distance, and the mechanical delay line should not introduce any angular shift in the THz beam direction, which would be amplified by the long propagation length. The dependence on the uncertainty with the sample thickness d is nicely studied in [7], where the reader will find additional information.

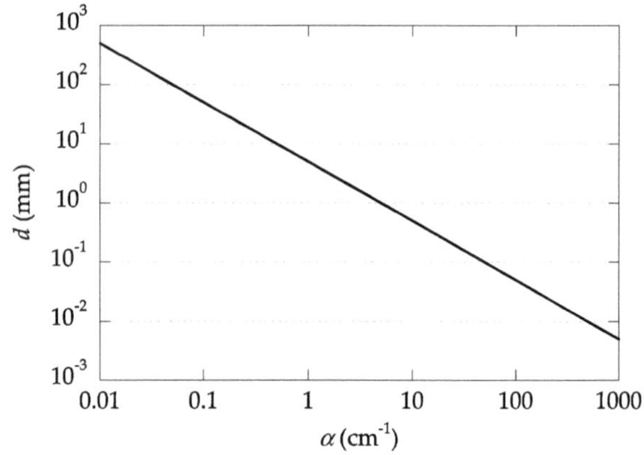

Figure 12.16 Optimum sample thickness d for a precise extraction of material parameters in THz-TDS.

Figure 12.17 presents the Δn and $\Delta\kappa$ maps versus n and κ, for the homemade setup at $f = 1$ THz and with $d = 1$ mm ($\omega d/c \approx 21$). Both Δn and $\Delta\kappa$ diverge when κ is large, because of the lack of transmitted signal at higher absorption (the shape of the maps is mostly governed by $\Delta t/t$, see Fig. 12.13). Below $\kappa \approx 0.16$ ($\alpha \approx 65$ cm^{-1}), Δn is smaller than 0.17%. The relative error on κ is minimum ($\Delta\kappa/\kappa \approx 0.12\%$) around $\kappa \approx 0.16$ and it increases when κ becomes smaller. Typically, it is smaller than 10%, except when κ tends towards zero. Thus, transmission THz-TDS is very convenient to measure the absorption of moderately opaque samples: at higher opacity, the error is large because there is almost no transmitted signal. For very transparent samples, the effect of absorption is hidden by both noise and the effect of Fresnel coefficients at the sample interfaces. We warn the reader that, when the noise-induced uncertainties are smaller than 1%, the overall precision on the parameters is limited by other causes of error, mostly a bad knowledge of the sample dimensions (thickness, flatness, etc.) or a bad adjustment of the THz-TDS system.

It could be useful, for experimentalists, to plot the noise-induced uncertainty versus the sample thickness d and the imaginary part κ of the refractive index. It could help in designing and manufacturing samples for a precise characterization. Figure 12.18 shows the

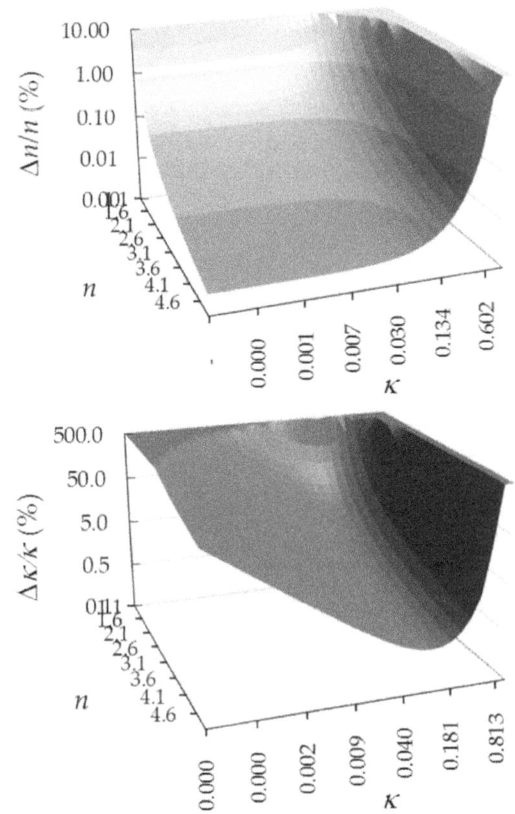

Figure 12.17 Relative errors $\Delta n/n$ (top) and $\Delta\kappa/\kappa$ (bottom) versus n and κ for the homemade setup at $f = 1$ THz, and with $d = 1$ mm.

relative uncertainty $\Delta n/n$ versus κ and the sample thickness d. This map has been calculated for the homemade THz-TDS setup, with $n = 2$ and at 1 THz. One observes the increase of the uncertainty when d becomes small, and when κ is large.

The experimental validation of these results has been first published by Duvillaret et al. [1]. Figure 12.19 shows the uncertainties Δn and $\Delta\alpha$ versus frequency for a 480 μm thick silicon wafer. Here the data were obtained by saving only the first THz pulse in the recorded waveform. The continuous line is calculated with Eq. 12.35, while open circles have been obtained from the analysis of 12 repetitive measurements. The agreement between theory and

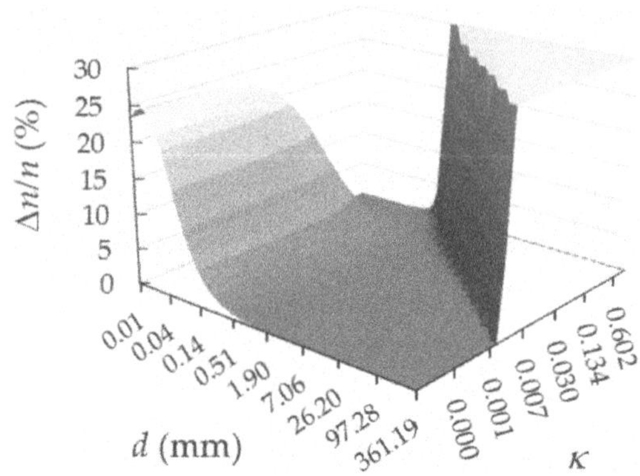

Figure 12.18 Relative error $\Delta n/n$ (left) versus κ and d for the homemade setup at $f = 1$ THz, and with $n = 2$.

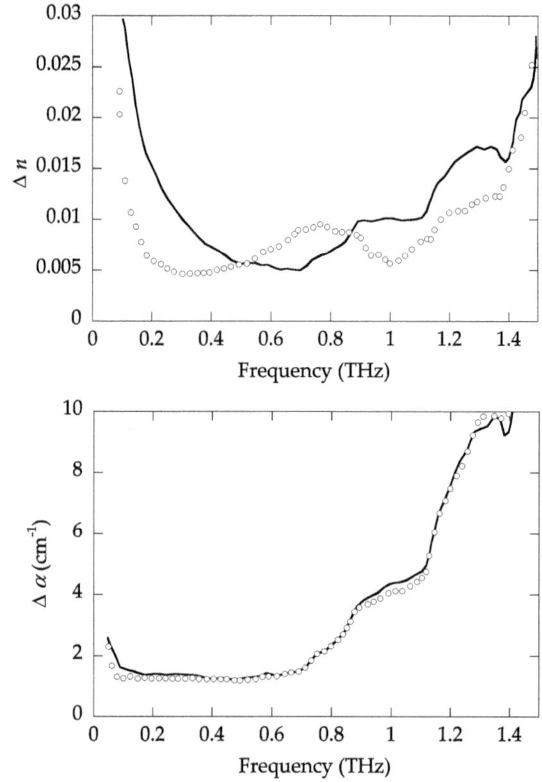

Figure 12.19 Measured (open circles) and calculated (continuous curves) Δn spectra for a 480 μm thick Si wafer. Adapted from Duvillaret et al. [1].

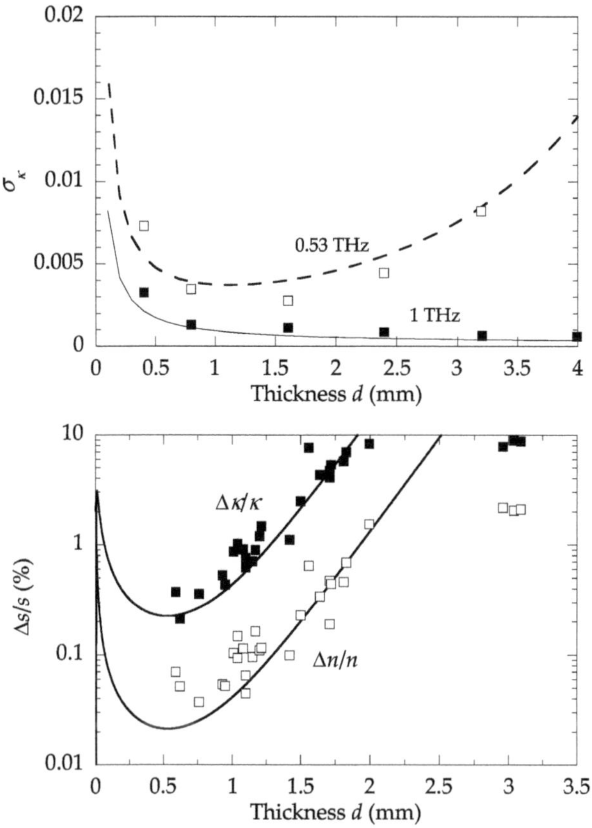

Figure 12.20 Top: Uncertainty $\sigma_\kappa = \Delta\kappa$ versus d for pellets of lactose-HDPE powder mixture at 0.57 and 1 THz. Squares are experimental data extracted from [7], continuous curves are fits with Eq. 12.35. Bottom: Uncertainties $\Delta n/n$ and $\Delta\kappa/\kappa$ versus d for slabs of bio-plastic at 2.25 THz. Squares are experimental data, and continuous curves are fits with Eq. 12.34 (extracted from Bernier et al. [4]).

experiment is excellent, especially for the absorption. Note that the authors used a Ti:Sa mode-locked laser pumped with an argon-ion laser, which was quite noisy. Even with such large noise, the relative precision on n is good ($\sim$3% in between 0.3–1.2 THz).

A complete study of the effect of the sample thickness has been published by Withayachumnankul et al. [7]. Their analysis and results agree with what has been described in this paragraph.

For example, Fig. 12.20 (top) presents the experimental uncertainty σ_κ versus the sample thickness. The samples are pellets of lactose powder mixed with HDPE powder in a 1:3 mass ratio. Samples with different thicknesses were prepared. Here we present experimental data, extracted from Fig. 11 of [7], together with fits calculated using relation 12.35. Either at 0.53 THz (high absorption regime due to a resonance) or at 1 THz (out of resonance), the agreement between theory and measurement is excellent.

Another study has been given in [4], from where Fig. 12.20 (bottom) is extracted. The samples are slabs of bio-plastic (Biomer P226). Averaged values and standard deviations are calculated from 27 sets of records. Both relative uncertainties $\Delta n/n$ and $\Delta\kappa/\kappa$ are plotted versus the sample thickness at $f = 2.25$ THz. Here again, the agreement between the modeling and the measurement is excellent. But for very thick samples ($\sim$3 mm), the model no longer fits the experimental data well, because transmission is too weak, and thus the uncertainty depends mostly on the noise parameter C, which is less precisely known than A and B.

Finally, the very weak uncertainty due to noise in transmission THz-TDS is depicted in Fig. 12.21, extracted from Bernier et al. [5]. Here the refractive index and its noise-induced uncertainty are plotted versus frequency. The material is a slab of hard Stycast, which is a glue used for cryogenic applications. Because Stycast is quite transparent, the noise-induced uncertainty is very small (less than 0.2%). In this case, the imprecision on n is mainly due to a poor knowledge of the sample thickness or to a bad alignment of the TDS system.

12.2.5 Reflection or Transmission THz-TDS Characterization?

It follows from the preceding paragraphs that transmission THz-TDS is well adapted to characterize transparent samples that are rather thick, while reflection THz-TDS is well suited for highly absorbing samples. For the same THz-TDS setup, the choice of the most appropriate technique is easy to determine, when taking into account only the uncertainties due to the noise. We suppose here again that rebounds in the sample can be properly

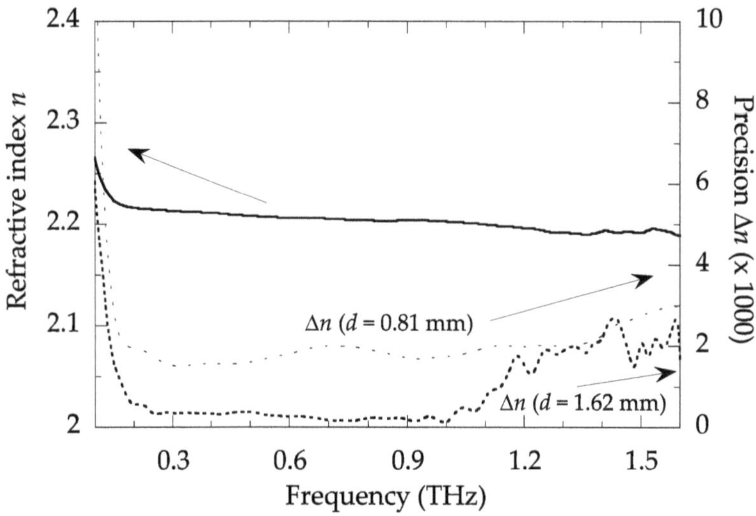

Figure 12.21 Refractive index n and its uncertainty Δn versus frequency for Stycast (adapted from Bernier et al. [5]). The uncertainties have been measured for two samples of different thicknesses.

filtered. Therefore, we compare the uncertainty expressions 12.15 (reflection) and 12.32 and 12.33 (transmission), using the same A, B, and C noise coefficients.

Figure 12.22 presents the ratio of the uncertainties calculated for transmission and reflection THz-TDS, i.e., $\Delta n_t/\Delta n_r$ (top plot) and $\Delta\kappa_t/\Delta\kappa_r$ (bottom plot), versus n and κ. Here the sample is 5-mm thick, the system is the homemade one, and the frequency is 1 THz. At large values of κ, both $\Delta n_t/\Delta n_r > 1$ and $\Delta\kappa_t/\Delta\kappa_r > 1$ because the transmitted signal is weak: therefore, it is preferable to employ reflection THz-TDS rather than transmission THz-TDS for a precise characterization of the sample. The same conclusion could be drawn for low refractive index n, because the amplitude and phase changes through such a sample are weak. However, the effect is less pronounced than for large κ, because the coefficient of reflection also decreases with n.

The limits $\Delta n_t = \Delta n_r$ and $\Delta\kappa_t = \Delta\kappa_r$, at which both techniques give similar precision, are shown by a black curve. Here the limits are located at $\kappa \approx 0.164$ ($\alpha \approx 66$ cm^{-1}) for any $n > 1.2$. Below

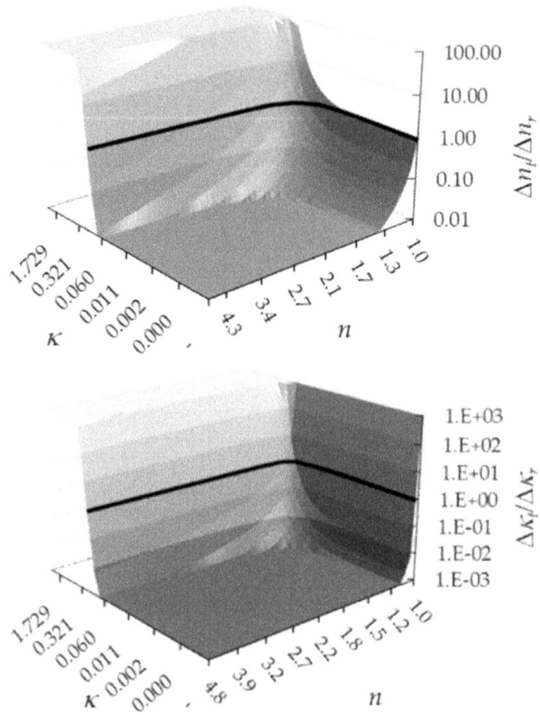

Figure 12.22 The ratio $\Delta n_t/\Delta n_r$ (top) and $\Delta\kappa_t/\Delta\kappa_r$ (bottom) versus n and κ calculated for the homemade system. The black lines indicate the limit between the reflection and transmission schemes for an optimized parameter extraction. Over these black lines, $\Delta x_t > \Delta x_r$ ($x = n$ or κ) and thus reflection THz-TDS is more precise than transmission THz-TDS.

these limits, for a 5-mm thick sample, transmission THz-TDS is more precise.

The limit, in the (n, κ) plane, between optimized reflection and transmission THz-TDS, depends both on the sample thickness d and the frequency f through the term $2\pi f\, d/c$. Figure 12.23 shows this limit in the (n, κ) plane for different values of $2\pi f\, d/c$. In the region over the limit, reflection THz-TDS is more precise than transmission THz-TDS. Except at low values of n, the κ-value of the limit does not depend so much on n. It varies typically as $1/(2\pi f\, d/c)$: as expected, the higher f or d, the smaller κ. For example, for the homemade system, a 0.5 mm thick sample must be characterized at 1 THz in

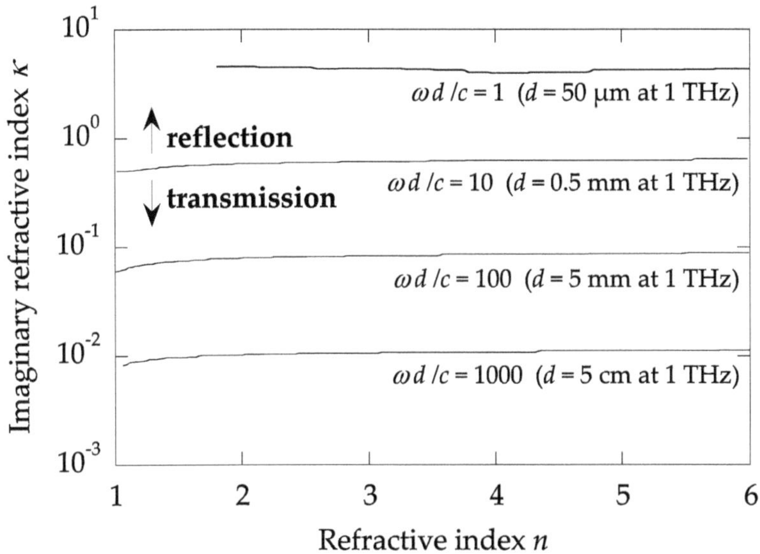

Figure 12.23 Limits, versus n and κ, for which reflection THz-TDS is more precise than transmission THz-TDS. These limits are calculated for the homemade system and for different values of $\omega\ d/c$.

reflection if $\kappa > 0.62$ ($\alpha > 250\ \text{cm}^{-1}$). The limit is $\kappa = 0.088$ ($\alpha = 35\ \text{cm}^{-1}$) if the sample is ten times thicker (5 mm).

12.3 Setup Misalignment

12.3.1 Effect of an Angular Tilt of the Sample

In most THz-TDS experiments, the sample is positioned normal to the THz beam (except in ATR and in same reflection schemes). An undesired angular tilt of the sample leads to several errors:

- In both transmission and reflection THz-TDS, the Fresnel coefficients at the sample faces are changed.
- In transmission THz-TDS, the effective thickness of sample that crosses the THz beam is enlarged.

- In both transmission and reflection THz-TDS, the THz beam transmitted or reflected by the sample is spatially deflected and/or shifted, and thus it does not totally reach the receiver.

12.3.1.1 Variation of the reflection or transmission coefficients

Let us first treat the change of the Fresnel coefficient and the longer effective sample thickness. The sample is tilted, such as the angle of incidence of the THz beam is θ instead of 0. The sample is located in air ($n_{\text{air}} = 1$). Without lack of generality, we suppose the sample is thick enough to allow one to select only the first transmitted or reflected THz pulse. Under this hypothesis, the complex transmission and reflection coefficients of the sample are written as

$$\tilde{r} = \frac{\tilde{b}\,\tilde{k}_{1z} - \tilde{k}_{2z}}{\tilde{b}\,\tilde{k}_{1z} + \tilde{k}_{2z}}, \quad \tilde{t} = \frac{4\tilde{b}\,\tilde{k}_{1z}\tilde{k}_{2z}}{\left(\tilde{b}\,\tilde{k}_{1z} + \tilde{k}_{2z}\right)^2} e^{-j\frac{\omega}{c}\tilde{n}\frac{d}{\cos\theta_2}}. \tag{12.37}$$

The subscripts 1 and 2 refer respectively to air and the sample material. d is the sample thickness. $\tilde{k}_{\text{iz}} = \sqrt{\tilde{n}_i^2 - \sin^2\theta}\,\omega/c$ is the wavevector component along z in medium i(z is the direction normal to the sample). Thus $\tilde{n}_1 = n_{\text{air}} = 1$ and $\tilde{n}_2 = \tilde{n}$. $\tilde{b} = \tilde{\mu}$ in the TE polarization case and $\tilde{b} = \tilde{n}^2$ in the TM case, where $\tilde{\mu}$ is the relative permeability of the sample material. In the TE case, the coefficients are related to the E field while, in the TM case, there are connected with the H field. Both TE and TM cases correspond to an angular tilt of the sample respectively in the plane of incidence and perpendicularly to it. In the TM case, the reflected and transmitted E fields are

$$\begin{pmatrix} E_{\text{rx}} \\ 0 \\ E_{\text{rz}} \end{pmatrix} = \tilde{r}_{\text{TM}} \begin{pmatrix} -E_{\text{ix}} \\ 0 \\ E_{\text{iz}} \end{pmatrix} = \tilde{r}_{\text{TM}} \begin{pmatrix} -\cos\theta \\ 0 \\ \sin\theta \end{pmatrix} E_i,$$

$$\begin{pmatrix} E_{\text{tx}} \\ 0 \\ E_{\text{tz}} \end{pmatrix} = \tilde{t}_{\text{TM}} \begin{pmatrix} E_{\text{ix}} \\ 0 \\ -E_{\text{iz}} \end{pmatrix} = \tilde{t}_{\text{TM}} \begin{pmatrix} \cos\theta \\ 0 \\ -\sin\theta \end{pmatrix} E_i.$$

In the limit of a small tilt angle, i.e., $\theta = \delta\theta$, and for non-magnetic materials, expressions 12.37 are approximated by

$$\tilde{r}\left(\theta \ll 1\right) = \tilde{r}\left(\delta\theta\right) = \delta\tilde{r}, \quad \tilde{t}\left(\theta \ll 1\right) = \tilde{t}\left(\delta\theta\right) = \delta\tilde{t} \Rightarrow$$

$$\begin{cases} \delta\tilde{r} = \tilde{r}\left(0\right)\dfrac{\tilde{b}\left(1-\tilde{n}^2\right)}{\tilde{b}^2-\tilde{n}^2}\frac{(\delta\theta)^2}{\tilde{n}}, \\ \delta\tilde{t} = \tilde{t}\left(0\right)\left(\left(\tilde{n}^2-1\right)\dfrac{\tilde{b}-\tilde{n}}{\tilde{b}+\tilde{n}} - j\frac{\omega}{c}\tilde{n}d\right)\frac{(\delta\theta)^2}{2\tilde{n}^2}, \end{cases} \tag{12.38}$$

with

$$\tilde{r}\left(0\right) = \frac{\tilde{b}-\tilde{n}}{\tilde{b}+\tilde{n}}, \quad \tilde{t}\left(0\right) = \frac{4\tilde{n}}{\left(1+\tilde{n}\right)^2}e^{-j\frac{\omega}{c}\tilde{n}d}. \tag{12.39}$$

Instead of measuring $\tilde{r}\left(0\right)$ or $\tilde{t}\left(0\right)$, one gets $\tilde{r}\left(\delta\theta\right)$ or $\tilde{t}\left(\delta\theta\right)$. Thus the recordings are altered by the error $\delta\theta$. This leads to an error $\delta\tilde{n}$ that is calculated as follows. Let us start with the reflection case. By inverting Eq. 12.39 using the erroneous value $\tilde{r}\left(0\right) + \delta\tilde{r}$, we get

$$\tilde{n} + \delta\tilde{n}_{\mathrm{r}} = \tilde{b}\frac{1-\tilde{r}\left(0\right)-\delta\tilde{r}}{1+\tilde{r}\left(0\right)+\delta\tilde{r}} \approx \tilde{b}\frac{1-\tilde{r}\left(0\right)}{1+\tilde{r}\left(0\right)}\left(1-\frac{2\delta\tilde{r}}{1-\tilde{r}\left(0\right)^2}\right),$$

$$\Rightarrow \quad \delta\tilde{n}_{\mathrm{r,TE}} = -\frac{2\delta\tilde{r}}{\left(1+\tilde{r}\left(0\right)\right)^2}, \quad \delta\tilde{n}_{\mathrm{r,TM}} = \frac{-2\tilde{n}^2\delta\tilde{r}}{\left(1+\tilde{r}\left(0\right)\right)^2}. \tag{12.40}$$

Finally, these errors are written as functions of $\delta\theta$ using Eq. 12.28:

$$\Rightarrow \quad \delta\tilde{n}_{\mathrm{r,TE}} = \delta\tilde{n}_{\mathrm{r,TM}} = \frac{\tilde{n}^2-1}{\tilde{n}}\frac{(\delta\theta)^2}{2},$$

i.e., for the relative real and imaginary parts,

$$\begin{aligned} \frac{\delta n_{\mathrm{r,TE}}}{n} &= \frac{\delta n_{\mathrm{r,TM}}}{n} = \frac{n^2+\kappa^2-1}{n^2+\kappa^2}\frac{(\delta\theta)^2}{2}, \\ \frac{\delta\kappa_{\mathrm{r,TE}}}{\kappa} &= \frac{\delta\kappa_{\mathrm{r,TM}}}{\kappa} = -\frac{n^2+\kappa^2+1}{n^2+\kappa^2}\frac{(\delta\theta)^2}{2}. \end{aligned} \tag{12.41}$$

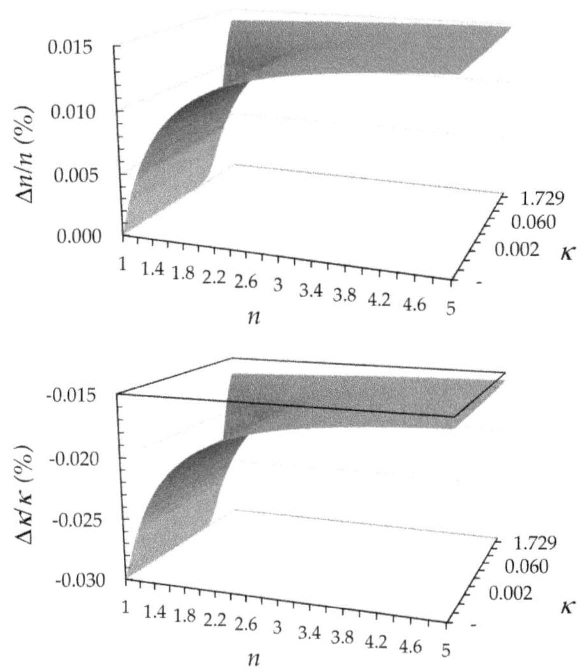

Figure 12.24 Relative errors $|\delta n/n|$ (top) and $|\delta\kappa/\kappa|$ (bottom) versus n and κ induced by an angular error $\delta\theta = 1°$, for reflection THz-TDS.

These errors are plotted in Fig. 12.24. Typically, for an angular tilt $\delta\theta = 1°$, they are of the order of a few 10^{-3}.

In the case of transmission THz-TDS, the expression of the error remains more complex, because it includes a propagation term:

$$\delta\tilde{n}_{\substack{\mathrm{TE}\\\mathrm{TM}}} = \pm\left(1 - \frac{1 \pm (\tilde{n}^2 - 1)}{1 - j\tilde{n}\left(\frac{1+\tilde{n}}{1-\tilde{n}}\right)\frac{\omega}{c}d}\right)\frac{\tilde{n}}{n^2}\frac{(\delta\theta)^2}{2}. \tag{12.42}$$

This expression is simplified if $\kappa \ll n$ and we get

$$\begin{cases} \delta n_{\mathrm{TE}} = n\left((n-1)^2 - \kappa n^2\left(\frac{\omega}{c}d\right) + n^2\left(\frac{n+1}{n-1}\right)\left(\frac{\omega}{c}d\right)^2\right)\frac{(\delta\theta)^2}{D}, \\ \delta n_{\mathrm{TM}} = n\left((n-1)^2 - \kappa\left(n^2 - 2\right)\left(\frac{\omega}{c}d\right) - n^2\left(\frac{n+1}{n-1}\right)\left(\frac{\omega}{c}d\right)^2\right)\frac{(\delta\theta)^2}{D}, \\ \delta\kappa_{\mathrm{TE}} = \left(\kappa(n-1)^2 + n^4\left(\frac{\omega}{c}d\right) + \kappa n^2\left(\frac{n+1}{n-1}\right)\left(\frac{\omega}{c}d\right)^2\right)\frac{(\delta\theta)^2}{D}, \\ \delta\kappa_{\mathrm{TM}} = \left(\kappa(n-1)^2 + n^2\left(n^2 - 2\right)\left(\frac{\omega}{c}d\right) - \kappa n^2\left(\frac{n+1}{n-1}\right)\left(\frac{\omega}{c}d\right)^2\right)\frac{(\delta\theta)^2}{D}, \end{cases} \tag{12.43}$$

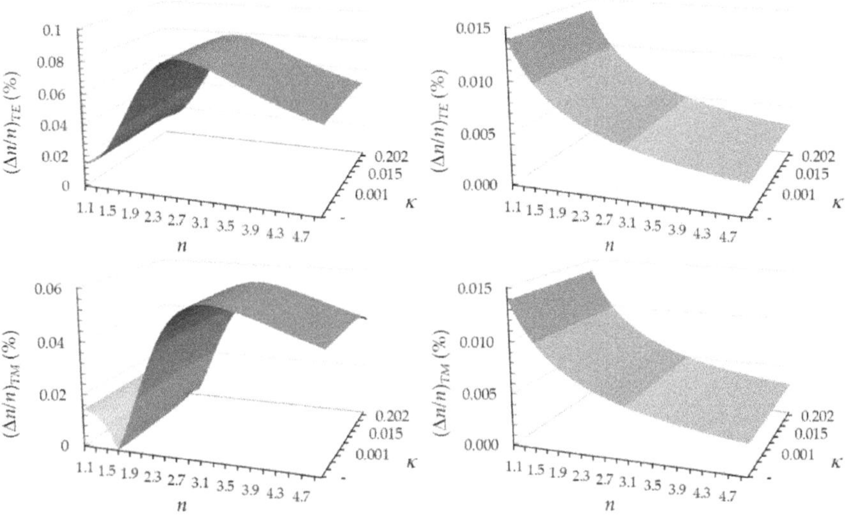

Figure 12.25 Absolute relative error $|\delta n/n|$ versus n and κ calculated with Eq. 12.43 for an angular error $\delta\theta = 1^\circ$, with $\omega d/c = 0.21$ (thin sample, left) and $\omega d/c = 21$ (thick sample, right), for transmission THz-TDS in TE (upper row) and TM (lower row) polarization.

with $D = 2n^2 \left(\left(\frac{1-n}{1+n} \right) + n^2 \left(\frac{n+1}{n-1} \right) \left(\frac{\omega}{c} d \right)^2 \right)$. The $\frac{\omega}{c} d$ terms originate in the longer path in the sample due to the oblique incidence, while other terms come from the variation of the Fresnel coefficients at the sample faces.

The errors given by expressions 12.43 are plotted versus n and κ in Fig. 12.25. For thin samples, i.e., $d\omega/c \ll 1$, expressions are simpler:

$$\left| \frac{\delta n}{n} \right| \approx \left| \frac{\delta \kappa}{\kappa} \right| \approx \frac{n^2 - 1}{2n^2} (\delta\theta)^2 , \tag{12.44}$$

while, for thick samples, i.e., $d\omega/c \gg 1$, we get

$$\left| \frac{\delta n}{n} \right| \approx \left| \frac{\delta \kappa}{\kappa} \right| \approx \frac{(\delta\theta)^2}{2n^2} . \tag{12.45}$$

It is noticeable that, in these two limit cases, the relative error does not depend on the polarization (TE or TM), on the material loss (κ) and on the sample thickness. The error is larger for thin samples, which is due to the change of the Fresnel coefficients.

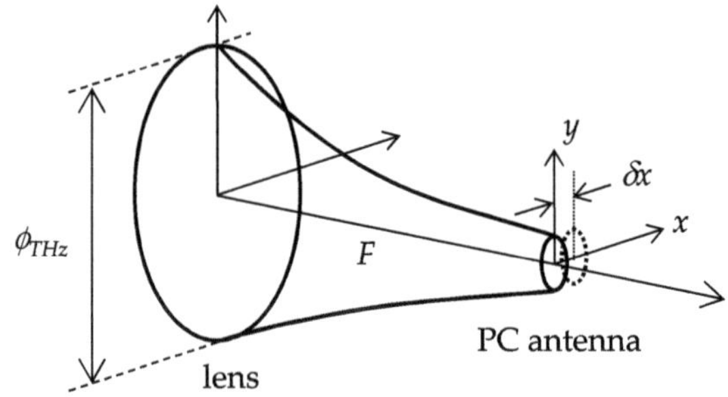

Figure 12.26 Scheme of a Gaussian beam that is focused onto a photo-conducting antenna by a lens of focal distance F.

For thick samples, both Fresnel coefficients and propagation effects compensate, making the error smaller. In any case, the error is less than 0.1%.

12.3.1.2 Deviation of the THz beam

The effect of the deviation of the THz beam is more difficult to treat, and it depends on the setup geometry. Let us give here a crude estimation of the induced errors in the case of a photo-conducting antenna. We consider that the THz beam exhibits a Gaussian shape and it is focused onto the receiver by a lens of focal length F (see Fig. 12.26), such as:

$$E_{\mathrm{THz}}(\rho) = A\, e^{-\left(\frac{\rho}{\rho_0}\right)^2},\ \rho_0 = \frac{4}{\pi}\frac{\lambda F}{\phi_{\mathrm{THz}}} = \frac{4}{\pi}\frac{cF}{\phi_{\mathrm{THz}} f}. \tag{12.46}$$

$E_{\mathrm{THz}}(\rho)$ is the THz field at the receiver, $\rho = \sqrt{x^2 + y^2}$ is the radial distance from the optical axis of the THz beam, and ϕ_{THz} is the diameter of the THz beam that illuminates the lens.

We suppose that the radiation pattern of the receiving antenna is also Gaussian, with the same shape as the incident Gaussian beam (Eq. 12.46). In other words, the coupling coefficient η_{antenna} is given by

$$\eta_{\mathrm{antenna}} = e^{-\left(\frac{\rho}{\rho_0}\right)^2}. \tag{12.47}$$

The signal delivered by the receiver, if the beam focused is shifted by δx at the receiver (see Fig. 12.26), is thus proportional to

$$S(\delta x) \propto \iint E_{\mathrm{THz}} \eta_{\mathrm{antenna}} dS = A \iint e^{-\frac{(x+\delta x)^2+y^2}{\rho_0^2}} e^{-\frac{x^2+y^2}{\rho_0^2}} dxdy$$
$$= A \iint e^{-\frac{x^2+(x+\delta x)^2+2y^2}{\rho_0^2}} dxdy,$$
$$\Rightarrow S(\delta x) = A\, e^{-\frac{\delta x^2}{2\rho_0^2}} \rho_0^2 \frac{\pi}{2}$$
$$= S(0)\, e^{-\frac{\delta x^2}{2\rho_0^2}} \approx S(0)\left(1 - \frac{\delta x^2}{2\rho_0^2}\right). \tag{12.48}$$

The shift δx originates in the angular tilt $\delta\theta$, namely $\delta x = F \times \delta\theta$. Hence,

$$\Delta S(\delta x) = S(\delta x) - S(0) = -S(0)\frac{\delta x^2}{2\rho_0^2} = -\frac{S(0)}{2}\left(\frac{\pi\, \phi_{\mathrm{THz}} f\, \delta\theta}{4c}\right)^2. \tag{12.49}$$

In reflection THz-TDS, this leads to an error Δr on the modulus of the coefficient of reflection r, which is determined as follows:

$$\Delta r = \frac{\Delta S(\delta x)}{S_{\mathrm{ref}}} = \frac{-\frac{S(0)}{2}\left(\frac{\pi\, \phi_{\mathrm{THz}} f\, \delta\theta}{4c}\right)^2}{S_{\mathrm{ref}}} = -\frac{r}{2}\left(\frac{\pi\, \phi_{\mathrm{THz}} f\, \delta\theta}{4c}\right)^2, \tag{12.50}$$

with $r = S(0)/S_{\mathrm{ref}}$. The induced phase variation, due to a bad positioning of the focus point, is

$$\Delta\varphi \approx \frac{\omega}{c} F\left(\frac{1}{\cos(\delta\theta)} - 1\right) \approx \frac{\omega}{c} F \frac{(\delta\theta)^2}{2}, \tag{12.51}$$

The error on $\tilde{n}$ is determined using the expression of $\tilde{r}$:

$$\tilde{n} = \frac{1-\tilde{r}}{1+\tilde{r}} \quad \Rightarrow \quad \Delta\tilde{n} = \frac{2\Delta\tilde{r}}{(1+\tilde{r})^2}$$
$$= 2e^{j\varphi}\frac{\Delta r + jr\Delta\varphi}{\left(1+r\,e^{j\varphi}\right)^2} = 2r\,e^{j\varphi}\frac{\frac{\Delta r}{r} + j\Delta\varphi}{\left(1+r\,e^{j\varphi}\right)^2}, \tag{12.52}$$

$$\Rightarrow \quad \Delta\tilde{n} = r\,e^{j\varphi}\frac{-\left(\frac{\phi_{\mathrm{THz}}}{2}\right)^2 \frac{2\pi f}{c} + jF\,\frac{2\pi f}{c}}{\left(1+r\,e^{j\varphi}\right)^2}(\delta\theta)^2. \tag{12.53}$$

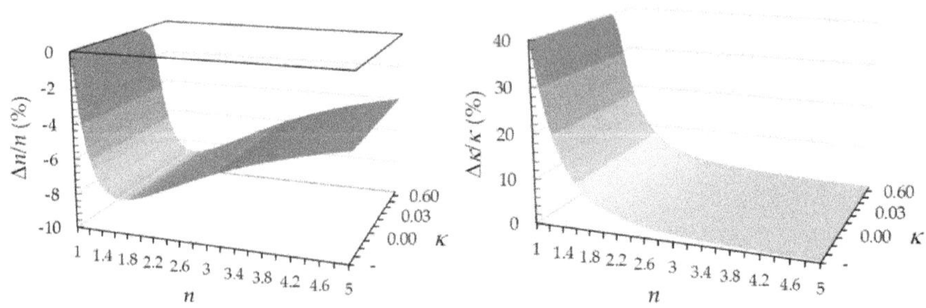

Figure 12.27 Relative errors $\Delta n/n$ and $\Delta\kappa/\kappa$ for a 0.1°angular tilt induced by a partial illumination of the receiver ($\phi_{\text{THz}} = 50$ mm, $F = 100$ mm, $f = 1$ THz).

The first term in the numerator of the ratio is caused by the change of field magnitude at the receiver, while the second one is produced by the phase variation. For typical actual geometrical parameters ($\phi_{\text{THz}} \approx$ a few centimeters, $F = 100$ mm, $f \sim 0.1 - 10$ THz), the phase variation is negligible, which leads to

$$\Delta n = -\left(\frac{\pi f}{2c}\phi_{\text{THz}}\right)^2 \frac{\left(n^2+\kappa^2\right)^2 - \left(n^2-\kappa^2\right)}{\left(n^2+\kappa^2\right)^2}(\delta\theta)^2, \qquad (12.54)$$

$$\Delta\kappa = \frac{2n\kappa}{n^2-\kappa^2-\left(n^2+\kappa^2\right)^2}\Delta n. \qquad (12.55)$$

In reflection THz-TDS, a small angular tilt of the sample leads to severe errors. Typically, with the above geometrical parameters ($\phi_{\text{THz}} = 50$ mm, $F = 100$ mm, $f = 1$ THz), a small tilt ($\delta\theta = 0.1°$) induces errors larger than 1% (see Fig. 12.27).

In transmission THz-TDS, the transmitted beam is only shifted laterally if the sample is angularly tilted, but there is no change in the transmitted beam direction (assuming the sample is a slab with parallel faces). If the focusing THz optics shows no aberration, a lateral shift of the incoming THz beam does not change the position of the focused beam. Therefore, the effect of beam deviation is negligible as compared to the propagation effect within the sample.

12.3.2 Effects due to a Bad Position of the Reference Mirror in Reflection THz-TDS

In the case of reflection THz-TDS, the main error source is the misalignment of the sample compared to the reference mirror [8, 9]. We treat here only an error in position. For the sake of simplicity, we suppose the sample is thick enough to allow us to neglect the Fabry–Perot rebounds. Let δ be the difference in position between the mirror and the sample. This position shift induces an erroneous phase difference $2\,\omega\,\delta/c$. Thus the measured complex coefficient of reflection is equal to the actual one times a phase error term:

$$\tilde{r}_{\text{meas}}\left(\omega,\tilde{n}\right)=\tilde{r}\left(\omega,\tilde{n}\right)\;e^{j2\frac{\omega}{c}\delta}=\frac{\tilde{n}-1}{\tilde{n}+1}e^{j2\frac{\omega}{c}\delta}, \qquad (12.56)$$

If one extracts the material parameters without taking into account this additional phase term, one will use the common expression $\tilde{r}\left(\omega,\tilde{n}\right)=(\tilde{n}-1)/(\tilde{n}+1)$ of the reflection coefficient, in which the refractive index will be shifted by $\Delta\tilde{n}$. One gets

$$\tilde{r}_{\text{meas}}\left(\omega,\tilde{n}\right)=\tilde{r}\left(\omega,\tilde{n}+\Delta\tilde{n}\right)\quad\Rightarrow\quad\frac{\tilde{n}-1}{\tilde{n}+1}e^{j2\frac{\omega}{c}\delta}=\frac{\tilde{n}+\Delta\tilde{n}-1}{\tilde{n}+\Delta\tilde{n}+1}, \qquad (12.57)$$

$$\Rightarrow\quad\Delta\tilde{n}=\left(\tilde{n}-1\right)\frac{1-e^{j2\frac{\omega}{c}\delta}}{\frac{\tilde{n}-1}{\tilde{n}+1}e^{j2\frac{\omega}{c}\delta}-1}. \qquad (12.58)$$

If the shift δ is small (as compared to the wavelength, i.e., $2\,\omega\,\delta/c\ll 1$), expression 12.58 simplifies into

$$\Delta\tilde{n}\approx j\frac{\omega}{c}\delta\left(\tilde{n}^2-1\right),$$

$$\Rightarrow\quad\frac{\Delta n}{n}\approx 2\frac{\omega}{c}\kappa\,\delta=\alpha\,\delta,\quad\Delta\kappa\approx\frac{\omega}{c}\delta\left(n^2-\kappa^2-1\right). \qquad (12.59)$$

The 2D map of these errors are plotted versus n and κ for $f=1$ THz and $\delta=1$ μm in Fig. 12.28.

The relative error on n is very weak (<1%) if the product $\alpha\left(\text{cm}^{-1}\right)\,\delta(\mu\text{m})$ does not exceed 100: For a transparent material (α < 5 cm^{-1}), the position error δ must be smaller than 20 μm to fulfill this condition. For strongly absorbing materials (α > 50 cm^{-1}), the mirror position must be precisely adjusted, if possible with δ < 2 μm. The error on κ is much bigger. Here (Fig. 12.28), for example, even

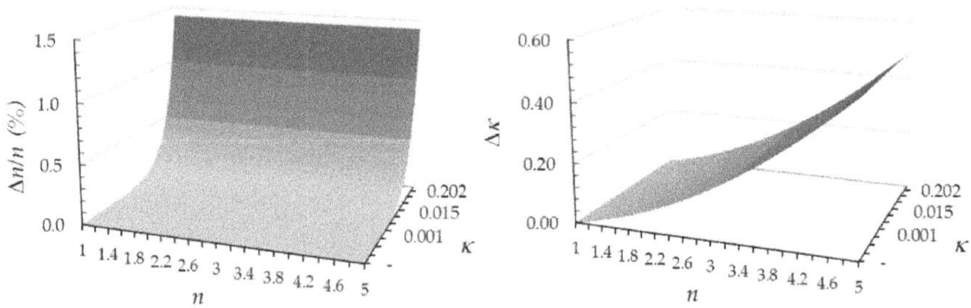

Figure 12.28 Errors $\Delta n/n$ (in %) and $\Delta\kappa$ versus n and κ for a 1-μm shift of the reference mirror position in a reflection THz-TDS experiment at $f =$ 1 THz.

for a shift as small as $\delta = 1$ μm, $\Delta\kappa \approx 0.06$ ($\alpha = 25$ cm^{-1}) for $n = 2$ whatever is κ. For transparent materials (i.e., $\alpha \ll 25$ cm^{-1}), the reflection method is too inaccurate and thus impractical to determine κ. For highly absorbing samples ($\alpha > 25$ cm^{-1}), the precision on κ is rather good, but the determination of n is poor!

12.4 Effect of the Sample Thickness Uncertainty

12.4.1 General Relations

This error affects mostly transmission experiments, as in reflection THz-TDS, the samples are usually thick enough to permit filtering the rebounds by a proper time-windowing, and thus the reflected signal does not depend on the sample thickness. Therefore, in a first stage, let us suppose the sample is thick enough to save only the first transmitted THz pulse:

$$\tilde{t} = \frac{4\tilde{n}}{(\tilde{n}+1)^2} e^{-j\frac{\omega}{c}(\tilde{n}-1)d}. \tag{12.60}$$

If the thickness of the sample is erroneous by an amount Δd, the refractive index $\tilde{n}$ is impacted by the error $\Delta\tilde{n}$ that obeys the following equation:

$$\frac{4\tilde{n}}{(\tilde{n}+1)^2} e^{-j\frac{\omega}{c}(\tilde{n}-1)(d+\Delta d)} = \frac{4(\tilde{n}+\Delta\tilde{n})}{(\tilde{n}+\Delta\tilde{n}+1)^2} e^{-j\frac{\omega}{c}(\tilde{n}+\Delta\tilde{n}-1)d},$$
$$\Rightarrow \quad \exp\left(j\frac{\omega}{c}d\left(\Delta\tilde{n}-(\tilde{n}-1)\frac{\Delta d}{d}\right)\right) = \frac{1+\frac{\Delta\tilde{n}}{\tilde{n}}}{\left(1+\frac{\Delta\tilde{n}}{\tilde{n}+1}\right)^2}. \tag{12.61}$$

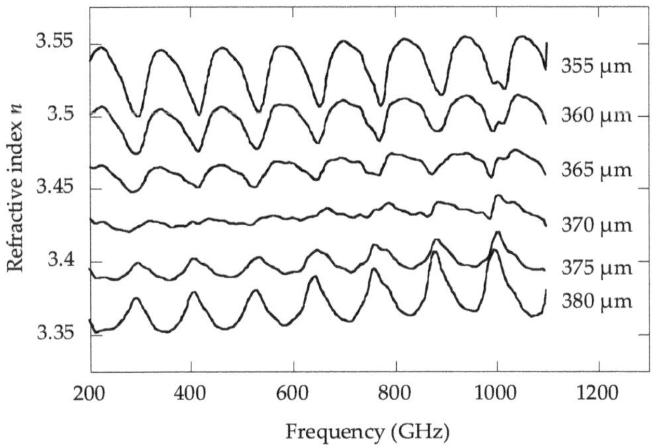

Figure 12.29 Refractive index of HR-Si versus frequency extracted using different values of the sample thickness (given on the right side). Adapted from Duvillaret et al. [2].

Therefore, because of the complex exponential in Eq. 12.61, the error $\Delta\tilde{n}$ induced by the thickness inaccuracy Δd shows an oscillatory behavior with frequency. This is clearly seen in Fig. 12.29, where the experimental refractive index spectrum of a HR-Si wafer is plotted for different values of the sample thickness. The oscillation magnitude decreases when the thickness value used in the extraction procedure tends towards the actual one (here 368 μm).

To get the relation between $\Delta\tilde{n}$ and Δd, (12.61) must be solved numerically. However, an analytical solution can be found, assuming the error $\Delta\tilde{n}$ remains small compared to $\tilde{n}$. We can take a series expansion of Eq. 12.61 and we get

$$\frac{\Delta\tilde{n}}{\tilde{n}} = \frac{j\frac{\omega}{c}d\,(\tilde{n}-1)}{j\tilde{n}\frac{\omega}{c}d + \frac{\tilde{n}-1}{\tilde{n}+1}}\frac{\Delta d}{d}. \tag{12.62}$$

In the limit cases of thick $(\omega d/c \gg 1)$ or thin $(\omega d/c \ll 1)$ samples, we obtain

$$\begin{cases} \frac{\omega}{c}d \gg 1: & \Delta n = (n-1)\frac{\Delta d}{d}, \quad \Delta\kappa = \kappa\frac{\Delta d}{d}, \\ \frac{\omega}{c}d \ll 1: & \Delta n = \frac{\omega}{c}\kappa\,(2n+1)\,\Delta d, \quad \Delta\kappa = -\frac{\omega}{c}\,(n+1)\,n\Delta d. \end{cases} \tag{12.63}$$

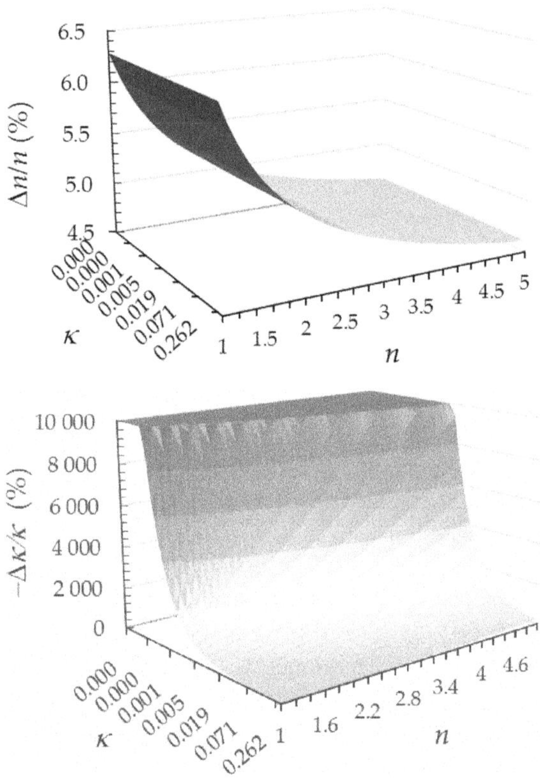

Figure 12.30 Errors $\Delta n/n$ and $\Delta\kappa$ versus n and κ for a 1-μm error on the sample thickness in a transmission THz-TDS experiment at $f = 1$ THz.

Typically, for thick samples, the relative error on n and κ is of the order of the relative error on the sample thickness. A common 1% error (thickness of 1 mm known with a 10 μm precision) leads to the same error on n and κ. For thin samples, the error also depends on frequency. Figure 12.30 shows the relative error on n and the error on κ versus n and κ when $\Delta d = 1$ μm and $f = 1$ THz. This estimation is valid for $\omega d/c \ll 1$, i.e., $d \ll 50$ μm. Typically, the error on n is around 6%, while the error on κ is negative (overestimating the thickness means that absorption occurs over a longer thickness, so the absorption coefficient is underestimated). This error on κ is amazingly big (in any case more than 100%).

12.4.2 Determining the Sample Thickness from THz-TDS

12.4.2.1 Initial method by Duvillaret et al. [2]

An elegant solution to improve the precision of the sample thickness and therefore the precision of the extracted parameters has been proposed by Duvillaret et al. [2]. Let us write again the coefficient of transmission (Eq. 5.3) in the general case:

$$\tilde{t} = \frac{4\tilde{n}}{(\tilde{n}+1)^2}\frac{e^{-j\frac{\omega}{c}(\tilde{n}-1)d}}{1-\left(\frac{\tilde{n}-1}{\tilde{n}+1}e^{-j\frac{\omega}{c}\tilde{n}d}\right)^2}$$
$$= \frac{4\tilde{n}}{(\tilde{n}+1)^2}e^{-j\frac{\omega}{c}(\tilde{n}-1)d}\sum_{p=0}^{\infty}\left(\frac{\tilde{n}-1}{\tilde{n}+1}e^{-j\frac{\omega}{c}\tilde{n}d}\right)^{2p}. \quad (12.64)$$

The summation over the integer p in Eq. 12.64 corresponds to the sum of the contribution of each rebound p within the sample. If the sample is thick enough, each rebounding pulse p can be isolated by a proper time-windowing (Fig. 12.31). Its spectrum is given by the p-th term in (12.64):

$$\tilde{t}_\mathrm{p} = \frac{4\tilde{n}}{(\tilde{n}+1)^2}\left(\frac{\tilde{n}-1}{\tilde{n}+1}\right)^{2p} e^{-j\frac{\omega}{c}(\tilde{n}(2p+1)-1)d}. \quad (12.65)$$

The modulus and phase of $\tilde{t}_\mathrm{p}$ are

$$\begin{cases} \left|\tilde{t}_\mathrm{p}\right| = \frac{4\sqrt{n^2+\kappa^2}}{\left((n+1)^2+\kappa^2\right)}\left(\frac{(n-1)^2+\kappa^2}{(n+1)^2+\kappa^2}\right)^p e^{-\frac{\omega}{c}\kappa(2p+1)d}, \\ \varphi_\mathrm{p} = -\frac{\omega}{c}\left(n\left(2p+1\right)-1\right)d + \arctan\frac{\kappa\left(n^2+\kappa^2-1\right)}{n(n+1)^2+(n+2)\kappa^2} - 2p\arctan\left(\frac{2\kappa}{n^2+\kappa^2-1}\right). \end{cases}$$

For rather transparent materials ($\kappa \ll 1$), the phase simplifies into

$$\varphi_\mathrm{p} \approx -\frac{\omega}{c}\left(n\left(2p+1\right)-1\right)d,$$

from which one easily determines the refractive index that we label with p (extracted from pulse p):

$$n_\mathrm{p} \approx \frac{1-\frac{c}{\omega d}\varphi_\mathrm{p}}{2p+1}. \quad (12.66)$$

If there is an error Δd on the thickness value, the related error on n_p depends on p:

$$\Delta n_\mathrm{p} \approx \frac{c}{\omega d^2}\frac{\varphi_\mathrm{p}}{2p+1}\Delta d = \left(\frac{1}{2p+1}-n\right)\frac{\Delta d}{d}. \quad (12.67)$$

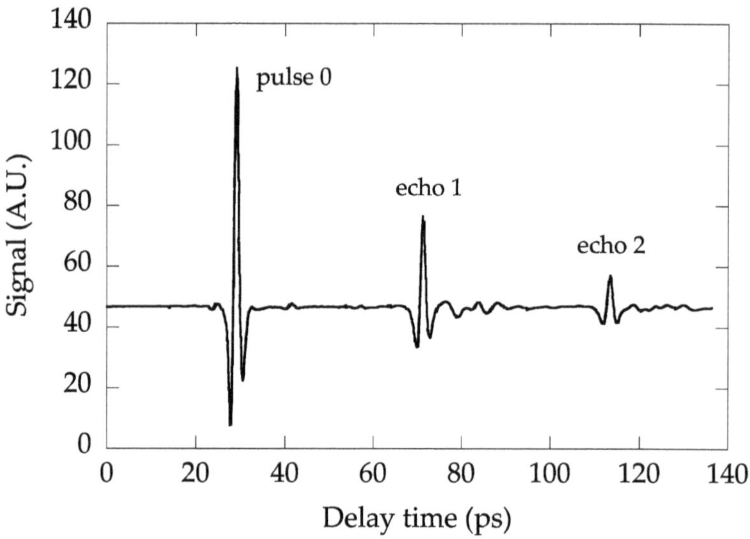

Figure 12.31 THz waveform transmitted by a 1.1 mm thick $LiNbO_3$ slab (THz polarization parallel to the slow axis of the crystal, i.e., transmission involves the ordinary index of the crystal). The directly transmitted pulse as well as the two first rebounds are clearly identified.

Here we have assumed that no error affects the phase (in fact, we will see later on that the phase could exhibit shifts or drifts mostly due to a bad running of the delay line). Therefore, if the sample thickness is erroneous, the refractive index extracted from the different time-windowed pulses depends on the pulse number p as expressed by Eq. 12.67. It follows that subtracting the refractive indices obtained with two different pulses leads directly to the thickness error Δd:

$$\frac{\Delta d}{d} = \frac{(2p+1)(2q+1)}{2(q-p)}\left(n_{\mathrm{p}} - n_{\mathrm{q}}\right). \tag{12.68}$$

From a practical point of view, the best estimation of Δd is achieved when $q - p$ is the largest, i.e., $p = 0$ and $q \rightarrow \infty$. On the other hand, the q-th pulse amplitude becomes smaller when q increases, and recording a high-q pulse requires a very long delay line, with possible shifts and variations of the THz signal. Thus, it is better to select the two first pulses, i.e., $p = 0$ and $q = 1$, for which $\Delta d/d = 1.5 \times (n_0 - n_1)$.

As an example, the top-side plot of Fig. 12.32 shows the ordinary index spectrum of $LiNbO_3$ extracted from experimental data using only pulse 0 (continuous lines) and echo 1 (dotted lines). Different sample thicknesses d are entered in the computation. One observes, as expected from (12.68), that the difference $n_1 - n_o$ turns out to be small when $d \approx 1.1$ mm. This is definitively verified in Fig. 12.32 (bottom-side plot). Here, $n_1 - n_o$ has been averaged over the whole experimental bandwidth (150–650 GHz) and then it is plotted versus d. The continuous line is the best fit with (12.64), which leads to the actual sample thickness $d = 1.100$ mm $\pm\, 0.003$ mm. The precision is amazing, i.e., $\Delta d \approx \lambda/200$. This is due to the large value of the ordinary index of $LiNbO_3$. Nevertheless, even for materials with a smaller refractive index, the so-achieved precision on the sample thickness is better than $\lambda/10$.

Thickness error in the case of thick samples can be treated by a proper analysis of the extracted refractive index spectra, like the ones in Fig. 12.29. We start again from the transmission expression 12.64. If $\tilde{n}_o$ and d are the actual values of the complex refractive index and of the thickness of the sample, extracting the refractive index from the measured $\tilde{t}_{meas}$ data with a bad thickness value, i.e., $d + \Delta d$, leads to incorrect $\tilde{n} = \tilde{n}_o + \Delta\tilde{n}$ values. Therefore, we have

$$\tilde{t}\,(\tilde{n}_o, d) = \tilde{t}\,(\tilde{n}, d + \Delta d) = \tilde{t}\,(\tilde{n}_o + \Delta\tilde{n}, d + \Delta d),$$

$$\Rightarrow \frac{\tilde{n}_o}{(\tilde{n}_o + 1)^2} \frac{e^{-j\frac{\omega}{c}(\tilde{n}_o - 1)d}}{1 - \left(\frac{\tilde{n}_o - 1}{\tilde{n}_o + 1} e^{-j\frac{\omega}{c}\tilde{n}_o d}\right)^2}$$

$$= \frac{\tilde{n}}{(\tilde{n} + 1)^2} \frac{e^{-j\frac{\omega}{c}(\tilde{n} - 1)(d + \Delta d)}}{1 - \left(\frac{\tilde{n} - 1}{\tilde{n} + 1} e^{-j\frac{\omega}{c}\tilde{n}(d + \Delta d)}\right)^2}. \qquad (12.69)$$

Then we separate the exponential terms in the numerators and we get

$$\exp\left[j\,\frac{\omega}{c}\left((\tilde{n}_o - 1 + \Delta\tilde{n})\,\Delta d + \Delta\tilde{n}\,d\right)\right] = \frac{\frac{(\tilde{n}+1)^2}{\tilde{n}}}{\frac{(\tilde{n}_o+1)^2}{\tilde{n}_o}} \frac{1 - \left(\frac{\tilde{n}-1}{\tilde{n}+1} e^{-j\frac{\omega}{c}\tilde{n}(d+\Delta d)}\right)^2}{1 - \left(\frac{\tilde{n}_o-1}{\tilde{n}_o+1} e^{-j\frac{\omega}{c}\tilde{n}_o d}\right)^2}. \qquad (12.70)$$

In the exponent on the left-hand side of (12.70), we first neglect the second order term, i.e., $\Delta\tilde{n}\,\Delta d$. Expression (12.70) is in the form of $\exp\left(j\,\tilde{\phi}\right) = \tilde{z}$, i.e., $\tilde{\phi} = \arg(\tilde{z}) - \ln|\tilde{z}|$, where $\tilde{\phi}$ and $\tilde{z}$ are complex

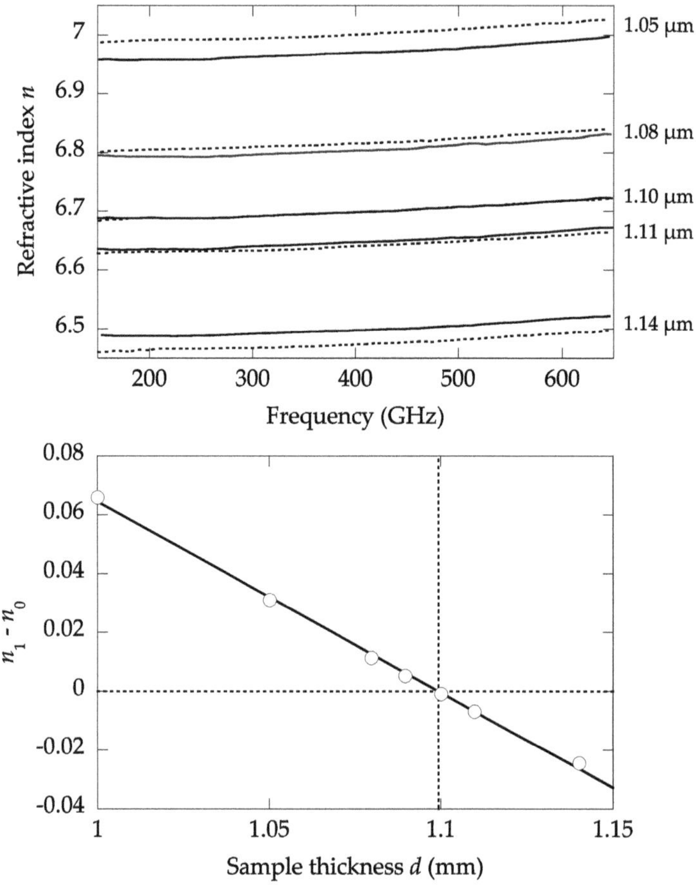

Figure 12.32 Top plot: Ordinary index of $LiNbO_3$ versus frequency extracted from pulse 0 (continuous lines) and echo 1 (dotted lines) for different values of the sample thickness. Bottom plot: Difference $n_1 - n_o$ versus the sample thickness. The dashed vertical line points out the actual value of the sample thickness.

numbers. Here the right-hand side of (12.70) is the product of two ratios whose moduli are close to 1. Thus $\ln|\tilde{z}| \approx \ln 1 \approx 0$ and we simply get

$$\Delta\tilde{n} = (1 - \tilde{n}_o)\,\frac{\Delta d}{d}$$
$$+\frac{c}{\omega d}\arg\left(\frac{\frac{(\tilde{n}+1)^2}{\tilde{n}}}{\frac{(\tilde{n}_o+1)^2}{\tilde{n}_o}}\right) + \frac{c}{\omega d}\arg\left(\frac{1-\left(\frac{\tilde{n}-1}{\tilde{n}+1}e^{-j\frac{\omega}{c}\tilde{n}(d+\Delta d)}\right)^2}{1-\left(\frac{\tilde{n}_o-1}{\tilde{n}_o+1}e^{-j\frac{\omega}{c}\tilde{n}_o d}\right)^2}\right). \qquad (12.71)$$

The first term on the right side of relation 12.71 is related to the contribution of propagation through the sample, the second one is due to the Fresnel coefficients at the sample faces, and the last one arises from the Fabry–Perot effect (rebounds within the sample). The first term is constant versus frequency: it induces a shift in the calculated refractive index when the thickness is erroneous. This shift is clearly seen in Fig. 12.29. The two other terms engender the oscillations seen in Fig. 12.29. Let us notice that the oscillation magnitude is always much smaller than the constant shift. Thus, one can estimate the oscillation magnitude by substituting in their expressions (two last right-side terms in 12.71) the first order expression of $\Delta\tilde{n}$ (first right-side term in 12.71, i.e., $\Delta\tilde{n} = (1-\tilde{n}_o)\,\Delta d/d)$. Moreover, we treat the case of weak absorption $n_o \gg \kappa_o$ (with $\tilde{n}_o = n_o - j\kappa_o$). It follows that

$$\frac{\frac{(\tilde{n}+1)^2}{\tilde{n}}}{\frac{(\tilde{n}_o+1)^2}{\tilde{n}_o}} \approx \frac{\left(1+\frac{\Delta\tilde{n}}{\tilde{n}_o+1}\right)^2}{1+\frac{\Delta\tilde{n}}{\tilde{n}_o}} \approx 1+\frac{\tilde{n}_o-1}{\tilde{n}_o+1}\Delta\tilde{n} \approx 1+\frac{n_o-1}{n_o+1}\Delta n.$$

This ratio is almost real, and thus its argument is null. The last term in (12.71) can also be simplified with

$$\frac{1-\left(\frac{\tilde{n}_o+\Delta\tilde{n}-1}{\tilde{n}_o+\Delta\tilde{n}+1}e^{-j\frac{\omega}{c}\tilde{n}(d+\Delta d)}\right)^2}{1-\left(\frac{\tilde{n}_o-1}{\tilde{n}_o+1}e^{-j\frac{\omega}{c}\tilde{n}_o d}\right)^2}$$
$$\approx \frac{1-\left(\frac{n_o-1}{n_o+1}\right)^2\left(1-\frac{4}{n_o+1}\frac{\Delta d}{d}\right)e^{-2j\frac{\omega}{c}n_o d}e^{-2j\frac{\omega}{c}\Delta d}}{1-\left(\frac{n_o-1}{n_o+1}\right)^2 e^{-2j\frac{\omega}{c}n_o d}}.$$

Therefore, the uncertainty is given by

$$\Delta\tilde{n} = (1-\tilde{n}_o)\frac{\Delta d}{d} + \frac{c}{\omega d}\phi, \tag{12.72}$$

with

$$\phi = \arg\left(\frac{1-(A+\Delta A)\,e^{-j\omega/\omega_1}e^{-j\omega/\omega_2}}{1-Ae^{-j\omega/\omega_1}}\right),$$

$$A = \left(\frac{n_o-1}{n_o+1}\right)^2, \quad \Delta A = -\frac{4A}{n_o+1}\frac{\Delta d}{d},$$

and $1/\omega_1 = 2n_o\frac{d}{c}, \quad 1/\omega_2 = 2\frac{\Delta d}{c}$.

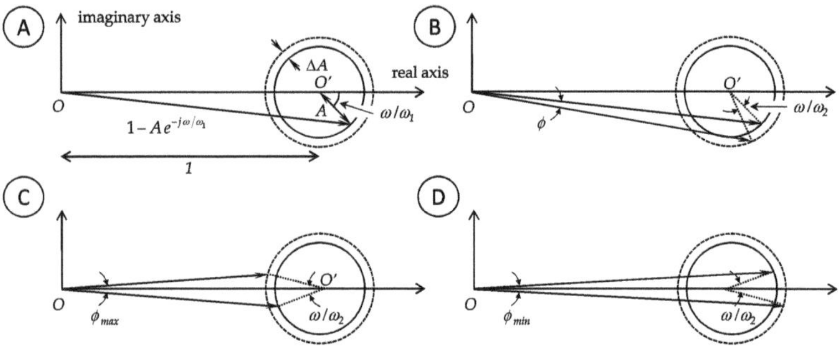

Figure 12.33 Representation of the oscillating term φ of $\tilde{n}$ (see relation 12.68) in the complex plane. Explanation of the A–D schemes is given in the text.

The role of the different terms in Eq. 12.72 can be simply explained by sketching this expression in the complex plane (Fig. 12.33).

In this complex plane, we define two vectors $\vec{A}$ and $\vec{A} + \Delta\vec{A}$ of respective lengths A and $A + \Delta A$, which are oriented at an angle ω/ω_1 and $\omega/\omega_1 + \omega/\omega_2$ with respect to the real axis. Drawing A in Fig. 12.33 shows the vector corresponding to $1 - Ae^{-j\omega/\omega_1}$, i.e., to the denominator in the definition of φ. When ω is varying, this vector is rotating around O'. Drawing B depicts both vectors corresponding to $1 - Ae^{-j\omega/\omega_1}$ and to $1 - (A + \Delta A)\, e^{-j\omega/\omega_1} e^{-j\omega/\omega_2}$. φ appears to be simply the angle between these two vectors. When ω is varying, φ oscillates between two limits, and thus $\tilde{n}$ exhibits an oscillatory behavior with ω. For a given value of ω/ω_2, the maximum and minimum values of φ are illustrated in (C) and (D) (note that $\varphi_{\max}$ is positive while $\varphi_{\min}$ is negative). A simple geometrical calculation leads to

$$\phi_{\substack{\text{mix}\\ \text{min}}} = \pm 2 \arctan\left(\frac{R}{1 \pm R} \sin\left(\omega/\omega_2\right)\right), \tag{12.73}$$

with

$$R = \frac{2A + \Delta A}{2} = \left(\frac{n_o - 1}{n_o + 1}\right)^2 \left(1 - \frac{2}{n_o + 1} \frac{\Delta d}{d}\right). \tag{12.74}$$

Therefore, the Δn_{osc} oscillation magnitude is

$$\Delta n_{osc} = \Delta n_{max} - \Delta n_{min} = \frac{8c}{\omega d} \arctan\left(\frac{2R\sin(\omega/\omega_2)}{1 - R^2 - R^2\sin^2(\omega/\omega_2)}\right)$$

$$\approx \frac{8c}{\omega d} \arctan\left(\frac{2R\sin(\omega/\omega_2)}{1 - R^2}\right). \qquad (12.75)$$

Validation of this analysis is made with the data recorded for a 368.5-μm thick HR-Si wafer, whose extracted refractive index is shown in Fig. 12.29. Figure 12.34 top shows the magnitude of the oscillation of the curves plotted in Fig. 12.29 versus the sample thickness value inputted in the extraction code. Here Δd is of the order of a few micrometers; thus, $\omega/\omega_2 \ll 1$ and $\Delta n_{osc} \approx (8c/\omega_2 d)\left(2R/1 - R^2\right)$ does not depend on frequency, as seen in Fig. 12.29. Therefore, the oscillation magnitude is deduced from averaging curves in Fig. 12.29 over the whole bandwidth. The continuous line is calculated from (12.72), with the thickness d as the only adjustable parameter. We see the excellent agreement with the experimental data. Here the thickness is obtained with a precision $\Delta d \approx 1$ μm. If the extraction procedure is performed with a thickness that is largely over or under estimated, i.e., Δd is rather big, the preceding approximation is no more correct, and the oscillations magnitude depends on frequency (Eq. 12.75).

In Fig. 12.34 bottom, we plot the magnitude of the oscillations when $\Delta d = 172$ μm ($\Delta d/d \approx 47\%$). The continuous line is calculated using Eq. 12.75 with d as the only variable parameter. The best fit is obtained for $d = 368.5$ μm. The agreement is very good, and the precision on the so-obtained thickness is excellent: indeed, if we use slightly different values of d (here 361 and 376 μm, i.e., 368.5 ±7.5 μm), the curves do not fit the experimental data so well. Thus, even if starting from a largely erroneous thickness, we end up with a very precise value (typically $\Delta d \approx 1$ μm): therefore, this method is robust.

12.4.2.2 Recent improved methods

Since this pioneering work by Duvillaret et al. [2], this initial method has been quite improved [10–14]. Li and coworkers [12] started from relation (12.66), which relates the refractive index determined

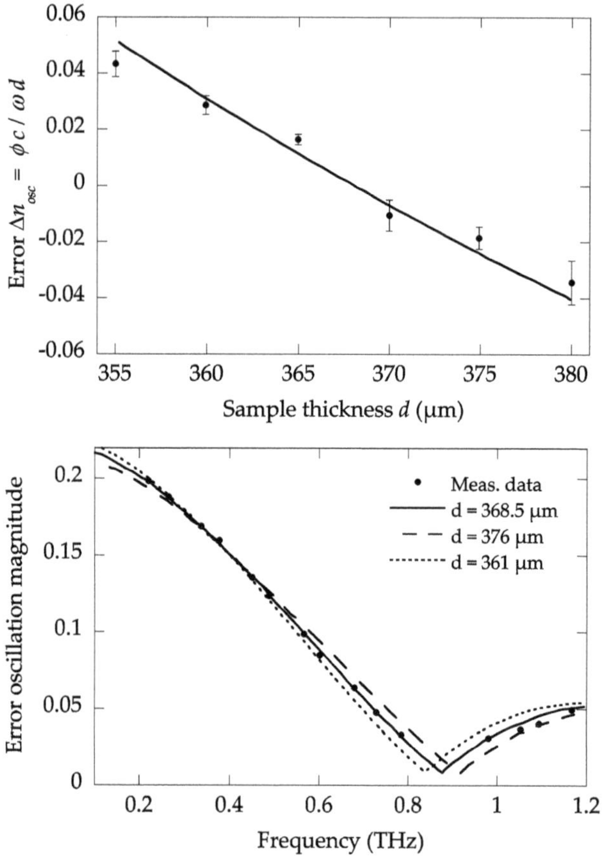

Figure 12.34 Top: Error $\Delta n_{\mathrm{osc}} = c\,\phi/\omega\,d$ (oscillatory term of (12.72)) versus sample thickness: black circles are values extracted from Fig. 12.8, the continuous line is calculated from relation 12.72 with d as the only adjustable parameter. Bottom: Magnitude of the oscillation of Δn versus frequency for $\Delta d = 172$ μm ($\Delta d/d \approx 47\%$): data are extracted from measurement, the continuous line is the best fit with (12.75), leading to a 368.5-μm thickness. Dotted and dashed lines are not-optimized curves ($d \neq 368.5$ μm).

from pulse p to the phase φ_{p}, and they write it, as well as (12.67), for $p = 1$:

$$n_1(\omega) = \frac{1}{3}\left(1 - \frac{c}{\omega\,d}\varphi_1(\omega)\right) \quad \Rightarrow \quad \Delta n_1(\omega) \approx \frac{c}{3\omega\,d^2}\varphi_1(\omega)\Delta d. \tag{12.76}$$

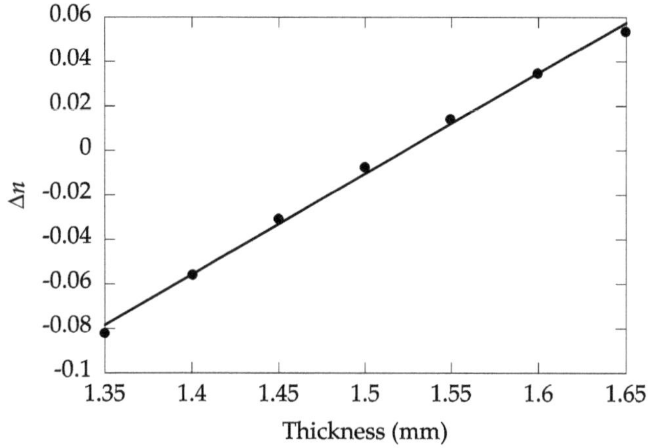

Figure 12.35 Error $\overline{\Delta n_1}$ versus the thickness d of a histidine pellet adapted from Li et al. [12]. Black circles are experimental values, and the continuous line is calculated from relation 12.77.

Then, they average $\Delta n_1(\omega)$ over the whole measured spectrum:

$$\overline{\Delta n_1} = \frac{1}{i_{\text{max}}} \sum_{i=1}^{i_{\text{max}}} \Delta n_1 (\omega_i) = \frac{1}{i_{\text{max}}} \sum_{i=1}^{i_{\text{max}}} \frac{c}{3\omega_i \, d^2} \varphi_1 (\omega_i) \, \Delta d = C \, \Delta d \,, \tag{12.77}$$

where C is a constant. The curve $\overline{\Delta n_1}$ versus the thickness d entered in the calculation is a straight line, which permits an easy determination of the actual thickness by extrapolating the line to 0. Figure 12.35 presents such a curve obtained for a pellet of amino acid (histidine). Experimental data are black circles, while the line is the best fit using relation 12.77. The sample thickness is determined with a great accuracy ($d = 1505.8 \pm 0.1$ μm). This allows the authors to extract the absorption coefficient of amino acid mixtures with improved precision.

Great progress in determining the sample thickness and correcting the induced errors was made by Dorney et al. [11]. As explained in Chapter 5, these authors extract the refractive index $\tilde{n}$ of the sample by minimizing an error function ER calculated over all the whole experimental spectrum. Let $\tilde{t}_{\text{exp}}(\omega)$ and $\tilde{t}(\omega)$ be respectively the measured complex coefficient of transmission and the theoretical one. The expression of this last one is given by

Eq. 12.64. ER is defined by

$$ER = \sum_i \left(\left|ER_{\text{mag}}(\omega_i)\right| + \left|ER_{\text{phase}}(\omega_i)\right|\right). \tag{12.78}$$

with $ER_{\text{mag}}(\omega_i) = t_{\text{exp}}(\omega_i) - t(\omega_i)$ and $ER_{\text{phase}} = \varphi_{\text{exp}}(\omega_i) - \varphi(\omega_i)$ (t and φ are respectively modulus and phase). The extraction procedure starts from a guess of the thickness. Then an estimation of $\tilde{n}$, independent of frequency, is obtained by considering the time delay between the first transmitted pulse and the reference pulse. Finally, this value is entered in Eq. 12.78 and ER is minimized through an iterative gradient descent algorithm. This leads to $\tilde{n}(\omega)$ for the initial input thickness. The iteration process is repeated over a given range of thicknesses, and the smallest achieved ER gives both the accurate values of $\tilde{n}(\omega)$ and d. Introducing a total variation metric TV instead of ER gives accurate results when the SNR of the THz signal is weak, or if the iteration on the sample thickness is too rough:

$$\text{TV} = \sum_i |D(\omega_i) - D(\omega_{i+1})|, \tag{12.79}$$

with $D(\omega_i) = |n(\omega_{i-1}) - n(\omega_i)| + |\kappa(\omega_{i-1}) - \kappa(\omega_i)|$.

This procedure is valid if the sample does not show sharp spectral features. When the right thickness is numerically obtained, TV shows a pronounced deepest local minimum versus d, as seen in Fig. 12.36: in this example, the so-estimated thickness of a GaAs sample is $d = 0.63$ mm, while the thickness measured with a caliper is 0.65 ± 0.2 mm. The method by Dorney et al. [11] is more general than the preceding one [2]. Pupeza et al. [13] used the same procedure as Dorney et al., but the same error function ER is minimized through a Nelder–Mead algorithm. Figure 12.37 shows the ER function experimentally obtained for two different samples, namely a Si wafer and a leu-gly-gly pellet. The pronounced deep minimum of the curves allows for the determination of the sample thickness with an accuracy of a few micrometers.

Very recently, van Mechelen and coauthors [14] extracted both thickness and refractive index of a layer in a multi-stack sample by performing a time-domain analysis, close to the method proposed by Li et al. [15] (see paragraph 9, Chapter 5). The basic idea is to

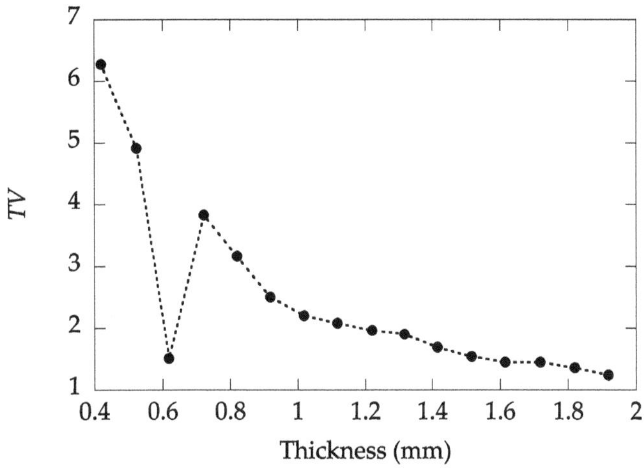

Figure 12.36 TV (see relation 12.77) versus d for a GaAs sample. A local deepest minimum is observed at $d = 0.63$ mm, while the thickness value measured with a caliper is 0.65 ± 0.02 mm. Adapted from Dorney et al. [11].

describe the dielectric constant of each layer k of the multi-stack by a Drude–Lorentz function:

$$\varepsilon_k(\omega) = \varepsilon_\infty + \sum_{l=1}^{m} \frac{\omega_{k,p,l}^2}{\omega_{k,o,l}^2 - \omega^2 - j\Gamma_{k,l}\omega}, \tag{12.80}$$

where $\omega_{k,p,l}$ is the plasma angular frequency of layer k, $\omega_{k,o,l}$ the characteristic frequency and $\Gamma_{k,l}$ its damping frequency for each resonance l. If the number of resonances is not so big, each layer is described by only four unknown parameters, i.e., $\omega_{k,p,l}$, $\omega_{k,o,l}$, $\Gamma_{k,l}$, and the thickness d_k. Expression 12.80 is entered in the transmission by a multi-stack (see A.33 in the annex). Transmission is then calculated for each frequency and subsequently is Fourier-transformed to obtain the transmitted waveform. This computed waveform is finally compared with the experimental one. van Mechelen et al. [14] have applied this procedure to a sheet of paper located over a copper reflector (Fig. 12.38). In this example, only the paper sheet is unknown as well as the distance between the sheet and the metal. Typically, the paper sheet thickness is obtained with a precision of the order of 0.1 μm.

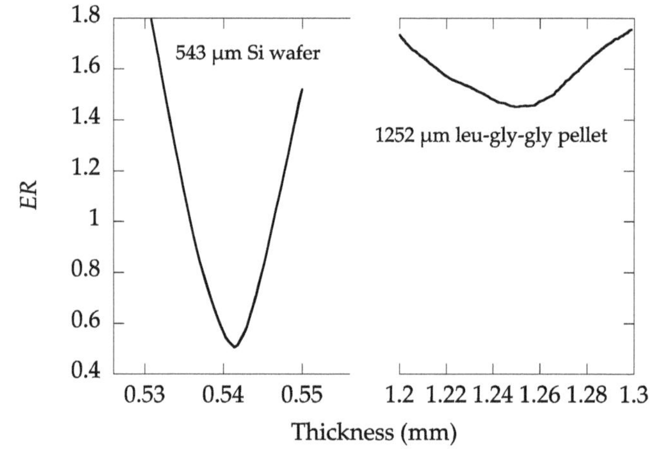

Figure 12.37 Experimental ER function versus sample thickness obtained for two different samples, namely a Si wafer (left) and a leu-gly-gly pellet (right). Adapted from Li et al. [13].

The uncertainty in THz parameters extraction due to a bad knowledge of the sample thickness has been meticulously treated by Krüger et al. [16]. As explained in Chapter 5, their extraction method is based on the search of the roots of relation 5.1 in the complex plane, which is divided into small domains. The size of the domains is made the smallest but with the constraint that each domain still contains a solution. The authors treat the uncertainty by Fourier-transforming the waveform standard deviation, and they show that

$$\mathrm{COV}\left(\vec{\Delta}\right) = J(\tilde{t})\mathrm{COV}\left(\vec{E}\right)J^{T}(\tilde{t}), \tag{12.81}$$

where $\vec{\Delta} = (t_R, t_I)$, $\vec{E} = (E_{tR}, E_{tI}, E_{iR}, E_{iI})$ and J is the Jacobian matrix (here any complex value is written as $\tilde{x} = x_R + jx_I$, and $\mathrm{COV}\left(\vec{X}\right)$ is the covariance matrix of the vector $\vec{X}$). They end up with the following simplified expressions of the errors:

$$\begin{cases} \Delta n = \frac{c}{\omega d}\sqrt{\left(\Delta \arg\left(\tilde{t}\right)\right)^2 + \left(\arg\left(\tilde{t}\right)\frac{\Delta d}{d}\right)^2}, \\ \Delta \kappa = \frac{c}{\omega d}\sqrt{\left(\frac{\Delta\left|\tilde{t}\right|}{\left|\tilde{t}\right|}\right)^2 + \left(\frac{\Delta d}{d}\ln\left|\tilde{t}\right|\right)^2}. \end{cases} \tag{12.82}$$

They determined the refractive index of capric acid with a precision of 0.005 (0.3%).

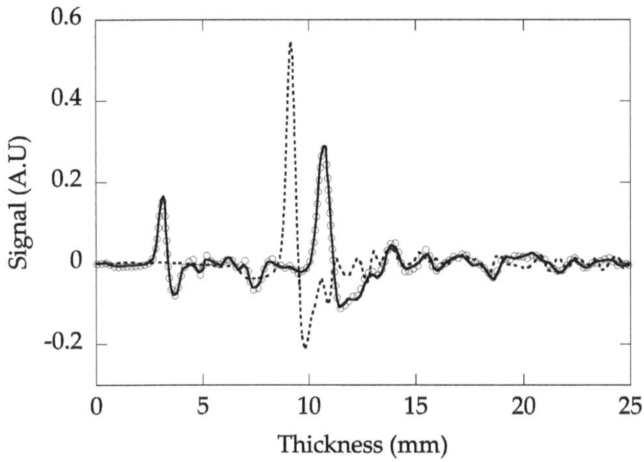

Figure 12.38 THz waveforms reflected by a copper plate: measured (open circles) and fitted (continuous line) with a paper sheet over the reflector. The dashed curve is the measured reference signal (without paper sheet). One can see the excellent agreement between the experiment and the fit. Adapted from van Mechelen et al. [14].

12.4.2.3 Applications of layer thickness determination by THz-TDS

Thickness issue is an important question in technology and industrial development and testing, for which THz pulsed techniques exhibit unique performance, especially for a nondestructive characterization of thin layers, like coats of paint, e.g., as used in the automotive industry [17]. It allows one to determine layer thickness, when other non-destructive techniques fail, because of opacity (visible or IR sensing), thickness (microwaves, acoustic waves), etc. Thus, this topic has been widely addressed [17–22]. Most of the published techniques are based on measuring the delay time Δt between a THz pulse transmitted through the layer and a reference one without a sample, or the delay time $\Delta t'$ between the pulses reflected by the upper and lower faces of the layer, if this layer is deposited on an opaque substrate. Assuming the sample shows no dispersion, the thickness is then given by

$$\begin{cases} d = c\frac{\Delta t}{n-1} \text{ (transparent substrate)}, \\ d = 2c\frac{\Delta t'}{n} \text{ (opaque substrate)}. \end{cases} \quad (12.83)$$

Therefore, determining d from the delay-time measurement, using this simplified relation 12.83, requires knowing the refractive index of the layer. The minimum measurable thickness $d_{\min}$ is roughly obtained when Δt or $\Delta t'$ are equal to the THz pulse duration Δt_{THz}. Typically, for a photo-conducting antenna with Gaussian-like current pulses, one gets

$$\begin{cases} d_{\min} = c\frac{\Delta t_{\mathrm{THz}}}{n-1} = \frac{1}{n-1}\frac{200}{f_{\mathrm{THz,max}}}, \text{ (transp. substrate)}, \\ d_{\min} = 2c\frac{\Delta t_{\mathrm{THz}}}{n} \approx \frac{1}{n}\frac{400}{f_{\mathrm{THz,max}}}, \text{ (opaque substrate)}. \end{cases} \tag{12.84}$$

Here $f_{\mathrm{THz,max}}$ is the bandwidth of the THz signal. For a 5 THz TDS set up and $n \approx 2$, $d_{\min}$ is of the order of a few tens of microns. This limit can be strongly improved ($0.1 \sim 0.2$ μm) by measuring the incoming THz waveform and fitting the transmitted or reflected one [14].

12.5 Effects due to Actual THz Beam Shape

In the preceding paragraphs, the investigation is made in the framework of the THz parallel beam, i.e., a plane-wave analysis. Real THz beams are spatially limited, at least by the finite size of the quasi-optical components, like mirrors, lenses, etc., but also by the radiation pattern of the antennas. These real THz beams are well described by Gaussian optics [23]. As compared to the plane wave analysis, a Gaussian study of the TDS signal will show two differences: the THz beam is slightly defocused by the sample and the phase of the detected signal is slightly different.

The rigorous study of the transmission of any Gaussian electromagnetic beam through a dielectric slab is not an easy task because each geometrical part of the beam is incident on the slab at a different angle. An elegant way of solving the problem is to write the incident beam as a sum of plane waves, through a Fourier transform from the real space to the wave vector space. Then each of these plane waves is transmitted through the sample: the whole transmission coefficient is easy to derive (see Annex). Finally, an inverse Fourier transform allows finding the transmitted Gaussian beam [24].

Here we aim only at deriving tendencies. Thus, we will neglect the rebounds inside the sample. Therefore, the shape of the transmitted beam, as well as its phase, are given by a simple Gaussian optics analysis [25]. We will consider the effect of the beam shape as compared to the plane wave analysis, versus the parameters of the sample (thickness d and refractive index n).

12.5.1 Gaussian Beam Approximation

In common THz-TDS setups, when components are perfectly aligned, the THz beam is almost Gaussian. Let us indeed consider a dipolar radiation, like the one emitted by a photo-conducting switch. The dipole far field is written as

$$E_{\mathrm{THz}}(\theta) \propto \frac{\sin\theta}{r}, \tag{12.85}$$

where r is the distance from the dipole and θ is the angle between the dipole and the direction of the considered radiation (Fig. 12.39, top). This radiation is made parallel by a component (silicon dome, parabolic mirror, refractive lens, etc.) of focal distance F. The THz power radiated within a $2\pi d\theta$ solid angle (ring in Fig. 12.39 top) propagates, after passing the lens, in a ring of area $2\pi\ dx$. Therefore, the THz field distribution after the lens is

$$(E_{\mathrm{THz}}(\theta))^2\, d\theta = (E_{\mathrm{THz}}(x))^2\, dx \quad \Rightarrow \quad E(x) \propto \frac{\sin\theta}{r}\sqrt{\frac{d\theta}{dx}},$$

$$\Rightarrow \quad E(x) \propto \left(x^2 + F^2\right)^{-3/2}. \tag{12.86}$$

For small values of x, this expression could be approximated by

$$E(x) \propto F^{-3}\left(1 + \frac{x^2}{F^2}\right)^{-3/2} \approx F^{-3}\left(1 - \frac{3}{2}\frac{x^2}{F^2}\right) \approx F^{-3}e^{-\frac{3}{2}\frac{x^2}{F^2}}. \tag{12.87}$$

Therefore, at least at the beam center, the Gaussian approximation describes rather well the beam's propagation properties. This is shown in Fig. 12.39 (bottom), where the dipolar field amplitude and the corresponding Gaussian field are plotted together.

Experimental measurements confirm the Gaussian-like shape of THz beams in common TDS systems. The easiest way to determine the beam shape is the "razor blade" experiment. A semi-infinite

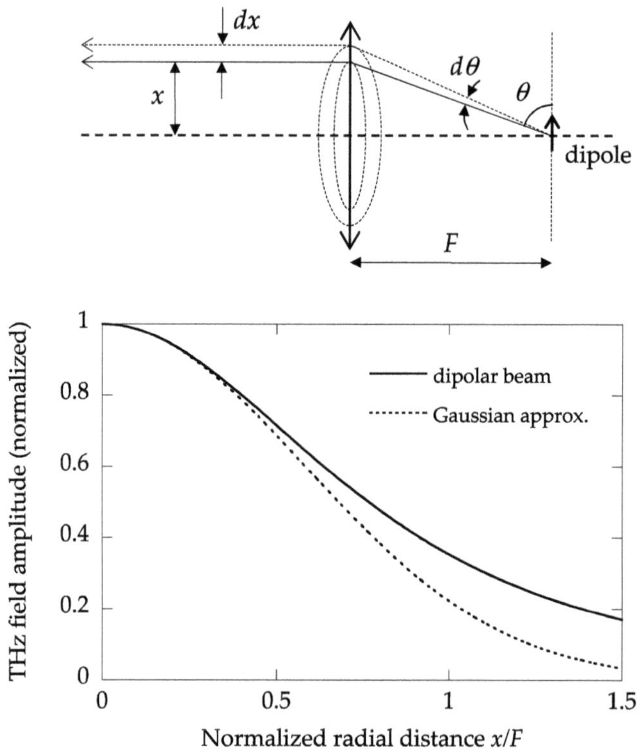

Figure 12.39 Top: THz dipole source with a collimating lens. Bottom: Radial dependence of the THz dipolar and Gaussian fields.

opaque plane is translated perpendicularly to the THz beam. If one takes into account only the beam power that is blocked by the plane and not diffraction, it is easy to show that the total power after the plane is

$$P(X) = \frac{P_i}{\pi w^2} \int_{x=X}^{\infty} \int_{y=-\infty}^{+\infty} e^{-\frac{x^2+y^2}{w^2}} dxdy = \frac{1}{2}\mathrm{erfc}\left(\frac{X}{w}\right) P_i, \quad (12.88)$$

with $\mathrm{erfc}(u) = \frac{2}{\sqrt{\pi}} \int_u^{\infty} e^{-t^2} dt$. erfc($u$) is the complementary error function. w is the beam waist, and X is the distance of the plane edge from the center of the beam.

Such a measurement is easy to perform. Figure 12.40 presents the recorded signal at 0.5 THz for the TDS system described by

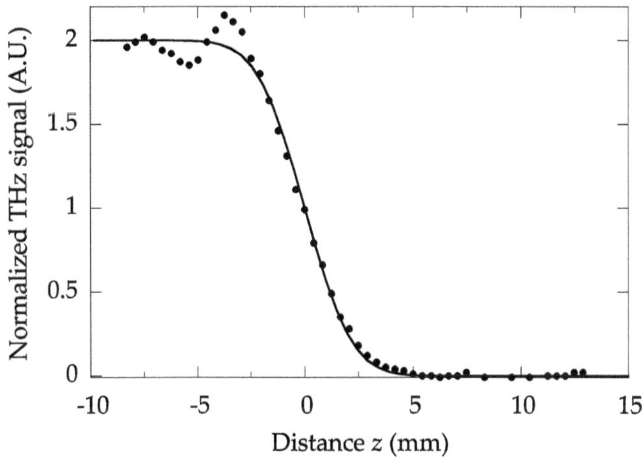

Figure 12.40 The detected THz signal in a razor blade experiment, versus the distance X between the blade edge and the optical axis. The continuous line is a fit with Eq. 12.88.

Siemion et al. [26]. The continuous line is a fit with Eq. 12.87, which gives $w = 5.9$ mm. At negative values of X, the recorded signal shows oscillations around the asymptotic value equal to 2. These oscillations are due to diffraction and request more sophisticated theories to describe them, e.g., the Kirchhoff–Fresnel one [27].

In the Gaussian beam theory, the field amplitude is given by [28]

$$E(x, y, z) = A\frac{w_o}{w(z)}e^{-\left(\frac{\rho}{w(z)}\right)^2}e^{-j(kz-\phi(z))}, \tag{12.89}$$

z is the direction of propagation, $\rho = \sqrt{x^2 + y^2}$ is the radial distance, i.e., the distance from the optical axis, A is the field amplitude in $z = 0$, k is the wavevector amplitude. $w(z)$ denotes the size (radius) of the beam cross section:

$$w(z)^2 = w_o^2 + \frac{4z^2}{w_o^2 k^2}. \tag{12.90}$$

Thus w_o is the minimum value of $w(z)$, which is observed in $z = 0$. w_o is generally named the waist of the beam. The phase $\phi(z)$ in Eq. 12.89 becomes

$$\phi(z) = \arctan\left(\frac{z}{z_{\text{Rayleigh}}}\right) - \frac{z}{z_{\text{Rayleigh}}}\frac{\rho^2}{w^2(z)}. \tag{12.91}$$

z_{Rayleigh} is the Rayleigh length, i.e., the distance z from 0 over which the waist is doubled:

$$w\left(z_{\text{Rayleigh}}\right) = 2w_{\text{o}} \quad \Rightarrow \quad z_{\text{Rayleigh}} = k\frac{w_{\text{o}}^2}{2} = \pi\frac{w_{\text{o}}^2}{\lambda}. \tag{12.92}$$

In Eq. 12.89, the term $w_{\text{o}}/w(z)$ in the expression of the field describes the variation of the field amplitude versus the propagation distance z, the next term depicts the Gaussian shape of the beam, and the last one is the propagation term. Compared to the plane wave case, its phase includes an additional term, namely $\phi(z)$, which is the sum of the Gouy phase, $\phi_{\text{Gouy}}(z) = \arctan\left(z/z_{\text{Rayleigh}}\right)$, and the radial phase $-\frac{z}{z_{\text{Rayleigh}}}\frac{\rho^2}{w^2(z)}$ of a quasi-spherical wave. One simplifies most of the preceding expressions by introducing the complex beam parameter $Q(z)$ such as

$$Q(z) = z + j\,z_{\text{Rayleigh}}. \tag{12.93}$$

It follows that

$$\frac{1}{Q(z)} = \frac{1}{R(z)} - j\frac{\lambda}{\pi\, w^2(z)}, \quad \tan\phi_{\text{Gouy}}(z) = \frac{\text{Re}(Q(z))}{\text{Im}(Q(z))}. \tag{12.94}$$

$R(z)$ is the radius of curvature of the Gaussian beam wavefront. The knowledge of $Q(z)$ is sufficient to describe the propagation properties of the Gaussian beam. Moreover, when the Gaussian beam crosses any optical component (Fig. 12.41), the change in $Q(z)$ obeys the ABCD law of classical optics:

$$Q_2 = \frac{A\,Q_1 + B}{C\,Q_1 + D}, \quad M = \begin{bmatrix} A & B \\ C & D \end{bmatrix}. \tag{12.95}$$

The matrix M is the same as in ray optics. For example,

$$\begin{cases} \text{- propagation over a distance } x\text{:} \quad M = \begin{bmatrix} 1 & x \\ 0 & 1 \end{bmatrix} \\ \text{- transmission through a } n_1/n_2 \text{ dioptre:} \quad M = \begin{bmatrix} 1 & 0 \\ 0 & n_1/n_2 \end{bmatrix} \\ \text{- transmission through a thin lens:} \quad M = \begin{bmatrix} 1 & 0 \\ -1/f & 1 \end{bmatrix} \end{cases} \tag{12.96}$$

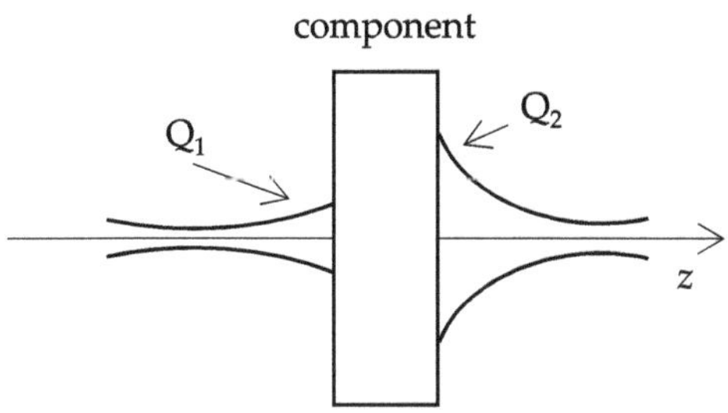

Figure 12.41 Transmission of a Gaussian beam through an optical system.

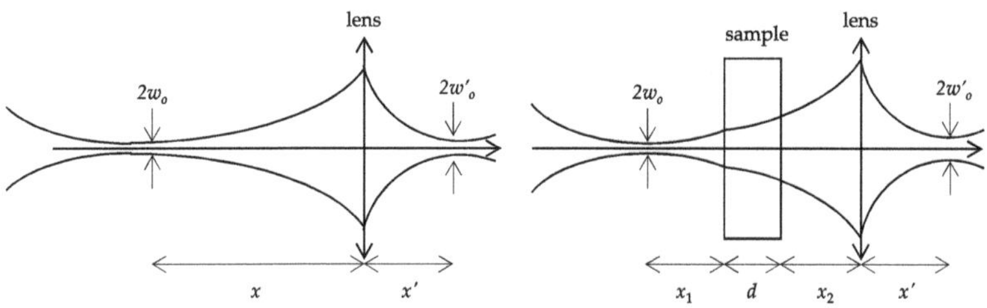

Figure 12.42 Simplified scheme of the Gaussian THz beam reaching the receiver in a transmission THz-TDS system: left, without sample; right, with sample.

12.5.2 Simple Analysis of the Influence of the Sample

We consider here a Gaussian THz beam which is incident on the optics that focuses it onto the receiver, and we look at the changes of the focused beam properties when a sample is placed in the beam before the optics (Fig. 12.42). This simplified scheme represents the detection part of the setup.

The waist of the impinging Gaussian beam is at a distance x from the optics, whose focal length is F, and its value is w_0. The M matrix of the system is

$$M = \begin{pmatrix} 1 & x' \\ 0 & 1 \end{pmatrix} \begin{pmatrix} 1 & 0 \\ -\frac{1}{F} & 1 \end{pmatrix} \begin{pmatrix} 1 & x \\ 0 & 1 \end{pmatrix} = \frac{1}{F} \begin{pmatrix} F - x' & F\left(x + x'\right) - xx' \\ -1 & F - x \end{pmatrix}. \tag{12.97}$$

It follows that the radius of curvature and the waist of the focused beam at a distance x' from the optics are

$$R\left(x'\right) = \frac{\left(F - x'\right)^2 Z_0^2 + \left(Fx - xx' + Fx'\right)^2}{\left(F - x\right)\left(Fx - xx' + Fx'\right) - Z_0^2\left(F - x'\right)}, \tag{12.98}$$

$$w^2\left(x'\right) = w_o^2 \frac{\left(F - x'\right)^2 Z_o^2 + \left(Fx - xx' + Fx'\right)^2}{Z_o^2 F^2}. \quad (12.99)$$

Here $Z_o = \pi\, w_o^2/\lambda$ the Rayleigh length of the incoming beam. The THz beam is focused at a distance $x' = x_{\text{Ro}}$ for which $R\left(x_{\text{Ro}}\right)$ tends towards infinity (its denominator is null):

$$R\left(x_{\text{Ro}}\right) \to \infty \quad \Rightarrow \quad x_{\text{Ro}} = F\frac{Z_o^2 - (F - x)\,x}{Z_o^2 + (F - x)^2}. \quad (12.100)$$

At $x' = x_{\text{Ro}}$, the Gouy phase is zero, and the size of the beam is

$$w^2\left(x_{\text{Ro}}\right) = w_o^2 \frac{F^2}{Z_o^2 + (F - x)^2}. \quad (12.101)$$

When placing a sample of thickness d and refractive index n at a distance x_2 in front of the lens (Fig. 12.42), the matrix M becomes

$$M = \begin{pmatrix} 1 & x' \\ 0 & 1 \end{pmatrix} \begin{pmatrix} 1 & 0 \\ -\frac{1}{F} & 1 \end{pmatrix} \begin{pmatrix} 1 & x_2 \\ 0 & 1 \end{pmatrix} \begin{pmatrix} 1 & 0 \\ 0 & n \end{pmatrix} \begin{pmatrix} 1 & d \\ 0 & 1 \end{pmatrix} \begin{pmatrix} 1 & 0 \\ 0 & \frac{1}{n} \end{pmatrix} \begin{pmatrix} 1 & x_1 \\ 0 & 1 \end{pmatrix}. \quad (12.102)$$

with $x_1 + x_2 + d = x$. The analytical form of $M$is

$$M = \frac{1}{F}\begin{pmatrix} F - x' & (F - x')(x - \Delta) + Fx' \\ -1 & F - (x - \Delta) \end{pmatrix}. \quad (12.103)$$

$\Delta = \frac{n-1}{n} d$ is simply the relative change in the x distance induced by the sample. It is noticeable that the matrix with sample (Eq. 12.103) is equal to the matrix without sample (Eq. 12.97), in which x has been substituted by $x - \Delta$. The position and the waist of the beam do not depend on the relative position x_2 of the sample in the beam! Hence, all the preceding expressions are still valid when making the same substitution. The position of the waist is now shifted by Δ':

$$\Delta' = x'_{\text{Ro}} - x_{\text{Ro}} = F^2 \frac{-Z_o^2 + (F - x + \Delta)(F - x)}{\left(Z_o^2 + (F - x + \Delta)^2\right)\left(Z_o^2 + (F - x)^2\right)}\Delta. \quad (12.104)$$

The Rayleigh length in the focus region is

$$Z' = \frac{\pi}{\lambda} w^2\left(x'_{\text{Ro}}\right) = \frac{\pi}{\lambda} w_o^2 \frac{F^2}{Z_o^2 + (F - x + \Delta)^2}. \quad (12.105)$$

At the former waist location $x' = x_{\text{Ro}}$, the variation of the beam waist and the Gouy phase are

$$\frac{\Delta w^2\left(x_{\text{Ro}}\right)}{w^2\left(x_{\text{Ro}}\right)} = \frac{w'^2\left(x_{\text{Ro}}\right) - w^2\left(x_{\text{Ro}}\right)}{w^2\left(x_{\text{Ro}}\right)} = -\frac{2(F - x) + \Delta}{Z_o^2 + (F - x + \Delta)^2}\Delta, \quad (12.106)$$

$$\tan\phi_{\text{Gouy}} = \frac{\Delta'}{Z'_o} = \frac{F^2}{Z_o}\frac{(F-x)(F-x+\Delta) - Z_o^2}{\left(Z_o^2 + (F-x+\Delta)^2\right)\left(Z_o^2 + (F-x+\Delta)^2\right)}\Delta. \tag{12.107}$$

If Δ is small as well as Z_o (tight focusing), one retrieves the result of ray optics:

$$\begin{cases} x_{\text{Ro}} \approx -\eta\, x, & \Delta' = x'_{\text{Ro}} - x_{\text{Ro}} \approx \eta^2 \Delta, \\ \frac{\Delta w^2(x_{\text{Ro}})}{w^2(x_{\text{Ro}})} \approx -F\eta\,\Delta, & \tan\phi_{\text{Gouy}} \approx \frac{\eta^2}{Z_o}\Delta, \end{cases} \quad \text{with } \eta = \left(1 - \frac{x}{F}\right)^{-1}. \tag{12.108}$$

Let us notice that these expressions are the same as the ones given in a reference paper on the influence of the Gaussian beam shape by P. Kuzel and colleagues [29]. In the paper, the authors also treat the effect of the rebounds in the sample.

When extracting the refractive index of the sample, the defocusing due to the sample leads to incorrect values if one uses a plane wave analysis. In a simplified model (the influence of Fresnel coefficients at the sample faces is neglected, as well as the rebounds), the phase at the detector is

$$\underbrace{\phi}_{\text{measured}} = \underbrace{\frac{2\pi f}{c} d\,(n-1) - \phi_{\text{Gouy}}}_{\text{Gaussian beam}} = \underbrace{\frac{2\pi f}{c} d\,(n + \Delta n - 1)}_{\text{Plane wave}},$$

$$\Rightarrow \quad \Delta n = -\frac{c}{2\pi f\, d}\phi_{\text{Gouy}} \approx -\frac{c}{2\pi f\, Z_o}\frac{1}{\left(1 - \frac{x}{F}\right)^2}\frac{n-1}{n}. \tag{12.109}$$

The error does not depend on the sample thickness. Typically, for $x = 2F$, $w_o = 1$ mm and $n = 2$, $\Delta n \approx 5 \times 10^{-3}$ at $f = 1$ THz, and $\Delta n \approx 5 \times 10^{-2}$ at $f = 300$ GHz.

The error on the absorption coefficient may be estimated as follows. We suppose that the focused THz beam, without sample, illuminates the whole surface of the receiver. When defocused by the sample, the detected energy varies as the ratio of the beam spot without and with the sample, which leads to

$$e^{-2(\alpha+\Delta\alpha)d} = \frac{w^2\left(x_{\text{Ro}}\right)}{w'^2\left(x_{\text{Ro}}\right)}e^{-2\alpha d},$$

$$\Rightarrow \quad \Delta\alpha = -\frac{1}{2d}\ln\left(1 + \frac{2(F-x) + \Delta}{Z_o^2 + (F-x)^2}\Delta\right) \approx \frac{1}{x-F}\frac{n-1}{n}. \tag{12.110}$$

Here again, the error does not depend on the sample thickness. Typically, for $x = 2F = 20$ cm and $n = 2$, $\Delta\alpha \approx 0.025$ cm^{-1}, which is negligible.

The validity of this modeling has been nicely checked by Kužel et al. [29] by comparison with experimental data.

12.5.3 Gaussian Beam in THz-TDS Setups

We consider now a simplified scheme of a whole classical THz-TDS setup that is used in transmission (Fig. 12.43). The beam radiated by the emitting antenna (E) is collected by a focusing component (either refractive lens or parabolic mirror). A second optical component, which is chosen to be the same as the first one for the sake of simplicity, focuses the beam onto the receiving antenna (R). The focal length of both optics is F.

The matrix of this system is

$$M = \begin{pmatrix} A & B \\ C & D \end{pmatrix} = \begin{pmatrix} 1 & x_R \\ 0 & 1 \end{pmatrix} \begin{pmatrix} 1 & 0 \\ -\frac{1}{F} & 1 \end{pmatrix} \begin{pmatrix} 1 & x_2 \\ 0 & 1 \end{pmatrix} \begin{pmatrix} 1 & 0 \\ 0 & n \end{pmatrix} \times \begin{pmatrix} 1 & d \\ 0 & 1 \end{pmatrix} \begin{pmatrix} 1 & 0 \\ 0 & \frac{1}{n} \end{pmatrix} \begin{pmatrix} 1 & x_1 \\ 0 & 1 \end{pmatrix} \begin{pmatrix} 1 & 0 \\ -\frac{1}{F} & 1 \end{pmatrix} \begin{pmatrix} 1 & x \\ 0 & 1 \end{pmatrix}. \tag{12.111}$$

Calculating the product of the matrixes in Eq. 12.111 leads to

$$\begin{cases} A = \frac{1}{F}\left(F - (x_s - \Delta) + x_R\left(-2 + \frac{x_s - \Delta}{F}\right)\right), \\ B = \frac{1}{F}\left(Fx + (F - x)(x_s - \Delta) + x_R\left(-\frac{1}{F}(F - x)(x_s - \Delta) + F - 2x\right)\right), \\ C = \frac{1}{F}\left(-2 + \frac{x_s - \Delta}{F}\right), \\ D = \frac{1}{F}\left(-\frac{1}{F}(F - x)(x_s - \Delta) + F - 2x\right). \end{cases} \tag{12.112}$$

with $x_s = x_1 + d + x_2$. Proceeding as previously leads to

$$x_{Ro} = F \frac{(x_s - F)(x_s - 2F) Z_E^2 + (Fx + (F - x)x_s)((F - x)x_s - F(F - 2x))}{(x_s - 2F)^2 Z_E^2 + (F(F - 2x) - (F - x)x_s)^2}. \tag{12.113}$$

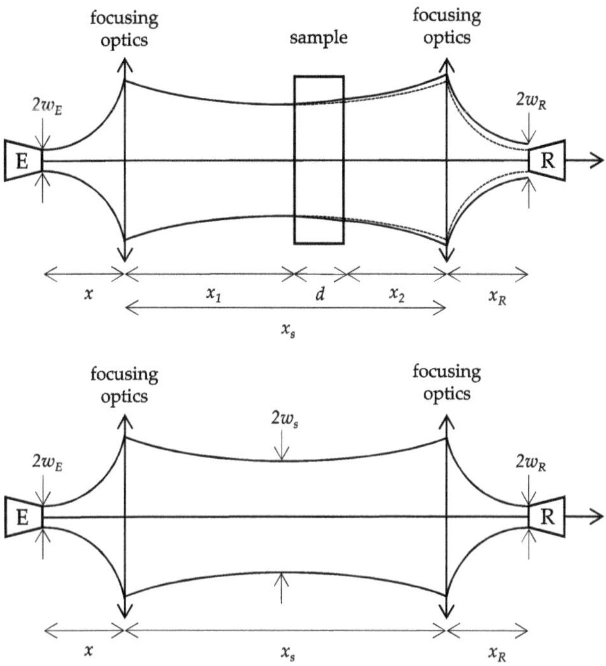

Figure 12.43 Gaussian THz beam in a simplified scheme of a transmission THz-TDS system without sample (upper scheme) and with a sample (lower scheme).

The system is achromatic, i.e., x_{Ro} does not depend on wavelength, if

$$F\,(F-2x)-(F-x)\,x_{\mathrm{s}}=0 \quad \Rightarrow \quad x_{\mathrm{s}}=F-F\frac{x}{F-x}, \tag{12.114}$$

$$\Rightarrow x_{\mathrm{Ro}}=F\frac{x_{\mathrm{s}}-F}{x_{\mathrm{s}}-2F}=x. \tag{12.115}$$

The receiver should be located at the same distance from the second optics than the distance from the emitter to the first optics. Usually, the emitter is located at a distance x from the first optics in such a way the beam after the first optics is almost parallel. A simple analysis of Eq. 12.100 shows that

$$x=F+Z_{\mathrm{E}}. \tag{12.116}$$

It follows that the system is symmetric regarding its geometrical center located at position $x_s/2$:

$$\frac{x_s}{2} = \frac{1}{2}\left(F - F\frac{(F+Z_E)}{F-(F+Z_E)}\right) = F\frac{F+2Z_E}{2Z_E} = F + \frac{F^2}{2Z_E}. \tag{12.117}$$

When inserting a sample in the setup, the focus is now given by expression 12.113 in which x_s is replaced by $x_s - \Delta$:

$$x_R = F\frac{(x_s - F - \Delta)(x_s - 2F - \Delta)Z_E^2 + [Fx + (F-x)(x_s-\Delta)][(F-x)(x_s-\Delta) - F(F-2x)]}{Z_E^2(x_s - \Delta - 2F)^2 + [(F-x)(x_s-\Delta) - F(F-2x)]^2}. \tag{12.118}$$

Two cases must be treated. First, we consider a system that is not optimized (relations 12.115 and 12.116 are not fulfilled). In the limit of the small value of Δ, Δx_R varies as Δ. Its expression is quite huge, but it could be simplified in the case of tight focusing ($Z_E \ll 1$):

$$x_R = F\frac{(Fx + (F-x)(x_s - \Delta))}{(F(F-2x) - (F-x)(x_s-\Delta))}. \tag{12.119}$$

Also, if Δ is small, we get

$$\Delta x_R = x_R - x_{Ro} = -\left(\frac{F(F-x)}{F(F-2x) - (F-x)x_s}\right)^2 \Delta. \tag{12.120}$$

It can be easily shown that expression 12.120 may be retrieved by a classical ray optics calculation.

The second case is related to an optimized THz-TDS system. We combine expressions 12.118, 12.115, and 12.114, and we obtain

$$x_R = \frac{F^4(Z_E + F) - 2Z_EF\left(F^2 + Z_EF - Z_E\Delta\right)\Delta}{\left(F^2 - Z_E\Delta\right)^2 + (Z_E\Delta)^2}. \tag{12.121}$$

As previously, the symmetrical design of the TDS setup makes the focal distance at the receiver equal to the one at the emitter when the system is empty ($\Delta = 0$ in Eq. 12.121):

$$x_{Ro} = F + Z_E. \tag{12.122}$$

In the limit of small Δ, the shift induced by the sample becomes

$$\Delta x_{\mathrm{R}} = x_{\mathrm{R}} - x_{\mathrm{Ro}} \approx -2\frac{Z_{\mathrm{E}}^{3}}{F^{4}}\Delta^{2}. \tag{12.123}$$

In this particular but common case, the linear term vanishes and one should consider the second term in Δ^2 of the development. At a position x' after the second optics, the Gouy phase is given by

$$\tan\phi_{\mathrm{Gouy}}\left(x'\right) = \frac{BD + AC\,Z_{\mathrm{E}}^{2}}{Z_{\mathrm{E}}\left(AD - CB\right)}$$
$$= \frac{x'\left(F^{2} + \left(2x - F + 2Z_{\mathrm{E}}\left(\frac{F-x_{\mathrm{s}}}{F}\right)\right)^{2}\right) - F\left(F^{2} + \left(2x - F + 2Z_{\mathrm{E}}\left(\frac{F-x_{\mathrm{s}}}{F}\right)\right)\left(x - Z_{\mathrm{E}} + 2Z_{\mathrm{E}}\left(\frac{F-x_{\mathrm{s}}}{F}\right)\right)\right)}{2Z_{\mathrm{E}}F^{2}}. \tag{12.124}$$

When inserting a sample, x_{s} must be changed into $x_{\mathrm{s}} - \Delta$ in (12.124). In the limit of a small Gouy phase, this leads to

$$\Delta\phi_{\mathrm{Gouy}}\left(x'\right) \approx \frac{2x'\left(\left(2x - F\right)F + 2Z_{\mathrm{E}}\left(F - x_{\mathrm{s}}\right)\right) - F\left(\left(3x - Z_{\mathrm{E}} - F\right)F + 4Z_{\mathrm{E}}\left(F - x_{\mathrm{s}}\right)\right)}{F^{4}}\Delta. \tag{12.125}$$

In an optimized system, this expression simplifies, at the receiver location ($x' = x_{\mathrm{Ro}}$), into

$$\tan\phi_{\mathrm{Gouy}}\left(x_{\mathrm{Ro}}\right) \approx \phi_{\mathrm{Gouy}}\left(x_{\mathrm{Ro}}\right) \approx 2\left(\frac{Z_{\mathrm{E}}}{F^{2}}\right)^{2}\Delta^{2}. \tag{12.126}$$

At the same receiver position, the waist is

$$w^{2}\left(x_{\mathrm{Ro}}\right) = \frac{\lambda}{\pi Z_{\mathrm{E}}}\frac{A^{2}Z_{\mathrm{E}}^{2} + B^{2}}{\left(AD - BC\right)} = w_{\mathrm{E}}^{2}\frac{\left(Z_{\mathrm{E}}\Delta\right)^{2} + \left(F^{2} + Z_{\mathrm{E}}\Delta\right)^{2}}{F^{4}}. \tag{12.127}$$

In the limit of small Δ, the waist tends to

$$w^{2}\left(x_{\mathrm{R}}\right) \approx w_{\mathrm{E}}^{2} + w_{\mathrm{E}}^{2}\frac{2Z_{\mathrm{E}}}{F^{2}}\Delta. \tag{12.128}$$

Let us also notice that at the center $x_{\mathrm{s}}/2$ of the TDS setup, the beam parameters are

$$\frac{x_{\mathrm{s}}}{2} = F\frac{F + 2Z_{\mathrm{E}}}{2Z_{\mathrm{E}}}, \quad w_{\mathrm{s}} = \frac{1}{\sqrt{2}}\frac{F}{Z_{\mathrm{E}}}w_{\mathrm{E}}, \quad Z_{\mathrm{s}} = \frac{F^{2}}{2Z_{\mathrm{E}}} = \frac{\lambda}{2\pi}\frac{F^{2}}{w_{\mathrm{E}}^{2}}. \tag{12.129}$$

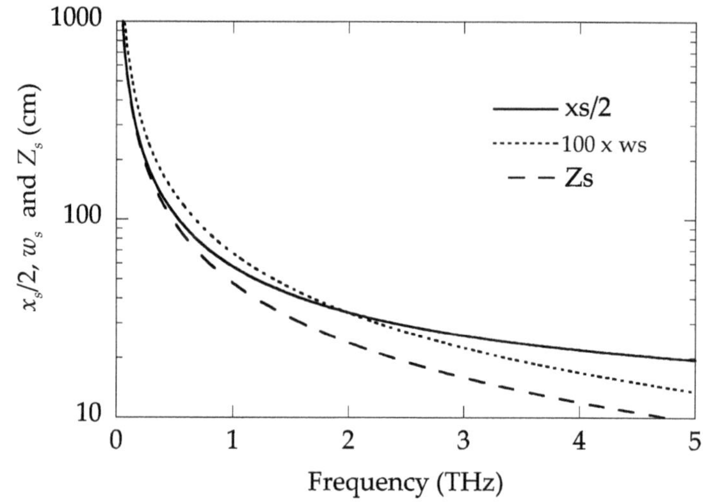

Figure 12.44 $x_s/2$, w_s and Z_s for a classical THz-TDS system with $F = 100$ mm and $Z_E = 1$ mm (the w_s values are multiplied by 100).

Typically, with both photo-conducting [23, 25, 29] and electro-optic emitting antennas [24], $w_E \approx 1 \sim 3$ mm, while $F \approx 100$ mm for common setups. Figure 12.44 presents $x_s/2$, w_s and Z_s versus frequency for a setup with an $F = 100$ mm optics and $w_E = 1$ mm. The curves decrease roughly inversely with frequency, i.e., they vary linearly with the THz wavelength. With the parameters of Fig. 12.43, one gets $x_s/2 \approx Z_s \approx 1600\,\lambda$ and $w_s \approx 22.5\,\lambda$. At 1 THz, $x_s/2 \approx Z_s \approx 500$ mm, $w_s \approx 6.75$ mm, while $x_s/2 \approx Z_s \approx 1600$ mm and $w_s \approx 22.5$ mm at 300 GHz.

Also, one should take care that the size of the studied samples is larger than the beam diameter at the waist. Otherwise, recorded signals will be perturbed by diffraction effects occurring at the edge of the sample. Moreover, because the higher frequencies are concentrated at the center of the beam while lower frequencies spread over longer distances from the beam axis, different regions of the sample will be characterized at lower and higher frequencies, which could lead to artefacts if the sample is not spatially homogeneous.

12.5.4 Errors Induced by a Plane-Wave Analysis Instead of a Gaussian One in an Optimized THz-TDS System

As already explained in the case of a simplified model (see relations 12.109 and 12.110), when performing the analysis of the recorded data with a plane wave modeling, two errors will affect the extracted parameters of a sample:

- The Gouy shift that does not exist for plane waves (the Gouy shift is related to the wave vector component perpendicular to the direction of propagation of the Gaussian beam) modifies the phase of the recorded signal: it will mostly alter the value of the refractive index.
- The geometrical shift of the focus will change the amount of THz light collected by the receiver, and thus it will lead mostly to an erroneous value of the coefficient of absorption of the tested material.

Let us first treat the Gouy shift. We extrapolate the result obtained in the simplified model case (Eq. 12.109) in which we input the expression of x_{Ro}:

$$\Delta n = -\frac{c}{2\pi f d}\Delta\phi_{\mathrm{Gouy}}\left(x_{\mathrm{Ro}}\right) \approx -\frac{c}{\pi f d}\left(\frac{Z_{\mathrm{E}}}{F^2}\right)^2 \Delta^2$$

$$\approx -\left(\frac{w_{\mathrm{E}}^2}{F^2}\frac{n-1}{n}\right)^2 \frac{\pi}{\lambda} d. \quad (12.130)$$

When inserting parameters of common THz-TDS setups ($w_{\mathrm{E}} \approx 1$ mm, $F = 100$ mm), relation 12.130 become

$$\Delta n \approx -\left(\frac{n-1}{n}\right)^2 3\times 10^{-7} d_{\mathrm{mm}} f_{\mathrm{THz}}.$$

Contrary to the simplified model, the optimization of the THz-TDS setup makes negligible the influence of the Gaussian beam on the extraction of n. Regarding the coefficient of absorption, we proceed as we did to get (Eq. 12.110)

$$\Delta\alpha = -\frac{1}{2d}\ln\frac{w^2\left(x_{\mathrm{Ro}}\right)}{w'^2\left(x_{\mathrm{Ro}}\right)}$$

$$= \frac{1}{2d}\ln\left(1 + 2\frac{Z_{\mathrm{E}}\Delta}{F^2} + \left(\frac{Z_{\mathrm{E}}\Delta}{F^2}\right)^2\right) \approx \frac{\pi}{\lambda}\left(\frac{w_{\mathrm{E}}}{F}\right)^2\left(\frac{n-1}{n}\right).$$

Here again, if the setup is ideally optimized, $\Delta\alpha$ is well below the sensitivity of the system.

12.6 Errors Related to the Pulse Train

Until now, the THz pulse emission and detection are supposed to be perfectly periodic and synchronous. Two phenomena may disrupt this hypothesis: the time delay between laser pulses in emission and detection is not constant, or its value is wrong due to some bad calibration; because of vibrations or thermal effects along the optical paths, the time delay exhibits a random noise.

12.6.1 Errors due to the Delay Line

12.6.1.1 Linear drift

The position of each stage $x_n = n\,\Delta x$ (n is an integer) of the delay line is shifted by an amount δx_n proportional to the stage position ($\delta x_n = a\,x_n = n\,a\Delta x$). Thus, the actual equivalent-time value is shifted by $\delta t_n = 2\Delta x_n/c = 2a\,x_n/c = a\,n\,T$ (T is the period in the equivalent-time scale). It follows that

$$S_{\text{rec,shift}}\left(t\right) = \text{III}\left(t\right) \otimes S\left(t\right) \quad \text{with} \quad \text{III}\left(t\right) = \sum_{n=-\infty}^{+\infty} \delta\left(t - n\left(1+a\right)T\right). \tag{12.131}$$

Consequently, the spectrum of this signal is

$$S_{\text{rec,shift}}(\omega) = \text{III}(\omega) S(\omega), \tag{12.132}$$

with $\text{III}(\omega) = \text{TF}\left[\text{III}\left(t\right)\right] = \frac{2\pi}{T(1+a)} \sum_{n=-\infty}^{+\infty} \delta\left(\omega - n\frac{2\pi}{T(1+a)}\right)$. The frequencies of the spectrum comb are also shifted:

$$\omega_{\text{shift},n} = n\frac{2\pi}{T\left(1+a\right)} = \omega_n\frac{1}{1+a} = \omega_n - \omega_n\frac{a}{1+a} \tag{12.133}$$

The THz spectrum is only linearly stretched. Typically, with commercial translation stages that are not worn, $\delta x/\Delta x = a < 10^{-3}$, $\Delta f/f \approx a < 10^{-3}$.

12.6.1.2 Errors induced by delay-line random vibrations

According to Jahn et al. [30], a random positioning error of any step at the delay line induces an SNR in the frequency domain equal to (see relation 11.96)

$$\mathrm{SNR}(\omega) = \frac{S^2(\omega)}{\sigma^2(\omega)} = 20\log\left(C_{\mathrm{DL}}\sigma_\tau\sqrt{\Delta\tau}\right) \equiv \frac{2}{A'}. \qquad (12.134)$$

$\Delta\tau$ is the corresponding time-equivalent average error and σ_τ its standard deviation, while C_{DL} is a proportionality constant. From Eq. 12.134, it is easy to obtain SNR for transmission:

$$\mathrm{SNR}\left(t(\omega)\right) = \frac{t^2(\omega)}{\sigma_\mathrm{t}^2(\omega)} = \frac{1}{A'} \quad \Rightarrow \quad \sigma_\mathrm{t}^2(\omega) = A't^2(\omega). \qquad (12.135)$$

It follows that the delay line noise contributes to the A coefficient of the ABC noise coefficients defined by (12.11). Jahn et al. [30] report an experimental value of $\mathrm{SNR}(\omega) = 2/A'$ of about 70 dB, i.e., $A' \approx 2 \times 10^{-6}$. This value is about 10 times and 1000 times smaller than the A coefficients of the Advantest and homemade systems, respectively (Fig. 12.2). For the latter, the delay line noise is negligible. For high dynamic systems like the Advantest one, the delay line is a major contributor to the noise through the coefficient A.

12.7 Comparison of the Sources of Error

The determination of the refractive index $\tilde{n}(\omega) = n(\omega) - j\kappa(\omega)$ from THz-TDS measurements is impacted by different error sources, the main ones being noise, sample position in the setup, and sample parameters. Some of them, like noise, depend on the refractive index. Others depend on the imprecision on some experimental parameters, like an angular tilt of the sample or poor system alignment. Also, the frequency plays a major role for example in the noise or focusing contribution. Therefore, the number of implied parameters is large, which makes almost impossible a general study valid for any sample. Hence, we will address here some common cases, either in reflection or transmission THz-TDS.

12.7.1 Reflection THz-TDS

12.7.1.1 Absorbing dielectric sample

We choose as a study case the characterization of a material with $n = 2$ and $\kappa = 0.1$ ($\alpha = 40$ cm^{-1} at 1 THz). Such a sample can be precisely characterized in reflection because it is quite absorbing. Figure 12.45 shows the calculated relative errors $\Delta n/n$ and $\Delta\kappa/\kappa$ versus frequency due to 1) an angular tilt (deviation of the reflected beam and change of the Fresnel coefficient) of 1°, 2) a shift (10 μm) of the reference mirror, and 3) the noise (we take the *ABC* values of the Advantest system). The deviation of the reflected beam is by far the main source of error. Here, with $\delta\theta = 0.1^\circ$ only, the errors are bigger than 100%, except at very low frequencies. Great care has to be taken when installing and aligning the sample and the reference mirror in the THz-TDS setup: to that purpose, the reflected detected signals must be maximized.

The second contribution of error in reflection THz-TDS is the bad position of the sample as compared to the reference mirror one. Typically, a 10 μm shift induces errors, on either n or κ, of a few tens of percent. The imprecision due to noise is about $10 \sim 100$ times smaller. Finally, the change of the Fresnel coefficient due to an angular tilt is negligible.

12.7.1.2 Conducting sample

Conducting materials, like metals or highly doped semiconductors, exhibit large and almost equal values of n and κ. This originates in their very big plasma frequency, compared to the THz frequencies. Here we choose $n = -40$ and $\kappa = 39$, i.e., a plasma frequency in the VIS-NIR. The corresponding coefficient of reflection r is bigger than 97%. Figure 12.46 depicts the calculated relative errors $\Delta n/n$ and $\Delta\kappa/\kappa$.

$\Delta n/n$ resembles the curves calculated for the dielectric sample (Fig. 12.25 left) but the noise contribution is weaker, thanks to a larger coefficient of reflection r which makes the detected signal stronger. The error $\Delta\kappa/\kappa$ due to the deflection of the reflected THz beam is colossal. Here again, this originates in the high

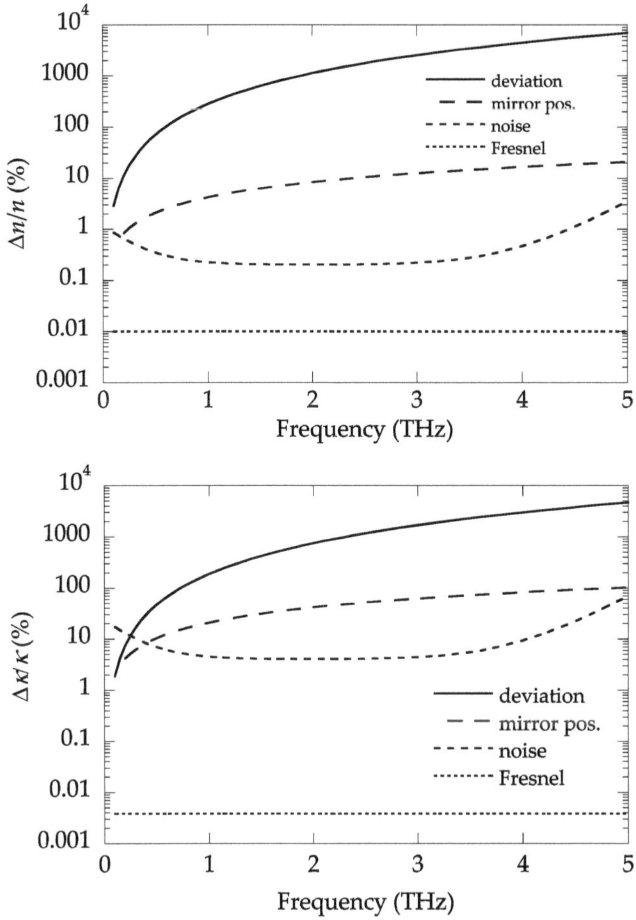

Figure 12.45 Relative errors $\Delta n/n$ (top) and $\Delta\kappa/\kappa$ (bottom) versus frequency induced by different error sources in reflection THz-TDS. The sample is a highly absorbing dielectric material (see text).

value of r. This makes a precise characterization of conducting material impossible if the sample and mirror are not perfectly aligned in the THz beam. Other sources of errors contribute at a much lower level, leading to a rather accurate determination of κ ($\Delta\kappa/\kappa < 1\%$).

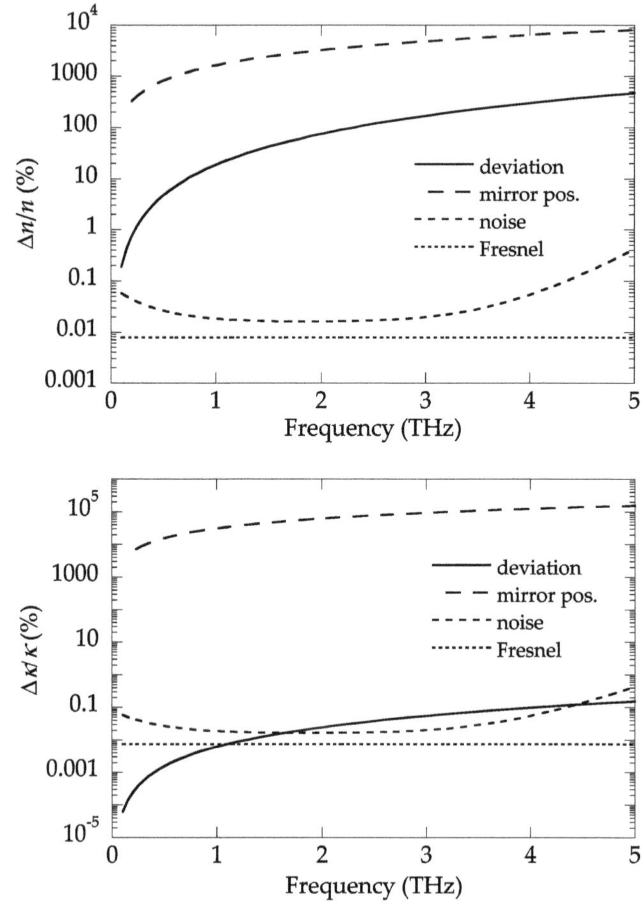

Figure 12.46 Relative errors $\Delta n/n$ (top) and $\Delta\kappa/\kappa$ (bottom) versus frequency induced by different error sources in reflection THz-TDS. The sample is a conducting material (see text).

12.7.2 Transmission THz-TDS

We treat three common cases:

(1) A wafer of moderate thickness (0.5 mm) and moderate absorption ($n = 2, \kappa = 3 \times 10^{-2}$, i.e., $\alpha = 6.3$ cm^{-1} at 1 THz).
(2) A rather thick (10 mm) transparent dielectric sample ($n = 2$, $\kappa = 10^{-3}$, i.e., $\alpha = 0.21$ cm^{-1} at 1 THz). Because absorption in

the sample is extremely weak, we already guess than absorption will be poorly determined.

(3) A very thin (10 nm) conducting layer with $n = -39$ and $\kappa = 40$. For a Drude-like conductor, these values correspond to a plasma frequency in the VIS-NIR, and to a collision time of several tens of femtoseconds. This could be an evaporated metallic layer.

12.7.2.1 Wafer of moderate thickness and moderate absorption

Here, the parameters of the study are: thickness 0.5 mm, refractive index $n = 2$ and $\kappa = 3 \times 10^{-2}$, error on thickness 10 μm, and angular tilt 1°. The noise is calculated for the Advantest system. For the Gouy shift calculation, the setup is supposed to be optimized, the waist at the emitter is 1 mm and the focal length of the optics is 100 mm. Figure 12.47 shows the different contributions to the relative errors $\Delta n/n$ (top) and $\Delta\kappa/\kappa$ (bottom) versus frequency, which are induced by an angular tilt, an error on the sample thickness, the Gouy shift, and the noise.

The main source of error on n is the noise, which limits the precision to a few percent. The error on the sample thickness ($\Delta d =$ 10 μm, i.e., $\Delta d/d = 2\%$) leads to an imprecision of about 1%. Gouy shift and angular tilt do not contribute effectively to the error. The determination of κ is mostly made imprecise by the error on thickness. Noise, in the frequency range of good SNR, and Gouy shift contribute 100 times less. The angular shift does not effectively affect the extraction of κ.

12.7.2.2 Thick transparent sample

The sample parameters of the study are: thickness 10 mm, refractive index $n = 2$ and $\kappa = 10^{-3}$, i.e., $\alpha = 0.21\ \text{cm}^{-1}$ at 1 THz and error on thickness 10 μm. All the other parameters are the same as in the preceding case. Figure 12.48 shows the different contributions to the relative errors $\Delta n/n$ (top) and $\Delta\kappa/\kappa$ (bottom). The results on $\Delta n/n$ behave similarly to the preceding case (Fig. 12.46). Here again, the main source of error is the noise. However, because the

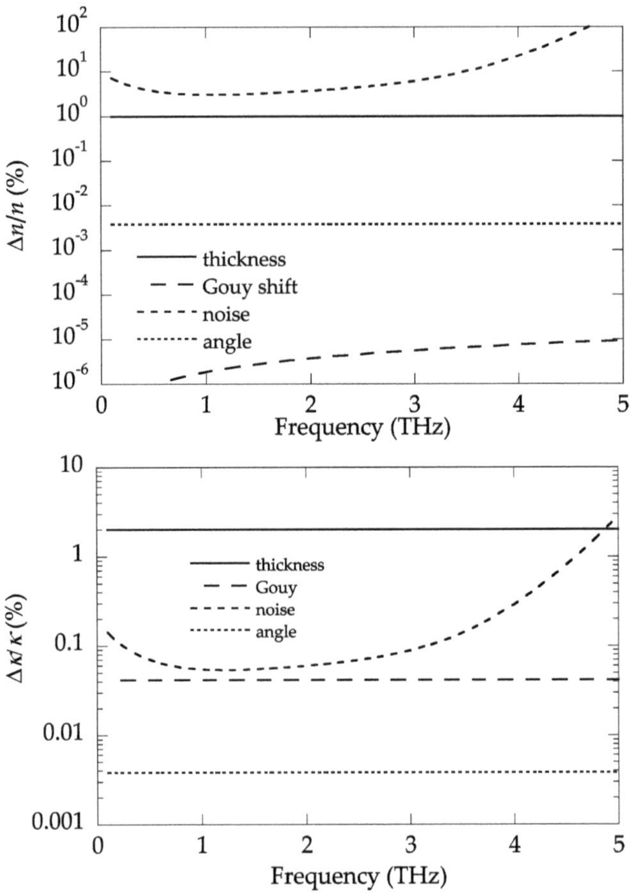

Figure 12.47 Relative errors $\Delta n/n$ (top) and $\Delta\kappa/\kappa$ (bottom) versus frequency induced by different error sources in transmission THz-TDS for a common sample of moderate thickness and absorption.

thickness is larger and the error on thickness is smaller, the Gouy shift and the thickness error contributions respectively increases and decreases. Regarding $\Delta\kappa/\kappa$, the Gouy shift is the main error term because the sample is thick and the material absorption is weak. Noise and thickness error contribute for about ten times less. The angular tilt of the sample is negligible.

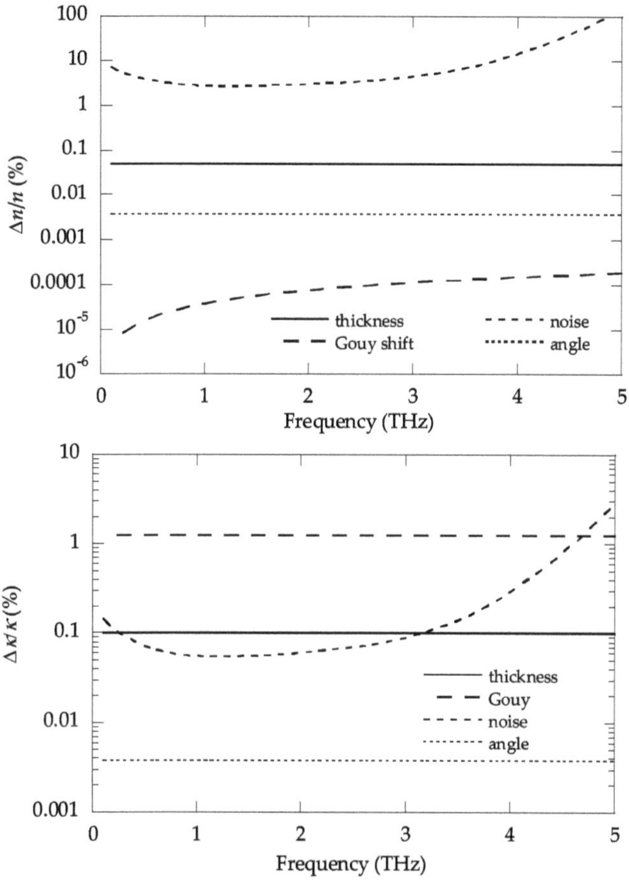

Figure 12.48 Relative errors $\Delta n/n$ (top) and $\Delta\kappa/\kappa$ (bottom) versus frequency induced by different error sources in transmission THz-TDS for a thick transparent sample.

12.7.2.3 Very thin conducting layer

The conducting layer is 10 nm thick, and the thickness uncertainty is 1 nm. Its refractive index is $n = -39$ and $\kappa = 40$, i.e., $\varepsilon \approx -80 - j\,3200$. Here the determination of n is more imprecise (typically 10%) than in the preceding cases of transparent samples. Noise is the main source of uncertainty. Due to the nanometer

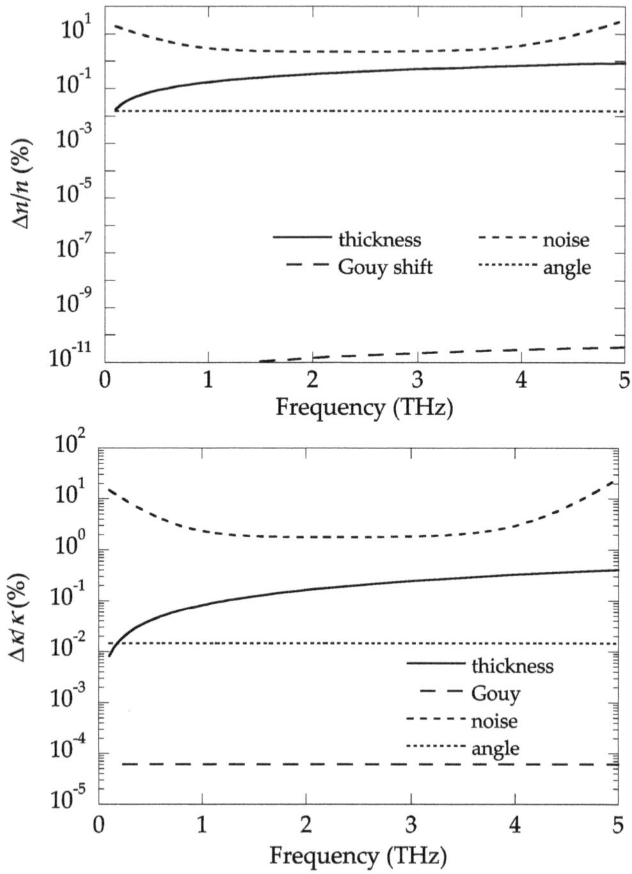

Figure 12.49 Relative errors $\Delta n/n$ (top) and $\Delta\kappa/\kappa$ (bottom) versus frequency induced by different error sources in transmission THz-TDS for a nanometer thick conducting layer.

thickness, the Gouy shift is negligible. κ is slightly more precisely extracted than n. Again, the Gouy shift does not contribute.

12.7.2.4 Experimental results

The contribution of the different sources of uncertainty, in transmission THz-TDS, has been clearly addressed by Withayachumnankul and co-workers [3]. They performed both an analysis close to one done in this chapter, and they validated it with an experimental study.

In their theoretical study, these authors [3] treated the same error sources like the ones explained in the above paragraphs, and they obtained similar results (e.g., relations 15 in [3] are comparable with Eq. 12.63). A few differences appear: they also studied the influence of (i) a random variation of the sample thickness within the sample area illuminated by the THz beam (its effect is extremely weak), (ii) an approximated transfer function (here we made use of a rigorous extraction without using approached formulas), and (iii) approached values for the physical constants, like the refractive index of air (the impact of these approximations are again very weak). They took into account the effect of the rebounds in the sample. Also, they considered that the sample thickness is measured several times and deduced the thickness uncertainty as the standard deviation of the measured values. Here, we took thickness uncertainty to be the precision of the instrument (Talystep, etc.). Finally, they did not take into account Gaussian beam effects and changes in the Fresnel coefficients under a weak oblique incidence.

The sample characterized by Withayachumnankul et al. [3] is a pellet of α-lactose powder diluted in a HDPE powder. This mixture exhibits a constant refractive index ($n \approx 1.46$) over the whole studied THz range. The thickness of the pellet is 1.85 mm $\pm$ 1 μm.

Figure 12.50 shows the relative uncertainties $\Delta n/n$ (top) and $\Delta\kappa/\kappa$ (bottom). In the spectral range where the THz signal is strong ($0.2 \sim 2.2$ THz), the main source of error for $\Delta n/n$ is the uncertainty on the sample thickness (here $\Delta d/d \approx 0.05\%$) and then the angular tilt (here $\delta\theta = \pm\ 2^{\circ}$) of the sample. Noise contributes a few times less but becomes dominant at smaller and higher frequencies. The sample roughness (random thickness uncertainty, labeled "delta" in Fig. 12.50) has a very small impact on $\Delta n/n$. We retrieve same tendencies as in Fig. 12.47, but here the noise contribution is weaker. The thickness uncertainty ($\Delta d/d \approx 0.05\%$) contributes more than in Fig. 12.50, simply because this plot has been computed with $\Delta d/d \approx 2\%$. Let us notice the very good accuracy of the determination of $n(\omega)$ since the total uncertainty is extremely weak ($\Delta n/n \approx 0.1\%$). Regarding $\Delta\kappa/\kappa$ (Fig. 12.50, bottom), the main

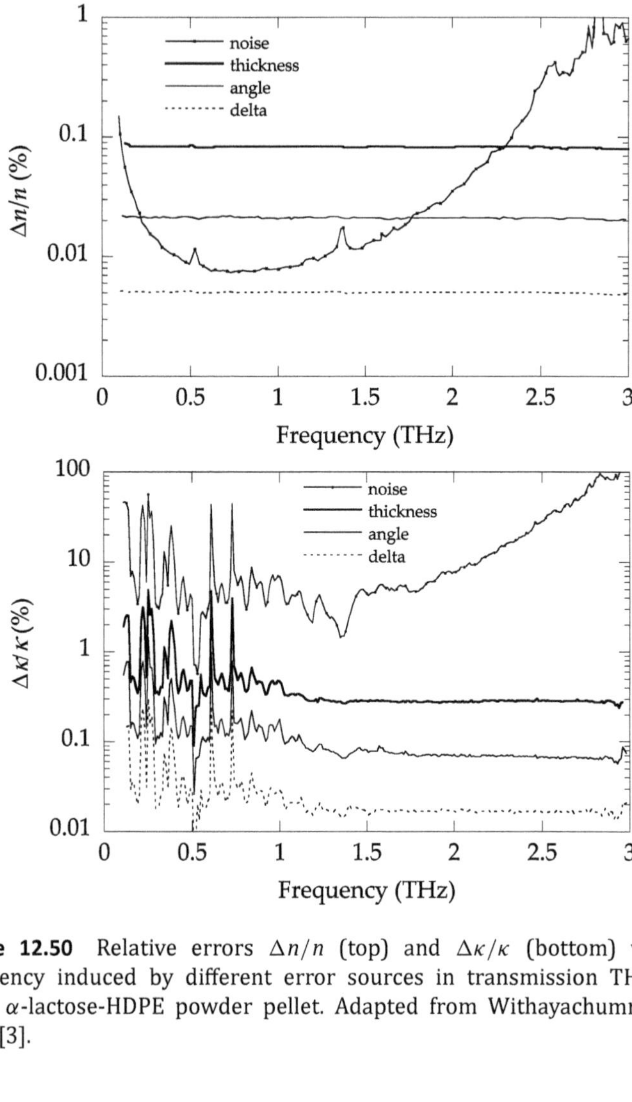

Figure 12.50 Relative errors $\Delta n/n$ (top) and $\Delta\kappa/\kappa$ (bottom) versus frequency induced by different error sources in transmission THz-TDS for a α-lactose-HDPE powder pellet. Adapted from Withayachumnankul et al. [3].

source of uncertainty is the noise. It limits the precision to about 10% in the central spectral range (0.2 $\sim$ 2.2 THz), and to higher values outside this range. Thickness uncertainty and angular tilt contributions are 10 $\sim$ 100 times weaker. Sample roughness effect is negligible. Compared to Fig. 12.50, again same tendencies are

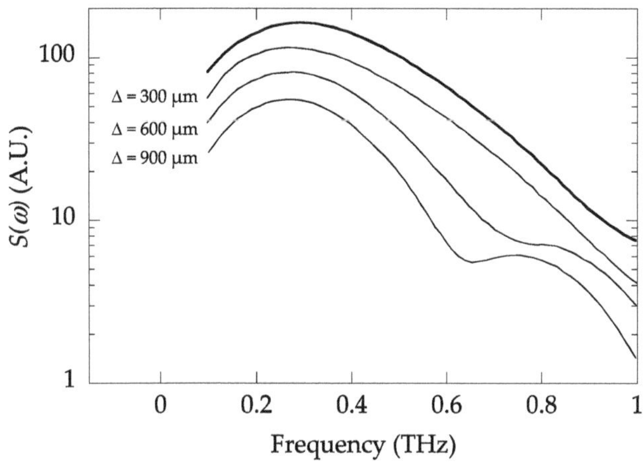

Figure 12.51 Spectra delivered by the same THz-TDS setup, showing the effect of a weak misalignment (lower plots) of the silicon dome attached to the emitting photo-conducting antenna. Adapted from Duvillaret et al. [31].

observed, but with a weaker contribution of the thickness error, because of the different values of $\Delta d/d$.

12.8 Conclusion

THz-TDS is a very accurate technique to determine the complex refractive index of materials when employed in optimal conditions. In this latter case, the precision on the real part $n(\omega)$ of the refractive index is of the order of a few percent, while it is a few times bigger for the imaginary part $\kappa(\omega)$.

When the sample is highly absorbing, reflection THz-TDS is preferable. In a reflection scheme, $n(\omega)$ is mostly determined from the modulus of the reflection coefficient, while $\kappa(\omega)$ is mostly obtained from the phase of the reflection coefficient. The relative positions of the sample and of a requested reference mirror are by far the main source of error in reflection THz-TDS. Moreover, the method is not suitable for characterizing rather transparent materials.

For transparent samples, transmission THz-TDS is more precise. Contrary to the reflection scheme, here $n(\omega)$ is mostly calculated from the phase of the transmission coefficient, while $\kappa(\omega)$ is mostly extracted from the modulus. The uncertainty on the actual sample thickness and the noise are the main causes of imprecision.

However, in this chapter, we did not address the main cause of error in either transmission or reflection THz-TDS. Hence, a setup that is not well aligned or a laser that is noisy leads to poor quality data. For example, if the laser beam does not arrive right at the center of photo-conducting THz antennas, the detected THz spectrum could exhibit unusual features, like dips, which could be interpreted to be interference like phenomena. These artefacts are clearly observed in Fig. 12.51, adapted from [31], where the silicon dome attached to the emitting antenna has been shifted by 300 μm from its optimized position.

References

1. Duvillaret, L., Garet, F., and Coutaz, J.-L. (2000). Influence of noise on the characterization of materials by terahertz time-domain spectroscopy, *J. Opt. Soc. Am. B*, **17**, p. 452.
2. Duvillaret, L., Garet, F., and Coutaz, J.-L. (1999). Highly precise determination of optical constants and sample thickness in terahertz time-domain spectroscopy, *Appl. Opt.*, **38**, p. 409.
3. Withayachumnankul, W., Fischer, B. M., Lin, H., and Abbott, D. (2008). Uncertainty in terahertz time-domain spectroscopy measurement, *J. Opt. Soc. Am. B*, **25**, pp. 1059–1072.
4. Bernier, M., Garet, F., Kato, E., Blampey, B., and Coutaz, J.-L. (2018). Comparative study of material parameter extraction using terahertz time-domain spectroscopy in transmission and in reflection, *J. Infrared Millimeter THz Waves*, **39**, pp. 349–366.
5. Bernier, M., Garet, F., and Coutaz, J.-L. (2017). Determining the complex refractive index of materials in the far-infrared from terahertz time-domain data, in *Terahertz Spectroscopy - Cutting Edge Technology*, edited by J. Uddin, chapter 7, pp. 119–141, IntechOpen Rijeka (Croatia).
6. Uhd Jepsen, P., and Fischer, B. M. (2005). Dynamic range in terahertz time-domain transmission and reflection spectroscopy, *Opt. Lett.*, **30**, pp. 29–31.
7. Withayachumnankul, W., Fischer, B. M., and Abbott, D. (2008). Material thickness optimization for transmission-mode terahertz time-domain spectroscopy, *Opt. Express*, **16**, pp. 7382–7396.

8. Pashkin, A., Kempa, M., Nemec, H., and Kuzel, P. (2003). Phase-sensitive time-domain terahertz reflection spectroscopy, *Rev. Sci. Instrum.*, **74**, pp. 4711–4717.
9. Khazan, M., Meissner, R., and Wilke, I. (2001). Convertible transmission-reflection time-domain terahertz spectrometer, *Rev. Sci. Instrum.*, **72**, pp. 3427–3430.
10. Scheller, M. (2014). Data extraction from terahertz time-domain spectroscopy measurements, *J. Infrared Millimeter THz Waves*, **35**, pp. 638–648.
11. Dorney, T. D., Baraniuk, R. G., and Mittleman, D. M. (2001). Material parameter estimation with terahertz time-domain spectroscopy, *J. Opt. Soc. Am. A*, **18**, pp. 1562–1571.
12. Li, Z., Zhang, Z., Zhao, X., Su, H., Yan, F., and Zhang, H. (2012). Method for improving terahertz band absorption spectrum measurement accuracy using noncontact sample thickness measurement, *Appl. Opt.*, **51**, pp. 4879–4884.
13. Pupeza, I., Wilk, R., and Koch, M. (2007). Highly accurate optical material parameter determination with THz time-domain spectroscopy, *Opt. Express*, **15**, pp. 4335–4350.
14. van Mechelen, J. L. M., Kuzmenko, A. B., and Merbold, H. (2014). Stratified dispersive model for material characterization using terahertz time-domain spectroscopy, *Opt. Lett.*, **39**, pp. 3853–3856.
15. Li, X., Hong, Z., He, J., and Chen, Y. (2010). Precisely optical material parameter determination by time domain waveform rebuilding with THz time-domain spectroscopy, *Opt. Commun.*, **283**, pp. 4701–4706.
16. Kruger, M., Funkner, S., Brundermann, E., and Havenith, M. (2011). Uncertainty and ambiguity in terahertz parameter extraction and data analysis, *J. Infrared Millimeter THz Waves*, **32**, pp. 699–715.
17. Su, K., Shen, Y.-C., and Zeitler, A. (2014). Terahertz sensor for non-contact thickness and quality measurement of automobile paints of various complexity, *IEEE Trans. THz Sci. Technol.*, **4**, pp. 432–439.
18. Mittleman, D., Hunsche, S., Boivin, L., and Nuss, M. (1997). T-ray tomography, *Opt. Lett.*, **22**, pp. 904–906.
19. Yasui, T., Yasuda, T., Sawakana, K., and Araki, T. (2005). Terahertz paintmeter for noncontact monitoring of thickness and drying process in paint film, *Appl. Opt.*, **44**, p. 6849.
20. Theuer, M., Beigang, R., and Grischkowsky, D. (2010). Highly sensitive terahertz measurement of layer thickness using a two-cylinder waveguide sensor, *Appl. Phys. Lett.*, **97**, p. 071106.

21. Mousavi, P., Haran, F., Jez, D., Santosa, F., and Dodge, J. S. (2009). Simultaneous composition and thickness measurement of paper using terahertz time-domain spectroscopy, *Appl. Opt.*, **48**, pp. 6541–6546.

22. Jonuscheit, J. (2016). Terahertz waves for thickness analyses, *Opt. Photonik*, **11**, pp. 30–33.

23. Uhd Jepsen, P., Jacobsen, R. H., and Keiding, S. R. (1996). Generation and detection of terahertz pulses from biased semiconductors, *J. Opt. Soc. Am. B*, **13**, pp. 2424–2436.

24. Hattori, T., Rungsawang, R., Ohta, K., and Tukamoto, K. (2002). Gaussian beam analysis of temporal waveform of focused terahertz pulses, *Jpn. J. Appl. Phys.*, **41**, pp. 5198–5204.

25. Reiten, M. T., Harmon, S. A., and Cheville, R. A. (2003). Terahertz beam propagation measured through three-dimensional amplitude profile determination, *J. Opt. Soc. Am. B*, **20**, pp. 2215–2225.

26. Siemion, Ag., Siemion, A., Makowski, M., Sypek, M., Hérault, E., Garet, F., and Coutaz, J.-L. (2011). Off-axis metallic diffractive lens for terahertz beams, *Opt. Lett.*, **36**, pp. 1960–1963.

27. Pearson, J. E., McGill, T. C., Kurtin, S., and Yariv, A. (1969). Diffraction of Gaussian laser beams by a semi-infinite plane, *J. Opt. Soc. Am.*, **59**, pp. 1440–1445.

28. Saleh, B. E. A., and Teich M. C. (1991). *Fundamentals of Photonics* (John Wiley & Sons, Inc., New York), Chapter 3, p. 80.

29. Kužel, P., Němec, H., Kadlec, F., and Kadlec C. (2010). Gouy shift correction for highly accurate refractive index retrieval in time-domain terahertz spectroscopy, *Opt. Express*, **18**, p. 15338.

30. Jahn, D., Lippert, S., Bisi, M., Oberto, L., Balzer, J. C., and Koch, M. (2016). On the influence of delay line uncertainty in THz time-domain spectroscopy, *J. Infrared Millimeter THz Waves*, **37**, pp. 605–613.

31. Duvillaret, L., Garet, F., and Coutaz, J.-L. (1999). Evidence of frequency dependent THz beam polarization in time-domain spectroscopy, in *THz spectroscopy and applications*, Sherwin, M. S., ed. *SPIE Proc.*, **3617**, pp. 30–37.

Chapter 13

Advanced Techniques

13.1 Combining Transmission and Reflection THz-TDS to Characterize Samples Exhibiting Strong Absorption Bands

13.1.1 Introduction

Some materials exhibit strong absorption peaks due to the excitation of molecular resonances, for example, rotational-vibrations in gases, or collective resonances in the solid state (phonons, surface plasmons, etc.). Also, artificial materials like photonic crystals and metamaterials display forbidden energy bandgaps, in which they are opaque. At these peaks or bandgaps, the transmission of the sample is weak and could be even below the noise level (saturated peaks or bandgaps). This means that the amplitude of the transmitted signal is almost zero and that its phase is lost. Therefore, for frequencies higher than the one of the first saturated region, extraction of the sample parameters cannot be precisely performed from transmission THz-TDS.

This is illustrated by the scheme in Fig. 13.1. The upper left plot shows a coefficient of transmission that is almost zero over part of the frequency range plotted. The corresponding phase (upper right

Principles of Terahertz Time-Domain Spectroscopy
Jean-Louis Coutaz, Frédéric Garet, and Vincent Wallace

ISBN 978-981-4774-56-7 (Hardcover), 978-0-203-70969-6 (eBook)
www.panstanford.com

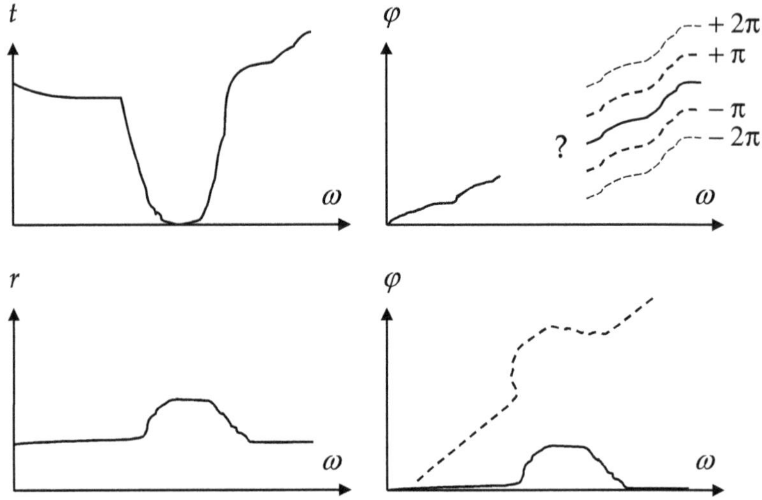

Figure 13.1 Upper plots: Typical transmission (modulus [left] and phase [right]) of a material showing a band of strong absorption. Lower plots: Same curves but for the reflection coefficient.

plot) is precisely known up to this forbidden range, where it cannot be determined. At higher frequency, the phase is known, but modulo π (remember that the phase is obtained by calculating the inverse tangent of the ratio of the imaginary and real parts of the Fourier-transformed waveform). A smart solution could be to perform reflection THz-TDS. However, the phase in reflection remains small, even in the regions of strong absorption, and the measurement could be perturbed by any bad positioning of the sample (dashed curve in the lower right plot of Fig. 13.1), as explained in Chapter 12.

A clever method to get rid of the loss of phase and the positioning error has been proposed by Bernier et al. [1]; it combines transmission and reflection THz-TDS data.

13.1.2 Principle of the Method

The basic idea behind the method is to perform a THz-TDS characterization of the sample both in transmission and in reflection.

Let us consider a sample which is transparent in the spectral regions 1 and 3 (Fig. 13.2a) and opaque in the intermediate region 2.

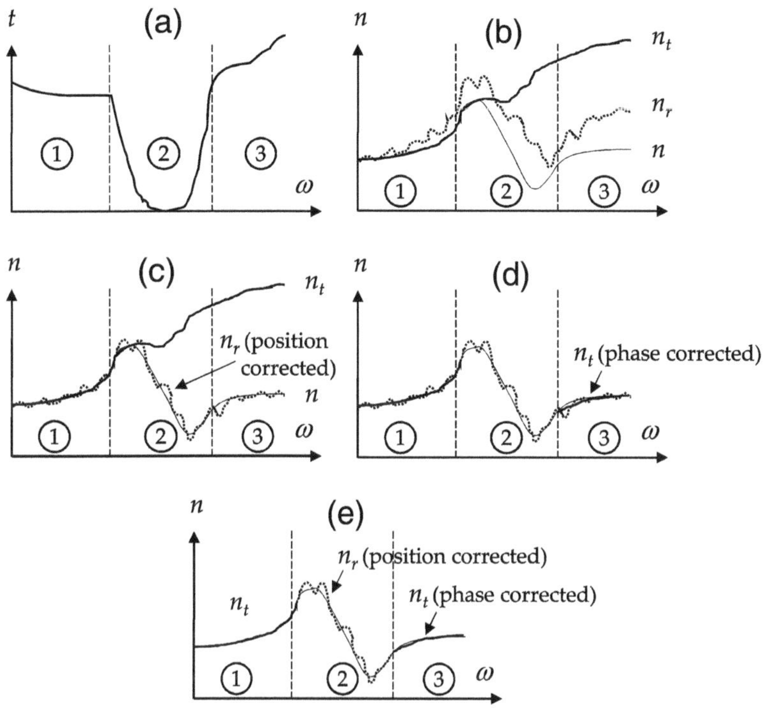

Figure 13.2 (a) Transmission of a sample showing an opaque spectral band; (b) refractive index n versus ω (thin curve: actual values; thick curve: transmission THs-TDS extraction; doted curve: reflection THs-TDS extraction); (c) with position corrected n_r; (d) with phase corrected n_t in region 3; (e) final result.

This sample is characterized both by transmission and reflection THz-TDS. A classical parameter extraction supplies both refractive index n and absorption coefficient α over the whole experimental frequency range. In region 1, because of transparency, the value $n_t(\omega)$ given by transmission THz-TDS is very close to the actual value $n(\omega)$ (see Fig. 13.2b), while, from the opaque spectral region, $n_t(\omega)$ differs from $n(\omega)$ because of the loss of phase in region 2. On the other hand, reflection THz-TDS delivers a noisier $n_r(\omega)$ curve, which may progressively differ with frequency from the actual value because of poor positioning of the sample. One corrects this frequency shift error by making $n_r(\omega)$ as close as possible as $n_t(\omega)$

in region 1: this is achieved by correcting the phase in reflection with the term $\omega\delta/c$, where δ is the error of the position of the reference mirror in reflection THz-TDS. Therefore, $n_r(\omega)$ is corrected over the whole frequency range (Fig. 13.2c), but with a weaker precision than transmission THz-TDS. In the second transparency region 3, one compares $n_t(\omega)$ and the corrected $n_r(\omega)$ curve. The difference arises from some unknown phase jump in the absorption region 2. This phase jump is corrected by making $n_t(\omega)$ as close as possible to $n_r(\omega)$. Therefore, the phase in transmission is now precisely known in region 3. A new extraction is performed with the phase corrected values, which leads to a precise estimation $n_t(\omega)$ of $n(\omega)$. Finally, one saves in region 1 the values extracted from a classical transmission THz-TDS measurement, one saves in region 3 the values extracted from the phase corrected transmission THz-TDS data, and in region 2, we keep the position-correction reflection data. This procedure is repeated as many times as the samples show opaque bands.

13.1.3 Example: Pellet of Maltose Powder

Figure 13.3 (top) shows the measured transmission spectrum (modulus and phase) of a pellet of maltose powder (the pellet was made by compressing the powder under 5 ton cm^{-2}). The pellet is 890 μm thick, which makes the transmitted signal below noise at the observed absorption peaks (~1.1, 1.6, 2.2 THz, etc.). Random phase jumps at the peaks are difficult to see, because the phase behavior is mostly dominated by the ω/c term.

Figure 13.3 (right) depicts the refractive index n_t and the absorption coefficient α_t versus frequency, as obtained using a common THz-TDS extraction code from the transmission data. The two first absorption peaks show a saturation, due to the lack of transmitted signal. At higher frequencies, noise leads to some overestimated α_t points. The refractive index curve exhibits an unexpected increase with frequency.

The phase of the reflected signal is plotted in Fig. 13.4 left (full circles). The measured value varies almost linearly in the range 0.5–1 THz, which is the signature of a bad positioning of the sample as regard to the reference mirror. The best linear fit is then subtracted to correct the positioning error (here the best fit gives a shift δ in

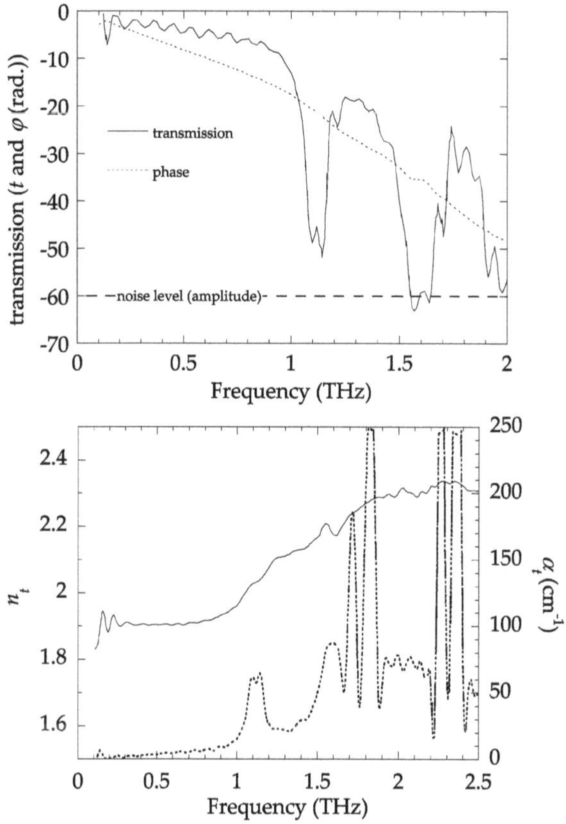

Figure 13.3 Top: Transmission spectrum of an 890 μm thick pellet of maltose powder. Bottom: $n_t(f)$ and $\alpha_t(f)$ extracted with a common THz-TDS code.

position equal to 5.1 μm, which corresponds to $\lambda/60$ at 1 THz). The so-corrected phase is plotted as open circles, and it is compared with the phase calculated from n_t and α_t. Between 0.5 and 1 THz, where transmission is weakly impacted by noise and by saturation of the peaks, the agreement is excellent, proving that the linear behavior of the measured phase is due to a bad positioning of the sample. The refractive index n_r extracted from reflection TDS data, with and without correction of the position, is plotted in Fig. 13.4 (bottom). The difference between corrected and uncorrected values increases with frequency, because of the $\delta\omega/c$ term. At low frequencies,

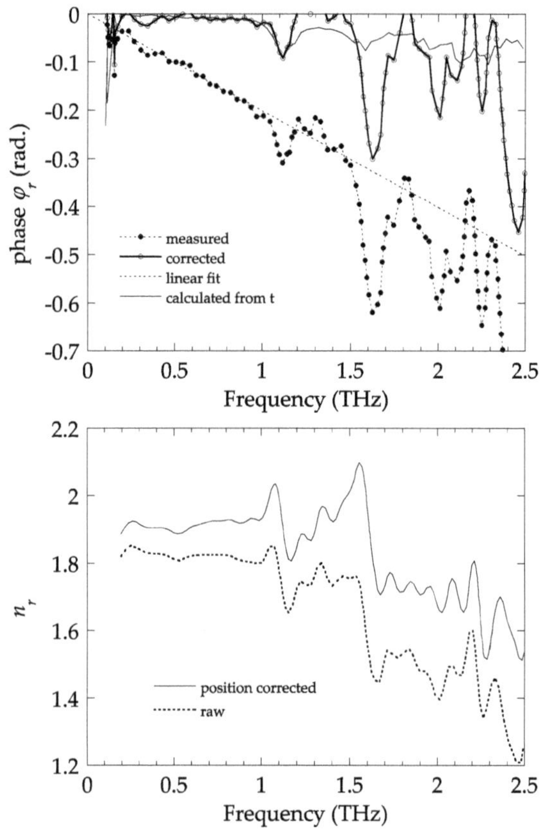

Figure 13.4 Top: Phase of the signal reflected by the maltose pellet. Bottom: Refractive index n_r deduced from the reflection data with and without the phase position error corrected.

the uncorrected refractive index is around 6% smaller than the corrected value, while this difference increases up to about 20% at 2 THz. Because of the error on n_r, the absorption coefficient α_r extracted from reflection measurements is also erroneous when the sample position effect is not corrected. This is observed in Fig. 13.5 top: the position shift adds an f^2-like contribution to α_r. This error is much bigger than α_r itself.

From the position-corrected n_r and α_r, the phase in transmission ϕ_t is calculated and compared with the value obtained from THz-

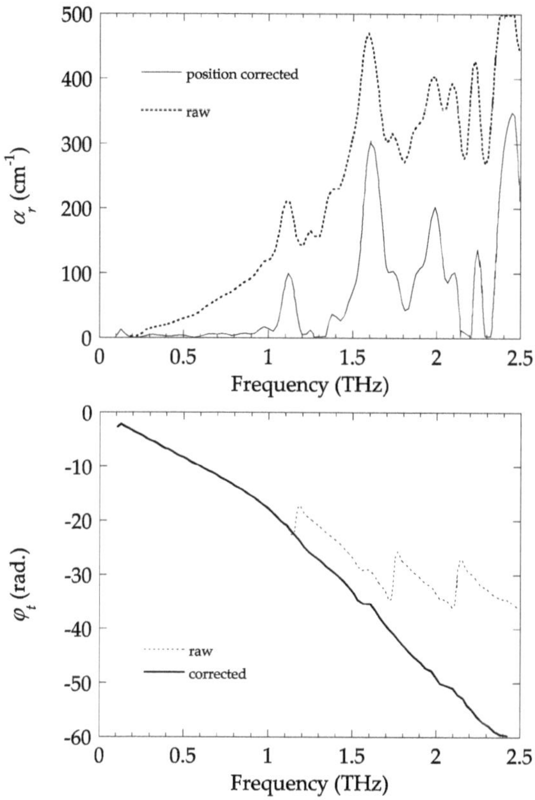

Figure 13.5 Top: Coefficient of absorption α_r deduced from the reflection data with and without the phase position error corrected. Bottom: Phase of the transmitted signal with and without phase jump correction.

TDS in transmission. Over the first saturated absorption peak, the difference is due to the random π-jump, which could be corrected. This is shown in Fig. 13.5 (bottom), where the phase jump correction is observed. At 1.1 THz, the phase correction is equal to 2π, while it is 3π at both 1.7 and 2.1 THz. Finally, the refractive index and the coefficient of absorption are calculated again from the phase-corrected transmission THz-TDS data, because transmission THz-TDS is more precise than reflection THz-TDS. This procedure gives n and α in the spectral regions of transparency. In the vicinity of the absorption peaks, transmission data cannot be exploited

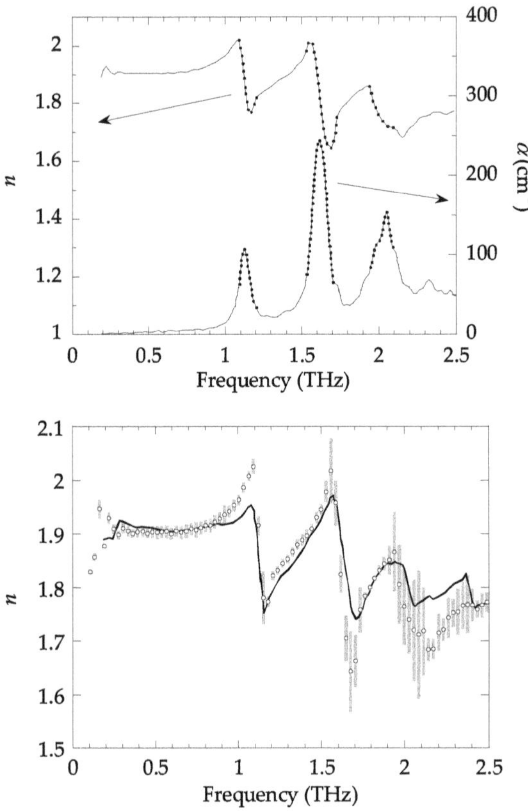

Figure 13.6 Top: Refractive index n and coefficient of absorption α versus frequency as extracted with the combined reflection/transmission method. The continuous curve is extracted from transmission THz-TDS, while the dotted ones are obtained from reflection data. Bottom: Refractive index n versus frequency extracted in transmission for a thin sample (325 μm) (continuous line) and for the thicker sample (890 μm) with the present reflection/transmission method (open circles, with error bars).

because the transmitted phase is unknown (one knows the value of the phase jumps, but we do not know at what frequency they happen). Thus reflection data must be saved. This is illustrated in Fig. 13.6, where the so-extracted n and α of the maltose pellet are plotted versus frequency. The continuous curves are obtained from phase-corrected transmission THz-TDS, while the dotted parts are extracted from position-corrected reflection THz-TDS.

To validate the method, n and α are determined from a thinner sample (325 μm thick), whose transmission does not exhibit any saturation: thus a classical precise extraction is performed using the transmission data. The two results, for the thin and thick samples, are plotted in Fig. 13.6 bottom. They are in very good agreement, especially if one takes into account the error bars.

13.2 Combining THz-TDS and Kramers–Kronig Analysis

13.2.1 Introduction

As explained in the Annex, the Kramers–Kronig (KK) transformation allows one to determine the real part of a complex function from the integration of its imaginary part (and vice versa). The complex function should be analytic in the upper half part of the complex plane, which means that, when time is the variable, the function obeys causality (the response of a material/device occurs after the excitation). When applied to electromagnetism, the KK relations permit to determine the refractive index n from the absorption coefficient α, and vice versa. The KK relations are derived in the Annex resulting in Eq. A.175, which is reproduced here for convenience:

$$n(\omega) = n_o + \frac{c}{\pi} PV \left(\int_0^{+\infty} \frac{\alpha(\omega')}{\omega'^2 - \omega^2} d\omega' \right), \qquad (13.1)$$

$$\alpha(\omega) = -\frac{4\omega^2}{\pi\, c} PV \left(\int_0^{+\infty} \frac{n(\omega') - 1}{\omega'^2 - \omega^2} d\omega' \right). \qquad (13.2)$$

PV means Cauchy principal value, i.e., the integration is performed for any ω' but $\omega' \neq \omega$. The first KK relation (13.1) is very often used in FTIR (Chapter 2) because FTIR delivers only $\alpha(\omega)$. Thus, it allows one to obtain also $n(\omega)$ from FTIR measurements. However, the big issue, when performing a KK transformation from experimental data, is to deal with the limited bandwidth of the signal. Therefore, the required integration from $\omega = 0$ to infinity is not possible, and the results are impacted by the so-called truncation error. Thus, the KK transform will give accurate results only if missing data outside the experimental bandwidth only weakly contribute to the mathematical integration. If the refractive index of the material is known for a given frequency, the impact of the truncation error may be minimized through smart methods, like the single subtractive KK transformation (SSKK, see [2]).

The big advantage of THz-TDS is that a KK analysis is not required to obtain both n and α simultaneously. This is often presented as the main advantage of THz-TDS compared to FTIR. Here we will show that both KK analysis and THz-TDS may be combined to improve the spectral characterization of given samples.

13.2.2 Comparison of Material Characterization by THz-TDS and FTIR

Before the advent of THz-TDS, FTIR was the main spectroscopic technique to investigate the properties of matter in the far infrared. Most of the studies were performed with the sample located outside the interferometer, just in front of the detector. As explained in Chapter 2, only the absorbance of a sample can be measured in this way (absorbance is defined as $-\ln T$, where T is the ratio of the transmitted and incident powers). However, if the Fresnel reflection at the sample surfaces is neglected, or if performing the measurements with several samples of different thicknesses, the coefficient of absorption of the studied material can be determined. Then the refractive index of this material is obtained with a KK transformation. Comparing the results extracted from FTIR and THz-TDS was initially a compulsory way of validating the THz-TDS method. This point was addressed by several teams [3, 4] in the early 2000s and even more recently [5].

As expected, FTIR and THz-TDS deliver very similar absorbance spectra. Figure 13.7 shows two examples of absorbance spectra recorded with FTIR and THz-TDS systems. Figure 13.7 (top) shows the absorbance spectrum of RDX, an explosive material, extracted from Huang et al. [4]. The FTIR and THz-TDS curves are very similar. The small vertical shift is due to a smaller RDX concentration for the THz-TDS measurements. Over 1.5 THz, the THz-TDS system cannot

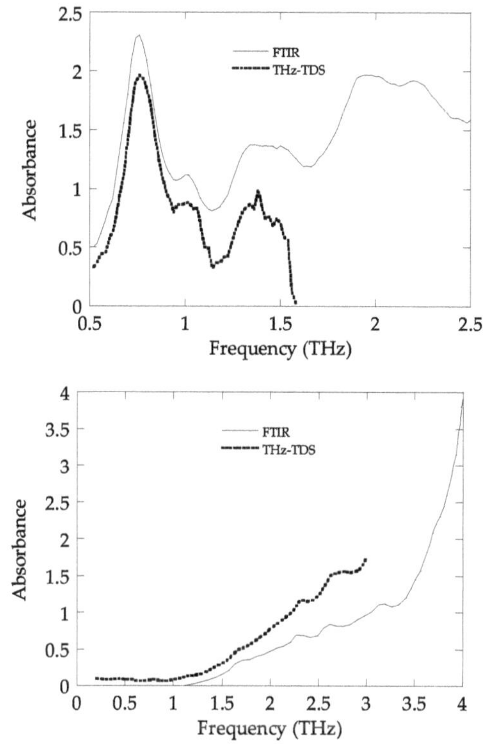

Figure 13.7 Top: Absorbance spectrum of an RDX sample, calculated with data extracted from Huang et al. [4]. Bottom: Absorbance spectrum of potassium fluoride with 1% molar concentration of potassium phosphate. Adapted from Ding et al. [6].

deliver useful results, due to the lack of THz signal. The absorbance spectrum of potassium fluoride with 1% molar concentration of potassium phosphate is shown in Fig. 13.7 (bottom). Here again, there is a good agreement between the results from the two FTIR and THz-TDS techniques. As seen in Fig. 13.7 (bottom), most FTIR setups cannot operate below 1 THz. When the refractive index of the sample is known, the absorption coefficient can be derived from absorbance. This is shown in Fig. 13.8. The absorption coefficient of a mixture of 75% maltose and 25% HDPE powders is plotted versus frequency [7]. Three curves are measured respectively with a Brucker Vertex FTIR system, and with THz-TDS setups (a

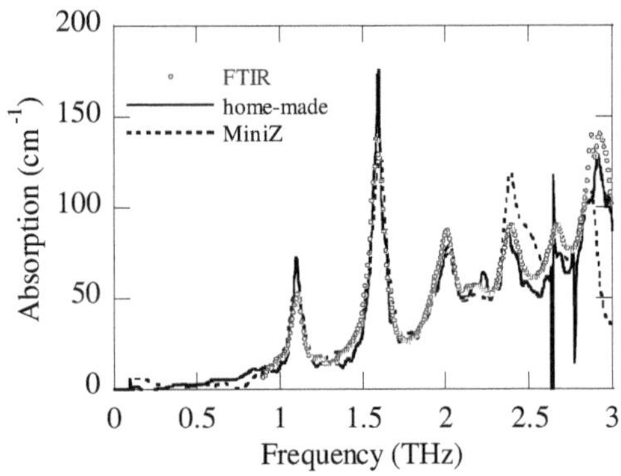

Figure 13.8 Absorption coefficient spectrum of a mixture of 75% maltose and 25% HDPE powders. Adapted from Coutaz et al. [7].

commercial Mini-Z Zomega instrument and a homemade system). The curves are very similar. The main differences arise from the THz beam that is strongly focused at the sample in the Mini-Z system. Here again, the FTIR spectrometer is limited to frequencies over 0.8 THz.

In conclusion, regarding absorbance or absorption, FTIR and TDS systems give similar results. With modern systems, the recording times are comparable. The main difference is the available bandwidth: most of the FTIR systems rarely operate below 1 THz, but on the contrary, they can explore the higher frequencies, i.e. over ten or some tens of THz. TDS systems exhibit complementary performances. Also, the source in FTIR systems radiates incoherent and not polarized beams, while the TDS signal is coherent and the TDS receiver records data over a limited time window centered at the THz pulse arrival. Between pulses, the signal is not recorded, which minimizes signal-independent noise. A good example is extracted from [8]: the authors recorded the same THz signal emitted by a photo-conducting antenna with both a triggered EO receiver and a Golay cell (energy detector). The experimental spectra are very similar (Fig. 13.9). However, the noise level with the Golay cell is about ten times bigger than the EO one.

This makes the FTIR signal a bit noisier than the THz-TDS one. Han et al. [3] reported that THz-TDS shows about five orders higher

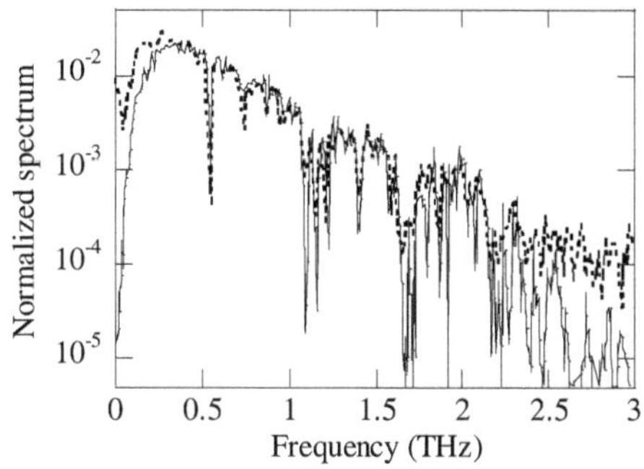

Figure 13.9 The THz spectrum of a signal emitted by a photo-conducting antenna, and recorded either with a triggered EO receiver (continuous line) or with a power meter (Golay cell, dotted line). Adapted from Naftaly et al. [8].

SNR than FTIR at less than 3 THz. This can be evaluated in terms of photon statistics. In common THz-TDS setups, the mean THz power is about 100 nW and the repetition rate is 10^8 Hz. This corresponds to about $N_{\text{THz}} \approx 2 \times 10^5$ THz photons per pulse. Because of coherence, the related noise obeys to a Poisson statistics, i.e., its fluctuation varies as $1/\sqrt{N_{\text{THz}}} \approx 1/450$. For thermal sources used in FTIR systems, like mercury lamps, the noise obeys the Bose–Einstein statistics, i.e., the fluctuation is proportional to $1/\left|e^{h\, f_{\text{THz}}/k_{\text{B}}T} - 1\right| \approx 1 \sim 10$ at THz frequencies. In a more recent paper where they used photo-induced plasma in gas for both emission and detection of THz pulses, Ho et al. [9] compare the relative performances of a classical THz-TDS system, the plasma system, and a commercial FTIR spectrometer (see Table 9.1, extracted from Ho et al. [9]).

To our knowledge, there exists no published result concerning the comparison of the refractive index extracted from THz-TDS and from a KK analysis of FTIR data.

13.2.3 Samples Exhibiting Opaque Spectral Bands [10]

13.2.3.1 Description of the problem

Let us consider a sample showing spectral peaks or bands of saturated absorption (i.e., in which there is no measurable transmitted signal; see the example shown in Fig. 13.10 [top], i.e.,

Table 13.1 Comparison of TDS systems and FTIR spectrometer [9]

	Plasma TDS	Traditional TDS	FTIR Bruker IFS 66v/S
Source	Dry nitrogen	ZnTe	Hg lamp
Detector	Dry nitrogen	ZnTe	Golay cell
Dynamic range (power)	$>10^6$	$>10^8$	$\sim$300
Bandwidth (THz)	0.5–10	0.1–3	150 GHz, UV
Peak power (W) with amplified laser	6×10^4	2×10^3	10^{-7}
Uniqueness	High peak power, broad bandwidth	High SNR	UV-FIR bandwidth

the transmission spectrum of a 790 μm thick DAST sample). Around 1.1 THz, the transmitted signal is below the noise level, and only transmission THz-TDS could be performed (this could be the case with some commercial setups).

As explained in the previous paragraph (combined reflection/transmission THz-TDS), absorption cannot be evaluated in the regions of opacity, and the refractive index could be strongly erroneous over the first absorption band, due to the unknown absolute phase of the transmitted spectrum.

Let us consider a numerical example. We start from known material data (e.g., RDX explosive material with parameters taken from the RIKEN database [11]; see Fig. 13.10 [bottom]). From these known $n_o(\omega)$ and $\alpha_o(\omega)$ values, we calculate the transmission of a 0.7-mm thick slab, and we simulate the noise as explained in chapter 11. The calculated transmission spectra (modulus and phase) with and without noise are presented in Fig. 13.11. At 6.7 THz, transmission is saturated, and phase is undetermined.

From this noisy transmission spectrum, we use a common THz-TDS extraction method to retrieve $n_o(\omega)$ and $\alpha_o(\omega)$ values. The so-obtained values $n_{TDS}(\omega)$ and $\alpha_{TDS}(\omega)$ are shown in Fig. 13.12. As anticipated, over the first absorption peak, $n_{TDS}(\omega)$ differs strongly from $n_o(\omega)$ due to unknown 2π phase jumps around the absorption peak. On the other hand, $\alpha_{TDS}(\omega)$ is almost equal to $\alpha_o(\omega)$ except in the vicinity of the peaks. Therefore, it is foreseen that a KK

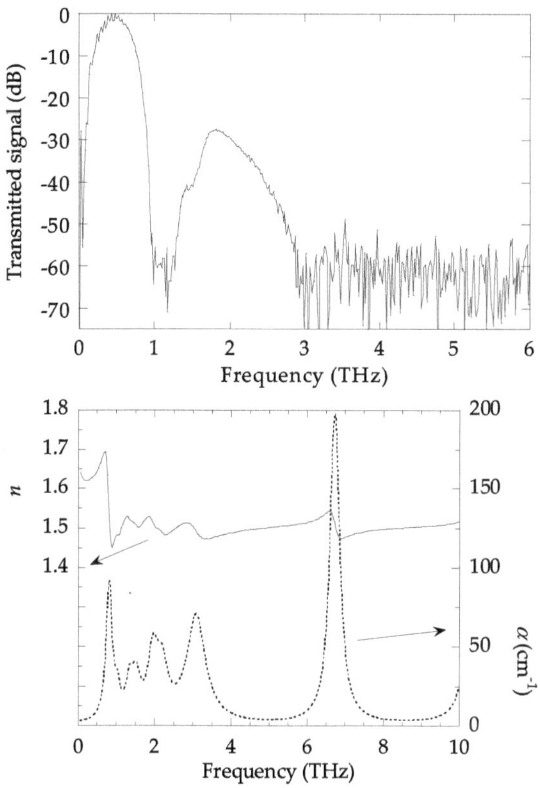

Figure 13.10 Top: Transmission (modulus) versus frequency of a 790-μm thick DAST sample. Bottom: Refractive index *n* and absorption coefficient *α* versus frequency of RDX (plotted from values taken from the RIKEN database [11]).

transformation could be performed using this more accurate value $\alpha_{\text{TDS}}(\omega)$. The KK calculation will only be impacted by the data missing at the absorption peaks and outside the experimental bandwidth, as well as by the slightly erroneous value of $\alpha_{\text{TDS}}(\omega)$ over the first saturated band (in fact, over this band, $n_{\text{TDS}}(\omega)$ is imprecisely known, which also induces an error on $\alpha_{\text{TDS}}(\omega)$).

The truncation error may be almost expunged by performing a modified KK transform, e.g., the single-subtractive KK transform (SSKK) [2].

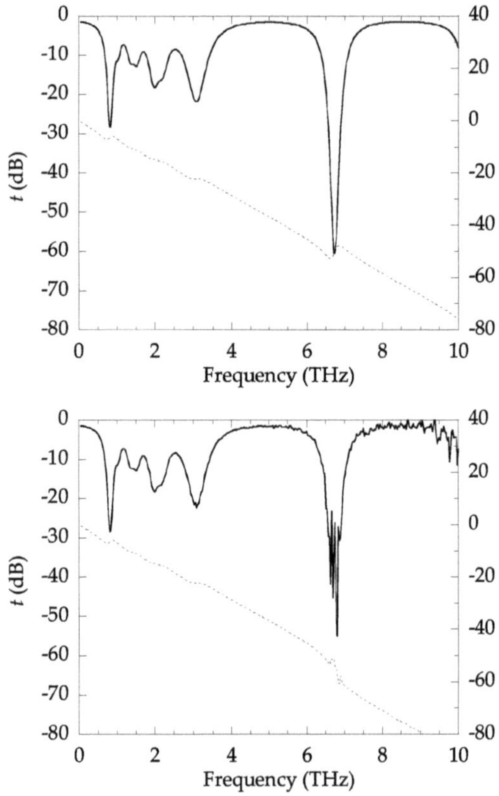

Figure 13.11 Calculated transmission spectra (modulus [continuous line] and phase [dotted line]) without (top) and with (bottom) noise.

13.2.3.2 Principle of the method

As explained above, the basic idea of the method is to perform a KK transformation of the absorption spectrum $\alpha_{\mathrm{TDS}}(\omega)$ that is obtained from a common THz-TDS extraction procedure. Thus, one obtains the refractive index $n_{\mathrm{KK}}(\omega)$:

$$n_{\mathrm{KK}}(\omega) = n_0 + \frac{c}{\pi} PV \left(\int_{\omega_{\min}}^{\omega_{\max}} \frac{\alpha_{\mathrm{TDS}}(\omega')}{\omega'^2 - \omega^2} d\omega' - \sum_{\text{peaks } i} \int_{\omega_i - \frac{\Delta\omega_i}{2}}^{\omega_i + \frac{\Delta\omega_i}{2}} \frac{\alpha_{\mathrm{TDS}}(\omega')}{\omega'^2 - \omega^2} d\omega' \right). \tag{13.3}$$

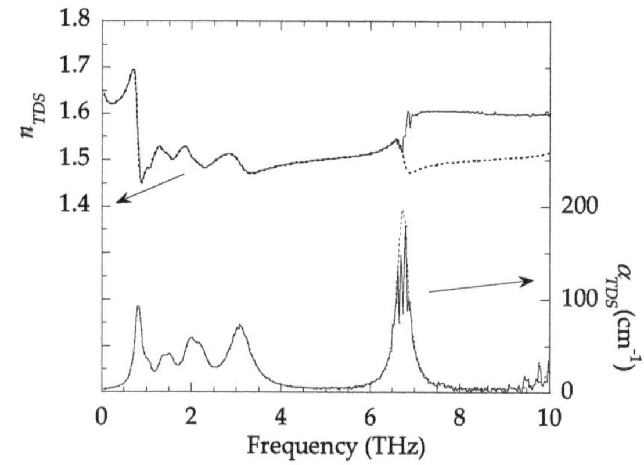

Figure 13.12 Spectra of the refractive index (upper curves) and of the coefficient of absorption (lower curves) of RDX. Dotted lines are values from the RIKEN database, continuous lines are determined by a common THz-TDS extraction code from a noisy calculated transmission spectrum.

The difference between the actual value and $n_{KK}(\omega)$ is

$$\Delta n(\omega) = n(\omega) - n_{KK}(\omega)$$
$$= \frac{c}{\pi} PV \left(\begin{array}{l} \int_0^{\omega_{\min}} \frac{\alpha_{TDS}(\omega')}{\omega'^2-\omega^2} d\omega' + \int_{\omega_{\max}}^{\infty} \frac{\alpha_{TDS}(\omega')}{\omega'^2-\omega^2} d\omega' \\ + \int_0^{\infty} \frac{\Delta\alpha(\omega')}{\omega'^2-\omega^2} d\omega' + \sum_{\text{peaks } i} \left(\int_{\omega_i - \frac{\Delta\omega_i}{2}}^{\omega_i + \frac{\Delta\omega_i}{2}} \frac{\alpha_{TDS}(\omega')}{\omega'^2-\omega^2} d\omega' \right) \end{array} \right). \tag{13.4}$$

The two first terms on the right side of Eq. 13.4 correspond to the truncation error induced by the missing data that are located outside the experimental bandwidth $[\omega_{\min}, \omega_{\max}]$. The third term is due to the difference $\Delta\alpha(\omega) = \alpha(\omega) - \alpha_{TDS}(\omega)$. $\alpha_{TDS}(\omega)$ is indeed different from the actual absorption because of noise and of the erroneous knowledge of $n_{TDS}(\omega)$ over the saturated peak frequency. Finally, the last term sums up all the missing data at the different saturated peaks.

The truncation error in KK transformation has been treated by many authors (see [2]). In the present study, the single subtractive method (SSKK) is well adapted to minimizing the truncation error because SSKK is based on the precise knowledge of the coefficient of absorption at a given frequency, namely the anchor frequency

ω_a. This is the case if the first saturated peak is not below the lowest experimentally reachable frequency: this permits to perform a common extraction and to get a precise value of both $n(\omega)$ and $\alpha(\omega)$ in the range below the first saturated peak frequency.

We will simply explain the SSKK method by starting from the KK expressions of $n(\omega)$ and $n(\omega_a)$:

$$n(\omega) = n_o + \frac{c}{\pi} PV \left(\int_0^{+\infty} \frac{\alpha(\omega')}{\omega'^2 - \omega^2} d\omega' \right),$$

$$n(\omega_a) = n_o + \frac{c}{\pi} PV \left(\int_0^{+\infty} \frac{\alpha(\omega')}{\omega'^2 - \omega_a^2} d\omega' \right).$$

The difference between both expressions leads to

$$n(\omega) - n(\omega_a) = \frac{c}{\pi} \left(\omega^2 - \omega_a^2\right) PV \left(\int_0^{+\infty} \frac{\alpha(\omega')}{\left(\omega'^2 - \omega^2\right)\left(\omega'^2 - \omega_a^2\right)} d\omega' \right). \tag{13.5}$$

In Eq. 13.5, the integrand at high frequencies (i.e., larger than ω_{max}) tends to

$$\frac{\alpha(\omega')}{\left(\omega'^2 - \omega^2\right)\left(\omega'^2 - \omega_a^2\right)} \underset{\omega' \to \infty}{\approx} \frac{\alpha(\omega')}{\omega'^4} \Rightarrow \int_{\omega_{max}}^{\infty} \ldots d\omega' \approx 0. \tag{13.6}$$

Consequently, the KK integral from ω_{max} to ∞ tends to zero (or is weak) because $\omega'^4 \gg \alpha(\omega')$. Similarly, at small frequencies, i.e., smaller than ω_{min}, we get

$$\frac{\alpha(\omega')}{\left(\omega'^2 - \omega^2\right)\left(\omega'^2 - \omega_a^2\right)} \underset{\omega' \to 0}{\approx} \frac{\alpha(\omega')}{\omega^2 \omega_a^2} \ll \frac{\alpha(\omega')}{\omega^2} \Rightarrow \int_0^{\omega_{min}} \ldots d\omega' \approx 0. \tag{13.7}$$

Therefore, expression 13.5 can be written as

$$n_{SSKK}(\omega) \approx n(\omega_a) + \frac{c}{\pi} \left(\omega^2 - \omega_a^2\right) PV \left(\int_{\omega_{min}}^{\omega_{max}} \frac{\alpha(\omega')}{\left(\omega'^2 - \omega^2\right)\left(\omega'^2 - \omega_a^2\right)} d\omega' \right). \tag{13.8}$$

The difference between $n_{SSKK}(\omega)$ and $n(\omega)$ is much smaller than the truncation error.

The combined procedure [10] follows the listed stages:

(1) One measures the complex transmission coefficient of a sample by TDS-TDS.
(2) One extracts $n_{\mathrm{TDS}}(\omega)$ and $\alpha_{\mathrm{TDS}}(\omega)$ from the TDS data.
(3) One sets $n(\omega) = n_{\mathrm{TDS}}(\omega)$ and $\alpha(\omega) = \alpha_{\mathrm{TDS}}(\omega)$ up to the first saturated absorption peak.
(4) One calculates $n_{\mathrm{SSKK}}(\omega)$ with an SSKK transform of $\alpha_{\mathrm{TDS}}(\omega)$, in which one chooses ω_{a} below the frequency of the first peak.
(5) One calculates, from $n_{\mathrm{SSKK}}(\omega)$ and $\alpha_{\mathrm{TDS}}(\omega)$, the phase $\varphi_{\mathrm{SSKK}}(\omega)$ in transmission beyond the first peak. The comparison between $\varphi_{\mathrm{SSKK}}(\omega)$ and $\varphi_{\mathrm{TDS}}(\omega)$ allows one to correct any phase jump in $\varphi_{\mathrm{TDS}}(\omega)$ occurring at the absorption peak.
(6) One performs again a classical extraction beyond the first peak with the corrected phase to get a precise estimation of $n_{\mathrm{TDS}}(\omega)$.

This procedure is repeated for any other saturated peak.

Let us now address the phase correction issue (stage 5 of the procedure). When the truncation error is minimized through the SSKK procedure, the resting error $\Delta n(\omega)$ is

$$\Delta n(\omega) = \frac{\pi}{c} PV \left(\begin{array}{l} \underbrace{\sum_{\text{peaks } i} \int_{\omega_i - \frac{\Delta\omega_i}{2}}^{\omega_i + \frac{\Delta\omega_i}{2}} \frac{\alpha_{\mathrm{TDS}}(\omega')}{\omega'^2 - \omega^2} d\omega'}_{\text{data missing at the peaks}} + \underbrace{\int_{\omega_{\min}}^{\omega_{\max}} \frac{\Delta\alpha_{\mathrm{noise}}(\omega)}{\omega'^2 - \omega^2} d\omega'}_{\text{bad estimation of } \alpha_{\mathrm{TDS}}(\omega) \text{ due to noise}} \\ + \underbrace{\int_{\omega_{\min}}^{\omega_{\max}} \frac{\Delta\alpha_{\pi}\text{-jumps}(\omega)}{\omega'^2 - \omega^2} d\omega'}_{\text{bad estimation } \alpha_{\mathrm{TDS}}(\omega) \text{ of due to phase jumps}} \end{array} \right).$$

It follows that the difference between the SSKK phase $\varphi_{\mathrm{SSKK}}(\omega)$ and the TDS phase $\varphi_{\mathrm{TDS}}(\omega)$ is simply written as

$$\Delta\varphi(\omega) = \varphi_{\mathrm{SSKK}}(\omega) - \varphi_{\mathrm{TDS}}(\omega) = \Delta\varphi_{\mathrm{noise}}(\omega) + m\pi + \sum_{\text{peaks } i} \Delta\varphi_i(\omega). \tag{13.9}$$

The phase jump ($m\pi$) could be exactly corrected if the additional phase error, i.e., $\Delta\varphi_{\mathrm{noise}}(\omega) + \sum_{\text{peaks } i} \Delta\varphi_i(\omega)$, is weaker than π. Otherwise, the number m will be overestimated. Fortunately, in the most intense region of the THz spectrum, $\Delta\varphi_{\mathrm{noise}}(\omega) \ll \pi$ and there exists a frequency ω_{oi} for which $\Delta\varphi_i\,(\omega_{\mathrm{oi}}) = 0$. Thus, at this frequency, m can be exactly valued. The determination of the value ω_{oi} of is rather long: the interested reader will find the complete explanation of this derivation in Bernier et al. [10].

13.2.4 Example

The method has been experimentally validated, as shown in the following example. A 790 μm thick sample of DAST [12] has been characterized using a TAS 7500TS THz-TDS system from Advantest. Figure 13.13 (top) shows the refractive index and absorption spectra as obtained with a common THz-TDS extraction. The refractive index curve exhibits an unusual long decay tail after the first resonance at 1.1 THz. The absorption curve definitively saturates at the first resonance around 1.1 THz, but also at the second one (around 3.1 THz). The refractive index curve obtained with the SSKK transform of the absorption is presented in Fig. 13.13 (bottom). This curve looks closer to the expected shape, i.e., the refractive index curve resembles the derivative of the absorption one. On the same plot, the frequencies f_{a} and f_{o1} are indicated.

The next stage of the method is illustrated in Fig. 13.14. In addition to the previous curve, the refractive index spectrum obtained from the phase-corrected THz-TDS is plotted. It gives accurate results, but in the absorption region (around 1.1 THz). Finally, in this region, we save the reflection data, even if they are a bit noisy. The so-achieved ultimate curve is plotted as a bold line in Fig. 13.14 (top). Moreover, we keep also the reflection data for $\alpha(\omega)$ in the absorption region (Fig. 13.14, bottom). The final results are in very good agreement with what is obtained with a much thinner sample of the same material, for which absorption does not saturate.

13.2.5 Improvement of Reflection THz-TDS

The comprehensive combination of KK analysis and reflection TDS technique has been widely addressed by K.-E. Peiponen and coworkers [2, 13, 14]. As already explained, in reflection THz-TDS, a misplacement of the sample compared to the reference mirror is the main cause of error in the extraction of the refractive index $\tilde{n}(\omega)$ of the material. The basic idea of the method by Peiponen et al. [13]

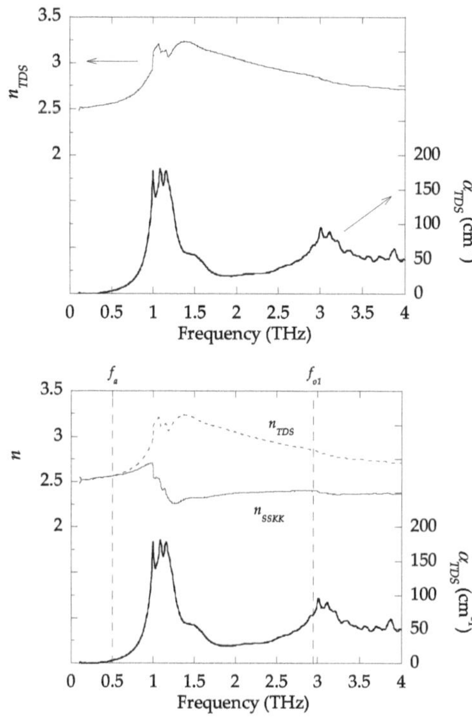

Figure 13.13 Top: $n_{\mathrm{TDS}}(\omega)$ and $\kappa_{\mathrm{TDS}}(\omega)$ curves of DAST obtained with a common THz-TDS extraction code. Bottom: Same plot together with the $n_{\mathrm{SSKK}}(\omega)$ curve.

is to compare the measured value of $\tilde{r}_{\mathrm{meas}}(\omega) = r_{\mathrm{meas}}(\omega)\, e^{j\varphi_{\mathrm{meas}}(\omega)}$ with the one calculated through an SSKK transform. Similarly to the derivation of Eq. 13.8, one can show that

$$\ln\left(\frac{r(\omega)}{r(\omega_{\mathrm{a}})}\right) = \frac{2}{\pi}\left(\omega^2 - \omega_{\mathrm{a}}^2\right) PV\left(\int_0^{\infty} \frac{\omega'\varphi(\omega')}{\left(\omega'^2 - \omega^2\right)\left(\omega'^2 - \omega_{\mathrm{a}}^2\right)} d\omega'\right), \tag{13.10}$$

$$\frac{\varphi(\omega)}{\omega} - \frac{\varphi(\omega_{\mathrm{a}})}{\omega_{\mathrm{a}}} = -\frac{2}{\pi}\left(\omega^2 - \omega_{\mathrm{a}}^2\right) PV\left(\int_0^{\infty} \frac{\ln(r(\omega'))}{\left(\omega'^2 - \omega^2\right)\left(\omega'^2 - \omega_{\mathrm{a}}^2\right)} d\omega'\right). \tag{13.11}$$

If the sample and the reference mirror are misplaced (i.e., not at exactly the same position when each was measured), the phase

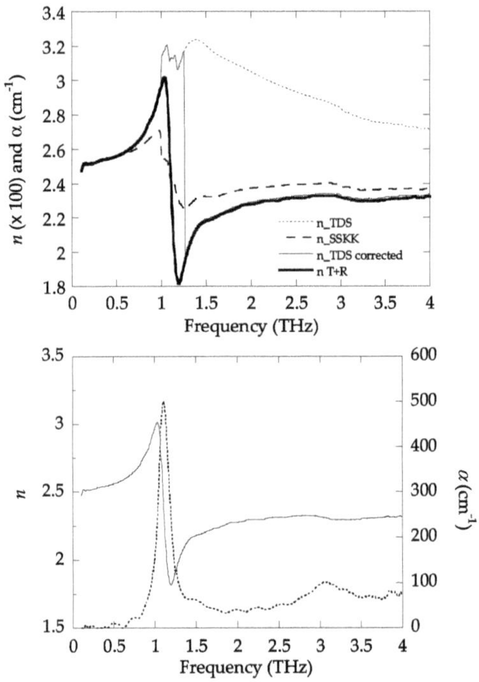

Figure 13.14 Top: Same plot as in Fig. 13.13 together with the phase-corrected $n_{\mathrm{TDS}}(\omega)$ and the final curve that combines reflection and transmission-corrected data. Bottom: Precise $n(\omega)$ and $\kappa(\omega)$ curves of DAST.

of the signal will be erroneous by the amount $\Delta\varphi(\omega) = 2\frac{\omega}{c}\delta$, where δ is the position shift. Peiponen et al. start their calculation with $r_{\mathrm{meas}}(\omega)$ and $\varphi_{\mathrm{corr}}(\omega) = \varphi_{\mathrm{meas}}(\omega) + 2\frac{\omega}{c}\delta$, and they perform an SSKK transform to obtain $r_{\mathrm{SSKK}}(\omega)$ from $\varphi_{\mathrm{corr}}(\omega)$ using Eq. 13.10 (stage 1). δ and ω_{a} are parameters that will be optimized in view of making the procedure converge. Then, stage 2, they calculate the mean value $r(\omega) = (r_{\mathrm{meas}}(\omega) + r_{\mathrm{SSKK}}(\omega))/2$, from which they deduce $\varphi_{\mathrm{SSKK}}(\omega)$ using (13.10) and then its mean value $\varphi(\omega) = (\varphi_{\mathrm{meas}}(\omega) + \varphi_{\mathrm{SSKK}}(\omega))/2$. The so-achieved value of $\tilde{r}(\omega) = r(\omega)\,e^{j\varphi(\omega)}$ is compared with the measured one. If the modulus of their differences is larger than a given value ε, $\tilde{r}(\omega)$ is entered in stage 2, and the loop is repeated as far as $|\tilde{r}(\omega) - \tilde{r}_{\mathrm{meas}}(\omega)| < \varepsilon$. This procedure is performed for different values of δ: the right value of δ is the one for which the results are the less sensitive to the value

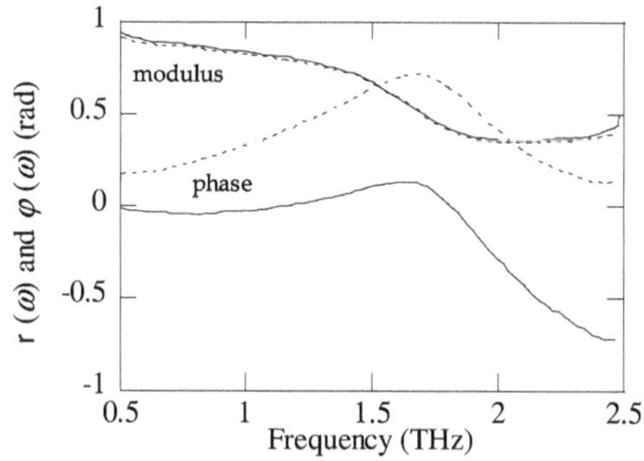

Figure 13.15 Modulus $r(\omega)$ and phase $\varphi(\omega)$ of the coefficient of reflection $\tilde{r}(\omega)$ for an intrinsic (100) InAs wafer. Continuous lines are measured THz-TDS values, while dashed lines are misplacement-corrected values. Adapted from Lucarini et al. [13].

of the anchor frequency ω_{a}. Figure 13.15 presents the measured and corrected spectra of $r(\omega)$ and $\varphi(\omega)$ for an intrinsic (100) InAs wafer. The values were measured by THz-TDS in reflection [15]. The misplacement δ of the sample was estimated to be around 53 μm. We see that the position error does not impact so much the modulus of $r(\omega)$, but it changes its phase dramatically.

Afterward, Herrmann et al. [16] have proposed and demonstrated a simplified method to extract values of $n(\omega)$ and $\kappa(\omega)$ from reflection THz-TDS, whose basic principle is based on the paper by Peiponen et al. [13]. First, they consider transmission THz-TDS. If neglecting the Fresnel coefficients at the sample surfaces, the modulus of the measured coefficient of transmission is simply described by the Beer–Lambert law, i.e.,

$$t(\omega) = \left|\tilde{t}(\omega)\right| = e^{-\frac{\omega}{c}\kappa d} \quad \Rightarrow \quad \alpha(\omega) = -\frac{2}{d}\ln\left(t(\omega)\right). \qquad (13.12)$$

Then, $n(\omega)$ can be calculated through a KK transform of Eq. 13.12. This method gives good results for low index materials, like PABA (4-aminobenzoic acid) diluted in a HDPE powder pellet [16].

When dealing with reflection THz-TDS of moderately absorbing samples ($\kappa \ll n$), the coefficient of reflection becomes

$$\tilde{r}(\omega) = r(\omega)\, e^{j\,\varphi(\omega)} = \frac{1 - n(\omega) + j\kappa(\omega)}{1 + n(\omega) - j\kappa(\omega)} \approx \frac{1 - n(\omega)^2 + 2j\kappa(\omega)}{(1 + n(\omega))^2}. \qquad (13.13)$$

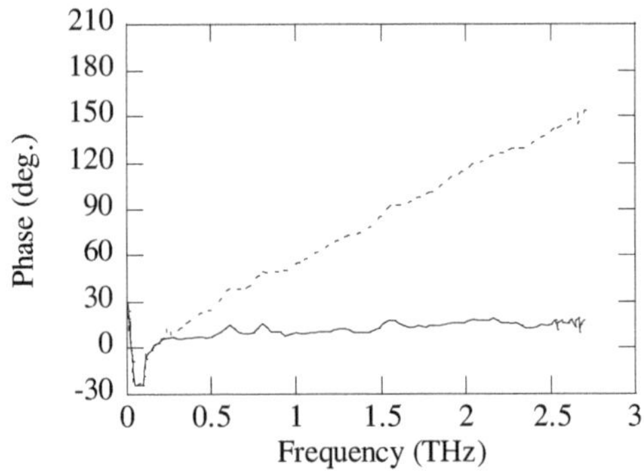

Figure 13.16 The phase of the reflected THz signal, measured by THz-TDS (dashed line) and correction with KK method (continuous line). Adapted from Herrmann et al. [16].

It follows that

$$n(\omega) \approx \frac{1 - r(\omega)}{1 + r(\omega)}, \; \kappa(\omega) \approx \frac{2r(\omega)}{(1 + r(\omega))^2} \tan \varphi(\omega). \tag{13.14}$$

The refractive index $n(\omega)$ does not depend on the phase, and thus it can be calculated from the modulus $r(\omega)$ of the coefficient of reflection. Then the extinction coefficient $\kappa_{\text{KK}}(\omega)$ is calculated through a KK transform of $n(\omega)$. It is compared to the experimental value $\kappa_{\text{exp}}(\omega)$ given by (13.14). If there is a position shift δ of the reference mirror, the experimental phase is erroneous by an amount $\Delta\varphi(\omega)$, such as $\varphi(\omega) + \Delta\varphi(\omega) = \varphi(\omega) + 2\frac{\omega}{c}\delta$. Hence, $\kappa_{\text{exp}}(\omega) \neq \kappa_{\text{KK}}(\omega)$. The difference leads to the error $\Delta\varphi(\omega)$, which is easily derived from Eq. 13.14:

$$\tan \Delta\varphi(\omega) = \frac{(1 + r(\omega))^2}{2r(\omega)} \frac{\kappa_{\text{exp}}(\omega) - \kappa_{\text{kk}}(\omega)}{\tan^2 \varphi(\omega) + 1} \tag{13.15}$$

Figure 13.16 presents the phase of the reflected signal obtained in a THz-TDS experiment (dashed line) and the corrected phase (continuous line) obtained with the help of the KK transform. Here, the sample and reference mirror were misaligned by 26 μm.

13.2.6 Improved Estimation of the Sample Thickness

A smart method to achieve a precise THz parameter extraction by improving the precision on the sample thickness, based on a KK transformation, has recently been proposed by Son and coworkers [17]. Let us suppose that the sample thickness d entered in the extraction code is erroneous by an amount Δd. Both refractive index and absorption coefficient will also by impacted by this error. We write

$$n_{\mathrm{TDS}}(\omega, \Delta d) = n(\omega) + \Delta n_{\mathrm{TDS}}(\omega, \Delta d), \tag{13.16}$$

$$\alpha_{\mathrm{TDS}}(\omega, \Delta d) = \alpha(\omega) + \Delta\alpha_{\mathrm{TDS}}(\omega, \Delta d). \tag{13.17}$$

Here $n(\omega)$ and $\alpha(\omega)$ are the actual values, while $n_{\mathrm{TDS}}(\omega, \Delta d)$ and $\alpha_{\mathrm{TDS}}(\omega, \Delta d)$ are the values obtained from the THz-TDS extraction using the wrong $d + \Delta d$ thickness. Then one performs a KK transformation of $n_{\mathrm{TDS}}(\omega, \Delta d)$:

$$\begin{aligned}\alpha_{\mathrm{KK}}(\omega, \Delta d) &= -\frac{4\omega^2}{\pi c} PV\left(\int_0^{+\infty} \frac{n_{\mathrm{TDS}}(\omega', \Delta d) - 1}{\omega'^2 - \omega^2} d\omega'\right)\\ &= \alpha(\omega) - \frac{4\omega^2}{\pi c} PV\left(\int_0^{+\infty} \frac{\Delta n_{\mathrm{TDS}}(\omega', \Delta d)}{\omega'^2 - \omega^2} d\omega'\right).\end{aligned} \tag{13.18}$$

The so-calculated $\alpha_{\mathrm{KK}}(\omega, \Delta d)$ differs from the actual value $\alpha(\omega)$ because of the error Δd. As well, it is different from $\alpha_{\mathrm{TDS}}(\omega, \Delta d)$. The difference is calculated by subtracting equations Eqs. 13.17 and 13.18:

$$\begin{aligned}&\alpha_{\mathrm{TDS}}(\omega, \Delta d) - \alpha_{\mathrm{KK}}(\omega, \Delta d)\\ &= \Delta\alpha_{\mathrm{TDS}}(\omega, \Delta d) + \tfrac{4\omega^2}{\pi c} PV\left(\int_0^{+\infty} \tfrac{\Delta n_{\mathrm{TDS}}(\omega', \Delta d)}{\omega'^2 - \omega^2} d\omega'\right).\end{aligned} \tag{13.19}$$

This difference tends to zero when $\Delta d = 0$, i.e., when the sample thickness is precisely known. Therefore, the method by Son et al. [17] is to minimize Eq. 13.19 by performing the calculation with different values of the thickness. The minimization is made by integrating over frequency the error on κ and making it smaller than a given value ε:

$$\int_0^{+\infty} (\kappa_{\mathrm{KK}}(\omega) - \kappa_{\mathrm{TDS}}(\omega))^2\, d\omega = \sum_{\omega_i} [\kappa_{\mathrm{KK}}(\omega_i) - \kappa_{\mathrm{TDS}}(\omega_i)]^2 < \varepsilon. \tag{13.20}$$

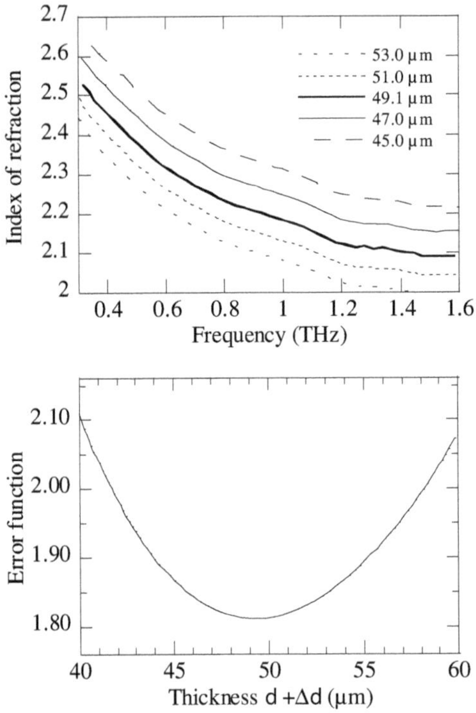

Figure 13.17 Top: Index of refraction of water calculated for different sample thicknesses. The bold curve shows the optimized values. Bottom: Error function (Eq. 13.20) versus sample thickness for a 49.1 μm thick water layer. Adapted from Son et al. [17].

Because of the already explained truncation error in the KK transformation of actual experimental data, Son et al. did use the SSKK method. The details of the procedure are explained by Son et al. in Ref. [17]. Figure 13.17 (top) shows the refractive index of a thin water layer, calculated from THz-TDS data for different values of the layer thickness. Figure 13.17 (bottom) presents the corresponding error function (Eq. 13.20). The error function exhibits a pronounced minimum at $d = 49.1$ μm. The thickness is thus obtained with a precision of about 1 μm. The corresponding imprecision on $n(\omega)$ is 0.03 (∼1.3%).

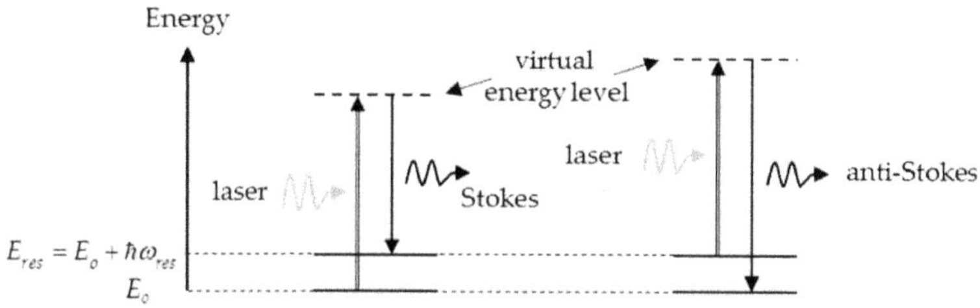

Figure 13.18 Raman scattering: in the Stokes phenomenon (left side), the excited matter radiates a photon at a frequency $\omega_{\text{Stokes}} = \omega_{\text{laser}} - \omega_{\text{res}}$, while in the anti-Stokes phenomenon, the radiated photon is at the frequency $\omega_{\text{anti-Stokes}} = \omega_{\text{laser}} + \omega_{\text{res}}$.

13.3 Comparison of Raman and THz-TDS Spectroscopies

13.3.1 Introduction

Raman spectroscopy, based on inelastic scattering, is another way of studying the far infrared response of matter. In Raman spectroscopy [18–20], one records the light inelastically scattered at a frequency ω_{Raman} by a sample excited by a laser beam at a frequency ω_{laser}, such as

$$\begin{cases} \omega_{\text{Raman}} \equiv \omega_{\text{Stokes}} = \omega_{\text{laser}} - \omega_{\text{res}}, \\ \omega_{\text{Raman}} \equiv \omega_{\text{anti-Stokes}} = \omega_{\text{laser}} + \omega_{\text{res}}. \end{cases} \tag{13.21}$$

ω_{res} is the frequency of a resonance of the molecule, while Stokes and anti-Stokes signals differ by a higher or lower scattered beam energy, i.e. adding or subtracting ω_{res} to or from ω_{laser}. These phenomena can be simply described in terms of energy conservation (see Fig. 13.18).

In the Stokes phenomenon, a photon from the pumping laser excites a molecule from the fundamental level of energy E_0 to a virtual level of energy $E = \hbar\omega_{\text{laser}}$. Then the molecule is de-excited by emitting a photon at a frequency ω_{Stokes}, not by returning to the fundamental level (Rayleigh scattering), but to a vibration/rotation level of energy $\hbar\omega_{\text{res}}$. The anti-Stokes signal is generated through a complementary phenomenon. A molecule, in the excited state of energy $\hbar\omega_{\text{res}}$, is pumped by the laser to a higher virtual level, and it returns to the fundamental level by emitting a photon at the frequency $\omega_{\text{anti-Stokes}}$. Thus, in a Raman scattering experiment, one always observes two lines symmetrically apart the laser frequency, namely the Stokes and anti-Stokes lines. At thermal equilibrium, the populations of the fundamental and vibration/rotation levels obey the Maxwell–Boltzmann statistics, and thus the Stokes and anti-Stokes intensities are related by

$$\frac{I_{\text{anti-Stokes}}}{I_{\text{Stokes}}} = e^{-\frac{\hbar\omega_{\text{res}}}{k_B T}}. \tag{13.22}$$

If $f_{\text{res}} = \omega_{\text{res}}/2\pi$ belongs to the THz range, Stokes and anti-Stokes signals are similarly intense, while for high-energy resonances (mid-infrared), the anti-Stokes signal is much weaker. The resonances involved in Raman scattering are molecular vibrations, rotations, or any more complex deformation of the molecule. In the solid state, collective resonances are involved, like optical phonons [21], magnon [22], spin waves [23], excitons [24], etc. Note that the Brillouin scattering, which occurs only in solids, is also an inelastic scattering like the Raman, but it involves phonons of weaker energy (acoustic phonons, molecular translations) rather than optical phonons. Therefore, Brillouin scattering results in frequency differences of the order of the GHz, while Raman scattering is within the THz range or above.

13.3.2 Basic Properties of Raman Scattering

When a light beam of E-field $\vec{E}(t) = \vec{E}_0\, e^{j\,\omega_{\text{light}}t}$ illuminates a molecule, it creates a dipole $\vec{p}(t)$ at the molecule due to the action of the Coulomb force on the positive and negative charges of the molecule (we have supposed here that multipolar excitations are very weak). Exciting E-field and induced dipole are related by

$$\vec{p}(t) = \overleftrightarrow{\alpha} : \vec{E}(t) \quad \Leftrightarrow \quad \vec{p}_0\, e^{j\,\omega_{\text{light}}t} = \overleftrightarrow{\alpha} : \vec{E}_0\, e^{j\,\omega_{\text{light}}t}. \tag{13.23}$$

$\overleftrightarrow{\alpha}$ is the polarizability of the molecule. For molecules, the polarizability could be a tensor (anisotropic polarizability), which is a requested property for the molecule to be Raman active (for atoms,

α is a scalar). $\overleftrightarrow{\alpha}$ depends on the positions $\vec{Q}_i$ of the atoms i of the molecule. At non-null temperatures, atoms move around their equilibrium positions, and thus the polarizability depends upon times, i.e., $\overleftrightarrow{\alpha} = \overleftrightarrow{\alpha}(t)$. These displacements are extremely small (smaller than about a tenth of the size of the atoms being studied), as far as we are concerned with room or colder temperatures. Thus we can approximate the polarizability by a Taylor series expansion:

$$\overleftrightarrow{\alpha}(t) = \overleftrightarrow{\alpha}_0 + \sum_i \frac{\partial \overleftrightarrow{\alpha}}{\partial Q_i} d Q_i(t). \tag{13.24}$$

At resonance, the displacement $d Q_i(t)$ of each atom i is oscillating at the resonance frequency ω_{res}. Therefore, we can rewrite relation 13.24 as

$$\overleftrightarrow{\alpha}(t) = \overleftrightarrow{\alpha}_0 + \sum_i \frac{\partial \overleftrightarrow{\alpha}}{\partial Q_i} Q_{\text{oi}}\, e^{j\omega_{\text{res}} t}. \tag{13.25}$$

It follows that the light-induced dipole moment (Eq. 13.23) is

$$\begin{aligned}\vec{p}(t) &= \overleftrightarrow{\alpha}(t) : \vec{E}_0\, e^{j\,\omega_{\text{light}} t}\\ &= \overleftrightarrow{\alpha}_0 : \vec{E}_0\, e^{j\,\omega_{\text{light}} t} + \left(\sum_i \frac{\partial \overleftrightarrow{\alpha}}{\partial Q_i} Q_{\text{oi}} : \vec{E}_0\right) e^{j\left(\omega_{\text{light}} \pm \omega_{\text{res}}\right)t}.\end{aligned} \tag{13.26}$$

The polarization $\vec{P}(t)$ of matter induced by the excitation light is the sum of all the dipole moments $\vec{p}(t)$ per volume unit. If there is no randomizing effect during the laser light period, $\vec{P}(t)$ obeys the same Eq. 13.26 as $\vec{p}(t)$, which is multiplied by the dipole volume density N:

$$\vec{P}(t) = N\overleftrightarrow{\alpha}_0 : \vec{E}_0\, e^{j\,\omega_{\text{light}} t} + N\left(\sum_i \frac{\partial \overleftrightarrow{\alpha}}{\partial Q_i} Q_{\text{oi}} : \vec{E}_0\right) e^{j\left(\omega_{\text{light}} \pm \omega_{\text{res}}\right)t}. \tag{13.27}$$

The first term in Eq. 13.27 gives the linear response of the matter and leads to the permittivity $\overleftrightarrow{\varepsilon}\left(\omega_{\text{light}}\right)$. The second term oscillates at frequencies $\omega_{\text{light}} \pm \omega_{\text{res}}$ and it renders for the inelastic process, i.e., Raman scattering. The $\pm$ sign originates from the complex notation we use here.[a] Relation 13.27 identifies the main difference, from a physical point of view, between Raman scattering and THz-TDS: THz-TDS involves the susceptibility of matter, i.e., the first term in Eq. 13.27, which is proportional to the polarizability $\overleftrightarrow{\alpha}_0$ of the molecules; Raman signals are related to the derivative, versus the local coordinates Q_i of the polarizability. Therefore, both techniques will give different and complementary information on the studied matter. Of course, if $\overleftrightarrow{\alpha}_0$ exhibits some spectral features due, for example, to resonances, its derivative will also show some features at the same frequencies. Thus we will observe similarities in both Raman and THz-TDS spectra.

However, the spectra could also strongly differ as it will be illustrated below. Mostly, susceptibility (THz-TDS) and Raman spectroscopies obey selection rules: for example, when dealing with vibrational energy levels, a vibration is IR-active if the dipole moment is changed during the vibration and is Raman-active if the polarizability is changed during the vibration [20]. It follows that the ν_1 vibration mode of the CO_2 molecule is IR-inactive, while it is Raman active: indeed, both oxygens move apart symmetrically from the central carbon atom. Thus the dipolar moment is not changed during the vibration. Selection rules are determined through a quantum mechanics study of the Raman effect. They depend on the involved resonances. In the simplest cases of rotational levels of linear molecules, $\Delta J = \pm 2$, and of vibrational levels of diatomic molecules, $\Delta \nu = \pm 1$. Figure 13.19 shows the rotational Raman spectrum of CO_2. The gas is excited by a discharge and is pumped by the 488 nm line of an Ar-ion laser. At zero frequency shift, the saturated signal is due to unfiltered pump laser light. Anti-Stokes and Stokes lines are seen respectively at positive and negative frequency shifts, and their overall shape is similar. The lines are periodically spaced as expected from the quantized rotational energy levels and the selection rule.

---

[a] We should note that the actual fields are real and therefore their oscillations must be described by a cosine function. Here the product $e^{j\,\omega_{\text{light}}t}e^{j\,\omega_{\text{res}}t}$ must be correctly written as a product of cosines:

$$\cos\left(\omega_{\text{light}}t\right)\cos\left(\omega_{\text{res}}t\right) = \cos\left(\left(\omega_{\text{light}} + \omega_{\text{res}}\right)t\right) + \cos\left(\left(\omega_{\text{light}} - \omega_{\text{res}}\right)t\right).$$

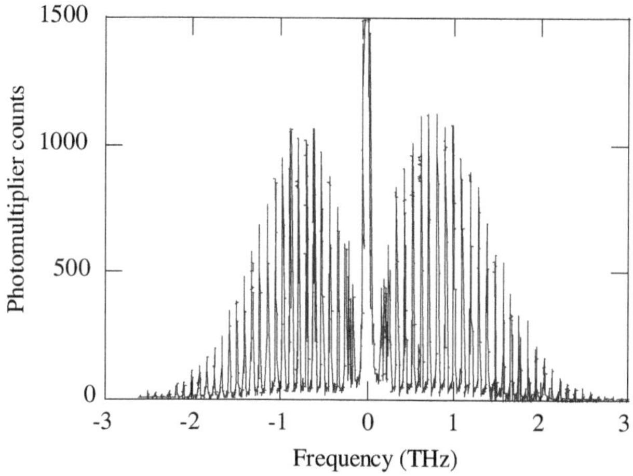

Figure 13.19 Raman spectrum of CO_2. The horizontal scale corresponds to the Raman frequency shift. The zero shift is the laser line frequency, 614.7 THz (i.e., $\lambda = 488$ nm), at which a large Rayleigh scattering contribution is observed. Adapted from Barrett and Weber [25].

Finally, we list the main advantages and disadvantages of Raman spectroscopy:

Advantages

- It does not require tunable laser. A common high-power laser is sufficient and, if possible, delivering a short wavelength because the Raman efficiency varies as $1/\lambda^4$.
- Several species can be measured simultaneously.
- Rotational and vibrational spectra can be achieved simultaneously.
- The Raman signal varies linearly with the pumping laser intensity, and no saturation effect is observed. The Raman intensity may indeed be written under the following expression:

$$I_{\text{Raman}} = I_{\text{laser}} \frac{\beta N L}{\lambda_{\text{laser}}^4}.$$

  N is the density of Raman-active molecules, L is the length of tested material, and β is proportional to the Raman cross-section. β is larger for molecules with extended π-electron systems as they are more easily polarized, and for molecules

with large, electron-rich atoms, such as S or I. As well, multiple bond stretches generally exhibit high cross sections. On the other hand, molecules with single C–H, C–O, and C–C bonds have small cross sections.
- Raman spectroscopy is well suited for major molecules like N_2, O_2, CO, CO_2, H_2O, etc.

Disadvantages

- Weak signal (about 1000 weaker than Rayleigh scattering).
- It requires a high-power laser, which can heat and degrade the sample.
- Optical pumping should be performed at a short wavelength (NIR, vis, or UV), which is very different from the THz range. This could be either an advantage or a disadvantage. For example, water solution can be easily characterized by Raman spectroscopy, because water is almost transparent at the laser and Stokes frequencies. THz spectroscopy of such water solutions is more difficult to perform because water absorbs the THz waves.
- Raman spectroscopy is sensitive to background fluorescence and emission, and stray light.

13.3.3 Comparison of Raman and THz-TDS Spectra

Numerous studies have been published on the combined characterization of materials using Raman spectroscopy, THz-TDS, and sometimes FTIR. They include organic materials [26, 27], biological materials [28], explosives [29], pharmaceuticals [30], polymers [31], crystals [32], glasses [33], and so on.

Figure 13.20 (top) shows the absorption coefficient, extracted from THz-TDS measurements, and the Raman intensity spectra of cysteine. The peaks are attributed to S–S bridges. We can observe the great similarity between TDS and Raman spectra. For example, the peak at 556 cm^{-1} (1.65 THz) is clearly seen on both curves. For the benzoic acid (Fig. 13.20, bottom), some spectral features are observed on both curves, but the similarity is weaker. Moreover, Raman and absorption curves differ strongly for crystalline indapamide (a medicine for the treatment of hypertension), as seen in Fig. 13.21.

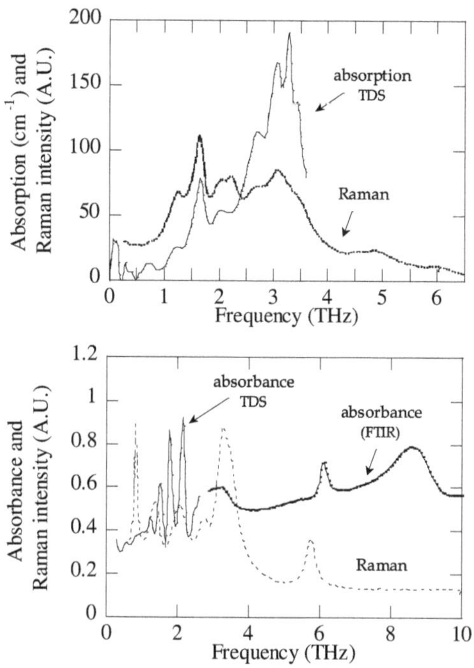

Figure 13.20 Top: Raman and absorption spectra cysteine (adapted from Brandt et al. [28]). Bottom: Absorbance and Raman spectra of benzoic acid (adapted from Kutuvantavida et al. [27]). Note that absorbance has been determined by THz-TDS below 2.6 THz and by FTIR over 2.6 THz.

13.4 Simultaneous Extraction of Permittivity and Permeability

Magnetic materials exhibit both relative permittivity $\tilde{\varepsilon}(\omega)$ and permeability $\tilde{\mu}(\omega)$ different from one. In actual materials, these values are complex because of electric and magnetic losses. Therefore, for a complete characterization of these materials requires the determination of four values, i.e., two real and two imaginary values. This cannot be achieved through a common THz-TDS measurement. In fact, this question was already addressed in the pioneering paper by Nicholson and Ross [31], who were the firsts to determine permittivity and permeability spectra from time-domain measurements. In their paper that was devoted to

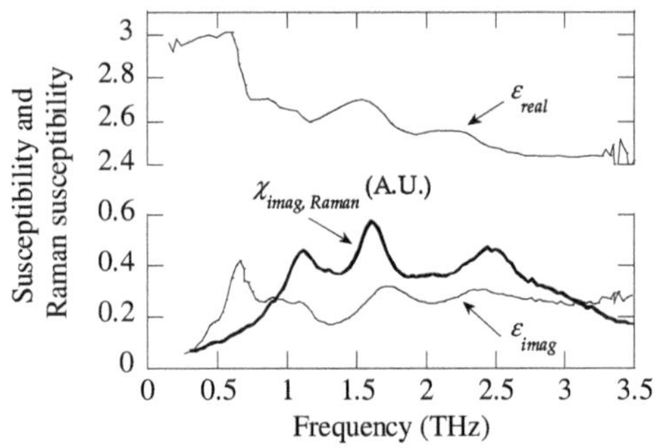

Figure 13.21 Raman and susceptibility spectra of crystalline indapamide (adapted from Kojima et al. [30]). Both real and imaginary parts of ε are given.

measurements in the GHz range using waveguides, the basic idea was to measure the impedance of a waveguide and the propagation factor of a wave propagating in the waveguide. Subsequently, this method was largely employed to characterize materials and devices in the microwave [35, 36] and optical domains [37].

When dealing with THz-TDS of magnetic materials, the procedure by Nicholson and Ross [34] is easy to adapt [38, 39]. Let us first remember the relations between the impedance $\tilde{Z}(\omega)$ and the refractive index $\tilde{n}(\omega)$ of a material, and $\tilde{\varepsilon}(\omega)$ and $\tilde{\mu}(\omega)$ (see Annex):

$$\tilde{n}(\omega) = \sqrt{\tilde{\varepsilon}(\omega)\tilde{\mu}(\omega)}, \quad \tilde{Z}(\omega) = \sqrt{\frac{\tilde{\mu}(\omega)}{\tilde{\varepsilon}(\omega)}}. \tag{13.28}$$

Determining $\tilde{Z}(\omega)$ and $\tilde{n}(\omega)$ from the experiment permits to get $\tilde{\varepsilon}(\omega)$ and $\tilde{\mu}(\omega)$:

$$\tilde{\varepsilon}(\omega) = \frac{\tilde{n}(\omega)}{\tilde{Z}(\omega)}, \quad \tilde{\mu}(\omega) = \tilde{n}(\omega)\tilde{Z}(\omega). \tag{13.29}$$

In transmission or reflection experiments, impedance $\tilde{Z}(\omega)$ appears only in the Fresnel coefficients at the sample faces, while $\tilde{n}(\omega)$ is only involved in the propagation factor through the sample (see Annex, relation A.26). Thus, let us write the coefficients of transmission (Eq. 5.3) and reflection (Eq. 5.4) through a slab located in air, as a function of $\tilde{Z}(\omega)$ and $\tilde{n}(\omega)$:

$$\tilde{t}(\omega) = \frac{4\tilde{Z}(\omega)}{\left(\tilde{Z}(\omega)+1\right)^2} \times \frac{e^{-j\frac{\omega}{c}(\tilde{n}-1)d}}{1-\left(\frac{\tilde{Z}(\omega)-1}{\tilde{Z}(\omega)+1}\right)^2 e^{-2j\frac{\omega}{c}\tilde{n}d}}, \tag{13.30}$$

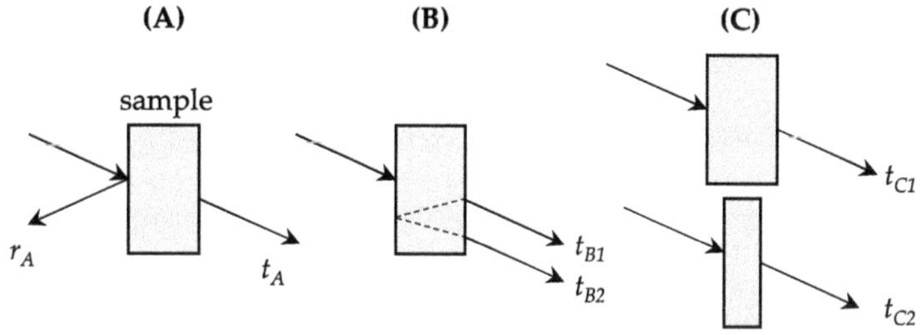

Figure 13.22 Different schemes for determining $\tilde{\varepsilon}(\omega)$ and $\tilde{\mu}(\omega)$ by THz-TDS.

$$\tilde{r}(\omega) = \frac{\tilde{Z}(\omega) - 1}{\tilde{Z}(\omega) + 1} - \frac{\tilde{Z}(\omega) - 1}{\left(\tilde{Z}(\omega) + 1\right)^3} \frac{4\tilde{Z}(\omega)e^{-2j\frac{\omega}{c}\tilde{n}d}}{1 - \left(\frac{\tilde{Z}(\omega)-1}{\tilde{Z}(\omega)+1}\right)^2 e^{-2j\frac{\omega}{c}\tilde{n}d}}. \quad (13.31)$$

$\tilde{Z}(\omega)$ and $\tilde{n}(\omega)$ are obtained by inverting Eqs. 13.30 and 13.31:

$$\tilde{Z}^2(\omega) = \frac{\tilde{t}(\omega)^2 e^{-2j\frac{\omega}{c}d} - (1 + \tilde{r}(\omega))^2}{\tilde{t}(\omega)^2 e^{-2j\frac{\omega}{c}d} - (1 - \tilde{r}(\omega))^2}, \quad (13.32)$$

$$\tilde{n}(\omega) = -1 + j\frac{c}{\omega\, d}\left[\ln\left(1 - \frac{\tilde{Z}(\omega) + 1}{\tilde{Z}(\omega) - 1}\tilde{r}(\omega)\right) - \ln\left(\tilde{t}(\omega)\right)\right]. \quad (13.33)$$

Other experimental schemes to determine $\tilde{Z}(\omega)$ and $\tilde{n}(\omega)$ from THz-TDS measurements have been carefully studied by Nemec et al. [40]. The goal is always to separate the contributions of the Fresnel coefficients and the propagation factor through the sample. The first scheme (A in Fig. 13.22) corresponds to the case studied above, including both transmission and reflection measurements. In the second scheme (B), to avoid to use two different setups for transmission and measurements, one looks at the first pulse directly transmitted by the sample and at the second one, which experiments a rebound in the sample. Comparing both signals permits us to separate the propagation factor in the sample from the Fresnel contribution. The last scheme (C) makes use of two samples of different thicknesses, here again, to directly determine the propagation factor. It appears that scheme A gives precise results if the sample thickness is optimized. Scheme B is good at low frequencies, but requires an optimized thickness that should be known with a great precision. Scheme C is good for low-impedance

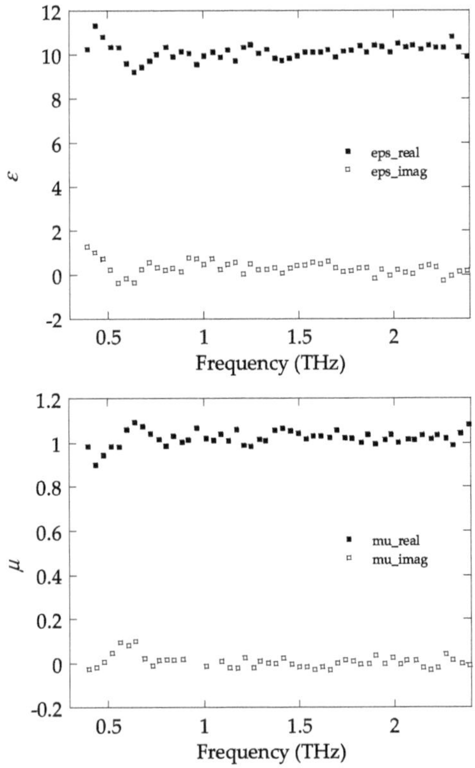

Figure 13.23 Real and imaginary parts of $\tilde{\varepsilon}(\omega)$ (bottom plot) and $\tilde{\mu}(\omega)$ (top plot) of ZnTe. Adapted from Nemec et al. [40].

materials and low frequencies, and the sample thickness must be precisely known.

Figure 13.23 shows the real and imaginary parts of $\tilde{\varepsilon}(\omega)$ and $\tilde{\mu}(\omega)$ of ZnTe determined by THz-TDS using method A, i.e., measuring the transmissivity and reflectivity of the sample. Here the crystal is not magnetic, therefore $\tilde{\mu}(\omega) \approx 1 + j \times 0$. The real part of $\tilde{\varepsilon}(\omega)$ is close to 10 and its imaginary part is almost null, i.e., the crystal is transparent with $n(\omega) \approx 3.2$.

Most of the related studies have been devoted to natural or artificial (meta) magnetic materials. In magnetic crystals, THz-TDS allows us to characterize the magnetic properties, like coercivity [41], and to measure magnetic resonances [42, 43]. Figure 13.24

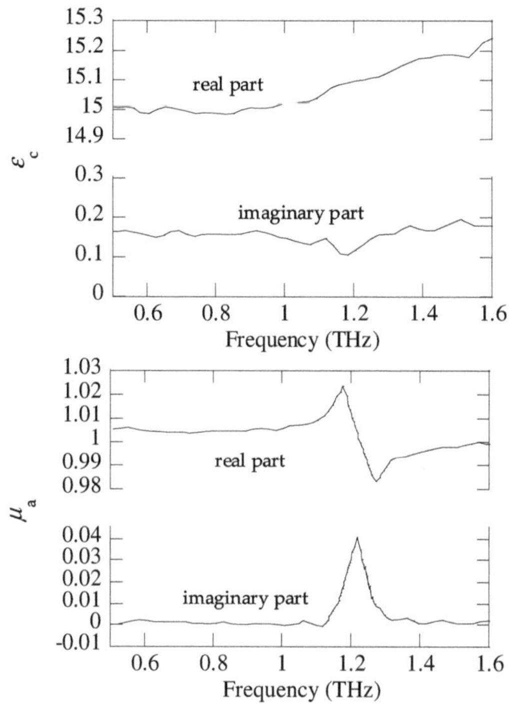

Figure 13.24 The real and imaginary parts of $\tilde{\varepsilon}(\omega)$ (bottom plot) and $\tilde{\mu}(\omega)$ (top plot) of $YMnO_3$, determined by THz-TDS performed at 10 K. Adapted from Kadlec et al. [42].

nicely illustrates an antiferromagnetic resonance (at 1.25 THz) in the permeability spectrum in $YMnO_3$, measured by THz-TDS at $T = 10$ K.

Magnetism in metamaterials has been the subject of numerous studies since this magnetism could be obtained without magnetic constituents [44], and the magnetic permeability could be large [45] and negative [46]. Magnetism arises mostly from conduction currents generated in the metallic parts of the metamaterials, and from related inductance effects. However, magnetism can also been observed in dielectric metamaterials near a Mie resonance [47, 48], for example in samples made of TiO_2 microspheres [49].

Figure 13.25 presents the real and imaginary parts of $\tilde{\varepsilon}(\omega)$ (top plot) and $\tilde{\mu}(\omega)$ (bottom plot) of a metamaterial made of phosphor

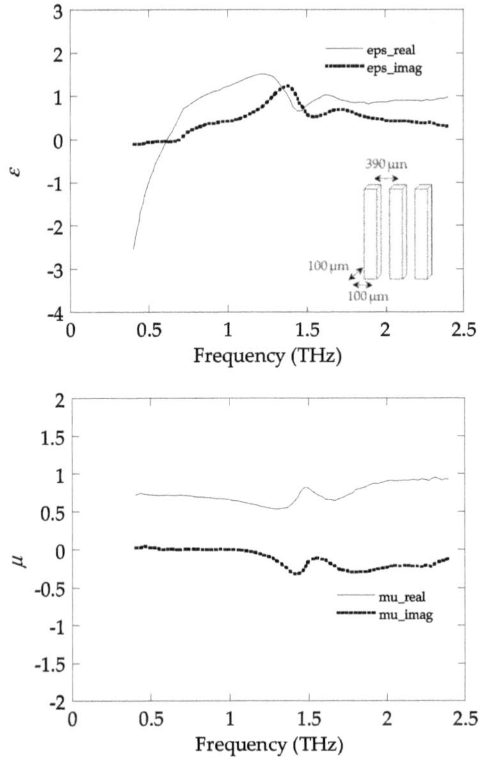

Figure 13.25 The real and imaginary parts of $\tilde{\varepsilon}(\omega)$ (top plot) and $\tilde{\mu}(\omega)$ (bottom plot) of a phosphor bronze wire grid (dimensions given in the inset), determined by THz-TDS. Adapted from Minowa et al. [38].

bronze wire grids [35]. The fact that the relative permeability is not equal to one, even if phosphor bronze is a nonmagnetic material, is explained by an inducting coupling between adjacent wires.

13.5 Comparison of THz-TDS and CW Techniques

When it was proposed in the 1980s, THz-TDS was showing unrivaled performance, in terms of achievable bandwidth, dynamics, fast records at room temperature, and phase/modulus measurements. Since those times, continuous-wave (CW) techniques have been strongly improved and, at least for the lower frequencies, are competing with THz-TDS. The modulus and phase of the signals are obtained, when using CW setups, by heterodyne techniques. Such instruments are named vector network analyzers (VNA). THz VNA delivers the S matrix of the studied sample. This could be performed either in free space, or in a waveguide configuration. Most of them are based on a microwave source that is frequency-multiplied to reach the THz range. Optoelectronic CW spectrometers are also available, in which the THz signal is generated and detected using the beating between two CW lasers of slightly different wavelengths.

13.5.1 THz VNA

To our knowledge, the first THz VNA was proposed by Philippe Goy and his colleagues [50]. Then P. Goy has created the society AB Millimetre, which was also the first to commercialize THz VNAs [51]. Nowadays, commercial systems up to 1.5 THz are supplied by a few companies, like Keysight Technologies (formerly Agilent), VDI, Anritsu, Rohde and Schwarz, and AB Millimetre. The initial scheme of Goy's THz VNA is still exploited. A stabilized microwave source, oscillating around $10 \sim 20$ GHz, is multiplied in frequency by Schottky diodes in a waveguide configuration. This signal, at a frequency f_{S}, is radiated by a horn antenna towards the sample, and the transmitted signal is collected by another horn antenna that feeds a mixer. This mixer mixes the detected signal with a reference one (local oscillator) at frequency f_{OL} delivered by a second source that is phase-locked to the first one. The mixer gives a signal at the intermediate frequency $f_i = f_{\mathrm{S}} - f_{\mathrm{OL}}$, whose magnitude and phase are read by common high-frequency electronics.

The advantages of VNAs are their high-frequency resolution and tunability (MHz and even sub-kHz [52]), and their high dynamics (higher than $80 \sim 100$ dB [51]). On the other hand, there are limited to frequencies below $1.1 \sim 1.5$ THz [53], they require a precise calibration [51, 54, 55] to know magnitude and phase at the input and output of the sample, and in a free-space configuration, they may be subject to standing wave perturbation. Because VNA measurements deliver magnitude and phase of the transmitted/reflected THz signals, extraction methods

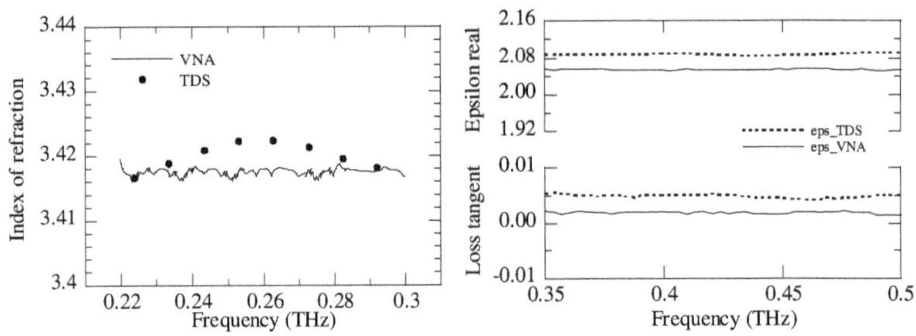

Figure 13.26 Left: Refractive index spectrum of HR-Si (adapted from Sun et al. [60]). Right: Real part of $\tilde{\varepsilon}(\omega)$ and loss tangent of PTFE (adapted from [61]).

for determining $n(\omega)$ and $\kappa(\omega)$, or $\varepsilon_r(\omega)$ and $\varepsilon_i(\omega)$ are rather similar to the ones used with THz-TDS data [56–60].

Several studies have been dedicated to the comparison of THz-TDS, FTIR and VNA characterization of materials and devices [51, 59–61]. Results from THz-TDS and VNA are in good agreement. Small differences arise from the calibration of the VNAs, and from the shape of the THz beam at the sample. Figure 13.26 (left) shows the refractive index of a wafer of high-resistivity Si. The VNA curve is obtained by taking into account the Fabry–Pérot oscillations in the sample. The TDS/VNA data differ only by 1.7%. Figure 13.26 (right) presents the spectral curves of the real part $\varepsilon_r(\omega)$ of the dielectric constant of PTFE and its loss tangent (defined as $\varepsilon_i(\omega)/\varepsilon_r(\omega)$). Here again, the agreement between the two techniques is excellent. $\varepsilon_r(\omega)$ differs by about 2%. The loss tangent is less precisely determined because its value is small due to the transparency of PTFE.

Recently, a detailed comparison of FTIR, THz-TDS, and VNA THz characterization has been published by Oberto et al. [63]. They conclude that the measurement of the refractive index can be performed easily with any of these techniques. However, in the spectral range attainable by all the three techniques, THz-TDS shows better performances, regarding uncertainty, while VNA is the less precise technique. This is seen in Fig. 13.27, where the THz-TDS, FTIR and VNA refractive index spectra of fused quartz are plotted, together with their error bars.

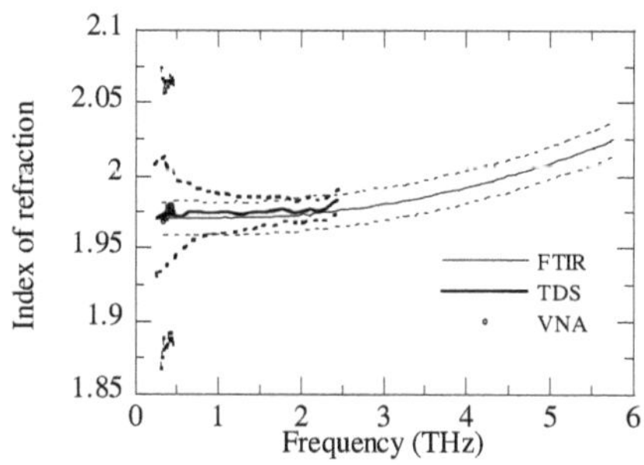

Figure 13.27 Refractive index spectrum of fused quartz. The dashed lines represent the error bars. Adapted from Sun et al. [60].

For transparent materials like fused quartz, FTIR gives the best results regarding the absorption coefficient. For less transparent materials like glass, THz-TDS is more precise.

13.5.2 CW THz Optoelectronic Setups

Continous-wave (CW) THz wave can be generated and detected through the optical beating of two CW laser sources, usually laser diodes. The beating of the two light beams is transformed into a THz signal in a nonlinear device. In most of the published or available CW spectrometers, the beating is performed in a photo-conducting switch or an ultrafast diode, like UTC diodes. Usually, one laser frequency is fixed, while the second one is tuned by varying the temperature of the laser diode. Let ω_1 and ω_2 be the pulsations delivered by the two laser diodes, whose intensities are supposed to be equal ($I_1 = I_2 = I_{\text{laser}}$). The photocurrent in the switch is proportional to the number of photo-generated carriers, i.e., to the total intensity of the two beams:

$$\begin{aligned} I_{\text{total}}(t) &= I_{\text{laser}}\left(\cos\left(\omega_1 t\right) + \cos\left(\omega_2 t\right)\right)^2 \\ &= I_{\text{laser}}\left[\cos^2\left(\omega_1 t\right) + \cos^2\left(\omega_2 t\right) + \cos\left(\left(\omega_1 + \omega_2\right) t\right)\right. \\ &\quad \left. + \cos\left(\left(\omega_1 - \omega_2\right) t\right)\right]. \end{aligned}$$

Because optical periods are much shorter than the photocarrier response time, the photocarrier density is proportional to the total intensity integrated over times longer than the optical periods but

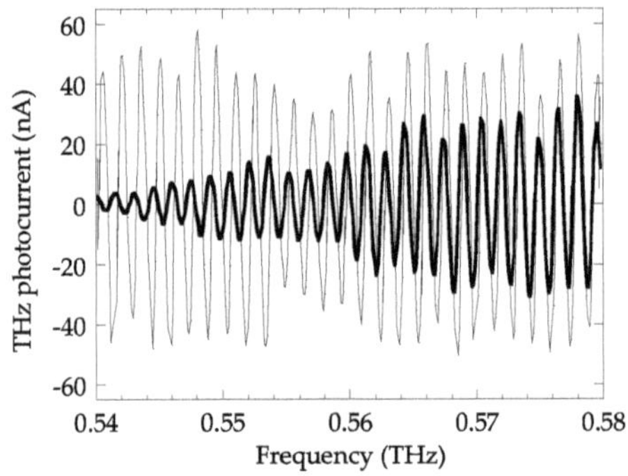

Figure 13.28 Recorded photocurrent in coherent continuous-wave THz spectroscopy system: empty system (thin line) and with a lactose sample (thick line). Adapted from Roggenbuck et al. [64].

shorter than the THz period:

$$\overline{I_{\text{total}}(t)} = I_{\text{laser}}\left[1 + \cos(\omega t)\right], \tag{13.34}$$

where $\omega = \omega_1 - \omega_2$ is the THz frequency. The radiated THz far field is proportional to the time derivative of the photocurrent:

$$E_{\text{THz}}(\omega, t) \propto I_{\text{laser}}\omega \sin(\omega t). \tag{13.35}$$

At the receiver, the current is proportional to the incoming THz field (Eq. 13.35) times the photo-generated carriers (which obey the same relation as Eq. 13.34):

$$i(\tau, t, \omega) \propto E_{\text{THz}}(\omega, t + \tau) \times \overline{I_{\text{laser}}(t)},$$

$$\Rightarrow \quad i(\tau, \omega) = \overline{i(\tau, t, \omega)} \propto I^2_{\text{laser}}\,\omega\, \sin(\omega\tau). \tag{13.36}$$

τ is the time delay between THz emission and detection. Thus the useful THz signal is varying sinusoidally with the time delay. If the difference of paths between the emitter and the receiver is $\Delta L = L_E + L_{\text{THz}} - L_D$ (L_E and L_D are the optical paths from the laser to respectively the emitter and the detector, and L_{THz} is the optical path from the emitting to the receiving THz antennas). Thus $\tau = \Delta L/c$. The spectral period of the interference is

$$\Delta f = c/\Delta L. \tag{13.37}$$

This interference pattern is observed in Fig. 13.28. Here $\Delta L = 20$ cm, leading to $\Delta f = 1.5$ GHz. Such interferential recorded signal forbids a continuous sweeping of the frequency when performing spectroscopy because the signal is null for $\omega\tau = m\pi$ (m is an integer). Usually, one saves data at the maxima of the recorded curve, i.e., with a period Δf given by Eq. 13.35. For ΔL equal to a few tens of centimeters, the periodicity of the recorded spectral data and thus the spectral resolution is comparable with common THz-TDS systems. A better resolution is obtained by enlarging the path difference ΔL.

As compared to THz-TDS, the dynamic range is better at lower frequencies ($80 \sim 100$ dB around 0.1 THz), but decreases strongly with the frequency. When dealing with THz photo-conducting antennas, this bandwidth is limited by both carrier lifetime and circuit constant time. Indeed, in the small signal approximation, the THz radiated power is [65]

$$P(\omega) = \frac{1}{2}\frac{(G_{\text{o}}V_{\text{B}})^2\,R_{\text{A}}}{\left(1+\omega^2\tau_e^2\right)\left(1+\omega^2\tau_{\text{RC}}^2\right)}. \tag{13.38}$$

V_{B} is the bias voltage, G_{o} is the switch photoconductance during illumination, R_{A} is the antenna-radiation resistance, and $\tau_{\text{RC}} = R_{\text{A}}C$, where C is the capacitance of the switch and τ_e is the photocarrier lifetime. Typically, $\omega\tau_e, \omega\tau_{\text{RC}} \gg 1$ and thus $P(\omega)$ varies as ω^{-4}. The largest reported THz bandwidths for such photo-mixing spectrometers are between 2 and 3 THz [66, 67].

Extracting the parameters of a sample from optical beating CW spectrometer data has been treated only in a few papers [64, 68–72]. In most of these publications, the contribution of Fresnel coefficients is neglected as well as the rebounds in the sample. The refractive index $n(\omega)$ and the absorption coefficient $\alpha(\omega)$ are directly determined from the phase and magnitude of the transmission coefficient. Let us write again the detected current (Eq. 13.36) versus the optical distance L between the emitting and receiving antennas:

$$\Rightarrow \quad i(\tau,\omega) \propto I^2_{\text{laser}}\,\omega\,\sin\left(\omega\frac{L_{\text{E}}+L-L_{\text{D}}}{c}\right). \tag{13.39}$$

The reference is measured without a sample between the antennas, so $L_{\text{THz}} = L \times n_{\text{air}}$. With the sample of thickness d and refractive index $n(\omega)$ in the setup, $L_{\text{THz}} = L \times n_{\text{air}} + (n(\omega) - n_{\text{air}})\,d$. The frequency

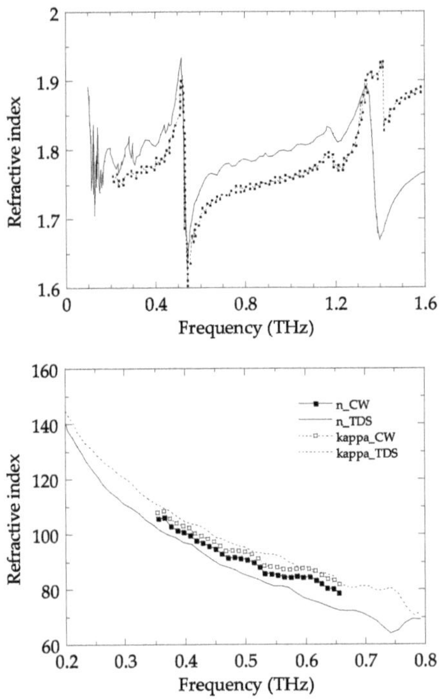

Figure 13.29 Top: Refractive index spectrum of α-lactose monohydrate measured using a CW spectrometer (continuous line). For comparison, the curve (dashed line) obtained from THz-TDS [73] is shown. Adapted from Roggenbuck et al. [64]. Bottom: $n(\omega)$ and $\kappa(\omega)$ of a 135 nm ITO layer deposited over a glass substrate. Both CW and TDS values are shown. Adapted from Moon et al. [71].

corresponding to a maximum m of the interference figure shifts from $f_{\text{ref,m}}$ to f_{m} when the sample is inserted in the setup, such as

$$f_{\text{m}} = m\frac{c}{L_{\text{E}} + l + (n(\omega) - 1)\,d - L_{\text{D}}}, \qquad f_{\text{ref,m}} = m\frac{c}{L_{\text{E}} + l - L_{\text{D}}}. \tag{13.40}$$

The refractive index $n(\omega)$ can be deduced from the frequency shift $\Delta f_{\text{m}} = f_{\text{m}} - f_{\text{ref,m}}$:

$$n(\omega) = 1 - m\frac{c}{d}\frac{\Delta f_{\text{m}}}{f_{\text{m}}\, f_{\text{ref,m}}} \approx 1 - m\frac{c}{d}\frac{\Delta f_{\text{m}}}{f_{\text{m}}^2}. \tag{13.41}$$

This procedure is straightforward as far as the frequency shift of a bright fringe is not bigger than the spectral interval between two consecutive fringes of the reference signal. This implies that

$$\Delta f_{\mathrm{m}} < f_{\mathrm{ref,m}} - f_{\mathrm{ref,m+1}} \quad \Rightarrow \quad d > d_{\min} = \frac{m}{m-1}\frac{1}{n(\omega)-1}\frac{c}{f_{\mathrm{ref,m}}}. \tag{13.42}$$

At 1 THz, with $L_{\mathrm{E}} + l - L_{\mathrm{D}} = 30$ cm, $m = 1000$ and $d_{\min} \approx 0.3/(n(\omega) - 1)$ mm. Otherwise, the integer m in (13.40) is not saved, and relation 13.39 is not valid. The transmittance of the sample is simply obtained by taking the ratio of the signal and reference detected currents. To avoid division by zero at the minima of the reference, one usually calculates the envelopes of the signal and reference by extrapolating the data between maxima.

An example, extracted from Roggenbuck et al. [64], is shown in Fig. 13.29: this is the refractive index of α-lactose monohydrate. Resonances at 0.46 and 1.4 THz are seen. The weak oscillations of the curve are due to the too much-simplified method to extract $n(\omega)$: these oscillations result from the contribution of the Fabry–Pérot effect, which has not been compensated. This curve is compared with the values obtained with the same material but using THz-TDS [68]. The agreement is good, but not excellent. The reason is maybe the simplified code used to extract $n(\omega)$ [64].

Another example, adapted from Moon et al. [71], is given in Fig. 13.29 (bottom). This is the refractive index of a 135 nm thick layer of ITO (indium tin oxide), which is a transparent (in the visible) conductive material. In the THz range, its high conductivity leads to high and comparable values of both $n(\omega)$ and $\kappa(\omega)$. The values achieved by CW spectroscopy are in excellent agreement with those from THz-TDS.

References

1. Bernier, M., Garet, F., and Coutaz, J.-L. (2013). Determination of the refractive index of samples showing low transmission bands by THz time-domain spectroscopy, *IEEE Trans. THz Sci. Technol.*, **3**, pp. 295–301.

2. Lucarini, V., Saarinen, J. J., Peiponen, K.-E., and Vartiainen, E. M. (2005). *Kramers-Kronig Relations in Optical Materials Research* (Springer Science & Business Media, Berlin), Chapter 4.

3. Han, P. Y., Tani, M., Usami, M., Kono, S., Kersting, R., and Zhang, X.-C. (2001). A direct comparison between terahertz time-domain spectroscopy and far-infrared Fourier transform spectroscopy, *J. Appl. Phys.*, **89**, pp. 2357–2359.
4. Huang, F., Schulkin, B., Altan, H., Federici, J. F., Gary, D., Barat, R., Zimdars, D., Chen, M., and Tanner, D. B. (2004). Terahertz study of 1,3,5-trinitro-s-triazine by time-domain and Fourier transform infrared spectroscopy, *Appl. Phys. Lett.*, **85**, pp. 5535–5537.
5. Palka, N., Szustakowski, M., Trzcinski, T., and Mozdzonek, M. (2011). Comparison of spectra of materials measured by time domain and Fourier transform spectroscopy in terahertz range, *Photon. Lett. Poland*, **3**, pp. 76–78.
6. Ding, T., Li, R., Zeitler, J. A., Huber, T. L., Gladden, L. F., Middelberg, A. P. J., and Falconer, R. J. (2010). Terahertz and far infrared Spectroscopy of alanine-rich peptides having variable ellipticity, *Opt. Express*, **18**, pp. 27431–27444.
7. Coutaz, J.-L., Bernier, M., Garet F., Joly, S., Miyake, Y., Minamide, H., Lheurette, E., and Lippens, D. (2014). Terahertz time-domain spectroscopy of absorbing materials and of metamaterials, *THz Sci. Technol.*, **7**, pp. 53–69.
8. Naftaly, M., and Miles, R. E. (2007). Terahertz time-domain spectroscopy for material characterization, *Proc. IEEE*, **95**, pp. 1658–1665.
9. Ho, I.-C., Guo, X., and Zhang, X.-C. (2010). Design and performance of reflective terahertz air-biased-coherent-detection for time-domain spectroscopy, *Opt. Express*, **18**, pp. 2872–2883.
10. Bernier, M., Garet, F., Coutaz, J.-L., Minamide, H., and Sato, A. (2016). Accurate characterization of resonant samples in the terahertz regime through a technique combining time-domain spectroscopy and Kramers-Kronig analysis, *IEEE Trans. THz Sci. Technol.*, **3**, pp. 295–301.
11. RIKEN Terahertz database (www.riken.jp/THzdatabase/).
12. Walther, M., Jensby, K., Keiding, S. R., Takahashi, H., and Ito, H. (2000). Far-infrared properties of DAST, *Opt. Lett.*, **25**, pp. 911–913.
13. Lucarini, V., Ino, Y., Peiponen, K.-E., and Kuwata-Gonokami, M. (2005). Detection and correction of the misplacement error in terahertz spectroscopy by application of singly subtractive Kramers–Kronig relations, *Phys. Rev. B*, **72**, p. 125107.
14. Peiponen, K.-E., Gornov, E., Svirko, Y., Ino, Y., Kuwata-Gonokami, M., and Lucarini, V. (2005). Testing the validity of terahertz reflection spectra by dispersion relations, *Phys. Rev. B*, **72**, p. 245109.

15. Vartiainen, E. M., Ino, Y., Shimano, R., Kuwata-Gonokami, M., and Peiponen, K.-E. (2004). Numerical phase correction method for terahertz time-domain reflection spectroscopy, *J. Appl. Phys.*, **96**, p. 4171.

16. Herrmann, M., Platte, F., Nalpantidis, K., Beigang, R., and Heise H. M. (2012). Combination of Kramers–Kronig transform and time-domain methods for the determination of optical constants in THz spectroscopy, *Vib. Spectrosc.*, **60**, pp. 107–112.

17. Son, H., Choi, D.-H., and Park, G.-S. (2017). Improved thickness estimation of liquid water using Kramers-Kronig relations for determination of precise optical parameters in terahertz transmission spectroscopy, *Opt. Express*, **25**, pp. 4509–4519.

18. Koningstein, J. A. (1972). *Introduction to the Theory of the Raman Effect* (D. Reidel Publishing Company, Dordrecht).

19. Colthup, N. B., Daly, L. H., and Wiberley S. E. (1990). *Introduction to Infrared and Raman Spectroscopy*, 3rd edn. (Academic Press, Inc., Boston).

20. Ferraro, J. R., Nakamoto, K., and Brown, C. W. (2003). *Introductory Raman Spectroscopy*, 2nd edn. (Academic Press, San Diego).

21. Merlin, R., Pinczuk, A., and Weber, W. H. (2000). Overview of phonon Raman scattering in solids, in *Raman Scattering in Materials Science*, Springer Series in Materials Science (Springer-Verlag Berlin Heidelberg), **42**, pp. 1–29.

22. Fleury, P. A., and Loudon, R. (1968). Scattering of light by one- and two-magnon excitations, *Phys. Rev.*, **166**, p. 514.

23. Pimenov, A., Mukhin, A. A., Ivanov, V. Y., Travkin, V. D., Balbashov, A. M., and Loidl, A. (2006). Possible evidence for electromagnons in multiferroic manganites, *Nat. Phys.*, **2**, pp. 97–100.

24. Peuker, K., and Enderlein, R. (1974). Theory of Raman scattering from excitons, *Phys. Stat. Sol. B*, **61**, pp. 247–255.

25. Barrett, J. J., and Weber, A. (1970). Pure-rotational Raman scattering in a CO_2 electric discharge, *J. Opt. Soc. Am.*, **60**, pp. 70–76.

26. Parrott, E. P. J., and Zeitler, J. A. (2015). Terahertz time-domain and low-frequency Raman spectroscopy of organic materials, *Appl. Spectrosc.*, **69**, pp. 1–25.

27. Kutuvantavida, Y., Williams, G. V. M., Pogson, E. M., Bhuiyan, D., Radhanpura, K., and Lewis, R. A. (2012). Material characterization at low frequencies using THz and Raman spectroscopy, *Proceed. 37th IRMMW-THz Conference*, pp. 1–3.

28. Brandt, N. N., Chikishev, A. Y., Kargovsky, A. V., Nazarov, M. M., Parashchuk, O. D., Sapozhnikov, D. A., Smirnova, I. N., Shkurinov, A. P., and Sumbatyan, N. V. (2008). Terahertz time-domain and Raman spectroscopy of the sulfur-containing peptide dimers: low-frequency markers of disulfide bridges, *Vib. Spectrosc.*, **47**, pp. 53–58.

29. Burnett, A. D., Fan, W. H., Upadhya, P. C., Cunningham, J. E., Edwards, H. G. M., Kendrick, J., Munshi, T., Hargreaves, M., Linfield, E. H., and Davies, A. G. (2007). Broadband terahertz time-domain and Raman spectroscopy of explosives, *Proc. SPIE*, **6549**, p. 654905.

30. Kojima, S., Mori, T., Shibata, T., and Kobayashi, Y. (2015). Broadband terahertz time-domain and low-frequency Raman spectroscopy of crystalline and glassy pharmaceuticals, *Pharm. Anal. Acta*, **6**, p. 1000401.

31. Ullah, R., Ud-Din Khan, S., Aamir, M., and Ullah, R. (2013). Terahertz time domain, Raman and Fourier transform infrared spectroscopy of acrylamide, and the application of density functional theory, *J. Spectrosc.*, **2013**, Article ID 148903.

32. Mori, T., Igawa, H., and Kojima, S. (2014). Progress of ultrafast terahertz time-domain spectroscopy: Raman inactive soft mode in quantum paraelectric $SrTiO_3$, *IOP Conf. Series: Materials Science and Engineering*, **54**, conference 1.

33. Kojima, S., Kitahara, H., Nishizawa, S., and Wada Takeda, M. (2004). Terahertz time domain spectroscopy of boson peak, *AIP Conf. Proc.*, **708**, p. 669.

34. Nicolson, A. M., and Ross, G. F. (1970). Measurement of the intrinsic properties of materials by time-domain techniques, *IEEE Trans. Instrum. Meas.*, **IM-19**, pp. 377–382.

35. Weir, W. B. (1974). Automatic measurement of complex dielectric constant and permeability at microwave frequencies, *Proc. IEEE*, **62**, pp. 33–36.

36. Smith, D. R., Schultz, S., Markos, P., and Soukoulis, C. M. (2002). Determination of effective permittivity and permeability of metamaterials from reflection and transmission coefficients, *Phys. Rev. B*, **65**, p. 195104.

37. Denton, R. E., Campbell, R. D., and Tomlin, S. G. (1972). The determination of the optical constants of thin films from measurements of reflectance and transmittance at normal incidence, *J. Phys. D: Appl. Phys.*, **5**, p. 852.

38. Minowa, Y., Fujii, T., Nagai, M., Ochiai, T., Sakoda, K., Hirao, K., and Tanaka, K. (2008). Evaluation of effective electric permittivity and

magnetic permeability in metamaterial slabs by terahertz time-domain spectroscopy, *Opt. Express*, **16**, pp. 4785–4796.

39. Wu, Z., Wang, L., Peng, Y., Young, A., Seraphin, S., and Xin, H. (2008). Terahertz characterization of multi-walled carbon nanotube films, *J. Appl. Phys.*, **103**, p. 094324.
40. Nemec, H., Kadlec, F., Kuzel, P., Duvillaret, L., and Coutaz, J.-L. (2006). Independent determination of the complex refractive index and wave impedance by time-domain terahertz spectroscopy, *Opt. Commun.*, **260**, pp. 175–183.
41. Namai, A., Yoshikiyo, M., Yamada, K., Sakurai, S., Goto, T., Yoshida, T., Miyazaki, T., Nakajima, M., Suemoto, T., Tokoro, H., and Ohkoshi, S.-I. (2016). Hard magnetic ferrite with a gigantic coercivity and high frequency millimetre wave rotation, *Nat. Commun.*, **3**, p. 1035.
42. Kadlec, C., Goian, V., Rushchanskii, K. Z., Kuzel, P., Lezaic, M., Kohn, K., Pisarev, R. V., and Kamba, S. (2011). Terahertz and infrared spectroscopic evidence of phonon-paramagnon coupling in hexagonal piezomagnetic $YMnO_3$, *Phys. Rev. B*, **84**, p. 174120.
43. Zhang, P., Su, F., Chen, X., Zhang, S., Mei, H., Yang, Z., Dai, J., and Pi, L. (2016). Observation of magnetic resonance and Faraday rotation in a spinel $ZnCr_2Se_4$ single crystal in the terahertz region, *Appl. Phys. Express*, **9**(10), p. 1.
44. Yen, T. J., Padilla, W. J., Fang, N., Vier, D. C., Smith, D. R., Pendry, J. B., Basov, D. N., and Zhang, X. (2004). Terahertz magnetic response from artificial materials, *Science*, **303**, pp. 1494–1496.
45. Choi, M., Lee, S. H., Kim, Y., Kang, S. B., Shin, J., Kwak, M. H., Kang, K.-Y., Lee, Y.-H., Park, N., and Min, B. (2011). A terahertz metamaterial with unnaturally high refractive index, *Nature*, **470**, p. 369.
46. Yen, T. J., Padilla, W. J., Fang, N., Vier, D. C., Smith, D. R., Pendry, J. B., Basov, D. N., and Zhang, X. (2004). Terahertz magnetic response from artificial materials, *Science*, **303**, pp. 1494–1496.
47. O'Brien, S., and Pendry, J. B. (2002). Photonic band-gap effects and magnetic activity in dielectric composite, *J. Phys. : Condens. Matter*, **14**, p. 4035.
48. Zhao, Q., Zhou, J., Zhang, F., and Lippens, D. (2009). Mie resonance-based dielectric metamaterials, *Mater. Today*, **12**, pp. 60–69.
49. Nemec, H., Kadlec, C., Kadlec, F., Kuzel, P., Yahiaoui, R., Chung U.-C., Elissalde, C., Maglione, M., and Mounaix, P. (2012). Resonant magnetic response of TiO_2 microspheres at terahertz frequencies, *Appl. Phys. Lett.*, **100**, p. 061117.

50. Goy, P., Gross, M., and Raimond, J. M. (1990). 8–1000 GHz vector network analyzer, *15th Int. Conf. on Infrared and Millimeter Waves* (IRMMW), Orlando, USA, Dec. 1990.

51. Goy, P., and Carooten, S. (2009). Sweep-able sub-millimeter sources and detectors for THz vector network analyzers and applications, *Workshop and Short Courses, 39th European Microwave Week*, Rome, Italy, 28 Sept. - 2 Oct. 2009.

52. George, D. K., Charkhesht, A., and Vinh, N. Q. (2015). New terahertz dielectric spectroscopy for the study aqueous solutions, *Rev. Sci. Instrum.*, **86**, p. 123105.

53. Koller, D., Durant, S., Rowland, C., Bryerton, E., and Hesler, J. (2016). Initial measurements with WM164 (1.1-1.5 THz) VNA extenders, *Paper H4D. 1, 41st IRMMW-THz Conference*, Copenhagen, Denmark, 25–30 Sept. 2016.

54. Hesler, J. L., Duan, Y., Foley, B., and Crowe, T. W. (2010). THz vector network analyzer measurements and calibration, *21st Int. Symp. Space THz Technol.*, Oxford, UK, 23–25 March 2010.

55. Hanning, J., Stenarson, J., Yhland, K., Sobis, P. J., Bryllert, T., and Stake, J. (2014). Single-flange 2-port TRL calibration for accurate THz-parameter measurements of waveguide integrated circuits, *IEEE Trans. THz Sci. Technol.*, **4**, pp. 582–587.

56. Hasar, U. C. (2009). A microwave method for noniterative constitutive parameters determination of thin low-loss or lossy materials, *IEEE Trans. Microwave Theory Tech.*, **57**, p. 1595.

57. Williams, T. C., Stuchly M. A., and Saville, P. (2003). Modified transmission reflection method for measuring constitutive parameters of thin flexible high-loss materials, *IEEE Trans. Microwave Theory Tech.*, **51**, p. 1560.

58. Hasar, U. C. (2011). Unique retrieval of complex permittivity of low -loss dielectric materials from transmission only measurements, *IEEE Geosci. Remote Sens. Lett.*, **8**, p. 562.

59. Yang, B., Wylde, R. J., Martin, D. H., Goy, P., Donnan, R. S., and Caroopen, S. (2010). Determination of the gyrotropic characteristics of hexaferrite ceramics from 75 to 600 GHz, *IEEE Trans. THz Sci. Technol.*, **58**, p. 3587.

60. Sun, W., Yang, B., Wang, X., Zhang, Y., and Donnan, R. (2013). Accurate determination of terahertz optical constants by vector network analyzer of Fabry–Pérot response, *Opt Lett.*, **38**, pp. 5438–5441.

61. Chang, T., Zhang, S., Yang, S., Sun, Z., and Cui, H.-L. (2017). Measurement of complex terahertz dielectric properties of polymers using an improved free-space technique, *Meas. Sci. Technol.*, **28**, p. 045002.
62. Naftaly, M., Ridler, N., Molloy, J., Shoaib, N., and Stokes, D. (2015). A comparison method for THz measurements using VNA and TDS, *40th IRMMW-THz conference*, Hong Kong, pp. 1–2.
63. Oberto, L., Bisi, M., Kazemipour, A., Steiger, A., Kleine-Ostmann, T., and Schrader, T. (2017). Measurement comparison among time-domain, FTIR and VNA-based spectrometers in the THz frequency range, *Metrologia*, **54**, pp. 77–84.
64. Roggenbuck, A., Schmitz, H., Deninger, A., Cámara Mayorga, I., Hemberger, J., Güsten, R., and Grüninger, M. (2010). Coherent broadband continuous-wave terahertz spectroscopy on solid-state samples, *New J. Phys.*, **12**, pp. 043017–043030.
65. Brown, E. R., Smith, F. W., and McIntosh, K. A. (1993). Coherent millimeter wave generation by heterodyne conversion in low-temperature-grown GaAs photoconductors, *J. Appl. Phys.*, **73**, pp. 1480–1484.
66. Mouret, G., Matton, S., Bocquet, R., Hindle, F., Peytavit, E., Lampin, J.F., and Lippens, D. (2004). Far-infrared CW difference-frequency generation a using vertically integrated and planar low-temperature grown GaAs photomixers: application to H_2S rotational spectrum up to 3 THz, *Appl. Phys. B: Lasers Opt.*, **79**, pp. 725–729.
67. Mangeney, J., Merigault, A., Zerounian, N., Crozat, P., Blary, K., and Lampin, J.F. (2007). Continuous wave terahertz generation up to 2 THz by photomixing on ion-irradiated $In_{0.53}Ga_{0.47}As$ at 1.55 μm wavelengths, *Appl. Phys. Lett.*, **91**, p. 241102.
68. Mouret, G., Matton, S., Bocquet, R., Bigourd, D., Hindle, F., Cuisset, A., Lampin, J. F., Blary, K., and Lippens, D. (2007). THz media characterization by means of coherent homodyne detection, results and potential applications, *Appl. Phys. B*, **89**(2–3), pp. 395–399.
69. Wilk, R. (2010). Continuous wave terahertz spectrometer with coherent detection, *Bull. Pol. Acad. Sci., Techn. Sci.*, **58**, pp. 713–716.
70. Jarzab, P. P., Nowak, K., Walczakowski, M. J., Augustyn, L., Mikulics, M., Wilk, R., Witkowski, J. S., and Plinski, E. F. (2012). Selected polymers on CW and pulsed THz investigations, *Opto-Electron. Rev.*, **20**, pp. 335–339.
71. Moon, K., Kim, N., Shin, J.-H., Yoon, Y.-J., Han, S.-P., and Park, K. H. (2014). Continuous-wave terahertz system based on a dual-mode laser for real-time non-contact measurement of thickness and conductivity, *Opt. Express*, **22**, pp. 2259–2266.

72. Lee, I. M., Kim, N., Lee, E. S., Han, S. P., Moon, K., and Park, K. H. (2015). Frequency modulation based continuous-wave terahertz homodyne system, *Opt. Express*, **23**, pp. 846–858.

73. Fischer, B. M. (2005). PhD thesis, Albert-Ludwigs University, Freiburg, Germany.

Annex

A.1 Electromagnetism

Let's consider a sample under study that is a slab of an isotropic material with flat and parallel faces, which is located in an isotropic surrounding medium. We will treat in the following paragraphs how an incident electromagnetic wave is transmitted and reflected by the sample. For the sake of simplicity, we will consider that the incident beam is a plane wave. First, we will recall how such a wave interacts with the interface between two different media. Then we will study the most common case of a slab with parallel faces. Finally, we will generalize this treatment to a multilayer sample.

A.1.1 Electromagnetic Fields and the Maxwell Equations

The electromagnetic field is known to correspond to an electric field, hereafter noted as $\vec{E}$, that propagates and oscillates in space and time together and simultaneously with a magnetic field noted $\vec{B}$. When these fields propagate into matter, they induce dipoles in the atoms/molecules and/or they move free electrical charges in the matter (here we do not consider the excitation of atoms/molecules to higher energy levels). The induced electrical effects tend to decrease the strength of $\vec{E}$, while the induced magnetic ones increase $\vec{B}$. One notes the vector $\vec{D}$, called the displacement field, that renders the effect of the induced dipoles in the matter and $\vec{H}$, the magnetic field strength, which accounts for the magnetic response of matter. Both $\vec{D}$ and $\vec{H}$ are functions of the exciting electric and magnetic fields $\vec{E}$ and $\vec{B}$, respectively. For weak exciting fields (as compared to the intra-atomic field), these functions are linear and are written as

$$\vec{D} = \varepsilon_o \varepsilon \, \vec{E}, \; \vec{B} = \mu_o \mu \, \vec{H}. \tag{A.1}$$

Equations A.1 are known as the constitutive relations. ε_o and μ_o are respectively the permittivity and permeability of vacuum.[a] ε and μ are the relative permittivity and permeability of the matter. In the international system of units, the strengths of $\vec{E}$, $\vec{D}$, $\vec{B}$, and $\vec{H}$ are respectively given in V/m, C/m^2, T and A/m. Let us notice that, in Eq. A.1, $\vec{E}$ and $\vec{H}$ play a similar role mathematically, while it should have been $\vec{E}$ and $\vec{B}$. This is simply due to historical reasons; researchers mixed up these notions at the beginning of the science of electromagnetism. At higher field strengths, the relations between $\vec{D}$ and $\vec{E}$ ($\vec{B}$ and $\vec{H}$) are no more linear due to the limited deformation of the nuclear or molecular clouds in the material, or to the nonlinear behavior of the Lorentz force applied to the free carriers. These functions can be extended as a series of the powers of $\vec{E}(\vec{H})$. These high exciting fields are easily produced by lasers, especially pulsed lasers. This explains why the new science of nonlinear optics started almost simultaneously with the invention of the lasers [1]. In anisotropic and nonlocal materials, $\vec{D}$ and $\vec{E}$ ($\vec{B}$ and $\vec{H}$) are not parallel, and thus ε or μ are tensors.

Electromagnetic fields are described by Maxwell's equations:

$$\begin{cases} \vec{\nabla} \times \vec{E} = -\frac{\partial \vec{B}}{\partial t}, \; \vec{\nabla} \times \vec{H} = \frac{\partial \vec{D}}{\partial t} + \vec{J}, \\ \vec{\nabla} \cdot \vec{D} = \rho, \; \vec{\nabla} \cdot \vec{B} = 0. \end{cases} \tag{A.2}$$

ρ and $\vec{J}$ are the charge density and the electrical current density in the medium where the electromagnetic field propagates. It is easy to derive the propagation equation of the electromagnetic field from Eqs. A.1 and A.2:

$$\Delta \vec{E} - \frac{\varepsilon \mu}{c^2} \frac{\partial^2 \vec{E}}{\partial t^2} = \mu_o \mu \frac{\partial \vec{J}}{\partial t}, \; \Delta \vec{H} - \frac{\varepsilon \mu}{c^2} \frac{\partial^2 \vec{H}}{\partial t^2} = -\vec{\nabla} \times \vec{J}. \tag{A.3}$$

Here the Laplace operator Δ stands for $\partial^2/\partial x^2 + \partial^2/\partial y^2 + \partial^2/\partial z^2$ and c is the speed of light in vacuum.[b]. The left-hand-side term of these equations is a second-order differential equation that

---

[a] $\varepsilon_o = 1/(36\pi \times 10^9) \approx 8.8 \times 10^{-12}$ F/m and $\mu_o = 4\pi \times 10^{-7} \approx 12.6 \times 10^{-7}$ H/m.
[b] $c = 299\ 792\ 458$ m/s, but in most of calculations, c can be approximated by $c \approx 3 \times 10^8$ m/s with an round error of 0.07 % Moreover, $\varepsilon_o \mu_o c^2 = 1$.

describes the propagation of the $\vec{E}$ and $\vec{H}$. The right-hand term is the source of the electromagnetic field. Thus the electromagnetic fields are generated by the temporal or spatial variations in the current density $\vec{J}$. Equations A.3 are the inhomogeneous Helmholtz equations. The homogeneous Helmholtz equations they are obtained when the waves propagate in media free of charges and currents, i.e., when the right-hand terms of A.3 are equal to zero:

$$\Delta\vec{E} - \frac{\varepsilon\,\mu}{c^2}\frac{\partial^2\,\vec{E}}{\partial\,t^2} = 0,\ \Delta\vec{H} - \frac{\varepsilon\,\mu}{c^2}\frac{\partial^2\,\vec{H}}{\partial\,t^2} = 0. \tag{A.4}$$

Equations A.4 exhibit an infinite number of solutions! Here, two are of interest: (i) the plane wave, which is a good approximation of any wave when observed far from the source, and (ii) the Gaussian wave that is delivered by monomode lasers. Gaussian waves constitute a good estimation of most of the spatially limited beams.

In this paragraph, we treat only the case of plane waves. A plane wave oscillates both in time and space. We simply define it as

$$\vec{E}\,(\vec{r},\,t) = \vec{E}_o\cos\left(\omega\,t - \vec{k}\cdot\vec{r}\right),\ \vec{H}\,(\vec{r},\,t) = \vec{H}_o\cos\left(\omega\,t - \vec{k}\cdot\vec{r}\right). \tag{A.5}$$

t is the time, $\vec{r}$ is the location in space, and ω is the pulsation (or angular frequency) of the fields and $\vec{k}$ is the wavevector. By substituting Eqs. A.5 into the propagation equation (Eq. A.4), we get

$$k = \pm\sqrt{\varepsilon\,\mu}\frac{\omega}{c} = \pm n\frac{\omega}{c}. \tag{A.6}$$

k is the magnitude of $\vec{k}$. The sign $\pm$ denotes forward (+) or backward (−) propagating waves, that obey the same propagation equation (Eq. A.4). The refractive index n of the medium is defined as $n = \sqrt{\varepsilon\,\mu}$. The speed of the wave, i.e., the speed of its maxima or minima, is the phase velocity that is equal to $v = c/n$. The frequency of the wave is $f = \omega/2\pi$ and its wavelength is $\lambda = c/f$. The direction of $\vec{k}$ is the one of the energy carried by the electromagnetic wave.

In many problems, it is easier to work with complex exponential functions rather than with trigonometric functions. Therefore, one uses the equality

$$\begin{aligned}\cos\left(\omega t - \vec{k}\cdot\vec{r}\right) &= \frac{1}{2}\left(e^{j\left(\omega t-\vec{k}\cdot\vec{r}\right)} + e^{-j\left(\omega t-\vec{k}\cdot\vec{r}\right)}\right)\\ &= \frac{1}{2}e^{j\left(\omega t-\vec{k}\cdot\vec{r}\right)} + \text{C.C.}\end{aligned} \tag{A.7}$$

C.C. means complex conjugate. As most electromagnetic problems are linear, the factor 1/2 in Eq. A.7 is forgotten, and one keeps only the first complex exponential term. Thus the fields A.5 become

$$\vec{E}\,(\vec{r},t) = \vec{E}\,e^{j\left(\omega t - \vec{k}\cdot\vec{r}\right)},\ \vec{H}\,(\vec{r},t) = \vec{H}\,e^{j\left(\omega t - \vec{k}\cdot\vec{r}\right)}. \tag{A.8}$$

This is not entirely correct but, as far as one is dealing with linear equations, the 1/2 factor may be simplified in the calculus, and the complex conjugates obey the same equation. When treating nonlinear equations, mostly concerning energy, the exact equation (A.7) must be employed. As soon as materials absorb a part of the energy of the electromagnetic waves, the wave amplitude decreases with the propagation distance in the material. Using complex exponential functions, the absorption is easily introduced in the propagation equation by using a complex refractive index $\tilde{n}$:

$$\tilde{n} = n - j\,\kappa. \tag{A.9}$$

n and κ are respectively the real and imaginary parts of the complex refractive index.[c] κ is also named extinction coefficient. People dealing with microwaves often use the loss tangent defined as

$$\tan\delta = \frac{\mathrm{Im}\,(\varepsilon)}{\mathrm{Re}\,(\varepsilon)} = \frac{2n\kappa}{n^2 - \kappa^2}. \tag{A.10}$$

For most materials, $n \gg \kappa$, which leads to the following simplified expression of the loss tangent:

$$\tan\delta \approx \frac{2\kappa}{n}. \tag{A.11}$$

For any vector having the form A.8, the curl is simply written as

$$\vec{A}\,(\vec{r},t) = \vec{A}\,e^{j\left(\omega t - \vec{k}\cdot\vec{r}\right)} \Rightarrow \vec{\nabla}\times\vec{A} = -j\,\vec{k}\times\vec{A}, \tag{A.12}$$

where $\times$ denotes the vectorial product of two vectors. The first two Maxwell equations (A.2) become

$$\vec{k}\times\vec{E} = \omega\,\mu_0\mu\,\vec{H},\ \vec{k}\times\vec{H} = -\omega\,\varepsilon_0\varepsilon\,\vec{E}. \tag{A.13}$$

[c] Let us notice that, instead of Eqs. A.8, one may have chosen

$$\vec{E}\,(\vec{r},t) = \vec{E}\,e^{-j\left(\omega t - \vec{k}\cdot\vec{r}\right)},\ \vec{H}\,(\vec{r},t) = \vec{H}\,e^{-j\left(\omega t - \vec{k}\cdot\vec{r}\right)}.$$

All the treatments presented here remain valid, but the sign "+" of the imaginary unit j must be changed into "−." Especially, the complex refractive index takes the form: $\tilde{n} = n + j\,\kappa$. Otherwise, a negative imaginary part will lead to gain instead of loss!

Thus $\vec{E}$, $\vec{H}$, and $\vec{k}$ are perpendicular. As $\vec{k}$ defines the direction of propagation of the wave, $\vec{E}$ and $\vec{H}$ are perpendicular to this direction of propagation. Multiplying each of equations A.13 by $\vec{k}\times$ leads to

$$\vec{k} \times \vec{k} \times \vec{E} = -\varepsilon\,\mu \frac{\omega^2}{c^2}\,\vec{E}, \vec{k} \times \vec{k} \times \vec{H} = -\varepsilon\,\mu \frac{\omega^2}{c^2}\vec{H}. \quad \text{(A.14)}$$

This is nothing but expression A.4 for a plane wave. The power per unit area carried by the electromagnetic wave is determined from the Poynting vector $\vec{P}$ defined as follows:

$$\vec{P}\,(\vec{r}, t) = \vec{E}\,(\vec{r}, t) \times \vec{H}\,(\vec{r}, t) \quad \text{(A.15)}$$

For a plane wave, using Eqs. A.4, we get the time-averaged value $\vec{P}_m$ of $\vec{P}$:

$$\vec{P}_m = \frac{1}{T}\int_0^{T=\frac{2\pi}{\omega}} \vec{P}\,dt = \frac{n}{2Z_o}E_o^2\,\vec{u}_k \quad \text{(A.16)}$$

$\vec{u}_k$ is the unit vector in the direction of $\vec{k}$. $Z_o = c\mu_o \approx 377\Omega$ is the impedance of vacuum. The magnitude of $\vec{P}_m$ is a power density, i.e., the power carried through a unit of section by the electromagnetic beam. Very often, this power density is called the intensity I of the beam:

$$I = \frac{n}{2Z_o}E_o^2 \quad \text{(A.17)}$$

Relations A.16 and A.17 are written for a transparent medium ($\kappa = 0$).

A.1.2 Reflection and Transmission at an Interface

From the four Maxwell equations (A.2), one can easily demonstrate [2] the rules of field conservation at an interface between two media noted 2 and 1:

$$\begin{cases} \vec{B}_2 - \vec{B}_1 = 0,\ \vec{N}_{12} \cdot \left(\vec{D}_2 - \vec{D}_1\right) = \sigma_S, \\ \vec{N}_{12} \times \left(\vec{E}_2 - \vec{E}_1\right) = 0,\ \vec{N}_{12} \times \left(\vec{H}_2 - \vec{H}_1\right) = \vec{J}_S. \end{cases} \quad \text{(A.18)}$$

$\vec{J}_S$ and σ_S are respectively the surface current density and the surface charge density. $\vec{N}_{12}$ is the unit vector perpendicular to the

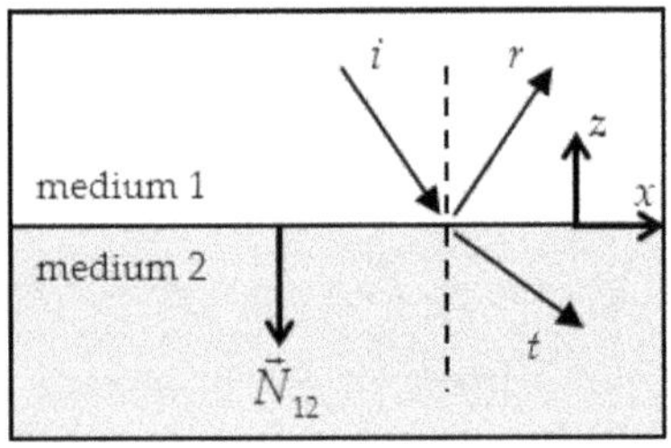

Figure A.1 Reflection and refraction at an interface.

surface, directed from medium 1 to 2. In the case of dielectric materials, the conductivity is null or small, so $\vec{J}_S \approx 0$ and thus the two last equalities express the conservation of the components of $\vec{E}$ and $\vec{H}$ perpendicular to $\vec{N}_{12}$, i.e., tangent to the interface.

Let us now consider the flat interface between media 2 and 1 of refractive index n_2 and n_1, as shown in Fig. A.1. These two media are semi-infinite, homogenous and isotropic. The problem shows an invariance of translation versus the direction y, so any derivation $\partial/\partial y$ is null. The common solution to Maxwell's equations in media 2 and 1 is a plane wave. We note $\vec{E}_r$ and $\vec{E}_t$ ($\vec{H}_i$, $\vec{H}_r$, and $\vec{H}_t$) the fields that are incident and reflected in medium 1, and transmitted in medium 2.

In the case $\vec{J}_S = 0$ and $\sigma_S = 0$, Eqs. A.18 lead to

$$\begin{cases} E_{ix} + E_{rx} = E_{tx} \\ E_{iy} + E_{ry} = E_{ty} \\ \varepsilon_1 E_{iz} + \varepsilon_1 E_{rz} = \varepsilon_2 E_{tz} \end{cases} \quad \begin{cases} H_{ix} + H_{rx} = H_{tx} \\ H_{iy} + H_{ry} = H_{ty} \\ \mu_1 H_{iz} + \mu_1 H_{rz} = \mu_2 H_{tz} \end{cases} \tag{A.19}$$

We have already taken into account the invariance along x, which imposes the conservation of the x component of $\vec{k}$. Because $\partial/\partial y = 0$, we write the x and z components of $\vec{E}$ ($\vec{H}$) versus H_y (E_y) from Eq. A.2, considering the plane-wave character of the fields. Thus, we end up with two independent sets of two equations:

$$\begin{cases} E_{iy} + E_{ry} = E_{ty} \\ E_{iy} - E_{ry} = \frac{k_{2z}}{k_{1z}}\frac{\mu_1}{\mu_2} E_{ty} \end{cases} \quad \begin{cases} H_{iy} + H_{ry} = H_{ty} \\ H_{iy} - H_{ry} = \frac{k_{2z}}{k_{1z}}\frac{\varepsilon_1}{\varepsilon_2} H_{ty} \end{cases} \tag{A.20}$$

Due to the symmetry of the problem ($\partial/\partial y = 0$), E_y and H_y can be independently calculated. Thus, we can project any incoming field of arbitrary polarization on the basis of TE (transverse electric,

$E_y \neq 0,\ E_x = E_z = 0$) and TM (transverse magnetic, $H_y \neq 0,\ H_x = H_z = 0$) polarized fields. Moreover, the two equations A.20 have a similar mathematical expression. Thus, we define

$$\begin{cases} F = E_y, & \alpha_{12} = \frac{\mu_1}{\mu_2}, \text{ (TE)}, \\ F = H_y, & \alpha_{12} = \frac{\varepsilon_1}{\varepsilon_2}, \text{ (TM)}. \end{cases} \tag{A.21}$$

Now the two cases can be solved simultaneously provided one makes use of definitions A.21. A.20 becomes

$$\begin{cases} F_i + F_r = F_t \\ F_i - F_r = \frac{k_{2z}}{k_{1z}} \alpha_{12} F_t \end{cases} \Leftrightarrow \begin{pmatrix} -1 & 1 \\ 1 & \frac{k_{2z}}{k_{1z}} \alpha_{12} \end{pmatrix} \begin{pmatrix} F_r \\ F_t \end{pmatrix} = \begin{pmatrix} 1 \\ 1 \end{pmatrix} F_i. \tag{A.22}$$

We then derive the ratio of the transmitted/reflected and incident fields:

$$\frac{F_t}{F_i} = \frac{2\, k_{1z}}{k_{1z} + k_{2z}\, \alpha_{12}}, \quad \frac{F_r}{F_i} = \frac{k_{1z} - k_{2z}\, \alpha_{12}}{k_{1z} + k_{2z}\, \alpha_{12}}. \tag{A.23}$$

The coefficients of reflection r and transmission t are usually defined for the E-field. Thus, Eq. A.23 gives these coefficients directly in the TE case. In the TM case, one should transform the H field appearing in Eqs. A.18–A.20 into the associated E-field. Here, for plane-wave beams, this relation, deduced from Eq. A.10, is simply

$$H_y = \frac{k}{\omega \mu_o \mu} E = \frac{1}{c \mu_o} \sqrt{\frac{\varepsilon}{\mu}} E.$$

Consequently, whatever is the polarization, the coefficients of reflection r and transmission t are

$$t = \sqrt{\frac{\mu_2}{\mu_1} \alpha_{12}} \frac{F_t}{F_i} = \sqrt{\frac{\mu_2}{\mu_1} \alpha_{12}} \frac{2\, k_{1z}}{k_{1z} + k_{2z}\, \alpha_{12}}, \quad r = \frac{F_r}{F_i} = \frac{k_{1z} - k_{2z}\, \alpha_{12}}{k_{1z} + k_{2z}\, \alpha_{12}}. \tag{A.24}$$

Let us notice that t^2 does not correspond to the coefficient T of transmission of the energy (or of the power noted p). In fact, because of expression A.17, the energy or the intensity in medium i is proportional to n_i. Moreover, when dealing with spatially limited beams, we must take into account the projection of the beam cross-section S onto the interface:

$$p_i = S_1 I_i = S_1 \frac{\sqrt{\varepsilon_1 \mu_1}}{2 Z_o} F_i^2, \ p_t = S_2 I_t = S_2 \frac{\sqrt{\varepsilon_2 \mu_2}}{2 Z_o} F_t^2, \ \frac{S_1}{\cos\theta_i} = \frac{S_2}{\cos\theta_t}.$$

This leads to

$$T = \frac{p_t}{p_i} = \frac{S_2 \frac{\sqrt{\varepsilon_2 \mu_2}}{2Z_o} F_t^2}{S_1 \frac{\sqrt{\varepsilon_1 \mu_1}}{2Z_o} F_i^2} = \frac{\cos\theta_t}{\cos\theta_i} \frac{\frac{\sqrt{\varepsilon_2 \mu_2}}{2Z_o} F_t^2}{\frac{\sqrt{\varepsilon_1 \mu_1}}{2Z_o} F_i^2} = \sqrt{\frac{\varepsilon_2 \mu_1}{\varepsilon_1 \mu_2}} \frac{\cos\theta_t}{\alpha_{12} \cos\theta_i} t^2. \tag{A.25}$$

Under normal incidence, we derive from Eq. A.23 the following expressions:

$$t = t_{\mathrm{TE}} = t_{\mathrm{TM}} = \frac{2\,Z_2}{Z_2 + Z_1}, \; r = r_{\mathrm{TE}} = -r_{\mathrm{TM}} = \frac{Z_2 - Z_1}{Z_2 + Z_1}. \tag{A.26}$$

$Z_n = \sqrt{\mu_n/\varepsilon_n}$ is the impedance of medium n. Of course, for a perfectly flat and isotropic sample, we can no more make the difference between the TE and the TM cases at normal incidence, as no direction in the plane is privileged. Thus the TE and TM transmission (reflection) coefficients equalize. The change of sign between r_{TE} and r_{TM} is only due to the definition of these coefficients: the first one is the ratio of the electric fields, while the second one is related to the magnetic field. Because, for a plane wave, the cross-product $\vec{E} \times \vec{H}$ points towards the direction of propagation, i.e., $\vec{k}$, the reflected beam that propagates along $-\vec{k}$ imposes the reflected $\vec{H}$ vector to be opposite in polarization to the incident one, as the electric field polarization is saved.

A.1.3 Reflection and Transmission by a Multilayer Stack

We generalize the single interface study to a stack of N parallel and homogeneous slabs, as drawn in Fig. A.2. The thickness of each layer labeled n is $d_n = z_{n-1} - z_n$, its permittivity is ε_n and its permeability μ_n. As each layer is made of a homogeneous material, the field F_n in each layer n is the sum of a plane wave propagating upward, whose magnitude is noted A_n, and of a plane wave propagating downward, noted B_n:

$$F_n = \left(A_n\, e^{j(\omega t - k_{nz} z)} + B_n\, e^{j(\omega t + k_{nz} z)}\right) e^{-j\, k_x x}. \tag{A.27}$$

We write the conservation of F and its z-derivative at z_n, i.e., at the interface between layers n and $n+1$:

$$\begin{cases} A_n\, e^{-j\, k_{nz} z_n} + B_n\, e^{j\, k_{nz} z_n} = A_{n+1}\, e^{-j\, k_{(n+1)z}\, z_n} + B_{n+1}\, e^{j\, k_{(n+1)z}\, z_n}, \\ A_n\, e^{-j\, k_{nz} z_n} - B_n\, e^{j\, k_{nz} z_n} = \frac{k_{(n+1)z}}{k_{nz}} \alpha_{n+1} \left(A_{n+1}\, e^{-j\, k_{(n+1)z}\, z_n} - B_{n+1}\, e^{j\, k_{(n+1)z}\, z_n}\right), \end{cases} \tag{A.28}$$

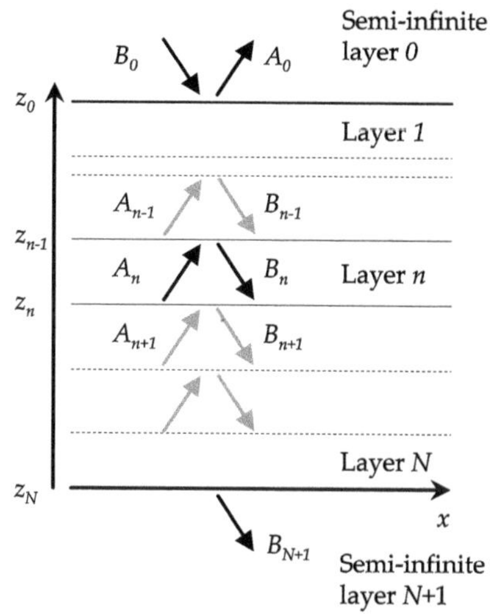

Figure A.2 Reflection and transmission by a multilayer structure.

where $\alpha_{n+1} = \mu_n/\mu_{n+1}$ (TE) and $\alpha_{n+1} = \varepsilon_n/\varepsilon_{n+1}$ (TM). Thus, we can obtain the fields in layer n from the fields in layer $n+1$.

We will employ hereafter a matrix formulation that makes the rest of the calculation easier. This transfer matrix between layer $n+1$ and layer n is named $M_{n,n+1}$. Thus we get from Eq. A.28

$$\begin{pmatrix} A_n \\ B_n \end{pmatrix} = M_{n,n+1} \begin{pmatrix} A_{n+1} \\ B_{n+1} \end{pmatrix}$$

$$= \frac{1}{t_{n,n+1}} \begin{pmatrix} e^{j\left(k_{nz}-k_{(n+1)z}\right) z_n} & r_{n,n+1} e^{j\left(k_{nz}+k_{(n+1)z}\right) z_n} \\ r_{n,n+1} e^{-j\left(k_{nz}+k_{(n+1)z}\right) z_n} & e^{-j\left(k_{nz}-k_{(n+1)z}\right) z_n} \end{pmatrix} \begin{pmatrix} A_{n+1} \\ B_{n+1} \end{pmatrix}. \tag{A.29}$$

$t_{n,n+1}$ and $r_{n,n+1}$ are the transmission and reflection coefficients from layer n to layer $n+1$, respectively defined similarly to Eq. A.24:

$$t_{n,n+1} = \frac{2\,k_{nz}}{k_{nz} + k_{(n+1)z}\,\alpha_{n+1}}, \quad r_{n,n+1} = \frac{k_{nz} - k_{(n+1)z}\,\alpha_{n+1}}{k_{nz} + k_{(n+1)z}\,\alpha_{n+1}} \tag{A.30}$$

The incident, reflected, and transmitted fields are linked by relation A.29 applied to each layer of the stack:

$$\begin{aligned}\begin{pmatrix} A_0 \\ B_0 \end{pmatrix} &= M_{0,1}\begin{pmatrix} A_1 \\ B_1 \end{pmatrix} = M_{0,1}\times M_{1,2}\begin{pmatrix} A_2 \\ B_2 \end{pmatrix} = \ldots \\ &= M_{0,1}\times M_{1,2}\times\ldots\times M_{N-1,N}\times M_{N,N+1}\begin{pmatrix} A_{N+1} \\ B_{N+1} \end{pmatrix} \\ &= M_{\mathrm{T}}\begin{pmatrix} A_{N+1} \\ B_{N+1} \end{pmatrix}\end{aligned} \tag{A.31}$$

The transfer matrix M_{T} of the whole stack is the product of each transfer matrix $M_{n,n+1}$:

$$M_{\mathrm{T}} = \prod_{n=0}^{N} M_{n,n+1} \tag{A.32}$$

Then, we relate the incident, reflected, and transmitted fields:

$$\begin{pmatrix} A_0 \\ B_0 \end{pmatrix} = \begin{pmatrix} A_r \\ B_i \end{pmatrix} = M_{\mathrm{T}}\begin{pmatrix} A_{N+1} \\ B_{N+1} \end{pmatrix} = M_{\mathrm{T}}\begin{pmatrix} 0 \\ B_t \end{pmatrix}$$

$$\Rightarrow \begin{cases} t = \sqrt{\dfrac{\mu_{n+1}}{\mu_0}\alpha_{0n+1}}\dfrac{B_t}{B_i} = \sqrt{\dfrac{\mu_{n+1}}{\mu_0}\alpha_{0n+1}}\dfrac{1}{M_{\mathrm{T}}(2,2)}, \\ r = \frac{A_r}{B_i} = \dfrac{M_{\mathrm{T}}(1,2)}{M_{\mathrm{T}}(2,2)}. \end{cases} \tag{A.33}$$

A.1.4 Reflection and Transmission by a Single Slab

Relation A.31 is greatly simplified in the case of a single slab of thickness d:

$$\begin{cases} \begin{pmatrix} A_0 \\ B_0 \end{pmatrix} = \frac{1}{t_{0,1}}\begin{pmatrix} e^{j(k_{0z}-k_{1z})z_0} & r_{0,1}e^{j(k_{0z}+k_{1z})z_0} \\ r_{0,1}e^{-j(k_{0z}+k_{1z})z_0} & e^{-j(k_{0z}-k_{1z})z_0} \end{pmatrix}\begin{pmatrix} A_1 \\ B_1 \end{pmatrix} \\ \begin{pmatrix} A_1 \\ B_1 \end{pmatrix} = \frac{1}{t_{1,2}}\begin{pmatrix} e^{j(k_{1z}-k_{2z})z_1} & r_{1,2}e^{j(k_{1z}+k_{2z})z_1} \\ r_{1,2}e^{-j(k_{1z}+k_{2z})z_1} & e^{-j(k_{1z}-k_{2z})z_1} \end{pmatrix}\begin{pmatrix} A_2 \\ B_2 \end{pmatrix} \end{cases}$$

$$\Rightarrow \begin{pmatrix} A_0 \\ B_0 \end{pmatrix} = \frac{1}{t_{0,1}t_{1,2}}\times$$

$$\left(\begin{matrix} \left[\begin{matrix} e^{j(k_{0z}z_0-k_{2z}z_1)}\times \\ \left(e^{-jk_{1z}(z_0-z_1)} + r_{0,1}r_{1,2}e^{jk_{1z}(z_0-z_1)}\right) \end{matrix}\right] & \left[\begin{matrix} e^{j(k_{0z}z_0+k_{2z}z_1)}\times \\ \left(r_{1,2}e^{-jk_{1z}(z_0-z_1)} + r_{0,1}e^{jk_{1z}(z_0-z_1)}\right) \end{matrix}\right] \\ \left[\begin{matrix} e^{-j(k_{0z}z_0+k_{2z}z_1)}\times \\ \left(r_{0,1}e^{-jk_{1z}(z_0-z_1)} + r_{1,2}e^{jk_{1z}(z_0-z_1)}\right) \end{matrix}\right] & \left[\begin{matrix} e^{-j(k_{0z}z_0-k_{2z}z_1)}\times \\ \left(r_{0,1}r_{1,2}e^{-jk_{1z}(z_0-z_1)} + e^{jk_{1z}(z_0-z_1)}\right) \end{matrix}\right] \end{matrix}\right)\begin{pmatrix} A_2 \\ B_2 \end{pmatrix}$$

From this result, we deduce the fields reflected/transmitted by the slab normalized to the incident fields:

$$\begin{cases} \dfrac{B_2}{B_0} = \dfrac{t_{0,1}t_{1,2}e^{j\left(k_{0z}z_0-k_{2z}z_1\right)}}{r_{0,1}r_{1,2}e^{-j\,k_{1z}\left(z_0-z_1\right)}+e^{j\,k_{1z}\left(z_0-z_1\right)}} \\ \\ \dfrac{A_0}{B_0} = \dfrac{\left(r_{1,2}e^{-j\,k_{1z}\left(z_0-z_1\right)}+r_{0,1}e^{j\,k_{1z}\left(z_0-z_1\right)}\right)e^{j2k_{0z}z_0}}{r_{0,1}r_{1,2}e^{-j\,k_{1z}\left(z_0-z_1\right)}+e^{j\,k_{1z}\left(z_0-z_1\right)}} \end{cases} \quad \text{(A.34)}$$

We set the origin of the z axis at the top of the stack ($z_0 = 0$); thus $z_1 = -d$. Equation A.34 simplifies into

$$\begin{cases} t = \sqrt{\dfrac{\mu_2}{\mu_0}}\alpha_{02}\dfrac{B_2}{B_0} = \sqrt{\dfrac{\mu_2}{\mu_0}}\alpha_{02}\,t_{0,1}t_{1,2}\dfrac{e^{j\left(k_{2z}-k_{1z}\right)d}}{1+r_{0,1}r_{1,2}e^{-2j\,k_{1z}d}} \\ \\ r = \dfrac{A_0}{B_0} = \dfrac{r_{0,1}+r_{1,2}e^{-2j\,k_{1z}d}}{1+r_{0,1}r_{1,2}e^{-2j\,k_{1z}d}} \end{cases} \quad \text{(A.35)}$$

At normal incidence, we get

$$\begin{cases} t = \dfrac{4\,Z_1Z_2e^{j\left(k_{2z}-k_{1z}\right)d}}{(Z_0+Z_1)(Z_1+Z_2)+(Z_0-Z_1)(Z_1-Z_2)e^{-2j\,k_{1z}d}} \\ \\ r = \dfrac{(Z_1-Z_0)(Z_1+Z_2)+(Z_0+Z_1)(Z_2-Z_1)e^{-2j\,k_{1z}d}}{(Z_0+Z_1)(Z_1+Z_2)+(Z_0-Z_1)(Z_1-Z_2)e^{-2j\,k_{1z}d}} \end{cases} \quad \text{(A.36)}$$

For non-magnetic materials, $\mu_n = 1$ and $Z_n = 1/\tilde{n}_n$, which leads to

$$\begin{cases} t = \dfrac{4\tilde{n}_0\tilde{n}_1\,e^{j\frac{\omega}{c}\left(\tilde{n}_2-\tilde{n}_1\right)d}}{(\tilde{n}_1+\tilde{n}_0)(\tilde{n}_1+\tilde{n}_2)-(\tilde{n}_0-\tilde{n}_1)(\tilde{n}_2-\tilde{n}_1)e^{-2j\,k_{1z}d}}, \\ \\ r = \dfrac{(\tilde{n}_0-\tilde{n}_1)(\tilde{n}_1+\tilde{n}_2)-(\tilde{n}_1+\tilde{n}_0)(\tilde{n}_2-\tilde{n}_1)e^{-2j\,\frac{\omega}{c}\tilde{n}_1d}}{(\tilde{n}_1+\tilde{n}_0)(\tilde{n}_1+\tilde{n}_2)-(\tilde{n}_0-\tilde{n}_1)(\tilde{n}_2-\tilde{n}_1)e^{-2j\,k_{1z}d}}. \end{cases} \quad \text{(A.37)}$$

In most of the practical cases, the sample is located in air, thus $\tilde{n}_0 = \tilde{n}_2 = 1$. We set $\tilde{n}_1 \equiv \tilde{n}$ and we obtain

$$\begin{cases} t = \dfrac{4\tilde{n}}{(1+\tilde{n})^2} \times \dfrac{e^{-j\,\frac{\omega}{c}(\tilde{n}-1)d}}{1-\left(\frac{\tilde{n}-1}{\tilde{n}+1}e^{-j\,\frac{\omega}{c}\tilde{n}d}\right)^2}, \\ \\ r = \dfrac{1-\tilde{n}}{1+\tilde{n}} \times \dfrac{1-e^{-2j\,\frac{\omega}{c}\tilde{n}d}}{1-\left(\frac{\tilde{n}-1}{\tilde{n}+1}e^{-j\,\frac{\omega}{c}\tilde{n}d}\right)^2}. \end{cases} \quad \text{(A.38)}$$

A.1.5 Ray-Tracing Approach of the Single Slab Response

The preceding treatment of the electromagnetic response of a single slab is based on a matrix approach. It is useful to consider another calculation procedure, in which the contribution of each wave propagating backward and forward in the slab is evidently taken into account. This method is close to a ray-tracing one, but we will make use of the wavy behavior of the electromagnetic fields. This study is well adapted to THz-TDS because it gives the magnitude and phase

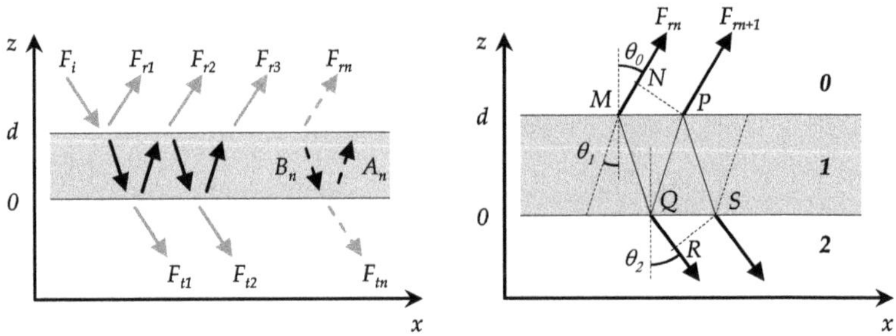

Figure A.3 Ray propagation through a slab: general scheme (left); zoom and geometrical parameters.

of each pulse transmitted or reflected by the slab after encountering several rebounds inside the slab.

We consider (see Fig. A.3) a plane wave impinging the slab and represented in Fig. A.3 as a ray noted F_i. Even if we only draw a ray, we have to remember that the plane wave illuminates the entire slab. The incident ray is reflected (signal F_{r1}) and transmitted into the slab (signal B_1). When B_1 reaches the second interface, one part (F_t) is transmitted in the medium below, and one part is reflected (A_1). The same phenomenon occurs for the rays (B_2, A_2) ... (B_n, A_n). The reflected and transmitted fields are the sums of the F_{rn} and F_{tn} waves, respectively. While performing this summation, we must take into account the phase difference between the different waves. The polarization (TE or TM) of the waves is introduced through the transmission and reflection coefficients (A.23 and A.21) at each interface.

The reflected signal becomes

$$F_r = r_{0,1}F_i + \frac{t_{0,1}t_{1,0}}{r_{1,0}}\left(r_{1,2}r_{1,0}e^{-j(k_1 MQP - k_0 MN)}\right)F_i$$
$$+ \ldots + \frac{t_{0,1}t_{1,0}}{r_{1,0}}\left(r_{1,2}r_{1,0}e^{-j(k_1 MQP - k_0 MN)}\right)^n F_i + \ldots$$
$$\Rightarrow \quad F_r = r_{0,1}F_i + \frac{t_{0,1}t_{1,0}}{r_{1,0}}F_i \sum_{n=1}^{\infty}\left(r_{1,2}r_{1,0}e^{-j(k_1 MQP - k_0 MN)}\right)^n. \tag{A.39}$$

The sum of the terms to the power of n is easy to determine:

$$S_N = \sum_{n=1}^{N} x^n = \frac{x\left(1 - x^N\right)}{1 - x} \underset{\substack{N\to\infty \\ 0<x<1}}{\longrightarrow} \frac{x}{1 - x}$$

Thus,

$$\begin{aligned} F_r &= r_{0,1}F_i + \frac{r_{1,2}t_{0,1}t_{1,0}e^{-j(k_1 MQP - k_0 MN)}}{1 - r_{1,2}r_{1,0}e^{-j(k_1 MQP - k_0 MN)}}F_i \\ &= \frac{r_{0,1} + r_{1,2}\left(t_{0,1}t_{1,0} - r_{0,1}r_{1,0}\right)e^{-j(k_1 MQP - k_0 MN)}}{1 - r_{1,2}r_{1,0}e^{-j(k_1 MQP - k_0 MN)}}F_i \end{aligned} \quad \text{(A.40)}$$

The propagation exponent $k_1 MQP - k_0 MN$ is equal to

$$k_1 MQP - k_0 MN = k_1 \frac{2d}{\cos\theta_1} - k_0 2d \tan\theta_1 \sin\theta_0 = 2k_{1z}d.$$

Also, using relations A.30, one easily demonstrates that $t_{0,1}t_{1,0} - r_{0,1}r_{1,0} = 1$. Therefore, we get

$$r = \frac{F_r}{F_i} = \frac{r_{0,1} + r_{1,2}e^{-j2k_{1z}d}}{1 - r_{1,2}r_{1,0}e^{-j2k_{1z}d}}, \quad \text{(A.41)}$$

which is exactly expression A.35 obtained with the electromagnetic calculation. In the same way, we obtain the transmitted signal:

$$\begin{aligned} F_t &= t_{0,1}t_{1,2}F_i e^{-jk_1 MQ} + t_{0,1}t_{1,2}r_{1,2}r_{1,0}e^{-jk_1 MQP}e^{jk_2 QR}e^{-jk_1 MQ}F_i \\ &\quad + \ldots + t_{0,1}t_{1,2}\left(r_{1,2}r_{1,0}e^{-jk_1 MQP}e^{jk_2 QR}\right)^n e^{-jk_1 MQ}F_i + \ldots \\ \Rightarrow F_t &= \frac{t_{0,1}t_{1,2}e^{-jk_1 MQ}}{1 + r_{0,1}r_{1,2}e^{j(k_2 QR - k_1 MQP)}}F_i. \end{aligned} \quad \text{(A.42)}$$

The propagation exponent in the denominator becomes

$$k_2 QR - k_1 MQP = k_2 2d \tan\theta_1 \sin\theta_2 - 2k_1 \frac{d}{\cos\theta_1} = -2k_{1z}d,$$

while

$$k_1 MQ = k_1 \frac{MQP}{2} = \frac{k_2 QR + 2k_{1z}d}{2} = k_{1z}d + k_2 \tan\theta_1 \sin\theta_2 d.$$

Finally, we write the transmission coefficient of the stack:

$$t = \frac{F_t}{F_i} = \frac{t_{0,1}t_{1,2}e^{-jk_{1z}d}e^{jk_2 \tan\theta_1 \sin\theta_2 d}}{1 + r_{0,1}r_{1,2}e^{-2jk_{1z}d}}. \quad \text{(A.43)}$$

As compared to Eq. A.35, Eq. A.43 is multiplied by a phase term $e^{jk_2 \tan\theta_1 \sin\theta_2 d}$. It corresponds to a change of the phase origin and is equal to the difference of phase with a beam that would have been propagating through the same thickness d in the surrounding medium. Therefore, Eqs. A.41 and A.43 are equivalent to Eq. A.35, but they are calculated from the contribution of each rebound inside the slab. In the case of pulsed signals, this calculation gives the amplitude of each pulse that makes a back and forth oscillation in the slab. This is very useful when the waveforms are time-windowed, i.e., only one or several temporal pulses are recorded.

A.1.6 Propagation in Anisotropic Media

In an anisotropic crystal, the permittivity is a tensor ($\overleftrightarrow{\varepsilon}$) of rank 2 ($3 \times 3$ matrix). When using as a referential the principal axes of the crystal, the tensor is diagonal:

$$\overleftrightarrow{\varepsilon} = \begin{pmatrix} \varepsilon_{xx} & 0 & 0 \\ 0 & \varepsilon_{yy} & 0 \\ 0 & 0 & \varepsilon_{zz} \end{pmatrix}. \tag{A.44}$$

Therefore, the fields $\vec{E}$ and $\vec{D}$ are not parallel, and the electromagnetic energy does not flow in the same direction as the wave. If we consider the propagation of a plane wave in a dielectric crystal (no free charges, no magnetic permeability), relation A.14 becomes

$$\vec{k} \times \vec{k} \times \vec{E} = \vec{k}\left(\vec{k} \cdot \vec{E}\right) - k^2 \vec{E} = \omega \mu_o \vec{k} \times \vec{H} = -\frac{\omega^2}{c^2} \overleftrightarrow{\varepsilon} : \vec{E}. \tag{A.45}$$

Then one sets

$$\vec{u} = \frac{\vec{k}}{k}, \quad \varepsilon_{ii} = n_i^2, \quad n = k\frac{c}{\omega},$$
$$\Rightarrow \quad E_i = \frac{n^2}{n^2 - n_i^2} u_i \left(\vec{u} \cdot \vec{E}\right), \quad \text{and} \quad D_i = \varepsilon_o n^2 \left(\vec{u} \cdot \vec{E}\right) \frac{n_i^2}{n^2 - n_i^2} u_i. \tag{A.46}$$

n is the refractive index at which the wave propagates. Solving Eq. A.46 for each component E_i of the field $\vec{E}$ leads to the Fresnel equation:

$$\sum_i \frac{n^2}{n^2 - n_i^2} u_i^2 = 1 \tag{A.47}$$

The Fresnel equation is quadratic in n, i.e., there are two solutions n for a plane wave propagating in the crystal. They correspond to two orthogonal polarizations of the $\vec{D}$ field, which are themselves perpendicular to the direction $\vec{u}$ of propagation (since $\vec{\nabla} \cdot \vec{D} = j\,\vec{k} \cdot \vec{D} = j\,k\vec{u} \cdot \vec{D} = 0$). The index ellipsoid is drawn by the extremity of vector $n\,\vec{u}$. From Eq. A.46, we get

$$\frac{D^2}{\varepsilon_o n^2} = \vec{E} \cdot \vec{D} = \sum_i E_i D_i = \sum_i \frac{D_i^2}{\varepsilon_o n_i^2} \qquad \text{(A.48)}$$

We set $\zeta = n\frac{D_\zeta}{D}$ $(\zeta = x, y, z)$ and we obtain the equation of the index ellipsoid:

$$\frac{1}{n^2} = \frac{x^2}{n_x^2} + \frac{y^2}{n_y^2} + \frac{z^2}{n_z^2} \qquad \text{(A.49)}$$

Therefore, a wave whose $\vec{D}$ field is polarized in the direction (x, y, z) propagates in the anisotropic crystal at speed c/n, where the index n is defined by relation A.49, which is the equation of an ellipsoid. In uniaxial crystals, $n_x = n_y \equiv n_o$ is the ordinary index, while $n_z \equiv n_e$ is the extraordinary index; z is the optical axis of the crystal. The ellipsoid shows a revolution symmetry around the optical axis. In isotropic materials, $n_x = n_y = n_z = n$ and the ellipsoid is a sphere. Table A.1 gives the crystal systems with their anisotropy properties, together with the ones that are centro-symmetric (in which second-order nonlinear phenomena are forbidden, like electro-optic effect and optical rectification).

Table A.1 Symmetry of optical materials

	Isotropic materials	Uniaxial crystals	Biaxial crystals
crystal systems	amorphous, cubic	hexagonal, trigonal, tetragonal	orthorhombic, triclinic, monoclinic
centrosymmetric materials	cubic: $m\bar{3}$, $m\bar{3}m$	hexagonal: $6/m$, $6/mmm$ trigonal: $\bar{3}$, $\bar{3}m$ tetragonal: $4/m$, $4/mmm$	orthorhombic: mmm triclinic: $\bar{1}$ monoclinic: $2/m$

A.2 Nonlinear Optics

A.2.1 Nonlinear Response of Matter

When an electromagnetic beam illuminates a dielectric crystal, it induces a dipole at each atom or molecule of the crystal, thanks to the interaction of the incident E-field and the electronic cloud of the atoms/molecules. We suppose here that the incoming photon energy is not equal to the difference in energy between two electronic levels of the atom/molecule. Also, we do not treat magneto-optical effects, and thus we suppose that the crystals are not magnetic (i.e., $\mu = 1$). When the incoming field is weak compared to the intra-atomic field, the electronic cloud deformation varies linearly with the applied field, and thus the response of the crystal is given by the constitutive relations A.1. When the field is strong, this deformation is limited by the neighboring atoms, and the saturation of the deformation makes the relation between $\vec{p}$ and $\vec{E}$ nonlinear. The induced dipole is a function of the E-field:

$$\vec{p} = \vec{g}\left(\vec{E}\right). \tag{A.50}$$

Since the nonlinearity of the function $\vec{g}$ is small, it is useful to write $\vec{g}$ as a series of the powers of $\vec{E}$:

$$\vec{p} = \overleftrightarrow{\alpha}^{L} : \vec{E} + \overleftrightarrow{\alpha}^{(2)} : \vec{E} \cdot \vec{E} + \overleftrightarrow{\alpha}^{(3)} : \vec{E} \cdot \vec{E} \cdot \vec{E} + \ldots \tag{A.51}$$

$\overleftrightarrow{\alpha}^{L}$ is the linear polarizability of the atoms/molecules and $\overleftrightarrow{\alpha}^{(n)}$ are the nonlinear polarizabilities. The polarizability is a tensor, which means that the exciting E-field and the induced dipole may be not parallel. Adding all the dipoles located in a unit volume of the crystal defines the polarizability $\vec{P}$ of the crystal:

$$\vec{P} = \frac{1}{V} \sum_{\text{volume } V} \vec{p} = \varepsilon_o \left(\overleftrightarrow{\chi}^{L} : \vec{E} + \overleftrightarrow{\chi}^{(2)} : \vec{E} \cdot \vec{E} + \overleftrightarrow{\chi}^{(3)} : \vec{E} \cdot \vec{E} \cdot \vec{E} + \ldots \right) \tag{A.52}$$

Here $\overleftrightarrow{\chi}$ is the susceptibility tensor, which is composed of both linear and nonlinear terms. Note that, for some authors, the ε_o term is entered in the definition of $\overleftrightarrow{\chi}$. For the reader who is not familiar with the tensorial notation, it may be useful to write down the expression of the component P_i $(i = x, y, z)$ of $\vec{P}$:

$$P_i = \varepsilon_o \sum_j \chi_{ij}^L E_j + \varepsilon_o \sum_j \sum_k \chi_{ijk}^{(2)} E_j E_k$$
$$+\varepsilon_o \sum_j \sum_k \sum_l \chi_{ijkl}^{(3)} E_j E_k E_l \ldots \tag{A.53}$$

We consider the case when the exciting optical field can be written as a plane wave of pulsation ω_{opt}, i.e.,

$$\vec{E} = \vec{E}_o e^{j\left(\omega_{\mathrm{opt}} t - \vec{k}_{\mathrm{opt}} \cdot \vec{r}\right)} + \mathrm{C.C.}$$

Here the complex notation makes the calculation simpler. Therefore, the polarization becomes

$$\vec{P} = \varepsilon_o \sum_n \overleftrightarrow{\chi}^{(n)} : \vec{E}_o^n \left(e^{j\left(\omega_{\mathrm{opt}} t - \vec{k}_{\mathrm{opt}} \cdot \vec{r}\right)} + \mathrm{C.C.}\right)^n . \tag{A.54}$$

If the optical field is constituted of several harmonics, the relative frequencies are added by the nonlinear process through the complex exponential term in Eq. A.54. Because of the power to n on the right side of Eq. A.54, new frequencies are created. Thus, by equaling the harmonics of the pumping field and the induced polarization, we get

$$\sum_{i=1}^{n} \hbar \omega_i = \sum_{i'=1}^{n'} \hbar \omega_{i'} . \tag{A.55}$$

This relation describes the conservation of energy in the nonlinear case. ω_i and $\omega_{i'}$ are the frequencies of the incoming and generated beams, respectively. n is the order of the nonlinearity. Here, we are mostly interested in the following second-order nonlinear phenomena involved in THz technology:

$$\begin{cases} \omega_{\mathrm{opt},1} - \omega_{\mathrm{opt},2} = \omega \ \text{difference of frequencies,} \\ \omega_{\mathrm{opt},1} + \omega = \omega_{\mathrm{opt},2} \ \text{sum of frequencies.} \end{cases} \tag{A.56}$$

$\omega_{\mathrm{opt},1}$ and $\omega_{\mathrm{opt},2}$ are the optical frequencies, while ω is the THz frequency. As $\omega_{\mathrm{opt},1} \approx \omega_{\mathrm{opt},2} = \omega_{\mathrm{opt}} \gg \omega$, Eq. A.56 may be approximated by

$$\begin{cases} \omega_{\mathrm{opt}} - \omega_{\mathrm{opt}} = 0 \leftrightarrow \ \text{optical rectification,} \\ \omega_{\mathrm{opt}} + 0 = \omega_{\mathrm{opt}} \leftrightarrow \ \text{Pockels effect.} \end{cases} \tag{A.57}$$

The second-order nonlinear tensor obeys some permutation symmetry rules known as the ABDP relations [3]:

$$\chi_{ijk}^{(2)} (\omega_1 = \omega_3 - \omega_2) = \chi_{kji}^{(2)} (\omega_2 = \omega_3 - \omega_1) = \chi_{jik}^{(2)} (\omega_3 = \omega_1 + \omega_2) . \tag{A.58}$$

Here, the tensors related to optical rectification and to the Pockels effect are thus linked by

$$\chi^{(2)}_{ijk}\left(\omega = \omega_{\text{opt}} - \omega_{\text{opt}}\right) = \chi^{(2)}_{kji}\left(\omega_{\text{opt}} = \omega_{\text{opt}} - \omega\right)$$
$$= \chi^{(2)}_{jik}\left(\omega_{\text{opt}} = \omega_{\text{opt}} + \omega\right). \quad \text{(A.59)}$$

Moreover, if the crystal is transparent at both the optical ω_{opt} and THz ω frequencies, the *ijk* indexes can be freely exchanged (Kleinmann's symmetry rules) and Eq. A.59 simplifies into

$$\overset{\leftrightarrow}{\chi}^{(2)}\left(\omega = \omega_{\text{opt}} - \omega_{\text{opt}}\right) = \overset{\leftrightarrow}{\chi}^{(2)}\left(\omega_{\text{opt}} = \omega_{\text{opt}} - \omega\right)$$
$$= \overset{\leftrightarrow}{\chi}^{(2)}\left(\omega_{\text{opt}} = \omega + \omega_{\text{opt}}\right). \quad \text{(A.60)}$$

In conclusion, optical rectification ($\omega = \omega_{\text{opt}} - \omega_{\text{opt}} \approx 0$) is driven by the same tensor as the Pockels effect ($\omega_{\text{opt}} = \omega_{\text{opt}} \pm \omega \approx \omega_{\text{opt}} + 0$).

In centrosymmetric materials, there exists an inversion symmetry with respect to the E-field, i.e., $\vec{P}\left(\vec{E}\right) = \vec{P}\left(-\vec{E}\right)$. This implies that all even order nonlinear processes, like optical rectification or electro-optic effect, cannot occur in centrosymmetric crystals and amorphous materials (see Table A.1).

A.2.2 Nonlinear Propagation

We treat the general case of the $\omega_3 = \omega_1 + \omega_2$ interaction, and then we will apply the achieve results to optical rectification and Pockels effect. The propagation of a plane wave in a nonlinear medium is described by the inhomogeneous Helmholtz equation (A.3), which can be written as

$$\Delta\vec{E}\left(\vec{r},\omega_3\right) - \frac{1}{c^2}\frac{\partial^2 \vec{E}\left(\vec{r},\omega_3\right)}{\partial t^2} = \mu_o\frac{\partial^2 \vec{P}\left(\vec{r},\omega_3\right)}{\partial t^2}$$
$$= \mu_o\frac{\partial^2 \vec{P}^{L}\left(\vec{r},\omega_3\right)}{\partial t^2} + \mu_o\frac{\partial^2 \vec{P}^{NL}\left(\vec{r},\omega_3 = \omega_1 + \omega_2\right)}{\partial t^2}. \quad \text{(A.61)}$$

We separate the time and space parts of the fields (e.g., $\vec{F}\left(\vec{r},\omega_i\right) = \vec{F}_i\left(\vec{r}\right)e^{j\,\omega_i t} + \text{C.C.}$). Hence,

$$\Delta\vec{E}_3\left(\vec{r}\right) + \varepsilon\left(\omega_3\right)\frac{\omega_3^2}{c^2}\vec{E}_3\left(\vec{r}\right) =$$
$$-\left(\frac{\omega_3}{c}\right)^2 \overset{\leftrightarrow}{\chi}^{(2)}\left(\omega_3 = \omega_1 + \omega_2\right) : \vec{E}_1\left(\vec{r}\right)\vec{E}_2\left(\vec{r}\right) \quad \text{(A.62)}$$

The solution of the inhomogeneous differential equation (A.62) is the sum of a free wave (right-side term of Eq. A.62 is null) and a forced wave (any particular solution of Eq. A.62). We chose plane waves as solutions, i.e., $\vec{E}_3(\vec{r}) = \vec{E}_{\text{free}}\, e^{-j\,\vec{k}_3\cdot\vec{r}} + \text{C.C.}$ for the free wave and $\vec{E}_3(\vec{r}) = \vec{E}_{\text{forced}}\, e^{-j\,(\vec{k}_1+\vec{k}_2)\cdot\vec{r}} + \text{C.C.}$ for the forced wave. Equation A.62 leads to

$$\vec{E}_3(\vec{r}) = e^{-j\,\vec{k}_3\cdot\vec{r}} \left[\left(\frac{\omega_3}{c}\right)^2 \frac{\overleftrightarrow{\chi}^{(2)} : \vec{E}_1 \vec{E}_2}{\left(\vec{k}_1 + \vec{k}_2\right)^2 - \vec{k}_3^2} e^{-j\,(\vec{k}_1+\vec{k}_2-\vec{k}_3)\cdot\vec{r}} + \vec{E}_{\text{free}} \right] + \text{C.C.} \quad \text{(A.63)}$$

Assuming the generated $\vec{E}_3(\vec{r})$ field is null at the entrance face of the crystal, and setting $\Delta\vec{k} = \vec{k}_1 + \vec{k}_2 - \vec{k}_3$, we finally obtain

$$\vec{E}_3(\vec{r}) = -2j \frac{\overleftrightarrow{\chi}^{(2)} : \vec{E}_1 \vec{E}_2}{\Delta\vec{k}\left(\vec{k}_1 + \vec{k}_2 + \vec{k}_3\right)} \left(\frac{\omega_3}{c}\right)^2 \sin\left(\frac{\Delta\vec{k}\cdot\vec{r}}{2}\right) e^{-\frac{j}{2}\,(\vec{k}_1+\vec{k}_2+\vec{k}_3)\cdot\vec{r}} + \text{C.C.} \quad \text{(A.64)}$$

$\vec{E}_3(\vec{r})$ is maximum when $\Delta\vec{k} = 0$ (phase-matching condition). Under this condition, $\left|\vec{E}_3(\vec{r})\right|$ increases linearly with the modulus r of $\vec{r}$.

A.2.3 Optical Rectification

Here,

$$\omega_3 = \omega, \quad \omega_1 = \omega_{\text{opt}} + \omega, \quad \omega_2 = -\omega_{\text{opt}}$$

$$\Rightarrow \quad \vec{k}_1 = \vec{k}_{\omega_{\text{opt}}+\omega}, \quad \vec{k}_2 = \vec{k}_{-\omega_{\text{opt}}} = -\vec{k}^*_{\omega_{\text{opt}}}, \quad \vec{k}_3 = \vec{k}_\omega$$

So the generated THz field is

$$\vec{E}_{\text{THz}}(\vec{r}) = -2j \frac{\overleftrightarrow{\chi}^{(2)} : \vec{E}_{\omega_{\text{opt}}+\omega} \vec{E}_{\omega_{\text{opt}}}}{\Delta\vec{k}\left(\vec{k}_{\omega_{\text{opt}}+\omega} - \vec{k}^*_{\omega_{\text{opt}}} + \vec{k}_\omega\right)} \left(\frac{\omega}{c}\right)^2 \times \sin\left(\frac{\Delta\vec{k}\cdot\vec{r}}{2}\right) e^{-\frac{j}{2}\left(\vec{k}_{\omega_{\text{opt}}+\omega} - \vec{k}^*_{\omega_{\text{opt}}} + \vec{k}_\omega\right)\cdot\vec{r}} + \text{C.C.} \quad \text{(A.65)}$$

The phase-matching condition writes $\Delta\vec{k} = \vec{k}_{\omega_{\text{opt}}+\omega} - \vec{k}^*_{\omega_{\text{opt}}} - \vec{k}_\omega = 0$. As $\omega_{\text{opt}} \gg \omega$ and the photons at frequencies $\omega_{\text{opt}} + \omega$ and ω are delivered by the same laser beam ($\vec{k}_{\omega_{\text{opt}}+\omega} // \vec{k}^*_{\omega_{\text{opt}}} // \vec{k}_\omega$), we get

$$\begin{cases} n_g = n_\omega, \\ \alpha_{\omega_{\text{opt}}+\omega} + \alpha_{\omega_{\text{opt}}} = \alpha_\omega, \end{cases} \quad \text{(A.66)}$$

where $n_g = c\frac{\partial k_{\omega_{\text{opt}}}}{\partial \omega}$ is the group index at optical frequencies.

For an optimized THz generation, the index n_ω at THz frequency must be equal to the group index n_g in the optical domain, and the THz intensity is given by

$$\vec{E}_{\text{THz}}(r) \approx \frac{-j}{2}\frac{\omega}{c}\frac{\overleftrightarrow{\chi}^{(2)} : \vec{E}_{\omega_{\text{opt}}}\vec{E}_{\omega_{\text{opt}}}}{n_\omega + j\kappa_\omega} r e^{-\frac{\alpha_\omega r}{2}} e^{-jn_\omega \frac{\omega}{c} r} + \text{C.C.} \quad \text{(A.67)}$$

A.2.4 Electro-optic Effect (Pockels Effect)

An electrical field applied to the crystal changes its refractive index, i.e., its permittivity tensor $\overleftrightarrow{\varepsilon}$ (Eq. A.44) as we will demonstrate below. The change is probed with a laser beam. We treat the case of a quasi-static applied field, i.e., its temporal variations are much slower than the period of the optical probe beam. Even if the applied field vibrates in the THz range, the THz period is about $100 \sim 1000$ times longer than the optical one, and the following treatment is valid.

The polarization of the crystal material under an applied field is derived from Eq. A.54 in which we save only the terms oscillating at the optical frequency:

$$\vec{P} = \varepsilon_o \overleftrightarrow{\chi}^L : \vec{E}_{\text{opt}} e^{j\left(\omega_{\text{opt}}t - \vec{k}_{\text{opt}}\cdot\vec{r}\right)} + \varepsilon_o \overleftrightarrow{\chi}^{(2)} : \vec{E}_{\text{opt}} \cdot \vec{E}_{\text{THz}} e^{j\left(\omega_{\text{opt}}t - \vec{k}_{\text{opt}}\cdot\vec{r}\right)} + \text{C.C.} \quad \text{(A.68)}$$

From Eq. A.68, we get $\vec{D}$:

$$D_i = \varepsilon_o \left(\varepsilon_{ii} E_{\text{opt},i} + \sum_{j,k} \chi_{ijk}^{(2)} E_{\text{opt},j} E_{\text{THz},k} \right) e^{j\left(\omega_{\text{opt}}t - \vec{k}_{\text{opt}}\cdot\vec{r}\right)} + \text{C.C.} \quad \text{(A.69)}$$

$$\Rightarrow \quad E_i = \frac{n^2}{n^2 - n_i^2} u_i \left(\vec{u}\cdot\vec{E}\right), \quad \text{and} \quad D_i = \varepsilon_o n^2 \left(\vec{u}\cdot\vec{E}\right) \frac{n_i^2}{n^2 - n_i^2} u_i.$$

Then we set $\zeta = \frac{D_\zeta}{D}$ $(\zeta = x, y, z)$ and proceed as we did to get Eqs. A.48 and A.49:

$$\frac{D^2}{n^2} = \sum_\zeta \frac{D_\zeta^2}{n_\zeta^2} = \sum_\zeta \frac{\left(D_\zeta + D_\zeta^{NL}\right)^2}{n_\zeta^2} \Rightarrow$$

$$\frac{1}{n^2} = \sum_\zeta \left(\frac{\zeta}{n_\zeta} + \sum_{j,k} \frac{\chi_{\zeta jk}^{(2)} j E_{\text{THz},k}}{n_\zeta n_j^2} \right)^2. \quad \text{(A.70)}$$

The ellipsoid is thus modified by the applied THz field. By gathering in Eq. A.70 the terms $\zeta^2$and ζj, we end up with

$$\frac{x^2}{n_x^2}\left(1+\sum_k \frac{\chi_{xxk}^{(2)}}{n_x^2} E_{\mathrm{THz},k}\right)+\frac{y^2}{n_y^2}\left(1+\sum_k \frac{\chi_{yyk}^{(2)}}{n_y^2} E_{\mathrm{THz},k}\right)$$
$$+\frac{z^2}{n_z^2}\left(1+\sum_k \frac{\chi_{zzk}^{(2)}}{n_z^2} E_{\mathrm{THz},k}\right)$$
$$+\sum_k \left[x\,y\frac{\chi_{xyk}^{(2)}+\chi_{yxk}^{(2)}}{n_x^2 n_y^2}+x\,z\frac{\chi_{xzk}^{(2)}+\chi_{zxk}^{(2)}}{n_x^2 n_z^2}+y\,z\frac{\chi_{yzk}^{(2)}+\chi_{zyk}^{(2)}}{n_y^2 n_z^2}\right] E_{\mathrm{THz},k}=\frac{1}{n^2}. \tag{A.71}$$

This equation is greatly simplified by setting

$$\Delta\left(\frac{1}{n^2}\right)_{ij}=\sum_k \frac{\chi_{ijk}^{(2)}}{n_i^2 n_j^2} E_{\mathrm{THz},k}. \tag{A.72}$$

Moreover, as the i and j indexes in Eq. A.72 can be freely permuted, one defines the reduced indexes:

$$ij \leftrightarrow l,\ xx \leftrightarrow 1,\ yy \leftrightarrow 2,\ zz \leftrightarrow 3,\ yz \leftrightarrow 4,\ xz \leftrightarrow 5,\ xy \leftrightarrow 6 \tag{A.73}$$

It follows that

$$\begin{cases} \frac{1}{n_l^2}=\frac{1}{n_l^2(0)}+\Delta\left(\frac{1}{n^2}\right)_l=\frac{1}{n_l^2(0)}+\sum\limits_k r_{lk}E_{\mathrm{THz},k},\ l=1,2,3. \\ \frac{1}{n_l^2}=\Delta\left(\frac{1}{n^2}\right)_l=\sum\limits_k r_{lk}E_{\mathrm{THz},k},\ l=4,5,6. \end{cases} \tag{A.74}$$

$r_{lk}=\frac{\chi_{ijk}^{(2)}}{n_i^2 n_j^2}$ $(ij \leftrightarrow l)$ is the electro-optic tensor and $n_l^2(0)=\varepsilon_l^L$ is the ij component of the susceptibility tensor without applied field. Several r_{lk} coefficients could be null depending on the symmetries of the crystallographic group of the material, as given in Fig. A.4. Equation A.71 becomes

$$x^2\left(\frac{1}{n_x^2}+\sum_k r_{1k}E_{\mathrm{THz},k}\right)+y^2\left(\frac{1}{n_y^2}+\sum_k r_{2k}E_{\mathrm{THz},k}\right)$$
$$+z^2\left(\frac{1}{n_z^2}+\sum_k r_{3k}E_{\mathrm{THz},k}\right)+2yz\sum_k r_{4k}E_{\mathrm{THz},k}$$
$$+2xz\sum_k r_{5k}E_{\mathrm{THz},k}+2xy\sum_k r_{6k}E_{\mathrm{THz},k}=\frac{1}{n^2} \tag{A.75}$$

For the sake of simplicity, relation A.75 is written using a matrix formalism:

$$\left(X\ Y\ Z \right) \begin{pmatrix} \frac{1}{n_1^2} & \frac{1}{n_6^2} & \frac{1}{n_5^2} \\ \frac{1}{n_6^2} & \frac{1}{n_2^2} & \frac{1}{n_4^2} \\ \frac{1}{n_5^2} & \frac{1}{n_4^2} & \frac{1}{n_3^2} \end{pmatrix} \begin{pmatrix} Z \\ Y \\ Z \end{pmatrix} = \left(X\ Y\ Z \right) [M] \begin{pmatrix} Z \\ Y \\ Z \end{pmatrix} = 1, \tag{A.76}$$

where $X = n\,x, Y = n\,y, Z = n\,z$. The eigenvalues λ and eigenvectors $\vec{V}$ of the matrix $[M]$ are a solution of

$$\det\left([M] - \lambda\,[I]\right) = 0, \quad \left([M] - \lambda\,[I]\right)\vec{V} = 0 \text{ with } \vec{V} = \begin{pmatrix} X \\ Y \\ Z \end{pmatrix}. \tag{A.77}$$

The axes of the index ellipsoid are equal to $1/\sqrt{\lambda}$ and their directions are the ones of the eigenvectors. In the general case, i.e., all the n_i^2 coefficients are not zero, the solutions of Eq. A.75 are complicated expressions of n_i^2. Let us treat here only a simple case as an example. We suppose the optical beam propagates along the x direction in the crystal. Thus its optical field has components only along y and z (we remember the reader that x, y, z are the principal axes of the crystal), and Eq. A.76 simplifies into the equation of an ellipse:

$$\left(Y\ Z \right) \begin{pmatrix} \frac{1}{n_2^2} & \frac{1}{n_4^2} \\ \frac{1}{n_4^2} & \frac{1}{n_3^2} \end{pmatrix} \begin{pmatrix} Y \\ Z \end{pmatrix} = 1. \tag{A.78}$$

The two eigenvalues and two eigenvectors are

$$\begin{aligned} \lambda_a &\approx \frac{1}{n_3^2} = \frac{1}{n_3^2(0)} + \sum_k r_{3k} E_{\mathrm{THz},k}, \\ \lambda_b &\approx \frac{1}{n_2^2} = \frac{1}{n_2^2(0)} + \sum_k r_{2k} E_{\mathrm{THz},k}. \\ \vec{V}_a &\approx \begin{pmatrix} 1 \\ \left(\frac{1}{n_3^2(0)} - \frac{1}{n_2^2(0)}\right) \sum_k r_{4k} E_{\mathrm{THz},k} \end{pmatrix}, \\ \vec{V}_b &\approx \begin{pmatrix} 1 \\ \dfrac{-1}{\left(\frac{1}{n_3^2(0)} - \frac{1}{n_2^2(0)}\right) \sum_k r_{4k} E_{\mathrm{THz},k}} \end{pmatrix} \end{aligned} \tag{A.79}$$

triclinic $\begin{pmatrix} r_{11} & r_{12} & r_{13} \\ r_{21} & r_{22} & r_{23} \\ r_{31} & r_{32} & r_{33} \\ r_{41} & r_{42} & r_{43} \\ r_{51} & r_{52} & r_{53} \\ r_{61} & r_{62} & r_{63} \end{pmatrix}$	monoclinic 2 x2 $\begin{pmatrix} 0 & r_{12} & 0 \\ 0 & r_{22} & 0 \\ 0 & r_{32} & 0 \\ r_{41} & 0 & r_{43} \\ 0 & r_{52} & 0 \\ r_{61} & 0 & r_{63} \end{pmatrix}$	monoclinic 2 x3 $\begin{pmatrix} 0 & 0 & r_{13} \\ 0 & 0 & r_{23} \\ 0 & 0 & r_{33} \\ r_{41} & r_{42} & 0 \\ r_{51} & r_{52} & 0 \\ 0 & 0 & r_{63} \end{pmatrix}$	monoclinic m x2 $\begin{pmatrix} r_{11} & 0 & r_{13} \\ r_{21} & 0 & r_{23} \\ r_{31} & 0 & r_{33} \\ 0 & r_{42} & 0 \\ r_{51} & 0 & r_{53} \\ 0 & r_{62} & 0 \end{pmatrix}$	monoclinic m x3 $\begin{pmatrix} r_{11} & r_{12} & 0 \\ r_{21} & r_{22} & 0 \\ r_{31} & r_{32} & 0 \\ 0 & 0 & r_{43} \\ 0 & 0 & r_{53} \\ r_{61} & r_{62} & 0 \end{pmatrix}$
orthorhombic 222 $\begin{pmatrix} 0 & 0 & 0 \\ 0 & 0 & 0 \\ 0 & 0 & 0 \\ r_{41} & 0 & 0 \\ 0 & r_{52} & 0 \\ 0 & 0 & r_{63} \end{pmatrix}$	orthorhombic $mm2$ $\begin{pmatrix} 0 & 0 & r_{13} \\ 0 & 0 & r_{23} \\ 0 & 0 & r_{33} \\ 0 & r_{42} & 0 \\ r_{51} & 0 & 0 \\ 0 & 0 & 0 \end{pmatrix}$			
tetragonal 4 $\begin{pmatrix} 0 & 0 & r_{13} \\ 0 & 0 & r_{13} \\ 0 & 0 & r_{33} \\ r_{41} & r_{52} & 0 \\ r_{51} & -r_{41} & 0 \\ 0 & 0 & 0 \end{pmatrix}$	tetragonal $\bar{4}$ $\begin{pmatrix} 0 & 0 & r_{13} \\ 0 & 0 & -r_{13} \\ 0 & 0 & 0 \\ r_{41} & -r_{42} & 0 \\ r_{42} & r_{41} & 0 \\ 0 & 0 & r_{63} \end{pmatrix}$	tetragonal 422 $\begin{pmatrix} 0 & 0 & 0 \\ 0 & 0 & 0 \\ 0 & 0 & 0 \\ r_{41} & 0 & 0 \\ 0 & -r_{41} & 0 \\ 0 & 0 & 0 \end{pmatrix}$	tetragonal $4mm$ $\begin{pmatrix} 0 & 0 & r_{13} \\ 0 & 0 & r_{13} \\ 0 & 0 & r_{33} \\ 0 & r_{42} & 0 \\ r_{42} & 0 & 0 \\ 0 & 0 & 0 \end{pmatrix}$	tetragonal $\bar{4}2m$ $\begin{pmatrix} 0 & 0 & 0 \\ 0 & 0 & 0 \\ 0 & 0 & 0 \\ r_{41} & 0 & 0 \\ 0 & r_{41} & 0 \\ 0 & 0 & r_{63} \end{pmatrix}$
cubic $\bar{4}3m$, 23 $\begin{pmatrix} 0 & 0 & 0 \\ 0 & 0 & 0 \\ 0 & 0 & 0 \\ r_{41} & 0 & 0 \\ 0 & r_{41} & 0 \\ 0 & 0 & r_{41} \end{pmatrix}$	cubic 32 $\begin{pmatrix} r_{11} & 0 & 0 \\ -r_{11} & 0 & 0 \\ 0 & 0 & 0 \\ r_{41} & 0 & 0 \\ 0 & -r_{41} & 0 \\ 0 & -r_{11} & 0 \end{pmatrix}$	trigonal $3m$ x1 $\begin{pmatrix} 0 & -r_{22} & r_{13} \\ 0 & r_{22} & r_{13} \\ 0 & 0 & r_{33} \\ 0 & r_{42} & 0 \\ r_{42} & 0 & 0 \\ -r_{22} & 0 & 0 \end{pmatrix}$	trigonal $3m$ x2 $\begin{pmatrix} r_{11} & 0 & r_{13} \\ -r_{11} & 0 & r_{13} \\ 0 & 0 & r_{33} \\ 0 & r_{42} & 0 \\ r_{42} & 0 & 0 \\ 0 & -r_{11} & 0 \end{pmatrix}$	
hexagonal 6 $\begin{pmatrix} 0 & 0 & r_{13} \\ 0 & 0 & r_{13} \\ 0 & 0 & r_{33} \\ r_{41} & r_{42} & 0 \\ r_{42} & -r_{41} & 0 \\ 0 & 0 & 0 \end{pmatrix}$	hexagonal $6mm$ $\begin{pmatrix} 0 & 0 & r_{13} \\ 0 & 0 & r_{13} \\ 0 & 0 & r_{33} \\ 0 & r_{42} & 0 \\ r_{42} & 0 & 0 \\ 0 & 0 & 0 \end{pmatrix}$	hexagonal 622 $\begin{pmatrix} 0 & 0 & 0 \\ 0 & 0 & 0 \\ 0 & 0 & 0 \\ r_{41} & 0 & 0 \\ 0 & -r_{41} & 0 \\ 0 & 0 & 0 \end{pmatrix}$		
hexagonal $\bar{6}$ $\begin{pmatrix} r_{11} & -r_{22} & 0 \\ -r_{11} & r_{22} & 0 \\ 0 & 0 & 0 \\ 0 & 0 & 0 \\ 0 & 0 & 0 \\ -r_{22} & -r_{11} & 0 \end{pmatrix}$	hexagonal $\bar{6}m2$ x1 $\begin{pmatrix} 0 & -r_{22} & 0 \\ 0 & r_{22} & 0 \\ 0 & 0 & 0 \\ 0 & 0 & 0 \\ 0 & 0 & 0 \\ -r_{22} & 0 & 0 \end{pmatrix}$	hexagonal $\bar{6}m2$ x2 $\begin{pmatrix} r_{11} & 0 & 0 \\ -r_{11} & 0 & 0 \\ 0 & 0 & 0 \\ 0 & 0 & 0 \\ 0 & 0 & 0 \\ 0 & -r_{11} & 0 \end{pmatrix}$		

Figure A.4 Electro-optic tensor. Adapted from Yariv [4].

To obtain these expressions, we have realistically supposed that $\sum_k r_{jk} E_{\mathrm{THz},k} \ll 1$. This also implies that we can take an approximate expression of n_y and n_z:

$$\begin{cases} n_y = n_2 = n_y(0) - \frac{1}{2} n_y^3(0) \sum_k r_{2k} E_{\mathrm{THz},k}, \\ n_z = n_3 = n_z(0) - \frac{1}{2} n_z^3(0) \sum_k r_{3k} E_{\mathrm{THz},k}. \end{cases} \quad \text{(A.80)}$$

A.3 Fourier Transform

A.3.1 Fourier Series

Joseph Fourier [5] stated as early as in 1807 that a periodic real function $f(t)$, of period T, can be written as the sum of discrete sinusoidal functions, now known as harmonics:

$$f(t) = a_o + \sum_{n=1}^{\infty} (a_n \cos(n\,\Omega\,t) + b_n \sin(n\,\Omega\,t)), \qquad \text{(A.81)}$$

with the pulsation Ω defined as

$$\Omega = \frac{2\pi}{T}. \qquad \text{(A.82)}$$

The harmonic of order n is

$$u_n(t) = a_n \cos(n\,\Omega\,t) + b_n \sin(n\,\Omega\,t) = A_n \cos(n\,\Omega\,t + \phi_n). \qquad \text{(A.83)}$$

This implies that

$$A_n = \sqrt{a_n^2 + b_n^2},\ \tan\phi_n = \frac{b_n}{a_n}\ (a_n \neq 0). \qquad \text{(A.84)}$$

The preceding relations can be written using a complex notation:

$$f(t) = \sum_{n=-\infty}^{\infty} c_n e^{jn\,\Omega\,t}. \qquad \text{(A.85)}$$

It is easily shown that

$$c_n = \frac{|A_n|}{2},\ \arg(c_n) = -\phi_n\ [2\pi]. \qquad \text{(A.86)}$$

The $e^{jn\,\Omega\,t}$ functions form a basis, i.e.,

$$\frac{1}{T}\int_{-T/2}^{T/2} e^{j(n-m)\,\Omega\,t} dt = \delta_{n,m}$$

We determine the coefficient c_n as follows. First, we multiply relation A.85 by $e^{-jm\Omega t}$ and then we integrate versus time over the T window:

$$f(t)\,e^{-jm\Omega t} = \sum_{n=-\infty}^{\infty} c_n e^{j(n-m)\,\Omega\,t}$$

$$\int_{-T/2}^{T/2} f(t)\,e^{-jm\Omega t} dt = \sum_{n=-\infty}^{\infty} c_n \underbrace{\int_{-T/2}^{T/2} e^{j(n-m)\,\Omega\,t} dt}_{\delta_{n,m}} = T\,c_m$$

$$\Rightarrow c_m = \frac{1}{T}\int_{-T/2}^{T/2} f(t)\, e^{-jm\Omega t} dt. \tag{A.87}$$

Usually, one prefers to write the c_n coefficients without the $1/T$ term, which is equivalent to divide the right-hand side of Eq. A.85 by T:

$$f(t) = \frac{1}{T}\sum_{n=-\infty}^{\infty} c_n e^{jn\Omega t}, \text{ with } \; c_n = \int_{-T/2}^{T/2} f(t)\, e^{-jn\Omega t} dt. \tag{A.88}$$

A.3.2 Handy Approach of the Fourier Transform

What is explained in the preceding paragraph is valid *only* for a periodic function. Treating a nonperiodic function is equivalent to making the period tend to infinity. As T is very large, the step $\Omega = 2\pi/T$ from the harmonic $n\Omega$ to the next harmonic $(n+1)\,\Omega$ is very small. Thus we need a very large number of harmonics to build $F(t)$, and these harmonic pulsations form an almost continuous function in the frequency domain. Thus we define the pulsation $\omega = n\Omega$, which is a continuous variable. Therefore, Eq. A.88 may be written as an integral:

$$f(t) = \frac{1}{T}\sum_{n=-\infty}^{\infty} c_n e^{jn\Omega t} = \frac{1}{2\pi}\sum_{n=-\infty}^{\infty} c_n e^{jn\frac{2\pi}{T}t}\frac{2\pi}{T}$$

$$\omega = n\frac{2\pi}{T},\; d\omega = \frac{2\pi}{T}$$

$$\sum_{n=-\infty}^{\infty} \cdots \frac{2\pi}{T} \rightarrow \int_{-\infty}^{+\infty} \ldots\, d\omega$$

$$f(t) = \frac{1}{T}\sum_{n=-\infty}^{\infty} c_n e^{jn\Omega t} \quad \rightarrow \quad f(t) = \frac{1}{2\pi}\int_{-\infty}^{+\infty} c(\omega) e^{j\omega t}\, d\omega, \tag{A.89}$$

with $c(\omega)$ calculated as in Eq. A.88:

$$c(\omega) = \int_{-\infty}^{+\infty} f(t) e^{-j\omega t}\, dt. \tag{A.90}$$

By extrapolating Eq. A.90, we define $c(\omega)$ as being the Fourier transform $F(\omega)$ of $f(t)$:

$$F(\omega) = \frac{1}{\sqrt{2\pi}} \int_{-\infty}^{+\infty} f(t) e^{-j\omega t} dt, \tag{A.91}$$

while, from Eq. A.89, we define the inverse Fourier transform of $F(\omega)$:

$$f(t) = \frac{1}{\sqrt{2\pi}} \int_{-\infty}^{+\infty} F(\omega) e^{j\omega t}\, d\omega. \tag{A.92}$$

In Eqs. A.91 and A.92, the constant coefficient is chosen as $1/\sqrt{2\pi}$ to make symmetrical the definitions. The Fourier transform allows one to turn from the temporal domain to the frequency domain.[d] Generally, even if $f(t)$ is real, $F(\omega)$ is a complex function. If $f(t)$ is real, we obtain

$$F^*(\omega) = \frac{1}{\sqrt{2\pi}} \int_{-\infty}^{+\infty} f(t) e^{j\omega t} dt = F(-\omega). \tag{A.93}$$

A.3.3 Useful Properties

A.3.3.1 Time reversal

The Fourier transform of the function $f(-t)$ is

$$\begin{aligned} F_-(\omega) &= \frac{1}{\sqrt{2\pi}} \int_{-\infty}^{+\infty} f(-t)\, e^{-j\omega t} dt = \frac{-1}{\sqrt{2\pi}} \int_{-\infty}^{+\infty} f(t')\, e^{-j\omega(-t')} dt' \\ &= \frac{-1}{\sqrt{2\pi}} \int_{-\infty}^{+\infty} f(t')\, e^{-j(-\omega)t'} dt' = -F(-\omega). \end{aligned} \tag{A.94}$$

[d] Here, we treat temporal and spectral functions, so the Fourier transform permits to turn from the time domain to the frequency one. More generally, the Fourier transform relates two domains of conjugated variables.

A.3.3.2 Plancherel's theorem

Plancherel's theorem indicates that the Fourier transform preserves the energy of the signal. Indeed, Plancheral's theorem states

$$\int_{-\infty}^{+\infty} |f(t)|^2\, dt = \int_{-\infty}^{+\infty} |F(\omega)|^2\, d\omega. \tag{A.95}$$

In electromagnetism, if $f(t) = E(t)$, where $E(t)$ is the amplitude of the electric field, the integral of $|E(t)|^2$ is nothing but the energy of the electromagnetic signal (see relations A.15–A.16).

A.3.3.3 Time and frequency shifting

We consider a temporal function $h(t)$, which is equal to a function $f(t)$ time-shifted by t_o:

$$h(t) = f(t - t_o).$$

Thus,

$$\begin{aligned} H(\omega) &= \frac{1}{\sqrt{2\pi}} \int_{-\infty}^{+\infty} h(t) e^{-j\omega t} dt \\ &= \frac{1}{\sqrt{2\pi}} \int_{-\infty}^{+\infty} f(t - t_o)\, e^{-j\omega t} dt \\ &= \frac{e^{j\omega t_o}}{\sqrt{2\pi}} \int_{-\infty}^{+\infty} f(u)\, e^{-j\omega u} du \\ \Rightarrow \quad & H(\omega) = e^{j\omega t_o} F(\omega). \end{aligned} \tag{A.96}$$

Now we consider the function $h(t)$, which is equal to a function $f(t)$ multiplied by $\exp(j\omega_o t)$:

$$h(t) = f(t) e^{j\omega_o t}.$$

Thus,

$$H(\omega) = \frac{1}{\sqrt{2\pi}} \int_{-\infty}^{+\infty} h(t) e^{-j\omega t} dt = \frac{1}{\sqrt{2\pi}} \int_{-\infty}^{+\infty} f(t) e^{-j(\omega - \omega_o)t} dt,$$

$$H(\omega) = F(\omega - \omega_o). \tag{A.97}$$

A.3.3.4 Differentiation

We consider a function $f(t)$, which is derivable for any value of t. We determine the Fourier transform of df/dt by integrating by parts:

$$\begin{aligned}F_{f'}(\omega) &= \frac{1}{\sqrt{2\pi}}\int_{-\infty}^{+\infty}\frac{d\,f(t)}{d\,t}e^{-j\omega t}dt\\ &= \frac{1}{\sqrt{2\pi}}\left[f(t)e^{-j\omega t}\Big|_{-\infty}^{+\infty} + j\omega\int_{-\infty}^{+\infty}f(t)e^{-j\omega t}dt\right]\\ &= \frac{f(t)e^{-j\omega t}\Big|_{-\infty}^{+\infty}}{\sqrt{2\pi}} + j\omega F(\omega).\end{aligned}$$

Assuming $f(\pm\infty) \to 0$, this gives

$$F_{f'}(\omega) = j\omega\, F(\omega). \tag{A.98}$$

A.3.3.5 Correlation and auto-correlation

The correlation product $g(t)$ (or simply speaking the correlation) of two functions $f(t)$ and $h(t)$ is defined as

$$g(t) = \int_{-\infty}^{+\infty} f(\tau + t)\,h(\tau)\,d\tau. \tag{A.99}$$

Its Fourier transform is

$$\begin{aligned}G(\omega) &= \frac{1}{\sqrt{2\pi}}\int_{-\infty}^{+\infty} g(t)e^{-j\omega t}dt\\ &= \frac{1}{\sqrt{2\pi}}\int_{-\infty}^{+\infty}\int_{-\infty}^{+\infty} f(\tau + t)\,h(\tau)\,e^{-j\omega t}d\tau\,dt\\ &= \int_{-\infty}^{+\infty} h(\tau)\left[\frac{1}{\sqrt{2\pi}}\int_{-\infty}^{+\infty} f(\tau + t)\,e^{-j\omega t}\,dt\right]d\tau\\ &= \int_{-\infty}^{+\infty} h(\tau)\,e^{j\omega\tau}\,F(\omega)d\tau,\end{aligned}$$

$$G(\omega) = \sqrt{2\pi}\,F(\omega)H(-\omega) = \sqrt{2\pi}\,F(-\omega)\,H(\omega). \tag{A.100}$$

Let us notice that in many textbooks, the Fourier transform is defined without the $\sqrt{2\pi}$ factor (Eq. A.91), but the inverse Fourier transform includes a factor $1/2\pi$. Within this definition, Eq. A.100 does not contain the $\sqrt{2\pi}$ factor. If $f(t)$ and $h(t)$ are real, using Eq. A.93, we get

$$G(\omega) = \sqrt{2\pi}\, F(\omega)H^*(\omega) = \sqrt{2\pi}\, F^*(\omega)H(\omega). \qquad \text{(A.101)}$$

When $h(t) = f(t)$, we deal with the autocorrelation of $f(t)$:

$$g(t) = \int_{-\infty}^{+\infty} f(\tau + t)\, f(\tau)\, d\tau,$$

$$\Rightarrow \quad G(\omega) = \sqrt{2\pi}\, F(\omega)F(-\omega). \qquad \text{(A.102)}$$

For a real function $f(t)$, the Fourier transform of its autocorrelation is

$$G(\omega) = \sqrt{2\pi}\, F(\omega)F^*(\omega) = \sqrt{2\pi}\, |F(\omega)|^2 \qquad \text{(A.103)}$$

This constitutes the Wiener-Kichine theorem.

A.3.3.6 Convolution

The convolution $s(t)$ of two functions $f(t)$ and $h(t)$ is similar to the correlation, but the time shift is now negative:

$$s(t) = \int_{-\infty}^{+\infty} f(\tau - t)\, h(\tau)\, d\tau \equiv f(t) \otimes h(t). \qquad \text{(A.104)}$$

Let us notice that the convolution is commutative:

$$s(t) = \int_{-\infty}^{+\infty} f(\tau)\, h(\tau - t)\, d\tau \equiv h(t) \otimes f(t). \qquad \text{(A.105)}$$

From the preceding paragraph, it follows simply that

$$TF\,[f(t) \otimes h(t)] = \sqrt{2\pi}\, F(\omega)H(\omega). \qquad \text{(A.106)}$$

This constitutes the convolution theorem. By reciprocity, we obtain

$$TF^{-1}\,[F(\omega) \otimes H(\omega)] = \sqrt{2\pi}\, f(t)h(t).$$

Hence,

$$TF\,[f(t)h(t)] = \frac{1}{\sqrt{2\pi}} F(\omega) \otimes H(\omega). \qquad \text{(A.107)}$$

A.3.4 Useful Examples

A.3.4.1 Rectangle

We consider a rectangle function, which is defined as follows:

$$f(t) = 1, \quad t \in \left[-\frac{a}{2}, \frac{a}{2}\right]; \quad f(t) = 0, \quad t \notin \left[-\frac{a}{2}, \frac{a}{2}\right]. \quad \text{(A.108)}$$

Its Fourier transform is

$$F(\omega) = \frac{1}{\sqrt{2\pi}} \int_{-a/2}^{+a/2} e^{-j\omega t} dt = \frac{a}{\sqrt{2\pi}} \mathrm{sinc} \frac{\omega a}{2}. \quad \text{(A.109)}$$

A.3.4.2 Gaussian functions

Let us consider a Gaussian function, whose FHWM width is $\Delta t = \tau 2\sqrt{\ln 2}$. We have

$$f(t) = e^{-\left(\frac{t}{\tau}\right)^2} \to F(\omega) = \frac{1}{\sqrt{2\pi}} \int_{-\infty}^{+\infty} e^{-\left(\frac{t}{\tau}\right)^2} e^{-j\omega t} dt. \quad \text{(A.110)}$$

We transform the integral to

$$F(\omega) = \frac{1}{2\pi} e^{-\frac{1}{4}\omega^2\tau^2} \int_{-\infty}^{+\infty} e^{-\left(\frac{t}{\tau} + \frac{1}{2} j\omega\tau\right)^2} dt.$$

Using the variable $u = t/\tau + j\omega\tau/2$ and performing a polar integration, we obtain the Fourier transform of a Gaussian function:

$$F(\omega) = \frac{\tau}{2\sqrt{\pi}} e^{-\left(\frac{\omega\tau}{2}\right)^2}. \quad \text{(A.111)}$$

Therefore, the Fourier transform of a Gaussian function is a Gaussian function, whose FHWM width is $\Delta\omega = 8\ln 2/\Delta t = 4\sqrt{\ln 2}/\tau \approx 1/(3\tau)$.

A.3.4.3 Dirac distribution and Dirac comb

We define the Dirac distribution $\delta(t)$, sometimes called the delta function or the delta Dirac function, by

$$\delta(t = 0) = 1, \quad \delta(t \neq 0) = 0. \quad \text{(A.112)}$$

A well-known and useful property of $\delta(t)$ is

$$\int_{-\infty}^{+\infty} f(t)\delta(t)dt = f(0) \rightarrow F_{\delta}(\omega) = \frac{1}{\sqrt{2\pi}}. \qquad \text{(A.113)}$$

Therefore, the Fourier transform of $\delta(t)$ is a constant. On the other hand, the Fourier transform of a constant is a delta function:

$$F(\omega) = \frac{1}{\sqrt{2\pi}} \int_{-\infty}^{+\infty} e^{-j\omega t} dt = \frac{1}{\sqrt{2\pi}} \lim_{\eta \to +\infty} \left[\frac{e^{j\omega\eta} - e^{-j\omega\eta}}{j\omega} \right],$$

$$\Rightarrow \quad F(\omega) = \frac{2}{\sqrt{2\pi}} \lim_{\eta \to +\infty} \left[\frac{\sin(\omega\eta)}{\omega} \right] = \delta(\omega). \qquad \text{(A.114)}$$

The Dirac comb is a series of Dirac distributions periodically spaced by a period T. We write it:

$$\text{III}(t) = \sum_{m=-\infty}^{+\infty} \delta(t - mT) = \sum_{m=-\infty}^{+\infty} \delta\left(\frac{t}{T} - m\right). \qquad \text{(A.115)}$$

Sometimes, $\text{III}(t)$ is named shah(t). Its Fourier transform is obtained by keeping in mind that the integral of a sum is the sum of integrals:

$$\begin{aligned}
\text{III}(\omega) = TF[\text{III}(t)] &= \frac{1}{\sqrt{2\pi}} \int_{-\infty}^{+\infty} \text{III}(t) e^{-j\omega t} dt \\
&= \frac{1}{\sqrt{2\pi}} \sum_{m=-\infty}^{+\infty} \left(\int_{-\infty}^{+\infty} \delta(t - mT) e^{-j\omega t} dt \right) \\
&= \frac{1}{\sqrt{2\pi}} \sum_{m=-\infty}^{+\infty} e^{j\omega mT} \int_{-\infty}^{+\infty} \delta(\tau) e^{-j\omega\tau} d\tau = \frac{1}{2\pi} \sum_{m=-\infty}^{+\infty} e^{-j\omega mT} \\
&= \frac{2\pi}{T} \sum_{k=-\infty}^{+\infty} \delta\left(\omega - k\frac{2\pi}{T}\right)
\end{aligned}$$

Therefore, the Fourier transform of a comb of period T is also a comb in the reciprocal domain, of period $2\pi/T$:

$$\begin{aligned}
TF[\text{III}(t)] &= TF\left[\sum_{m=-\infty}^{+\infty} \delta(t - mT) \right] \\
&= \frac{2\pi}{T} \sum_{k=-\infty}^{+\infty} \delta\left(\omega - k\frac{2\pi}{T}\right) = \text{III}(\omega). \qquad \text{(A.116)}
\end{aligned}$$

A.3.4.4 Discrete time Fourier transform (DTFT)

In many experiments, a signal is not recorded continuously, but only at selected points via a computer. Therefore, the recorded data are a series of periodically spaced discrete values. Let us treat a temporal signal $S(t)$ that is recorded only at discrete times $t = nT$. T is the period of the data acquisition, and n is an integer. The whole recorded signal is

$$S_{\text{rec}}(t) = \sum_{n=-\infty}^{+\infty} \delta\,(t - nT)\,S(t). \tag{A.117}$$

Because $\sum_{n=-\infty}^{+\infty} \delta\,(t - nT) = \text{III}(t)$ is the Dirac comb, (A.117) can be transformed into

$$S_{\text{rec}}(t) = \text{III}(t) \otimes S(t). \tag{A.118}$$

The Fourier transform of Eq. A.118 is easy to calculate: First, the Fourier transform of the convolution product is the product of the Fourier transforms. Second, the Fourier transform of a comb is also a comb (A.116). Consequently, we obtain

$$\begin{aligned} S_{\text{rec}}(\omega) = TF\,[S_{\text{rec}}(t)] &= \sum_{n=-\infty}^{+\infty} \delta\left(\omega - n\frac{2\pi}{T}\right) \times S(\omega) \\ &= \sum_{n=-\infty}^{+\infty} S\left(\omega - n\frac{2\pi}{T}\right). \end{aligned} \tag{A.119}$$

The spectrum of the measured signal is the spectrum of the actual signal repeated with a period $\Delta\omega = 2\pi/T$ or $\Delta f = 1/T$. If the bandwidth of the actual signal exceeds $\Delta\omega$, two succeeding spectra $(n,\, n+1)$ overlap and the overlapping spectral region becomes useless: this is the aliasing phenomenon. Therefore, the maximum reachable frequency bandwidth, in time-domain spectroscopy techniques, is inversely proportional to the time delay between two successive temporal recorded data:

$$\Delta\omega_{\max} = \frac{2\pi}{2T} = \frac{\pi}{T} \quad \Leftrightarrow \quad \Delta f_{\max} = \frac{1}{2T}. \tag{A.120}$$

The factor 1/2 originates in the fact that the spectra are even functions of the frequency, i.e., the repeated spectra spread towards increasing and decreasing frequencies around $n\Delta\omega$. (A.120) is another way of writing down the Shannon–Nyquist sampling theorem.

A second practical limitation arises from the actual record, which does not allow one to perform the exact Fourier transform of the signal. Indeed, waveforms are measured over a limited time scale. Let us call N the total number of recorded data in the time domain. The total duration, over which the record is performed, is $\Delta T = N \times T$. Any Fourier transform is not performed over an infinite time scale, but only over ΔT. We set $W(t)$ the window function, which is equal to 1 for $0 \leq t \leq \Delta T$, and to 0 elsewhere. Using Eqs. A.107 and A.108, the Fourier transform of the recorded signal is expressed by

$$\begin{aligned} S_{\text{rec}}(\omega) &= \frac{1}{\sqrt{2\pi}} \int_0^{\Delta T} S_{\text{rec}}(t) e^{-j\omega t} dt \\ &= \frac{1}{\sqrt{2\pi}} \int_{-\infty}^{+\infty} W(t) S(t) e^{-j\omega t} dt = T\,F\,[W(t)S(t)] \\ \Rightarrow \quad S_{\text{rec}}(\omega) &= \frac{W(\omega) \otimes S(\omega)}{\sqrt{2\pi}} = \frac{\Delta T}{2\pi} \text{sinc}\left(\frac{\omega\,\Delta\,T}{2}\right) \otimes S \end{aligned} \quad \text{(A.121)}$$

The experimental spectrum is the convolution product of the actual spectrum and of the Fourier transform of the window function, i.e., a sinc function. Let us imagine that the actual spectrum shows an ideal peak, whose width is null, i.e., a Dirac delta, at frequency ω_p. The recorded pulse is

$$S(\omega) = \delta_{\omega_p} \quad \Rightarrow \quad S_{\text{rec}}(\omega) = \frac{\Delta T}{2\pi} \text{sinc}\left(\frac{1}{2}\left(\omega - \omega_p\right)\,\Delta T\right) \quad \text{(A.122)}$$

The sinc function is the impulse frequency response of the experiment. This function (Fig. A.5) includes a strong bump at ω_p (peak value equal to 1) and weaker additional secondary peaks (the first maximum at $\omega \approx \omega_p \pm 2.46 \times \frac{2\pi}{\Delta T}$ is about 0.13). The question is now to know if we can observe in the experimental spectrum a second peak separated from the first one by $\delta\omega$. The Rayleigh criterion says that this observation is possible if the sinc curve of the second peak crosses the first one at half the maximum value

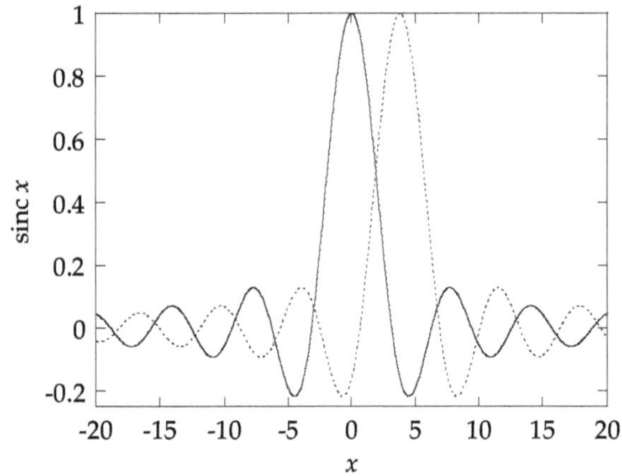

Figure A.5 Impulse spectral function sinc(x) (continuous curve). The dotted curve is shifted by $x = 3.79$ compared to the continuous curve (see text). According to the Rayleigh criterion, the continuous and dotted curves are the closest ones that can be identified separately.

(Fig. A.5), i.e., at 0.5. As $\text{sinc}\,(1.895) = 0.5$, it means that the second peak should be shifted by $2x \approx 3.79 \approx 1.2\,\pi$, which also corresponds to the FHWM of the sinc function.

In a first approximation, one chooses $2x = \pi$, i.e., the maximum of the sinc function of the second peak corresponds to the first zero of the first peak curve. In the case of the experimental spectrum (Eq. A.122), this leads to

$$\delta\omega = \frac{2\pi}{\Delta T} \quad \Leftrightarrow \quad \delta f = \frac{1}{\Delta T} \tag{A.123}$$

The frequency resolution of a spectrum obtained by Fourier transform a temporal signal recorded over a time-window T is $1/\Delta T$.

In conclusion, when recording discrete data over a limited time duration, the maximum frequency of the corresponding spectrum is $f_{\max} = 1/2T$, where T is the period of the temporal data acquisition (also called the time sampling period), and the frequency resolution is $\delta f = 1/\Delta T$, where $\Delta T = N \times T$ is the duration of the recording time window, and N is the number of recorded temporal data. Therefore, the number of significant frequency values in the spectrum is also equal to N ($f_{\max}/\delta f = N\Delta T/2\Delta T = N/2$).

This is illustrated in Fig. A.6. Figure A.6 (top) is the plot of a Gaussian pulse versus time. The horizontal scale is without units: it means that if one chooses picosecond as unity, the scale of the spectrum will be given in THz, and as well nanosecond will correspond to GHz. The FHWM is 1. The continuous line is the calculated curve, while black dots and open squares are data selected respectively every $T = 0.1$ and $T = 0.4$, over the same total duration $\Delta T = 12.8$. The respective numbers of recorded data are 128 and $128/4 = 32$. Figure A.6 (bottom) shows the corresponding spectra (only the moduli). The continuous curve corresponds to the 128 temporal data record. It is periodic with a spectral period equal to 10, i.e., $1/T$. As clearly seen, from $f = 5$ to 10, the spectrum is an image of the actual spectrum obtained with $S_{\text{rec}}(\Delta f - f) = S_{\text{rec}}(\Delta f - 1/T) = S(f)$, which illustrates why the experimental spectra are only useful up to $f_{\max} = 1/2T$. The doted curve is the spectrum of the 32 recorded temporal data. Now, because T has been multiplied by 4 as compared to the 128 data record, the period of the spectrum and the useful spectral range are divided by 4. In this case, we are at the limit for which the actual spectrum spreads over $f_{\max}$. Thus information on higher frequencies of the actual spectrum are lost: one says the signal is under sampled. This could be solved by reducing the time delay T between two successive records.

In this example, the frequency resolution of the spectra is the same, as the recording window ΔT is the same. This is clearly observed in Fig. A.7, which is an enlarged view of Fig. A.6 for small frequencies. Points obtained with the 128 and 32 temporal data are located at the same frequencies. The aliasing effect is seen for the 32-data spectrum. The increase of the spectrum over $f = 1.25$ is an artefact, because of under sampling.

To make the spectral curves better, one can add data equal to zero to the recorded temporal data, which artificially increase the recording window ΔT. This procedure is called "zero padding." However, one has to take care that it does not increase the resolution of the spectrum. To demonstrate this point, we start from Eqs. A.117 and A.121, i.e., from a set of discrete temporal data recording during

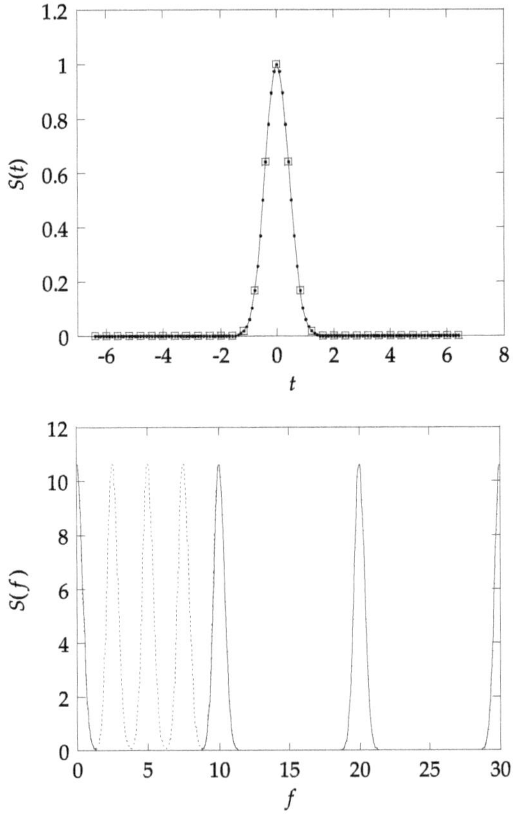

Figure A.6 (Top) Gaussian pulse versus time (continuous line). The time scale is of arbitrary unity. The pulse FHWM is 1. Recorded data are open squares ($N = 32$) and black dots ($N = 128$); (bottom) Fourier transform (modulus) of the recorded temporal data (continuous curve $\leftrightarrow N = 128$; dotted curve $\leftrightarrow N = 32$).

a limited time:

$$S_{\text{rec}}(t) = \sum_{n=0}^{N-1} \delta\,(t - nT)\,S(t) \quad \Rightarrow \quad S_{\text{rec}}(\omega) = \sum_{n=0}^{N-1} S\,(nT)\,e^{-j\omega nT} \tag{A.124}$$

If we add $\Delta N = P - N$ additional temporal data, such as $S\,(pT) = 0$ $(N < p \leq P)$, expression A.124 becomes

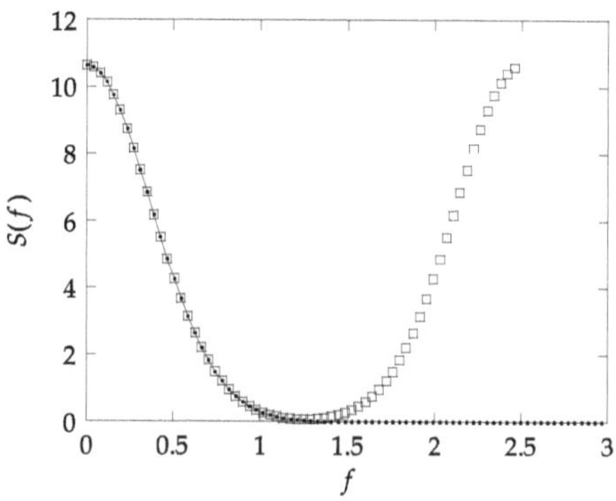

Figure A.7 Zoom of Fig. A.6. The continuous line is calculated with the analytical expression of the Fourier transform of the Gaussian pulse. Dots and open squares are determined from the recorded temporal data (respectively $N = 128$ and $N = 32$).

$$\begin{aligned} S_{\text{rec}}^{P}(\omega) &= \sum_{n=0}^{P-1} S(nT)\, e^{-j\omega nT} \\ &= \sum_{n=0}^{N-1} S(nT)\, e^{-j\omega nT} + \sum_{n=N}^{P-1} \underbrace{S(nT)}_{=0} e^{-j\omega nT} \\ &= S_{\text{rec}}^{N}(\omega) = S_{\text{rec}}(\omega). \end{aligned} \tag{A.125}$$

Adding temporal points of zero weight does not change the achieved spectrum. However, this could be surprising as we have shown in Eqs. A.121–A.123 that the spectral resolution is the width of the main peak of the sinc function, which is inversely proportional to the recording window duration $\Delta T = N \times T$. Here, as N has been substituted by $P > N$, we expect a better resolution. In fact, the misinterpretation could come to a poor definition of the $W(t)$ function. We write Eq. A.121 again including the additional data:

$$\begin{aligned} S_{\text{rec}}^{P}(\omega) = &\frac{1}{\sqrt{2\pi}} \int_{0}^{\Delta T = N\,T} S_{\text{rec}}(t) e^{-j\omega t} dt \\ &+ \frac{1}{\sqrt{2\pi}} \int_{0}^{\Delta T' = \Delta N\,T} \underbrace{S_{\text{rec}}(t)}_{=0}\, e^{-j\omega t} dt \end{aligned} \tag{A.126}$$

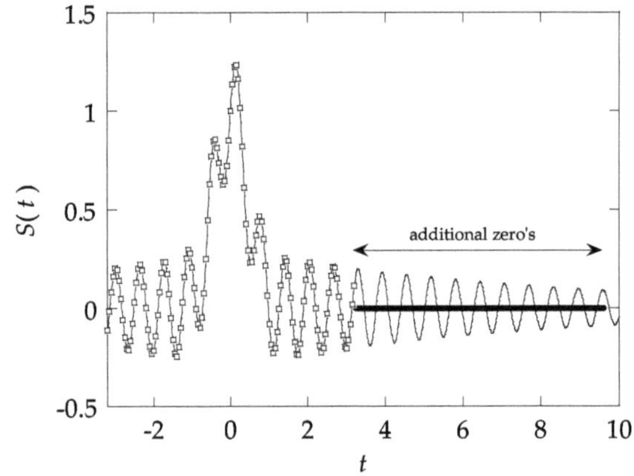

Figure A.8 An example of the waveform (continuous line), with recorded discrete points (open squares), and additional zeros (dots).

It is tempting to write this expression as

$$S_{\mathrm{rec}}^{P}(\omega) = \frac{1}{\sqrt{2\pi}} \int_{-\infty}^{+\infty} W'(t) S_{\mathrm{rec}}(t) e^{-j\omega t} dt,$$

with

$$W'(t) = (N + \Delta N)\, T = W(t) + \Delta N \times T = W(t) + \Delta W(t).$$

In fact, this is not correct because only the number of significant data counts in the calculation, as shown in Eq. A.126. However, when adding zeros, the number of temporal data is $N + \Delta N$. Therefore, the number of calculated discrete values in the spectrum is also $N + \Delta N$. The spectrum curve is plotted with more points, so it appears smoother than without adding zeros. However, its resolution is not improved.

Figures A.8 and A.9 present an example of the influence of adding zeros. The calculated waveform (continuous line) is the sum of a Gaussian pulse of width 1 and a sine function (pulsation $\omega = 10$) that decays exponentially. The sine will induce a resonant peak in the spectrum of the waveform, at the frequency $f = 5/\pi \approx 1.59$. Open squares represent recorded temporal data: their number is 128. The recording time window extends from –3.2 to 3.2, i.e., $\Delta T = 6.4$. 128

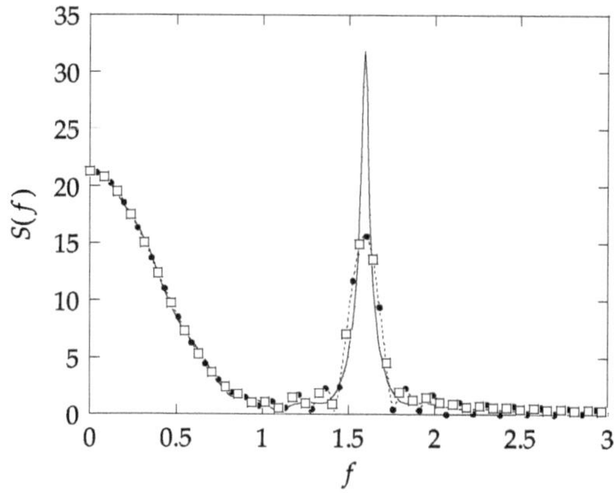

Figure A.9 Spectra of the waveforms shown in Fig. A.8. The continuous line is the exact spectrum, open squares, and black dots are calculated respectively without and with additional zeros. The dashed line is the spectrum calculated with the actual time window.

additional zeros are added, making the artificial time window $W'(t)$ twice longer than the actual one $W(t)$.

The corresponding spectra (modulus) are plotted on Fig. A.9. The continuous line is the exact spectrum. Open squares and black dots are computed respectively without and with the additional zeros. The dashed line is a guide to the eye, which corresponds to the experimental spectra, i.e., obtained from recorded temporal data. Because of the limited time windowing, the spectral resolution makes the experimental resonance at $f = 1.59$ slightly wider than the actual one (see continuous and dashed lines). Data calculated with and without additional zeros are along the dashed curve, and thus the additional zeros do not improve the resolution, i.e., make the calculated resonance narrower. They only define more precisely the dashed line.

Most of the numerical Fourier transform methods are based on the fast Fourier transform (FFT) algorithm. The spectrum $S_{\text{rec}}(\omega)$ is given by Eq. A.124, and the spectral resolution $\delta\omega$ by (A.123). It is worthless to calculate spectral data with a periodicity smaller than $\delta\omega$ (oversampling). Thus, $S_{\text{rec}}(\omega)$ should be computed at discrete frequencies given by $\omega_k = k \times \delta\omega$ $(k = 0, \ldots, N-1)$. Injecting

ω_k in (A.124) gives

$$S_{rec,k} = S_{\text{rec}}(\omega_k) = \sum_{n=0}^{N-1} S(nT)\, e^{-j2\pi\frac{kn}{N}} = \frac{T}{N}\sum_{n=0}^{N-1} S_n M^{nk},$$

$$M = e^{-j\frac{2\pi}{N}}. \quad \text{(A.127)}$$

Equation A.127 can be written using the matrix formalism:

$$\widehat{S_{\text{rec}}} = \widehat{M} \cdot \widehat{S}, \quad \text{(A.128)}$$

with

$$\widehat{M} = \begin{pmatrix} 1 & 1 & 1 & \dots \\ 1 & M & M^2 & \dots \\ 1 & M^2 & M^4 & \dots \\ \dots & \dots & \dots & \dots \end{pmatrix}, \quad \widehat{S} = \begin{pmatrix} S_0 \\ S_1 \\ \dots \\ S_{N-1} \end{pmatrix}, \quad \widehat{S_{\text{rec}}} = \begin{pmatrix} S_{rec,0} \\ S_{rec,1} \\ \dots \\ S_{rec,N-1} \end{pmatrix}.$$

It is easy to show that the matrix $\widehat{M}$ is orthogonal ($\widehat{M}\widehat{M}^T = \widehat{I}$). Therefore, one can deduce the spectrum, i.e., $\widehat{S_{\text{rec}}}$, from the measured temporal data, i.e., $\widehat{S}$:

$$\widehat{S} = \widehat{M}^T \cdot \widehat{S_{\text{rec}}}. \quad \text{(A.129)}$$

Because $\widehat{M}^T$ is an N-rank square matrix, resolving (A.129) requires a number of mathematical operations proportional to N^2. The basic routine in FFT consists of separating the S_n temporal data into a set of $N/2$ even terms ($S_n^a = S_{2n}$) and a set of $N/2$ odd terms ($S_n^b = S_{2n+1}$). This implies that N should be even:

$$S_{rec,k} = \sum_{n=0}^{\frac{N}{2}-1} S_n^a W^{2nk} + W^k \sum_{n=0}^{\frac{N}{2}-1} S_n^b W^{2nk} = S_{rec,k}^a + S_{rec,k}^b. \quad \text{(A.130)}$$

Then for each $\widehat{S_{\text{rec}}^a}$ and $\widehat{S_{\text{rec}}^b}$, we could proceed as previously, i.e., solving the corresponding (A.129) matrix system, which is faster because $\widehat{S_{\text{rec}}^a}$ and $\widehat{S_{\text{rec}}^b}$ are ($N/2$)-dimensions vectors. However, it is smarter to separate again each $\widehat{S_{\text{rec}}^a}$ and $\widehat{S_{\text{rec}}^b}$ in two sets of even and odd terms, and to repeat the procedure until the last sets of records include only 1 term. This implies that N should be a power of 2, let us set $N = 2^m$. Each step of separation rearranges the N values of recorded data (N operations) and there are m steps, i.e., the total number of operations is $m \times N = N \ln N / \ln 2 = N \log_2 N$. As compared to solving (A.129), the gain in calculation time is $m \times N/N^2 = (\log_2 N)/N$. For $N = 2^{10} = 1024$, the gain is 102.

A.3.5 Kramers–Kronig Relations

We suppose that the function $g(t')$ describes the effect of a phenomenon that occurs at time t' (e.g., the generation of free carriers in a semiconductor by a laser beam excitation). The function $f(t-t')$ describes how the effect of the phenomenon evolves with time (for example the decrease of the free carrier population). At time $t > t'$, the evolution of the effect depends only on the delay $\tau = t - t'$ between the time of observation t and the time t' of occurrence of the effect. Thus the product $f(t-t') \times g(t')$ corresponds to the remaining part, at time t, of the effect that occurs at time t'. If the phenomenon happens during a given time interval, we must add the contribution of all the different times t' and we get

$$s(t) = \int_{-\infty}^{+\infty} f\left(t-t'\right) g\left(t'\right) dt'. \tag{A.131}$$

In physics, an event cannot occur before its cause: this is the principle of causality induced by the always positive variation of time. Therefore, the time delay $\tau = t - t'$ must always be positive or null. We now use τ as the variable instead of t' in Eq. A.131:

$$s(t) = \int_{0}^{+\infty} f(\tau)\, g(t-\tau)\, d\tau \tag{A.132}$$

We can express the causality principle mathematically by simply multiplying $f(t)$ with the Heaviside step function $h(t)$ ($h(t \geq 0) = 1$, $h(t < 0) = 0$) in the calculation

$$s(t) = \int_{-\infty}^{+\infty} f(\tau)\, h(\tau)\, g(t-\tau)\, d\tau \tag{A.133}$$

This is, in fact, the definition (Eq. A.104) of a convolution. The Fourier transform of $s(t)$ is obtained using Eq. A.105:

$$S(\omega) = \sqrt{2\pi}\, F(\omega) G(\omega), \tag{A.134}$$

where $F(\omega)$ and $G(\omega)$ are the Fourier transforms of $f(t)h(t)$ and $g(t)$, respectively. Using the convolution theorem (A.107) leads to

$$F(\omega) = TF\left[f(t)\, h(t)\right] = \frac{1}{\sqrt{2\pi}} F(\omega) \otimes H(\omega) \tag{A.135}$$

Then, to determine the Fourier transform $F(\omega) = TF\,[f(t)h(t)]$, we must use the Fourier transform $H(\omega)$ of the Heaviside function, whose derivation is quite tricky, but may be found in books of mathematics [6]:

$$TF\,[h(t)] = H(\omega) = \frac{1}{\sqrt{2\pi}}\left[\pi\delta(\omega) + \frac{j}{\omega}\right] \tag{A.136}$$

We substitute this expression in Eq. A.135:

$$F(\omega) = \frac{1}{\sqrt{2\pi}} F(\omega) \otimes H(\omega) = \frac{1}{\sqrt{2\pi}} \int_{-\infty}^{+\infty} F(\omega') H(\omega' - \omega)\, d\omega'$$

$$= \frac{1}{2\pi} \int_{-\infty}^{+\infty} F(\omega') \left(\pi\delta(\omega' - \omega) + \frac{j}{\omega' - \omega}\right) d\omega'$$

$$\Rightarrow \quad F(\omega) = \frac{F(\omega)}{2} + \frac{j}{2\pi} \int_{-\infty}^{+\infty} \frac{F(\omega')}{\omega' - \omega} d\omega' \tag{A.137}$$

In fact, the Fourier transform of $1/(\omega' - \omega)$ in (A.137) cannot be calculated for $\omega' = \omega$. So this integral is calculated everywhere, but not in $\omega' = \omega$. This is the definition of the Cauchy principal value (PV):

$$\mathrm{PV}\left(\int_{-\infty}^{+\infty} \frac{F(\omega')}{\omega' - \omega} d\omega'\right) = \lim_{a \to 0}\left[\int_{-\infty}^{\omega - a} \frac{F(\omega')}{\omega' - \omega} d\omega' + \int_{\omega + a}^{+\infty} \frac{F(\omega')}{\omega' - \omega} d\omega'\right] \tag{A.138}$$

So we can derive $F(\omega)$ from Eq. A.137:

$$F(\omega) = \frac{j}{\pi}\mathrm{PV}\left(\int_{-\infty}^{+\infty} \frac{F(\omega')}{\omega' - \omega} d\omega'\right) \tag{A.139}$$

The great result of this study is obtained when looking at the real and imaginary part of $F(\omega)$. As $\omega - \omega'$ is real, Eq. A.139 can be transformed into

$$\mathrm{Re}\{F(\omega)\} = \frac{1}{\pi}\mathrm{PV}\left(\int_{-\infty}^{+\infty} \frac{\mathrm{Im}\{F(\omega')\}}{\omega - \omega'} d\omega'\right) \tag{A.140}$$

$$\mathrm{Im}\{F(\omega)\} = -\frac{1}{\pi}\mathrm{PV}\left(\int_{-\infty}^{+\infty} \frac{\mathrm{Re}\{F(\omega')\}}{\omega - \omega'} d\omega'\right) \tag{A.141}$$

These two equations are known as the Kramers–Kronig relations [7]. In physics, they are not relevant because the integration spreads over negative frequencies, which is unrealistic. Nevertheless, we can limit the integration to positive frequencies thanks to the fact that $f(t)$ is a real function. Thus $F^*(\omega) = F(-\omega)$ (see Eq. A.93), which leads to

$$\operatorname{Re}\{F(\omega)\} - j\operatorname{Im}\{F(\omega)\} = \operatorname{Re}\{F(-\omega)\} + j\operatorname{Im}\{F(-\omega)\}$$
$$\Leftrightarrow \quad \begin{cases} \operatorname{Re}\{F(-\omega)\} = \operatorname{Re}\{F(\omega)\}, & \operatorname{Re}\{F(\omega)\} \text{ is even} \\ \operatorname{Im}\{F(-\omega)\} = -\operatorname{Im}\{F(\omega)\}, & \operatorname{Im}\{F(\omega)\} \text{ is odd} \end{cases} \tag{A.142}$$

We apply these properties to the first Kramers–Kronig relation:

$$\operatorname{Re}\{F(\omega)\} = \frac{1}{\pi}\mathrm{PV}\left(\int_{-\infty}^{+\infty} \frac{\operatorname{Im}\{F(\omega')\}}{\omega-\omega'}d\omega'\right) = \frac{1}{\pi}\mathrm{PV}\left(\int_{-\infty}^{+\infty} \frac{\operatorname{Im}\{F(\omega')\}}{\omega-\omega'}\frac{\omega+\omega'}{\omega+\omega'}d\omega'\right)$$
$$= \mathrm{PV}\left(\underbrace{\frac{\omega}{\pi}\int_{-\infty}^{+\infty} \frac{\operatorname{Im}\{F(\omega')\}}{\omega^2-\omega'^2}d\omega'}_{=0 \text{ because } \operatorname{Im}\{F(\omega)\} \text{ is odd}} + \frac{1}{\pi}\int_{-\infty}^{+\infty} \frac{\operatorname{Im}\{F(\omega')\}}{\omega^2-\omega'^2}\omega' d\omega'\right)$$

As $\operatorname{Im}\{F(\omega)\}$ is odd, the integral from $-\infty$ to $+\infty$ is twice the one for positive frequencies:

$$\operatorname{Re}\{F(\omega)\} = \frac{2}{\pi}\mathrm{PV}\left(\int_{0}^{+\infty} \frac{\operatorname{Im}\{F(\omega')\}}{\omega^2-\omega'^2}\omega' d\omega'\right) \tag{A.143}$$

We treat likewise the second Kramers–Kronig relation:

$$\operatorname{Im}\{F(\omega)\} = -\frac{1}{\pi}\mathrm{PV}\left(\int_{-\infty}^{+\infty} \frac{\operatorname{Re}\{F(\omega')\}}{\omega-\omega'}d\omega'\right)$$
$$= -\frac{\omega}{\pi}\mathrm{PV}\left(\int_{-\infty}^{+\infty} \frac{\operatorname{Re}\{F(\omega')\}}{\omega^2-\omega'^2}d\omega' + \underbrace{\int_{-\infty}^{+\infty} \frac{\operatorname{Re}\{F(\omega')\}}{\omega^2-\omega'^2}\omega' d\omega'}_{=0 \text{ because is } \operatorname{Re}\{F(\omega)\} \text{ even}}\right)$$
$$\Rightarrow \quad \operatorname{Im}\{F(\omega)\} = -\frac{2\omega}{\pi}\mathrm{PV}\left(\int_{0}^{+\infty} \frac{\operatorname{Re}\{F(\omega')\}}{\omega^2-\omega'^2}d\omega'\right) \tag{A.144}$$

Equations A.143 and A.144 are the useful expressions of the Kramers–Kronig relations. We can now apply them in electromagnetism to determine the relations between the real part $n(\omega)$ and the imaginary part $\kappa(\omega)$ of the refractive index (or, even better, the absorption coefficient which writes $\alpha(\omega) = 2\omega\kappa(\omega)/c$). We start from the relation A.1, in the frequency domain, between the exciting field $\vec{E}(\omega)$ and the response $\vec{D}(\omega)$ of the matter:

$$\vec{D}(\omega) = \varepsilon_o \varepsilon(\omega)\, \vec{E}(\omega) = \varepsilon_o\, (1 + \chi(\omega))\, \vec{E}(\omega). \tag{A.145}$$

The susceptibility $\chi(\omega)$ denotes the capacity of the exciting field to create dipoles in the matter. It obeys to the causality principle, as these induced dipoles cannot be created before the exciting field impinges the matter! We will apply the Kramers–Kronig relations to $\chi(\omega)$, to determine the relation between the real and imaginary part of the dielectric constant $\varepsilon(\omega)$:

$$\begin{cases} \chi_r(\omega) = \frac{2}{\pi}\mathrm{PV}\left(\int\limits_0^{+\infty} \frac{\chi_i(\omega')}{\omega^2-\omega'^2}\omega' d\omega'\right) \\ \chi_i(\omega) = -\frac{2\omega}{\pi}\mathrm{PV}\left(\int\limits_0^{+\infty} \frac{\chi_r(\omega')}{\omega^2-\omega'^2} d\omega'\right) \end{cases} \tag{A.146}$$

If $\chi_r(\omega),\ \chi_i(\omega) \ll 1$, we obtain

$$\begin{aligned} \varepsilon(\omega) = 1 + \chi(\omega) = 1 + \chi_r(\omega) + j\chi_i(\omega) = (n(\omega) - j\kappa(\omega))^2 \\ \Rightarrow n(\omega) - j\kappa(\omega) \approx 1 + \tfrac{\chi(\omega)}{2} = 1 + \tfrac{\chi_r(\omega)}{2} + j\tfrac{\chi_i(\omega)}{2} \end{aligned}$$

We substitute $(\chi_r(\omega),\ \chi_i(\omega))$ by $(n(\omega),\ \alpha(\omega))$ in Eq. A.145:

$$\begin{cases} n(\omega) = 1 + \frac{c}{\pi}\mathrm{PV}\left(\int\limits_0^{+\infty} \frac{\alpha(\omega')}{\omega'^2-\omega^2} d\omega'\right) \\ \alpha(\omega) = -\frac{4\omega^2}{\pi c}\mathrm{PV}\left(\int\limits_0^{+\infty} \frac{n(\omega')-1}{\omega'^2-\omega^2} d\omega'\right) \end{cases} \tag{A.147}$$

Practically, relations A.147 are very useful to derive the refractive index spectrum from the absorption one, which is quite easy to measure. However, measurements are always performed over a limited bandwidth, and thus the integrals in Eq. A.147 cannot be performed from 0 to infinity. This leads to a truncation error $\Delta n(\omega)$:

$$\Delta n(\omega) = \frac{c}{\pi}\mathrm{PV}\left(\int\limits_0^{\omega_{\min}} \frac{\alpha(\omega')}{\omega'^2 - \omega^2} d\omega' + \int\limits_{\omega_{\max}}^{+\infty} \frac{\alpha(\omega')}{\omega'^2 - \omega^2} d\omega'\right). \tag{A.148}$$

In the general case, $\Delta n(\omega)$ depends on the measurement bandwidth and on the sample under test. If $\omega_{\text{min}} \ll \omega$ and $\omega_{\text{max}} \gg \omega$, i.e., if the experimental bandwidth is sufficiently large and/or if one is interested in the central part of the spectrum, one gets

$$\Delta n(\omega) \approx \frac{c}{\pi}\left(-\frac{1}{\omega^2}\int_0^{\omega_{\text{min}}} \alpha\left(\omega'\right) d\omega' + \int_{\omega_{\text{max}}}^{+\infty} \frac{\alpha\left(\omega'\right)}{\omega'^2} d\omega'\right) \approx -\frac{A}{\omega^2}, \tag{A.149}$$

where A is a constant. Thus the truncation error is bigger at the lower frequencies of the spectrum.

References

1. Bloembergen, N. (1965), *Nonlinear Optics* (W. A. Benjamin, Inc., New York).
2. Born, M., and Wolf, E. (1999). *Principles of Optics*, 7th ed. (Cambridge University Press, Cambridge).
3. Armstrong, J. A., Bloembergen, N., Ducuing, J., and Pershan, P. S. (1962). Interactions between light waves in a nonlinear dielectric, *Phys. Rev.*, **127**, pp. 1918–1939.
4. Yariv, A. (1989). *Quantum Electronics*, 3rd ed. (John Wiley and Sons, New York).
5. Fourier, J. B. J. (1822). *Théorie analytique de la chaleur* (Firmin Didot, père et fils, Paris). The title of this book may be translated from French as *Analytical Theory of Heat.*
6. Arfken, G. (1985). *Mathematical Methods for Physicists*, 3rd ed. (Academic Press, Orlando).
7. Lucarini, V., Saarinen, J. J., Peiponen, K.-E., and Vartiainen, E. M. (2005). *Kramers–Kronig Relations in Optical Materials Research* (Springer-Verlag, Berlin).

Index

For Product Safety Concerns and Information please contact our EU representative GPSR@taylorandfrancis.com
Taylor & Francis Verlag GmbH, Kaufingerstraße 24, 80331 München, Germany

www.ingramcontent.com/pod-product-compliance
Ingram Content Group UK Ltd.
Pitfield, Milton Keynes, MK11 3LW, UK
UKHW021513250926
475702UK00043B/263
9789814774567